The Evolution of National Wildlife Law

THE EVOLUTION OF NATIONAL WILDLIFE LAW

Third Edition

Michael J. Bean and
Melanie J. Rowland

A Project of the Environmental Defense Fund
and World Wildlife Fund–U.S.

Westport, Connecticut
London

Library of Congress Cataloging-in-Publication Data

Bean, Michael J., 1949–
The Evolution of national wildlife law / Michael J. Bean and Melanie J. Rowland.—3rd ed.
p. cm.
"A project of the Environmental Defense Fund and World Wildlife Fund–U.S."
Rev. ed. of: The evolution of national wildlife law. Rev. & expanded ed. c1983.
Includes bibliographical references and index.
Rev. ed. of: The evolution of national wildlife law
ISBN 0–275–95988–0 (alk. paper).—ISBN 0–275–95989–9 (pbk. : alk. paper)
1. Wildlife conservation—Law and legislation—United States.
I. Rowland, Melanie J. II. Bean, Michael J., 1949– Evolution of national wildlife law. III. Environmental Defense Fund. IV. World Wildlife Fund (U.S.) V. Title.
KF5640.B4 1997
346.7304'69516—dc21 97–8860

British Library Cataloguing in Publication Data is available.

Library of Congress Catalog Card Number: 97–8860
ISBN: 0–275–95988–0
0–275–95989–9 (pbk.)

First published in 1997

Praeger Publishers, 88 Post Road West, Westport, CT 06881
An imprint of Greenwood Publishing Group, Inc.

Printed in the United States of America

The paper used in this book complies with the Permanent Paper Standard issued by the National Information Standards Organization (Z39.48–1984).

10 9 8 7 6 5 4 3 2 1

To Emily Starr Bean
—Michael J. Bean

To my mother, Marjorie Rowland, and to my husband, Randy Brook
—Melanie J. Rowland

CONTENTS

PART III
WILDLIFE, LAND, AND WATER

PART IV
INTERGOVERNMENTAL WILDLIFE CONSERVATION

FOREWORD

When *The Evolution of National Wildlife Law* first appeared in 1977, it filled an empty niche in legal scholarship. It was the first thorough and authoritative analysis of wildlife conservation law in the United States. Remarkably, two decades later it remains the standard reference, the essential companion, for anyone seeking to understand the numerous and complex statutes, regulations, and court decisions on wildlife law. Since the second edition was published in 1983, the field of wildlife law has grown by leaps and bounds. Those who have come to rely on this indispensable work will be grateful for the third edition. It completely revises and updates the earlier editions to reflect current legal perspectives on conservation of wildlife and biological diversity.

Like many other areas of environmental law, wildlife law stirs strong passions and intense controversy. Few who have written on the subject have been able to separate their own views of what the law ought to be from what it is. What distinguishes this book from much of the rest of the literature in this field is its balanced, objective scholarship. While the authors make no attempt to conceal their belief that the goal of conserving wildlife is a worthy one, their legal analysis is straightforward and impartial. It is not surprising that, in 1995, when the United States Supreme Court considered the *Sweet Home* case, a highly contentious Endangered Species Act dispute, both sides cited this work in their briefs.

Though this is a rigorous work of scholarship, its authors are not ivory tower academics. Michael J. Bean has for two decades been a lawyer for the Environmental Defense Fund in Washington, D.C., and one of the most innovative and influential forces in the country for wildlife conservation. He is Chair of EDF's Wildlife Program and is especially noted for his expertise on endangered species conservation. Melanie J. Rowland, who joins Michael as co-author for the third edition, has been Senior Counsel with The Wilderness Society and a faculty member at the University of Washington. She is now an attorney with the National Oceanic and Atmospheric Administration's Office of General Counsel in Seattle, where her practice concentrates on recovery efforts for threatened and endangered salmon.

The subject of this book is one about which they care deeply and know a great deal. Like the two that preceded it, the new edition of *The Evolution of National Wildlife Law* is a monumental achievement that will serve lawmakers, administrators, educators, conservationists, and scholars for years to come.

Bruce Babbitt

PREFACE

A project of this magnitude does not become a reality without the help and support of a great number of individuals and organizations. Thanks are due first to World Wildlife Fund–U.S., without whose help and encouragement this work would not have been undertaken. It will surprise no one who knows him that the impetus for this project within World Wildlife Fund was Don Barry, who has since rejoined the Interior Department as the Deputy Assistant Secretary for Fish, Wildlife and Parks. For two decades Don has been one of the most respected and influential figures in Washington in wildlife conservation policy. He has been an Interior Department lawyer, a congressional staffer, a public advocate, and a key policymaker. Those with a keen eye can find at least one of Don's fingerprints on just about every important wildlife conservation initiative at the national level in recent years. They are on this book, just as they were on the previous edition. For bringing World Wildlife Fund to this project, for a friendship now approaching a quarter century, and for a truly singular role in conserving wildlife, a heartfelt thanks is due to Don.

The role of keeping this project on track after Don left World Wildlife Fund fell first to David Hoskins, now of the Senate Environment and Public Works Committee staff, and then to Chris Williams. David and Chris are two of the real standouts among the next generation of conservation leaders. For their help, encouragement, and collegiality, a real debt of gratitude is owed. Finally, Kathryn S. Fuller, World Wildlife Funds's distinguished President, and James P. Leape, its Senior Vice President, warrant thanks for their unflagging confidence and support.

The National Fish and Wildlife Foundation provided critical financial support for this project. Amos S. Eno, the Foundation's executive director, and Whitney C. Tilt, its director of conservation programs, have been key players in providing funding for conservation efforts. We are honored that this project was worthy of their support. The Munson Foundation provided crucial funding as well. Wolcott Henry deserves thanks for the Foundation's generosity. Thanks go also to William Y. Brown, formerly of Waste Management, Inc., who arranged a matching grant from that company to the National Fish and Wildlife Foundation.

Melanie Rowland's work on this book was made possible by generous support from the University of Washington in Seattle, where she served as a Visiting Scholar at the former Institute for Environmental Studies and the School of Law during preparation of the manuscript. Special thanks are due to the then-members of the Institute's administration and conservation biology faculty: Dr. James Karr, Director; Vim Wright, Assistant Director; and Drs. Dee Boersma, Estella Leopold, and Gordon Orians. Their enthusiasm, encouragement, and insights were invaluable. Sincere thanks are due to the School of Law's former Dean Wallace Loh, Dean Roland Hjorth, former Associate Dean Robert Aronson, and Associate Dean Thomas Andrews, who provided the opportunity to work on this project at the law school. The law school provided access to Westlaw, an invaluable research tool for a work of this scope. Thanks also to law librarian Penny Hazelton and the law library's outstanding reference librarians. Many thanks to the environmental law faculty members who contributed freely of their time and expertise: Daniel Bodansky, William Burke, Gregory Hicks, Ralph Johnson, and William Rodgers, Jr. Finally, thanks to Dr. David Fluharty of the university's School of Marine Affairs, a member of the North Pacific Fishery Management Council, for his help in unravelling the mysteries of the Magnuson Act.

Many people helped in a variety of other ways. Margaret McMillan, an outstanding administrative assistant at the Environmental Defense Fund, did a huge amount of formatting, proofreading, cite checking, and other tasks too numerous to list. Randy Brook, an attorney, computer expert, and Melanie Rowland's husband, saved the day on countless occasions when the technological challenges of producing a manuscript of this length sorely tested one author's patience. Randy also helped a great deal with editing and proofreading. Allan Clark, a former Environmental Defense Fund legal intern and now a lawyer with a major Seattle firm, kindly offered not only to review, but also to help research and write, portions of this edition. Charles Carson, a Harvard Law School student, helped with cite checking and other tasks, as did Elizabeth Kaplan of Stanford University, and Steve Greetham of Northeastern University School of Law.

Finally, we owe thanks to a long list of attorneys and conservation professionals who shared their expertise, provided important leads or copies of hard-to-find documents, or reviewed drafts of chapters, and who contributed mightily to the quality of this book: Michael Anderson, Pamela

Baldwin, Adam Berger, Lynne Corn, Robert Dreher, Mark Eames, Pamela Pride Eaton, John Fitzgerald, Alix Foster, Albert Gidari, Rosemarie Gnam, Nancy Green, Robert Irvin, Jeffrey Kopf, Thomas Lustig, Ruth Musgrave, Dr. Elliott Norse, Suzy Sanders, Paul Schmidt, Victor Sher, William Snape, Todd True, Johanna Wald, and most especially, Dr. David Wilcove, an outstanding ecologist on the staff of the Environmental Defense Fund. We deeply appreciate all of their help. If we made errors despite their able assistance, the responsibility is entirely our own.

Michael J. Bean
Melanie J. Rowland

Part I

OVERVIEW

The three chapters that comprise this part introduce the reader to the scope of the work and examine the historical antecedents and legal foundations for today's panoply of wildlife conservation laws. They set the stage for the later in-depth examination of the key statutes that comprise the body of federal wildlife law in the United States. They also explore the first major federal undertaking to conserve wildlife, the regulation of commerce in wildlife.

An understanding of contemporary wildlife laws and programs requires familiarity with the historical and constitutional context in which they arose. No less than in other areas of governmental action, the arena of wildlife conservation has been a crucible in which the relationship of federal and state authority has been continually refined. The United States Supreme Court has often been called upon to resolve disputes over the respective roles of the state and federal governments. In fact, it is surprising how many of the cases that a student of constitutional law encounters concern federal regulation of wildlife. Chapter 2 examines these cases and traces the substantial expansion of federal wildlife regulation that has occurred in this century.

As the reader will see in Chapter 2, the states initially assumed responsibility for conserving wildlife, and there was serious doubt as to the scope of federal authority over wildlife. Consequently, the federal government's first foray into the field of wildlife regulation, the Lacey Act of 1900, was a modest one. It was based on the federal government's unquestioned authority to regulate interstate commerce. Its purpose was simply to aid enforcement of state wildlife laws, most of which related to commercial and sport hunting. Chapter 3 examines the Lacey Act, which has remained a key tool in the federal government's efforts to conserve wildlife.

Chapter 3 also discusses more recent federal wildlife conservation efforts based on the authority to regulate commerce. These are aimed at controlling the spread of injurious wildlife. They include Executive Order 11987, the Federal Noxious Weed Act, and the Nonindigenous Aquatic Nuisance Prevention and Control Act of 1990. These laws are assuming greater importance, and will continue to do so in the future, as the intentional and accidental introduction of non-native species becomes one of the leading threats to the survival of native species throughout the United States and the world.

Chapter **1**

INTRODUCTION

> It is interesting to contemplate a tangled bank, clothed with many plants of many kinds, with birds singing on the bushes, with various insects flitting about, and with worms crawling through the damp earth, and to reflect that these elaborately constructed forms, so different from each other, and dependent upon each other in so complex a manner, have all been produced by laws acting around us.
>
> Charles Darwin, *The Origin of Species*

The laws to which Darwin referred were the laws of nature, not of human society. Yet few would doubt that our laws have also had a major impact on the tangled bank of life that Darwin described. Many of our laws have served only to disentangle and destroy it; others have been intended to protect and conserve it. This book examines some of these protective laws, the federal laws to conserve wildlife.

Major federal wildlife legislation was enacted as early as 1900, and the body of federal wildlife law is now voluminous and complex. Yet until recent decades it had largely escaped the notice of legal scholars. Prior to publication of the first edition of this work in 1977, there was no single comprehensive text addressing wildlife law as a distinct segment of environmental law. Nor was there much critical scrutiny of the subject in legal journals.

The very term "wildlife law" was novel in 1977. Few had seen fit to distinguish such a body of law from the broader categories of "environmental law" or "natural resources law." By the time the second edition of this book appeared in 1983, that had clearly changed. The United States Department of Justice had established a new Wildlife Section within its Land and Natural Resources Division in 1979. This section, since expanded and renamed the Wildlife and Marine Resources Section, represents the United States in all litigation concerning the federal wildlife laws described in this book. Law schools began to offer courses in wildlife law, and for many of these the first or second edition of this book has served as the basic text.

The original work played a role in stimulating these developments. It probably also contributed significantly to what one scholar described as "[a]n efflorescence of scholarship" on wildlife law.[1] Moreover, to a surprising degree, the original work proved immensely popular. It exhausted several printings and reached an audience that included lawyers and nonlawyers alike.

The intense interest in wildlife law since publication of the first edition has been matched by a great increase in the amount of wildlife-related litigation. Since 1977 the Supreme Court has decided more than a dozen cases presenting important questions of wildlife law. These cases have addressed such fundamental issues as the very nature of federal and state interests in wildlife. The lower federal courts have also handled a growing number of cases brought by the government, conservation groups, and commercial interests. The resulting body of new court decisions, as well as frequent congressional action, has created a need for this third edition.

As in earlier editions, this book explores the most significant federal laws and cases relating to wildlife conservation. Though its scope is broad, it is not truly comprehensive. Part of the problem in providing a focused analysis of federal wildlife law is that its boundaries are so uncertain. If the umbrella of "wildlife law" encompasses all federal laws that could have a significant impact on the tangled bank of life, directly or indirectly, then one can exclude hardly any environmental law. This book makes no claim to be that comprehensive. Rather, the laws and programs examined here have regulation of human use of wildlife or its habitat as one of their principal aims.

If it is hard to say precisely what constitutes federal wildlife law, the authors take some comfort in the fact that it has also been hard for federal lawmakers to say precisely what constitutes wildlife. The term "wildlife" has not had a fixed meaning in federal law. Indeed, its steady evolution and expansion over time is one of the important trends in the development of federal wildlife law. For purposes of this book, the term "wildlife" has the broadest meaning recognized in federal law, encompassing all nonhuman and nondomesticated animals. This book also discusses unique aspects of plant conservation under federal wildlife law. Although there is no compelling biological reason to exclude plants from the def-

[1] T. Lund, American Wildlife Law 2 (1980).

inition of wildlife, plants have occupied a conceptual status in law different from that of animals.[2]

The question of what is embraced by the term "wildlife" became even more complex after the Supreme Court's 1980 decision in *Diamond v. Chakrabarty* that microorganisms produced by genetic engineering may be subject to patent protection.[3] Patentability, in the Court's view, depends not on whether the object in question is living or inanimate, but on whether it occurs in nature or is "the result of human ingenuity and research."[4] This distinction may be useful when considering microbes produced in a laboratory, but it is less so as applied to the myriad results of human ingenuity and research in animal husbandry and wildlife management. The notion that new breeds and varieties of "wild" animals could be "owned" by a patent holder would add to the complexity of defining what wildlife is[5] and who owns it.

The discussion of wildlife law in this edition is arranged differently from the earlier works, and a few subjects treated in those editions have been dropped altogether. The new format itself reflects an evolution in what is thought to be most important in wildlife law.

The first part examines the historical and constitutional contexts in which American wildlife law has developed. This part also looks closely at the first major federal wildlife law, the Lacey Act of 1900, and its strategy of conserving wildlife through controls on commerce.

The second part examines the laws that approach conservation from a species-management perspective. There are several federal initiatives aimed at protecting particular categories of wildlife: birds, marine mammals, ocean fish, and endangered species. This part explores the federal conservation programs for each of these groups, the statutes that established them, and the litigation they have spawned.

The third part focuses on wildlife habitat. This subject has assumed ever greater importance as the pace of human alteration of natural habitat has accelerated. The discussion of wildlife conservation on federal lands is considerably expanded from prior editions, reflecting both the growing role that public lands play in habitat conservation and the heightened

[2]A useful discussion of these differing legal concepts, as well as the bases for state and federal protection of rare plants, is found in Linda McMahan, *Legal Protection for Rare Plants*, 29 Am. U.L. Rev. 515 (1980).

[3]447 U.S. 303 (1980).

[4]*Id.* at 313.

[5]To illustrate the bizarre results that can sometimes happen when courts consider such questions, see the Canadian case of Regina v. Ojibway, 8 Criminal Law Quarterly 137 (1965–66) (Op. Blue, J.), described in United States v. Byrnes, 644 F.2d 107, 112 n.9 (2d Cir. 1981). The Canadian court concluded that a pony saddled with a down pillow was a "bird" within the meaning of a statute defining the term as a "two legged animal covered with feathers." The court reasoned that the two legs were the statutory minimum and that the feather covering need not be natural.

controversy over human use of these lands. This edition's more detailed examination of "multiple-use" federal lands administered by the U.S. Forest Service and the Interior Department's Bureau of Land Management was prompted by the recent heavy volume of litigation testing the obligations of federal land managers toward wildlife under legislation adopted in the mid-1970s. This part also includes two new chapters. One surveys federal programs to promote wildlife conservation on private lands, and the other focuses on water resource development and wildlife conservation.

The final part includes two topics: wildlife and Native Americans, and international wildlife law. Most federal relations with Native American tribes are subject to formal treaties. One chapter explores the nature of Native American treaty rights as they relate to federal wildlife conservation. International treaties similarly involve the challenge of cooperation among sovereign nations to advance the shared objective of wildlife conservation. The chapter on international wildlife law primarily examines the most important treaties to which the United States is a party.

A quick appraisal of the topics covered in this book reveals that federal wildlife law is largely statutory. Apart from the nineteenth-century doctrine of state ownership of wildlife and some recent cases concerning this doctrine as a basis for civil damage actions, the common law has played a minor role in wildlife conservation. One consequence is that the legislative programs Congress has created require governmental bureaucracies to implement them. How government officials have implemented wildlife conservation programs, and how the public and the courts have responded to their actions, is the principal subject of this book.

Chapter 2

THE LEGAL FRAMEWORK FOR THE DEVELOPMENT OF FEDERAL WILDLIFE LAW

The ultimate source of authority for all governmental action in the United States is the Constitution. This document sets forth the respective powers and duties of the three branches of the federal government. It also expressly prohibits the states and the federal government from the exercise of certain powers. All powers not delegated to the federal government nor prohibited to the states, however, are "reserved to the States respectively, or to the people" by virtue of the Constitution's Tenth Amendment. Accordingly, it is often said that the federal government is a government of enumerated powers, or a "limited" government.

The Constitution's enumeration of federal powers, however, is often couched in the most general language. Because of this generality, the courts, and especially the United States Supreme Court, play a pivotal role in the process of constitutional interpretation, the process by which the respective legal authorities of the state and federal governments are demarcated.

This chapter examines the constitutional bases for federal authority over wildlife. It also examines the rise and fall of the very important doctrine of state ownership of wildlife, a judicially created doctrine that furnished the basis for repeated challenges to the exercise of federal authority. Before those topics are addressed, however, the ancient origins of both state and federal wildlife regulation will be explored.

HISTORICAL ANTECEDENTS

In the history of Western thought, there is an almost unbroken tradition, starting at least as early as the Roman Empire, in which wild animals (or animals *ferae naturae*, as they were called) were regarded as occupying a nearly unique status. The law considered wild animals in their natural state to be like the air and the oceans, in that they were the property of no one. Yet unlike the air and the oceans, wild animals could become the property of anyone who captured or killed them. The only legal restriction in Rome on the right to acquire property in wildlife in this way was that a private landowner had the exclusive right to reduce to possession the wildlife on his property. This restriction, however, was apparently more "a recognition of the right of ownership in land than an exercise by the state of its undoubted authority to control the taking and use of that which belonged to no one in particular, but was common to all."[1]

Governmental regulation of the right to take wildlife was soon evident in feudal Europe. According to Sir William Blackstone, the eighteenth-century English legal scholar whose writings influenced Anglo-American law more than those of any other individual, the origins of game regulation in Europe were the very same as those of the feudal system itself. That is, to retain the fruits of their conquest, the feudal kings and barons of Europe sought to keep weapons out of the hands of those whom they had conquered. According to Blackstone:

> Nothing could do this more effectually than a prohibition of hunting and sporting; and therefore it was the policy of the conqueror to reserve this right to himself and such on whom he should bestow it; which were only his capital feudatories, or greater barons. And accordingly we find, in the feudal constitutions, one and the same law prohibiting the *rustici* in general from carrying arms, and also proscribing the use of nets, snares, or others engines for destroying the game.[2]

So too in England restrictions on hunting and land uses detrimental to wildlife were imposed quite early and for similar reasons. With the Saxon invasion of England around 450 A.D., land began to be parcelled out to the nobility. The lands that were not parcelled out became known as "royal forests" in which the king alone had the right to hunt and in which

[1]Geer v. Connecticut, 161 U.S. 519, 523 (1896).

[2]3 W. Blackstone, Commentaries *413. Blackstone's intense animosity toward royal privilege is evident in his discussion of feudal game laws. Consider the following passage, which immediately follows the passage quoted in the text:

> This exclusive privilege well suited the martial genius of the conquering troops, who delighted in a sport which in its pursuit and slaughter bore some resemblance to war. . . . And indeed, like some of their modern successors, they had no other amusement to entertain their vacant hours; despising all arts as effeminate, and having no other learning, than was couched in such rude ditties, as were sung at the solemn carousels which succeeded these antient huntings.

Id. *414.

private landowners were "required to retain adequate vegetation for wildlife forage and cover."[3]

The Norman Conquest in 1066 brought with it a great expansion of the royal forests. William Nelson, a contemporary of Blackstone, wrote that "William the Conqueror laid waste thirty-six Towns in Hampshire to make a Forest."[4] As the royal forests expanded, so too did the "Forest Jurisdiction," an elaborate system of royal forest laws, together with special courts and officials charged with administering them.[5] For example, all forest land "was subject to an easement for the benefit of wildlife" that allowed forest officials to enter private land and remove vegetation needed for wildlife.[6]

The Norman Conquest also witnessed an expansion beyond the forests of the king's exclusive authority to hunt. Indeed, the king soon claimed the sole right to pursue game or to take fish anywhere in the kingdom, though he frequently bestowed limited parts of that prerogative upon the favored nobility by means of various royal franchises. These included the franchises of "park" and "chase," which granted to the holder the right to pursue such prized beasts as deer, fox, and marten across his own land, in the case of the former, or across the lands of others, in the case of the latter. Yet another franchise was that of "free-warren," which authorized its holder to kill less-prized beasts, such as fowl and hares, in a particular area, as long as he prevented others from doing so. Similar franchises applied to fishing. A right of "free fishery" gave its holder an exclusive right to fish in a particular river. A right of "several fishery" was essentially the same, except that it applied only in those waters adjoining the lands of the franchise holder. A "common of piscary" was a nonexclusive right to fish in particular waters.[7]

As England's political system evolved, so too did its system of wildlife regulation. By the thirteenth century, so many franchises of free fishery and several fishery had been granted that the great numbers of private weirs placed in England's rivers impeded navigation. So significant was this problem that in 1215 the Magna Charta directed removal of the weirs

[3]T. Lund, American Wildlife Law 16 (1980).

[4]William Nelson, The Laws Concerning Game (1762), quoted in W. Sigler, Wildlife Law Enforcement 3 (3d ed. 1980).

[5]T. Lund, *British Wildlife Law before the American Revolution: Lessons from the Past,* 74 Mich. L. Rev. 49, 60–61 (1975). This article appears as Chapter 2 in T. Lund, American Wildlife Law (1980). Other chapters provide an interesting discussion of the adaptation by the new American states of English legal principles regarding wildlife.

[6]Lund, *supra* note 3, at 16.

[7]The descriptions of the various royal franchises discussed in the text are all taken from 3 W. Blackstone, Commentaries *38–40. Even Blackstone conceded that the distinctions between some of the franchises were "very much confounded in our law-books." *Id.* *40.

throughout England, a directive which was later "judicially expanded to bar the king from granting private fisheries in tidal waters."[8]

Royal power over wildlife gradually gave way to Parliament. The slow transition of authority did not signal any great democratization of rights to wildlife, however. Rather, the principal mechanism of parliamentary control was enactment of the so-called "qualification statutes." These statutes prohibited the taking of game by anyone not "qualified," in the sense of having the requisite amount of wealth or land prescribed in the statute. Thus, the qualification statutes merely perpetuated a pervasive system of class discrimination and at the same time kept weapons out of the hands of those considered unfriendly, or potentially so, to those in power.[9]

All of these features of English wildlife law were still current at the time of the first settlement of the New World, although the Forest Jurisdiction had fallen into disuse. The qualification statutes, on the other hand, were very much alive and would continue to be so until several decades after American Independence. Stripped of its many formalities, the essential core of English wildlife law on the eve of the American Revolution was the complete authority of the King and Parliament to determine what rights others might have with respect to the taking of wildlife. The following section explores what became of that central doctrine in the United States after ties with the mother country were severed.

DEVELOPMENT OF THE STATE OWNERSHIP DOCTRINE

The Carryover of British Wildlife Law and the "Public Trust" Doctrine

The first case to come before the United States Supreme Court concerning the relationship of government and citizen with respect to wildlife was *Martin v. Waddell* in 1842.[10] The facts of this case were ideally suited for an examination of the extent to which British wildlife law carried over to the independent United States.

At issue in *Martin v. Waddell* was the right of a riparian landowner to exclude all others from taking oysters from certain mudflats in New Jersey's Raritan River. The landowner claimed to own both the riparian and submerged lands, tracing his title to a grant in 1664 from King Charles

[8]MacGrady, *The Navigability Concept in the Civil and Common Law: Historical Development, Current Importance, and Some Doctrines That Don't Hold Water,* 3 Fla. St. U. L. Rev. 513, 555 (1975).

[9]Lund, *supra* note, 3 at 8–10. The precise relationship of the qualification statutes to the royal franchises previously discussed is not altogether clear. Blackstone asserts that the qualification statutes did not affirmatively authorize anyone to take game, that right remaining dependent upon the grant of a royal franchise, but suggests that the statutes were not commonly understood to mean this. *See* 3 W. Blackstone, Commentaries *417–18.

[10]41 U.S. (16 Pet.) 367 (1842).

II to the Duke of York that purported to convey "all the lands, islands, soils, rivers, harbours, mines, minerals, quarries, woods, marshes, waters, lakes, fishings, hawkings, huntings and fowlings" within described metes and bounds.

In the view of the inventive Chief Justice Roger Taney, the legal question presented was more than the interpretation of a mere deed of title, for the original deed from the King to the Duke "was an instrument upon which was to be founded the institutions of a great political community."[11] Accordingly, Taney deemed it necessary to consider first "the character of the right claimed by the British crown"[12] and second, whether the character changed when title to the lands passed from the King to the Duke and ultimately to the plaintiff. Addressing these fundamental issues, Taney declared that "dominion and property in navigable waters, and in the lands under them [were] held by the King as a public trust" and that it "must be regarded as settled in England against the right of the King since Magna Charta," to make a private grant of such lands and waters.[13] That is, by virtue of his public trust responsibilities, the King was without power to abridge "the public common of piscary."[14]

If the public trust character of navigable waters and their submerged lands survived a grant by the King of his proprietary interest in them, there was still the question whether it also survived the American Revolution. Taney declared that it did:

> [W]hen the people of New Jersey took possession of the reins of government, and took into their own hands the powers of sovereignty, the prerogatives and regalities which before belonged either to the crown or the parliament, became immediately and rightfully vested in the state.[15]

In so declaring, Taney seemed to place the states in the role of successors to Parliament and the crown, thus laying the groundwork for the later development of the doctrine of state ownership of wildlife. As will become apparent, however, that development largely ignored Taney's important

[11] *Id.* at 412. Chief Justice Taney is best known in history as the author of the 1857 Dred Scott decision, which gave impetus to the Civil War. Taney's opinion held that because slaves were property, the property rights of a slave owner could not be nullified by Congress.

[12] *Id.* at 409.

[13] *Id.* at 410, 411.

[14] *Id.* at 412. One commentator, in a provocative article, argued persuasively that Taney's decision vastly expanded the Magna Charta's limited prohibition against the creation of private fishing rights into a much broader prohibition against the creation of any sort of private rights in the submerged lands of navigable waters. *See* MacGrady, *supra* note 8, at 554–55 and 589–91. Nonetheless, the "public trust doctrine," as it has come to be called, is now well established in American law. *See* Sax, *The Public Trust Doctrine in Natural Resource Law: Effective Judicial Intervention*, 68 Mich. L. Rev. 471 (1970).

[15] 41 U.S. at 416.

qualifier that the powers assumed by the states were "subject . . . to the rights since surrendered by the Constitution to the general government."[16]

Until the turn of the century, there were few occasions to consider the scope of the rights surrendered by the states, because prior to 1900 the only federal wildlife legislation was limited in scope and relatively insignificant in impact.[17] There was, however, a steady growth in the regulation of wildlife at the state and territorial levels. The few Supreme Court cases that considered the validity of this regulation uniformly upheld it.

In *Smith v. Maryland*, for example, the Supreme Court considered the validity of a Maryland law that prohibited the taking of oysters from the state's waters by means of scoop or drag.[18] The defendant shipowner, whose vessel was licensed by the United States to engage in the coasting trade, contended that the state law interfered with the exclusive federal power to regulate interstate commerce.[19] The Supreme Court rejected the defendant's argument and held that the state's ownership of the soil conferred upon it the authority to regulate the taking of oysters from that soil.[20] The Court emphasized the limited scope of its holding, however:

> The law now in question . . . does not touch the subject of the common liberty of taking oysters, save for the purpose of guarding it from injury, to whomsoever it may belong, and by whomsoever it may be enjoyed. Whether this liberty belongs exclusively to the citizens of the State of Maryland, or may lawfully be enjoyed in common by all citizens of the United States; whether this public use may be restricted by the State to its own citizens, or a part of them, or by force of the Constitution of the United States must remain common to all citizens of the United States; whether the national government, by a treaty or act of congress, can grant to foreigners the right to participate therein; or what, in general, are

[16] *Id.* at 410. Technically, the holding in Martin v. Waddell applied only to the original thirteen states. It was soon held to apply as well to all later admitted states in Pollard v. Hagan, 44 U.S. (3 How.) 212 (1845). Legislation admitting new states to the nation has traditionally included an "equal footing" clause, which grants to each new state the same rights and powers held by existing states. For an application of the equal footing doctrine to the question of state authority over wildlife, see Ward v. Race Horse, 163 U.S. 504 (1896), discussed in Chapter 13 at text accompanying notes 3–5.

[17] *See, e.g.*, Act of July 27, 1868, ch. 273, § 6, 15 Stat. 241 (*repealed* 1944), prohibiting the killing of certain fur-bearing animals in the territory and waters of Alaska; Act of February 9, 1871, 16 Stat. 593 (*repealed* 1964), creating the Office of the United States Commissioner of Fish and Fisheries "for the Protection and Preservation of the Food Fishes of the Coast of the United States"; Act of February 28, 1887, ch. 288, 24 Stat. 434, regulating the importation of mackerel into the United States; and Act of May 7, 1894, ch. 72, 28 Stat. 73, prohibiting hunting in Yellowstone National Park.

[18] 59 U.S. (18 How.) 71 (1855).

[19] "The Congress shall have power . . . To regulate Commerce with foreign Nations, and among the several States, and with the Indian Tribes." U.S. Const. art. I. § 8, cl. 3.

[20] 59 U.S. at 75.

the limits of the trust upon which the State holds this soil, or its power to define and control that trust, are matters wholly without the scope of this case, and upon which we give no opinion.[21]

Some of the many questions left unanswered by the Court's *Smith* opinion were answered relatively soon thereafter. Others remained unanswered for more than a century.[22]

McCready v. Virginia gave the Court its first opportunity to answer one of the questions left open in *Smith.*[23] There, the Court upheld a Virginia statute prohibiting citizens of other states from planting oysters in Virginia's tidewaters. In so doing, however, it substantially expanded the narrow holdings of *Martin v. Waddell* and *Smith v. Maryland* by declaring that the state owned not only the tidewaters, but also "the fish in them, so far as they are capable of ownership while running."[24] Accordingly, Virginia could regulate the planting or taking of oysters in those tidewaters, even to the point of excluding altogether the citizens of other states, because such a regulation "is in effect nothing more than a regulation of the use by the people of their common property."[25]

Fifteen years later, in *Manchester v. Commonwealth of Massachusetts,* the Court upheld a Massachusetts statute prohibiting the use of purse seines to take menhaden in Buzzard's Bay.[26] The Court's opinion signalled a more cautious approach to the question of ultimate authority for the state's regulation. Rather than base that authority on any concept of "ownership" of the fish, the Court emphasized that inasmuch as the Bay was a body of navigable water within the state's territorial jurisdiction, "the subject is one which a state may well be permitted to regulate within its territory, in the absence of any regulation by the United States."[27] Moreover, the Court felt compelled to point out that the regulation in question served a valid public purpose, the preservation of menhaden, which, "although they are not used as food for human beings, but as food for other fish, which are so used, is for the common benefit."[28] The quoted language is significant because it reflects a fundamental nineteenth-century conception of the purpose of wildlife law, the preservation of a food supply.

In still other respects the *Manchester* decision revealed some uncertainty about the breadth of the principles previously announced in *McCready.*

[21] *Id.*
[22] *See* text accompanying notes 121–153 *infra.*
[23] 94 U.S. 391 (1876).
[24] *Id.* at 394.
[25] *Id.* at 395.
[26] 139 U.S. 240 (1891).
[27] *Id.* at 265.
[28] *Id.*

Thus, the Court observed that, since the Massachusetts statute did not discriminate in favor of the citizens of that state, there was no need to consider whether there existed "a liberty of fishing for swimming fish in the navigable waters of the United States common to the inhabitants or the citizens of the United States."[29] The Court's implication that a state's powers with respect to "swimming fish" may be less than those with respect to sedentary shellfish was later to become the basis for virtually eliminating the *McCready* decision's vitality.[30]

The Landmark Case of *Geer v. Connecticut*

It was against this background of developing principles regarding state ownership of submerged land and state control over fisheries that the Supreme Court in 1896 decided the case which eloquently articulated a general theory of state "ownership" of wildlife, *Geer v. Connecticut.*[31] In that case, the defendant Geer appealed his conviction under state law for possessing game birds with the intent to ship them out of Connecticut. The birds had been lawfully killed there; only Geer's intent to transport them outside the state was unlawful. So the legal question presented was the same as that considered in *Smith*: whether the statute improperly interfered with Congress's power to regulate interstate commerce.

In the view of Justice Edward White, writing for the majority, the answer required a thorough examination "of the nature of the property in game and the authority which the state had a right lawfully to exercise in relation thereto."[32] Tracing the history of governmental control over the taking of wildlife from Greek and Roman law through the civil law of the European continent and the common law of England, White concluded that the states had the right "to control and regulate the common property in game," which right was to be exercised "as a trust for the benefit of the people."[33] As an incident of this right of control, the states could affix conditions on the taking of game, and, most importantly, those conditions would remain with the game even after being killed.

Addressing the narrow legal question of whether the conditions affixed by Connecticut improperly impeded interstate commerce, White offered three alternative grounds on which to uphold the state law. First, without deciding the question, he asserted that in view of the "peculiar nature" of the state's ownership of game, "it may well be doubted whether commerce is created" by the killing and subsequent sale of such game.[34] Sec-

[29] *Id.*
[30] See text accompanying notes 113–116 *infra.*
[31] 161 U.S. 519 (1896).
[32] *Id.* at 522.
[33] *Id.* at 528–529.
[34] *Id.* at 530.

ond, even if it did constitute commerce, it was at most only intrastate commerce.[35] Finally, even if interstate commerce were impeded, the "duty of the state to preserve for its people a valuable food supply" authorized the exercise of the state's police power to that end so long as interstate commerce was only "remotely and indirectly affected."[36]

Because of the breadth of its language, the *Geer* opinion soon came to be regarded, and was long regarded, as the bulwark of the state ownership doctrine. Yet the precise legal issue decided in *Geer* was quite narrow. Moreover, *Geer* itself recognized that the power that it found in the states could continue to exist only "in so far as its exercise may be not incompatible with, or restrained by, the rights conveyed to the federal government by the Constitution."[37] Nevertheless, by intermixing questions of state authority to regulate the taking and disposition of wildlife with such technical property concepts as "ownership," *Geer* sparked a long and continuing debate about the respective powers of the state and federal governments over wildlife. Taken to its extreme, as some of its proponents would try to do, the state ownership doctrine would render impossible the development of a body of federal wildlife law. Yet only four years after *Geer*, federal wildlife law took its first major step with passage of the Lacey Act of 1900.[38]

THE CONSTITUTION AS A SOURCE OF FEDERAL AUTHORITY FOR WILDLIFE REGULATION

The Lacey Act was a cautious first step in the field of federal wildlife regulation. Founded upon the Constitution's grant to Congress of the power to regulate commerce between the states,[39] the Act's central provision prohibited the interstate transportation of "any wild animals or birds" killed in violation of state law.[40] Thus, the Act's principal thrust was to enlist the aid of the federal government, through its powers over interstate commerce, in the enforcement of state game laws.

The Lacey Act went even further in bolstering the states' regulatory authority over wildlife. It included a provision taken almost verbatim from legislation designed to permit "dry" states to block the importation of alcohol, stating that whenever dead wildlife was imported into a state, it

[35]*Id.* at 530–31. This was obviously a "bootstrap" argument because it was the state's own prohibition of export, the very matter under dispute, that confined the commerce in game to intrastate commerce.

[36]*Id.* at 534.

[37]*Id.* at 528.

[38]Ch. 553, 31 Stat. 187 (1900) (current version at 16 U.S.C. §§ 3371–3378 and 18 U.S.C. § 42).

[39]*See* note 19 *supra*.

[40]Ch. 553, § 3, 31 Stat. 188 (1900) (current version at 16 U.S.C. § 3372(a)).

was subject to the state's laws as if it were killed there.[41] Thus, whereas *Geer* had upheld a state's authority to prohibit the export of game lawfully killed within the state, the Lacey Act sanctioned a state's prohibition of the import of game lawfully killed in other states. Accordingly, in the view of some early courts, the Act was tantamount to an abdication of federal powers to regulate interstate commerce.[42]

Other provisions of the Lacey Act had implications for the scope of federal power, though their importance was not much noted at the time. One provision prohibited the importation of named animals, including starlings and English sparrows, and "such other birds or animals as the Secretary of Agriculture may . . . declare injurious to the interest of agriculture or horticulture."[43] The source of this exercise of federal authority, the congressional power over foreign commerce, has never been seriously challenged.[44]

Finally, in direct response to the decimation of the passenger pigeon and the depletion of several other bird species, the Lacey Act authorized the Secretary of Agriculture to adopt all measures necessary for the "preservation, distribution, introduction, and restoration of game birds and other wild birds," subject, however, to the laws of the various states and territories.[45] In this manner, the federal government began to inch slowly toward direct wildlife management.

The cautious approach embodied in the Lacey Act appeared to be eminently justified when, twelve years later, the Supreme Court decided "*The Abby Dodge*."[46] In that action, the United States brought suit against a vessel for its alleged violation of a federal statute prohibiting the taking of sponges from the Gulf of Mexico or the Straits of Florida by means of diving apparatus.[47] The vessel owner contended that the sponges were taken in Florida's territorial waters and that, if the federal statute applied to such waters, it was unconstitutional because the taking of sponges there was a matter "exclusively within the authority of the states."[48] Chief Justice White, author of the *Geer* opinion sixteen years earlier, agreed. Although he upheld the federal statute's validity, he did so by construing it to apply only beyond Florida's territorial waters. He offered the following rationale:

[41]Ch. 553, § 5, 31 Stat. 188–189 (1900) (*repealed* 1981).

[42]*See, e.g.*, State v. Shattuck, 96 Minn. 45, 104 N.W. 719 (1905); People v. Bootman, 180 N.Y. 1, 72 N.E. 505 (1904). *Cf.* New York *ex rel.* Silz v. Hesterberg, 211 U.S. 31 (1908).

[43]Ch. 553, § 2, 31 Stat. 188 (1900) (current version at 18 U.S.C. § 42).

[44]The Congress shall have power . . . To regulate Commerce with Foreign Nations. . . . U.S. Const. art. I, § 8, cl. 3.

[45]Ch. 553, § 1, 31 Stat. 187 (1900) (current version at 16 U.S.C. § 701).

[46]223 U.S. 166 (1912).

[47]Act of June 20, 1906, ch. 3442, 34 Stat. 313 (*repealed* 1914).

[48]223 U.S. at 173.

In view of the clear distinction between state and national power on the subject, long settled at the time the act was passed . . . we are of opinion that its provisions must be construed as alone applicable to the subject within the authority of Congress to regulate, and, therefore, be held not to embrace that which was not within such power.[49]

Thus, Justice White's *Abby Dodge* decision was the first—and last—Supreme Court statement that the state ownership doctrine actually precluded federal wildlife regulation. Since then the Court's decisions have made clear that there are at least three separate sources of constitutional authority for federal wildlife regulation: the treaty-making power, the property power, and the commerce power.

The Federal Treaty-Making Power

With the *Abby Dodge* decision only a year old, the outlook was hardly auspicious when Congress enacted the Migratory Bird Act of 1913.[50] The Act was a part of the Appropriations Act for the Department of Agriculture. It declared all migratory game and insectivorous birds "to be within the custody and protection of the government of the United States" and prohibited their being hunted except pursuant to federal regulations.

The Act's constitutionality was considered in two federal district court cases, *United States v. Shauver*[51] and *United States v. McCullagh*,[52] and in each case it was found wanting. It appears that the government made a rather feeble effort to support its law on the basis that the migratory character of the birds made them subject to Congress's power to regulate interstate commerce. Both courts rejected the government's argument as foreclosed by *Geer*. The government asserted a more vigorous claim that the law was supported by the Constitution's property clause, which authorized Congress to make all "needful Rules and Regulations" concerning the property of the United States.[53] The courts likewise rejected that argument on the basis that *Geer* had placed "property" in game in the states.

The *Shauver* case was appealed to the Supreme Court and argued twice, initially before a bench of only six Justices Apparently fearful of an adverse decision, the Department of Agriculture urged the Department of State to conclude a treaty with Great Britain (on behalf of Canada) for

[49] *Id.* at 175.

[50] Act of March 4, 1913, ch. 145, 37 Stat. 828, 847 (*repealed* 1918).

[51] 214 F. 154 (E.D. Ark 1914), *appeal dismissed*, 248 U.S. 594 (1919).

[52] 221 F. 288 (D. Kan. 1915).

[53] "The Congress shall have power to dispose of and make all needful Rules and Regulations respecting the Territory or other Property belonging to the United States." U.S. Const. art. IV, § 3.

the protection of migratory birds.[54] The treaty was signed on August 16, 1916.[55] After passage in 1918 of implementing legislation, the Migratory Bird Treaty Act,[56] the Supreme Court dismissed the government's appeal in *Shauver* and thus never decided the constitutionality of the 1913 Act.

The question of the 1918 Act's constitutionality was very soon before the Supreme Court when the state of Missouri filed a bill in equity seeking to restrain Ray Holland, a United States Game Warden, from enforcing the Act within the state. The United States contended that the Treaty and its implementing legislation took precedence over any conflicting state power of regulation by virtue of the Constitution's supremacy clause.[57] The Court's landmark decision in *Missouri v. Holland*[58] dealt a stunning blow to those who believed the state ownership doctrine was a bar to federal wildlife regulation. Justice Oliver Wendell Holmes, writing for a seven-member majority that included Justice Louis Brandeis and even Chief Justice White, easily disposed of Missouri's ownership argument:

> The State . . . founds its claim of exclusive authority upon an assertion of title. . . . No doubt it is true that as between a State and its inhabitants the State may regulate the killing and sale of such birds, but it does not follow that its authority is exclusive of paramount powers. To put the claim of the State upon title is to lean upon a slender reed. Wild birds are not in the possession of anyone; and possession is the beginning of ownership. . . .
>
> . . .
>
> But for the treaty and the statute there soon might be no birds for any powers to deal with. We see nothing in the Constitution that compels the Government to sit by while a food supply is cut off and the protectors of our forests and our crops are destroyed. It is not sufficient to rely upon the States. The reliance is vain.[59]

Missouri v. Holland established beyond question the supremacy of the federal treaty-making power as a source of authority for federal wildlife regulation.[60] More importantly, it forcefully rejected the contention that

[54]Comment, *Treaty-Making Power as Support for Federal Legislation*, 29 Yale L.J. 445 (1920).

[55]Convention for the Protection of Migratory Birds, Aug. 16, 1916, United States-Great Britain (on behalf of Canada), 39 Stat. 1702. T.S. No. 628. See Chapter 4 for a discussion of this treaty and its implementing legislation.

[56]Ch. 128, 40 Stat. 755 (1918) (current version at 16 U.S.C. §§ 703–711).

[57]"This Constitution, and the Laws of the United States which shall be made in Pursuance thereof; and all Treaties made, or which shall be made, under the Authority of the United States, shall be the Supreme Law of the Land." U.S. Const. art. VI.

[58]252 U.S. 416 (1920).

[59]*Id.* at 434–35.

[60]Coggins & Hensley, *Constitutional Limits on Federal Power to Protect and Manage Wildlife: Is the Endangered Species Act Endangered?*, 61 Iowa L. Rev. 1099, 1124–25 (1976), discusses various asserted limitations on the federal government's authority to regulate wildlife through the treaty-making power, and concludes that they are illusory. Palila v. Hawaii Dep't of Land and Natural Resources, 471 F. Supp. 985 (D. Ha. 1979), *aff'd*, 639 F.2d 495 (9th Cir. 1981),

the doctrine of state ownership of wildlife barred federal wildlife regulation, and it invited the question of what further sources of federal power might be used in developing a body of federal wildlife law.

The Federal Property Power

The scope of the property clause, rejected by lower federal courts as a source of authority for federal wildlife regulation in *Shauver* and *McCullagh*, was considered by the Supreme Court only eight years after *Missouri v. Holland*. In fact, the federal government had been exercising this power for a considerable time, for in 1894 it had prohibited all hunting in Yellowstone National Park.[61] In addition, in 1906 it prohibited the hunting of birds "on any lands of the United States which have been set apart or reserved as breeding grounds for birds by any law, proclamation, or Executive order," except under regulations of the Secretary of Agriculture, thus asserting regulatory authority over the taking of wildlife on the newly created federal wildlife refuges.[62] There are no reported cases challenging these assertions of authority, probably because the federal government was assumed to have the same right as any other landowner to prohibit hunting on its land. When the government sought to remove wildlife from these lands without complying with state law, however, several states challenged this action as being outside the scope of the powers conferred by the property clause.

The first of several cases to present the issue was the 1928 Supreme Court case of *Hunt v. United States*.[63] In that case the Secretary of Agriculture directed the removal of excess deer in Kaibab National Forest because of their threatened harm to the forest through overbrowsing. State officials arrested people carrying out the Secretary's directive, and

---

might have furnished the basis for an interesting inquiry into the limits of the treaty-making power. In that action the state contested the constitutionality of the Endangered Species Act as applied to a nonmigratory bird found only on state-owned lands within a single state. With only cursory discussion of the treaty issue, the district court upheld the law under both the treaty and commerce powers of the federal government. 471 F. Supp. at 993–94.

[61]Act of May 7, 1894, ch. 72, 28 Stat. 73.

[62]Act of June 28, 1906, ch. 3565, § 1, 34 Stat. 536 (current version at 18 U.S.C. § 41). Pelican Island Refuge, established in 1903, is often regarded as the first federal wildlife refuge. *See, e.g.*, Department of the Interior, United States Fish and Wildlife Service, Final Environmental Statement, Operation of the National Wildlife Refuge System, app. F (Nov. 1976). President Harrison's proclamation some eleven years earlier reserving Alaska's Afognak Island may in fact deserve that distinction. Harrison's proclamation reserved the island "in order that salmon fisheries in the waters of the Island, and salmon and other fish and sea animals, and other animals and birds . . . may be protected and preserved unimpaired." Proclamation No. 39, 27 Stat. 1052 (1892). See discussion of the National Wildlife Refuge System in Chapter 8.

[63]278 U.S. 96 (1928).

the United States brought suit to enjoin the state from enforcing its game laws with respect to the removal program. The state relied upon *Geer* and the other ownership cases. Without even mentioning those cases, however, the Supreme Court ruled that "the power of the United States to thus protect its lands and property does not admit of doubt . . . the game laws or any other statute of the state . . . notwithstanding."[64] Twelve years later, the *Hunt* holding was extended to acquired national forest lands in *Chalk v. United States*.[65]

Despite the decisions in *Hunt* and *Chalk*, disagreement concerning the federal government's authority to regulate wildlife on its own lands remained. On December 1, 1964, the Office of the Solicitor for the Department of the Interior issued a memorandum opinion in response to a request from the Fish and Wildlife Service for a determination of the Secretary's authority to promulgate hunting and fishing regulations for lands within the National Wildlife Refuge System. The Solicitor's opinion went beyond the narrow question put to him and declared that the United States "has constitutional power to enact laws and regulations controlling and protecting . . . [its] lands, including the . . . resident species of wildlife situated on such lands, and that this authority is superior to that of a State."[66]

The Solicitor's opinion touched off a storm of controversy. Even the comprehensive study of fish and wildlife resources on the public lands undertaken for the Public Land Law Review Commission concluded that the Solicitor had been overzealous and that at the very least the cases required a "clear showing of damage to Federal property before action in violation of State Law is sanctioned."[67]

The asserted distinction relied upon in the Commission's study was promptly rejected in *New Mexico State Game Commission v. Udall*, in which the Secretary of the Interior directed the killing of some deer in Carlsbad Caverns National Park solely for research purposes, without compliance with state game laws and without any showing of depredation by the deer.[68] The concerns of many of the state game agencies were expressed through the amicus brief of the International Association of Game, Fish, and Conservation Commissioners:

> [T]his occurrence is but one in a series of recent endeavors by the Department of the Interior to enter the field of game management, a role which has been historically and competently fulfilled by the States. . . . Interior Department admin-

[64] *Id.* at 100.

[65] 114 F.2d 207 (4th Cir. 1940).

[66] Quoted in G. Swanson, Fish and Wildlife Resources on the Public Lands 15 (1969).

[67] *Id.* at 32.

[68] 410 F.2d 1197 (10th Cir.), *cert. denied sub nom.* New Mexico State Game Comm'n v. Hickel, 396 U.S. 961 (1969).

istrators increasingly claim that, on federally owned lands, the federal government has the right to manage and control wildlife, including the right to take and dispose of such game as the Department deems appropriate, and the State may not interfere.[69]

Similarly, the state of Michigan in its amicus brief argued that the federal government's "next logical step will be to use this new power for the purpose of regulating hunting and fishing on these lands and charging a license for such privilege."[70] Notwithstanding these fears, the Court upheld the research program because of the Secretary's necessary power to determine which animals "*may be detrimental* to the use of the park."[71]

Although the *Udall* decision obviated the need for a "clear showing of damage" to federal land before wildlife removal would be permitted, there nevertheless appeared to remain some required nexus between protecting land and regulating wildlife. Indeed, it was on this basis that a three-judge federal court in 1975 struck down the Wild Free-Roaming Horses and Burros Act, thus setting the stage for the Supreme Court's decision in *Kleppe v. New Mexico.*[72] The *Kleppe* decision is the Court's most recent, and probably definitive, pronouncement on the property clause as a basis for federal authority to regulate wildlife.

The Wild Free-Roaming Horses and Burros Act[73] was enacted in 1971 to protect all unbranded and unclaimed horses and burros on the public lands of the United States as "living symbols of the historic and pioneer spirit of the West." The Act declared these animals to be "an integral part of the natural system of public lands"[74] and directed the Secretaries of the Interior and Agriculture "to protect and manage [them] as components of the public lands."[75] The *Kleppe* case began when, at the request of a federal grazing permittee, New Mexico authorities removed some wild burros from federal land and sold them at auction. The federal Bureau of Land Management demanded that New Mexico recover and return them. Instead, the state sued the Secretary of the Interior to have the federal act declared unconstitutional. The lower court agreed with the state, distinguishing *Hunt, Chalk,* and *Udall* on the grounds that in those cases the federal efforts were lawful only because they served to protect

[69] *Id.* at 1201 n. 6.

[70] *Id.* Coggins & Hensley, *supra* note 60, at 1150, argue that the states' fear of federal wildlife regulation is probably "founded less on loss of sovereignty . . . than on loss of revenue."

[71] 410 F.2d at 1201 (emphasis in original).

[72] 426 U.S. 529 (1976).

[73] 16 U.S.C. §§ 1331–1340. For further discussion of the Wild Free-Roaming Horses and Burros Act, see Chapter 10 at text accompanying notes 165–229.

[74] 16 U.S.C. § 1331.

[75] *Id.* § 1333(a).

the federal lands, whereas here the Act was designed solely to protect the animals.[76]

The Supreme Court unanimously reversed the lower court on the grounds that protection of federal land is a sufficient, but not a necessary, basis for action under the property clause. While noting that the "furthest reaches of the power granted by the Property Clause have not yet been definitively resolved," the Court declared that the power "necessarily includes the power to regulate and protect the wildlife living there."[77] The Court's holding was a solid ratification of the controversial view expressed by Interior's Office of the Solicitor twelve years earlier. Moreover, adverting to issues raised but not resolved fifty years earlier in *Shauver* and *McCullagh,* the Court observed that "it is far from clear . . . that Congress cannot assert a property interest in the regulated horses and burros superior to that of the State."[78]

The Supreme Court has not had further occasion to consider the scope of federal authority over wildlife under the property clause. Several lower courts, however, have addressed the questions left unresolved in *Kleppe.* In particular, *United States v. Brown,* a case involving a challenge to a National Park Service prohibition against hunting on state waters within (but not a part of) the National Park System, purported to answer affirmatively the question "whether the Property Clause empowers the United States to enact regulatory legislation protecting federal lands from interference occurring on non-federal public lands, or, in this instance, waters."[79] The court upheld the prohibition as a proper exercise of property clause powers, finding that it was needed to protect wildlife and visitors on the federal lands.[80]

In *Palila v. Hawaii Dep't of Land and Natural Resources,* the district court carried *Kleppe*'s suggestion of a federal ownership interest in wildlife a step further.[81] In *Palila,* the court upheld the Endangered Species Act, as applied to nonmigratory species found on state lands, on the basis of the treaty power and commerce clause. It nonetheless suggested that the "importance of preserving such a national resource [as an endangered species] may be of such magnitude as to rise to the level of a federal property interest."[82]

[76]New Mexico v. Morton, 406 F. Supp. 1237 (D. N.M. 1975).

[77]426 U.S. at 539, 541.

[78]*Id.* at 537.

[79]552 F.2d 817, 822 (8th Cir. 1977).

[80]The court's alternative holding was that the waters on which hunting was prohibited were in fact a part of the National Park System. *Id.* at 821. Among other non-wildlife cases exploring the limits of property clause authority over non-federal lands are Leo Sheep Co. v. United States, 440 U.S. 668 (1979), United States v. Lindsey, 595 F.2d 5 (9th Cir. 1979), and Minnesota v. Bergland, 499 F. Supp. 1253 (D. Minn. 1980).

[81]*See* text accompaning note 78, *supra.*

[82]471 F. Supp. at 995 n. 40.

The Federal Commerce Power

The *Kleppe* decision and *Missouri v. Holland* clearly establish the property clause and the treaty-making power as sound sources of authority for federal wildlife law, notwithstanding the state ownership doctrine. A third source of federal authority, the power to regulate interstate commerce, has had a more uncertain relationship to the state ownership doctrine. This uncertainty stemmed largely from *Geer*, which cast considerable doubt on the federal power under the commerce clause. The *Geer* case, however, involved a question of the validity of state action rather than federal action. Moreover, the holding in *Geer* was narrowed considerably in *Foster-Fountain Packing Company v. Haydel*, in which the court held that a state terminates its absolute control over the use of wildlife once it permits part or all of the wildlife to enter the stream of commerce.[83] Nevertheless, until 1977 there was no Supreme Court decision spelling out the scope of federal wildlife regulatory authority conferred by the commerce clause.

In *Douglas v. Seacoast Products, Inc.*, the Supreme Court held that federal licenses granted to vessels pursuant to the Enrollment and Licensing Act conferred upon licensees a right to fish in coastal waters.[84] Rejecting arguments that this authority was beyond Congress's power, the Court declared that, while "at earlier times in our history there was some doubt whether Congress had power under the Commerce Clause to regulate the taking of fish in state waters, there can be no question today that such power exists where there is some effect on interstate commerce."[85] Although the fish involved in *Douglas* were migratory, the Court did not base its finding of federal authority on that fact. Rather, "[t]he movement of vessels from one State to another in search of fish, and back again to processing plants, is certainly activity which Congress could conclude affects interstate commerce."[86]

Two decisions in 1979 further reinforced the *Douglas* conclusion. The Court in *Hughes v. Oklahoma*, which concerned the validity of a state statute discriminating against interstate commerce rather than the validity of a federal statute enacted pursuant to the commerce clause, held that "[t]he definition of 'commerce' is the same when relied on to strike down or restrict state legislation as when relied on to support some exertion of

[83]278 U.S. 1 (1928).

[84]431 U.S. 265 (1977).

[85]*Id.* at 281–282

[86]*Id.* at 282. *Cf.* Brown v. Anderson, 202 F. Supp. 96 (D. Alas. 1962) (Alaska fishing law an unlawful burden on interstate commerce because many of the fishermen it affected were from other states and moved in interstate commerce). *See also* Thornton v. United States, 271 U.S. 414 (1926) (grazing cattle that wander freely across state lines are subject to federal regulation).

federal control or regulation.''[87] Citing *Hughes*, the Supreme Court later that same year, in *Andrus v. Allard*,[88] held that the Migratory Bird Treaty Act, the constitutionality of which had been upheld under the treaty power in *Missouri v. Holland*, was equally valid as an exercise of federal power under the commerce clause.[89]

Whether the commerce clause furnished an adequate basis for federal regulation of a nonmigratory species that did not move from state to state as part of any commercial activity was answered affirmatively in *Palila v. Hawaii Dep't of Land and Natural Resources*.[90] The *Palila* court reasoned that ''a national program to protect and improve the natural habitats of endangered species preserves the possibilities of interstate commerce in these species and of interstate movement of persons, such as amateur students of nature or professional scientists who come to a state to observe and study these species.''[91]

An even more sweeping statement of federal authority under the commerce clause is found in *United States v. Helsley*,[92] a case upholding the constitutionality of the Airborne Hunting Act.[93] Although the court was prepared to sustain the act simply on the basis of federal authority over air space, it went on to assert, *arguendo*, that even if the dominant purpose of the statute had been the regulation of game management,

> [C]ongressional regulation is not thwarted by arguments that the incidental connection between commerce and the regulation is used merely as an expedient to justify the law. . . . Congress may find that a class of activities affects interstate commerce and thus regulate or prohibit all such activities without the necessity of demonstrating that the particular transaction in question has an impact which is more than local.[94]

In light of these decisions, it is clear that federal authority to regulate wildlife under the commerce clause is of equal stature to that conferred by the property clause. Accordingly, federal regulation of wildlife pursuant

[87] 441 U.S. 322, 326 n. 2 (1979). This issue was expressly left open in Douglas, 431 U.S. at 282 n. 17, but resolved the following year in Philadelphia v. New Jersey, 437 U.S. 617, 621–623 (1978).

[88] 444 U.S. 51 (1979).

[89] 444 U.S. 63, note 19. In so declaring, the Court implicitly confirmed the correctness of the views expressed in the early cases of Cochrane v. United States, 92 F.2d 623 (7th Cir. 1937) and Cerritos Gun Club v. Hall, 96 F.2d 620 (9th Cir. 1938), and in the initial edition of this work at pages 32–33.

[90] 471 F. Supp. 985 (D. Ha. 1979), *aff'd on other grounds*, 639 F.2d 495 (9th Cir. 1981).

[91] 471 F. Supp. at 995. The court relied upon the analysis in the initial text at pages 32–33 in support of this conclusion. *Id.*, note 39.

[92] 615 F.2d 784 (9th Cir. 1979).

[93] 16 U.S.C. § 742j–l.

[94] 615 F.2d at 787.

to the commerce clause is unrestrained by the state ownership doctrine. In fact, the contention that state ownership bars federal wildlife regulation has received no authoritative judicial support since the 1912 decision in *The Abby Dodge*, a decision that, though never overruled, has been given a quiet interment.[95]

In sum, it is clear that the Constitution, in its treaty, property, and commerce clauses, provides support for the development of a comprehensive body of federal wildlife law and that, to the extent federal law conflicts with state law, it takes precedence over the latter. That narrow conclusion, however, does not automatically divest the states of any role in wildlife regulation or imply a preference for a particular allocation of responsibilities between the states and the federal government. It does affirm, however, that an allocation can be designed without serious fear of constitutional hindrance. There is little question that for reasons of policy, pragmatism, and political comity, the states will continue to play an important role in wildlife conservation, either as a result of federal forbearance or through creation of opportunities for states to participate in implementing federal wildlife programs.

Wildlife and the Federal Authority to Regulate Land Use under the Commerce Clause

As discussed earlier, the commerce clause of the Constitution offers an expansive basis for federal wildlife regulation. There is a growing body of authority that the presence of wildlife, particularly migratory birds, creates a commerce clause basis for federal regulation of land use. This issue has arisen in the context of federal regulations restricting the filling of non-navigable waters or so-called "isolated wetlands" that are neither part of nor adjacent to navigable or interstate waterways.

One of the first cases of this sort was *State of Utah by and through Division of Parks v. Marsh.*[96] The state sought to restrain the Army Corps of Engineers from regulating the discharge of fill material into Utah Lake, the largest freshwater lake in Utah. Because the lake had no navigable connection to waters used for interstate commerce, Utah argued that it was beyond the reach of the commerce clause. The court of appeals rejected this argument, citing a number of factors that established a nexus with interstate commerce, including the fact that "the lake is on the flyway of

[95]Coggins & Hensley, *supra* note 60, at 1139–43, suggest a fourth, implied basis for federal regulation of wildlife, that being the inherent federal power to protect those wildlife species having "symbolic" value to the nation. Although this source of federal authority has not yet been tested in the courts, Congress relied upon it in enacting such wildlife legislation as the Wild Free-Roaming Horses and Burros Act (see text accompanying note 73 *supra*) and the Bald and Golden Eagle Protection Act (see discussion in Chapter 4).

[96]740 F.2d 799 (10th Cir. 1984).

several species of migratory waterfowl which are protected under international treaties.''[97]

The *Utah* decision was not especially noteworthy, given the many other factors that provided an independent basis for concluding that the commerce clause extended to Utah Lake. However, later cases applied its holding in more tenuous factual circumstances. For example, in *Leslie Salt Co. v. United States*,[98] the court upheld Army Corps of Engineers jurisdiction over certain artificially created, seasonal wetlands that may have been used by migratory birds and endangered species. The court took note of Corps criteria defining the waters under its jurisdiction to include those that "are or would be used as habitat by birds protected by the Migratory Bird Treaties," by other migratory birds that cross state lines, or by endangered species.[99] The court found that the wetlands in question fell within the regulations because "[t]he record showed . . . that migratory birds . . . and one endangered species may have used the property as habitat," and that "[t]he commerce clause power . . . is broad enough to extend the Corps' jurisdiction to local waters which may provide habitat to migratory birds and endangered species."[100] For the last proposition, the court cited *Utah* and *Palila v. Hawaii Dep't of Land and Natural Resources.*

The idea that the mere possibility that a wetland site might be used by a migratory bird conferred federal regulatory authority over that site was too much for the Seventh Circuit Court of Appeals in *Hoffman Homes, Inc. v. EPA*.[101] There the court considered whether a one-acre wetland, unconnected via surface or groundwater with any other body of water, was within the Army Corps' jurisdiction. There was no evidence of actual use of the site by any migratory bird, only expert opinion to the effect that such use could reasonably be expected. Without evidence of actual use, the court was unwilling to uphold the Corps' jurisdiction. Judge Manion, in a concurring opinion, reasoned that even if evidence of actual use by a migratory bird had been offered, that would not have established Corps jurisdiction over an isolated wetland. "The commerce power as construed by the courts is indeed expansive," he conceded, "but not so expansive as to authorize regulation of puddles merely because a bird traveling interstate might decide to stop for a drink."[102] Whether, as Judge Manion urges, there is a limit to using the presence of migratory wildlife as a constitutional justification for extending the reach of federal regulation, the clear weight of authority is that such presence, if documented and

[97] *Id.* at 804.
[98] 896 F.2d 354 (9th Cir. 1990).
[99] *Id.* at 360.
[100] *Id.*
[101] 999 F.2d 256 (7th Cir. 1993).
[102] *Id.* at 263.

regularly occurring, will provide an adequate constitutional foundation for federal regulation.

THE CONSTITUTION AS A LIMITATION ON STATE AND FEDERAL REGULATION OF WILDLIFE

The Constitution not only serves as a source of governmental authority to conserve wildlife but also restricts the manner in which authority may be exercised. The concluding section of this chapter examines several key constitutional provisions that may limit the states and the federal government in their wildlife conservation efforts.

The Equal Protection and the Privileges and Immunities Clauses

As observed earlier in this chapter, in the year of the nation's centennial the Supreme Court in *McCready v. Virginia* upheld the validity of a state statute totally excluding noncitizens of the state from planting oysters in its tidewaters.[103] Despite the suggestion a few years later in *Manchester v. Massachusetts* that the Court had some doubts about the breadth of its holding in *McCready*, in the early decades of the twentieth century the Court upheld two other plainly discriminatory statutes. In the first of these, *Patsone v. Pennsylvania*, the Court upheld a Pennsylvania statute prohibiting unnaturalized foreign-born residents from killing wild game, and, in furtherance of this ban, prohibiting the possession by these residents of shotguns and rifles.[104] A convicted defendant challenged the statute's constitutionality under the equal protection clause of the Fourteenth Amendment.[105] In an opinion noteworthy by today's standards for its reluctance to probe the legislature's motives, the Court simply considered whether "the protection of wild life . . . warrants the discrimination" and concluded that it could not deem the assumption "that resident unnaturalized aliens were the peculiar source of the evil that it desired to prevent" to be "manifestly wrong."[106]

Ten years later, in *Haavik v. Alaska Packers' Association*, the Alaskan territorial legislature had imposed a $5 tax on each nonresident fisherman in the state.[107] The Court found the tax not to be prohibited by the priv-

[103] *See* text accompanying notes 23–25 *supra.*

[104] 232 U.S. 138 (1914).

[105] "No State shall . . . deny to any person within its jurisdiction the equal protection of the laws." U.S. Const. amend. XIV, § 1.

[106] 232 U.S. at 143.

[107] 263 U.S. 510 (1924).

ileges and immunities clause of the Constitution,[108] because it was uniformly applied to all nonresidents and nothing in the Constitution prohibited favoring local residents.

In two decisions in 1948, the Supreme Court dramatically changed course and held that the privileges and immunities clause and the equal protection clause of the Fourteenth Amendment impose strict limits on state wildlife regulation. In *Takahashi v. Fish and Game Commission*, the Court had before it a California statute, which, when originally passed during the period of Japanese evacuation, denied commercial fishing licenses to alien Japanese.[109] After the war, the law was amended to apply to all persons ineligible for United States citizenship, a classification that still affected the Japanese most heavily.

Mr. Takahashi, who had fished the waters of California pursuant to state license from 1915 to 1942, sued in the California state courts to compel the state to issue him another license. Although initially successful, Takahashi failed in the Supreme Court of California, which upheld the law on the basis of California's "proprietary interest in fish in the ocean waters within 3 miles of the shore."[110] When Takahashi sought review in the Supreme Court of the United States, he was supported by some of the best legal talent in the country. Amicus briefs on his behalf were filed by the Attorney General and the Solicitor General of the United States, Thurgood Marshall for the National Lawyers Guild, and numerous others; his case was argued before the Court by Dean Acheson. In reversing the California court, the Supreme Court, through Justice Black, quoted the "slender reed" passage of Justice Holmes's opinion in *Missouri v. Holland*,[111] and stated:

> We think that same statement is equally applicable here. To whatever extent the fish in the three-mile belt off California may be "capable of ownership" by California, we think that "ownership" is inadequate to justify California in excluding any or all aliens who are lawful residents of the State from making a living by fishing in the ocean off its shores while permitting all others to do so.[112]

Only Justices Reed and Jackson dissented.

At issue in the Court's other 1948 decision, *Toomer v. Witsell*, were sev-

[108]"The Citizens of each State shall be entitled to all Privileges and Immunities of Citizens of the several States." U.S. Const. art. IV, § 2. In 1952, the Supreme Court characterized the Haavik decision as premised on the erroneous assumption that Congress had intended to relieve the Alaskan territorial legislature from constitutional restrictions applicable to the states. *See* Mullaney v. Anderson, 342 U.S. 415, 419–20 (1952).

[109]334 U.S. 410 (1948).

[110]*Id.* at 414.

[111]252 U.S. 416, 434–35 (1920). The passage is quoted in full at text accompanying note 59 *supra*.

[112]334 U.S. at 421.

eral South Carolina statutes governing commercial fishing in the three-mile zone off its coast.[113] Principal among these was a provision imposing on nonresident commercial shrimp harvesters a license fee of $2,500, which was one hundred times greater than the fee charged residents. A group of Georgia citizens brought suit, charging that the fee differential, which the Court found to be "so great that its practical effect is virtually exclusionary," contravened the guarantees of the privileges and immunities clause.[114]

In upholding the Georgians' claim, the Court distinguished many of its precedents enunciating or relying upon the state ownership doctrine. Thus, *McCready* was distinguished on the basis that it concerned sedentary shellfish in inland waters, *Haavik* on the basis that it concerned the power of Congress rather than the power of a state, and *Patsone* on the basis (hardly believable) that "persuasive independent reasons justifying the discrimination" were advanced there.[115] The Court summed up its view of the ownership doctrine:

> The whole ownership theory, in fact, is now generally regarded as but a fiction expressive in legal shorthand of the importance to its people that a State have power to preserve and regulate the exploitation of an important resource. And there is no necessary conflict between that vital policy consideration and the constitutional command that the State exercise that power . . . so as not to discriminate without reason against citizens of other States.[116]

The Court in *Toomer* also struck down a statute requiring owners of shrimp boats fishing within three miles of South Carolina to dock at a South Carolina port, unload, pack, and stamp their catch before shipping it to another state. The state argued that if *Geer* permitted a state to prohibit altogether the shipment of its wildlife to other states, then surely it could impose lesser restrictions on export. The Court rejected the state's contention, based on *Foster-Fountain Packing Company v. Haydel.*[117] Significantly, Justices Frankfurter and Jackson, who in a separate opinion expressed the view that the state's "technical ownership" of its wildlife resources exempted its regulation of wildlife from the privileges and immunities clause, nevertheless concurred with the majority on the basis that the state statutes were offensive to the commerce clause.

Toomer and *Takahashi* thus placed broad limitations on state regulatory authority over wildlife. Nevertheless, significant opportunities for distin-

[113]334 U.S. 385 (1948).
[114]*Id.* at 396–97.
[115]*Id.* at 400.
[116]*Id.* at 402.
[117]278 U.S. 1 (1928). The Haydel case is described briefly at text accompanying note 83 *supra.*

guishing either decision on factual grounds remained. The Court in *Toomer* itself carefully noted that a differential fee structure that "merely compensate[d] the State for any added enforcement burden [nonresidents] may impose or for any conservation expenditures from taxes which only residents pay," would be permissible.[118] Despite myriad state efforts to avoid their essential holdings, the courts were largely unsympathetic.[119]

In recent years, however, the Supreme Court has given some indication that it might be inclined to narrow the scope of its earlier rulings. In *Reetz v. Bozanich*, for example, the Court refrained from ruling on a challenge to an Alaska statute limiting commercial net gear salmon-fishing licenses to persons who previously held such a license for specific salmon registration areas or who held any other commercial fishing license for a period of three years in a particular registration area.[120] Because the action alleged a violation of a provision of the Alaska Constitution that had never been construed by an Alaska court, the Supreme Court determined that the matter should first be considered by the state courts.

In 1976, it appeared that the Supreme Court might be ready to reconsider its holdings in *Toomer* and *Takahashi* when it agreed to review a lower court decision striking down Virginia laws restricting the rights of aliens and nonresidents to engage in certain commercial fisheries within the state's coastal waters. However, the Court's decision in *Douglas v. Seacoast Products, Inc.* avoided the constitutional issues and focused instead on the narrower question of whether the Virginia laws conflicted with federal legislation.[121] The precise legal question was whether Virginia's laws were preempted by the federal Enrollment and Licensing Act, a statute of 1793 vintage under which the plaintiffs' vessels had been licensed to engage in coastal fisheries.

The state sought to escape the Court's conclusion that its laws were void because they conflicted with the federal law by arguing that it could exclude federal licensees from taking resources that the state owned.[122] The Court replied that "it is pure fantasy to talk of 'owning' wild fish, birds, or animals. . . . The 'ownership' language of cases such as [*Geer* and its

[118]334 U.S. at 399.

[119]*See, e.g.*, Mullaney v. Anderson, *supra* note 108; Massey v. Apollonio, 387 F. Supp. 373 (D. Me. 1974) (three-year residency requirement for a Maine lobster license), discussed in Note, *Massey v. Apollonio: Is Residency an Impermissible Conservation Device?*, 6 Envt'l L. 543 (1975); Brown v. Anderson, 202 F. Supp. 96 (D. Alas. 1962) (Alaskan law permitting the emergency closure of certain salmon fishing areas to nonresidents); Edwards v. Leaver, 102 F. Supp. 698 (D.R.I. 1952) (Rhode Island law restricting commercial menhaden licenses to residents). *Cf.* Lynden Transport, Inc. Alaska, 532 P.2d 700, 710 (Alas. 1975).

[120]397 U.S. 82 (1970).

[121]431 U.S. 265 (1977).

[122]The state's argument rested not only upon Geer and its predecessors, but also section 3(a) of the Submerged Lands Act, 43 U.S.C. § 1311(a). *See* 431 U.S. at 283–84.

predecessors] must be understood as no more than a 19th-century legal fiction expressing 'the importance to its people that a State have power to preserve and regulate the exploitation of an important resource.''[123] Thus, while conceding the importance of state power to preserve and regulate exploitation of wildlife, the Court flatly rejected any claim that this power derives from a state's "ownership" of wildlife.

Two members of the *Douglas* court filed a separate opinion concurring with the majority's judgment but differing with part of its reasoning. In particular, they thought "that the States' substantial regulatory interests" had not been "given adequate shift" by the majority.[124] The Justices conceded that states do not "own" wildlife "in any conventional sense of that term," but they nonetheless contended that "the States have a substantial proprietary interest . . . in the fish and game within their boundaries."[125] The states' regulatory interests, according to the two Justices, "are of substantial legal moment, whether or not they rise to the level of a traditional property right."[126] Indeed, in their view, the states' regulatory interests in wildlife were so substantial that "only a direct conflict with the operation of federal law . . . will bar the state regulatory action . . . no matter how 'peripatetic' the objects of the regulation or however 'Balkanized' the resulting pattern of commercial activity."[127] Two years later, however, in *Hughes v. Oklahoma*, the Supreme Court explicitly rejected precisely that characterization of the states' interest.[128]

Before *Hughes*, however, the Court was presented with yet another challenge to a discriminatory state wildlife statute, a Montana law imposing upon out-of-state hunters a substantially higher fee for hunting elk than that imposed upon state residents. Unlike *Douglas*, this case involved no conflicting federal statute to preempt state law. Thus, the only issue before the Court in *Baldwin v. Fish and Game Commission of Montana*[129] was whether the state statute conflicted with the privileges and immunities clause of article IV, section 2, or the equal protection clause of the Fourteenth Amendment to the Constitution. In a five-member majority opinion, the Court held that it did not.

On the privileges and immunities claim, the plaintiffs contended that the case was controlled by *Toomer v. Witsell*. The Court disagreed, however, distinguishing *Toomer* on the basis that it involved licensing an activity for a "commercial livelihood." Recreational sport hunting, on the other

[123] *Id.* at 284, quoting Toomer v. Witsell, 334 U.S. 385, 402 (1948).
[124] 431 U.S. at 288 (Rehnquist, J., concurring).
[125] *Id.* at 287–88.
[126] *Id.* at 288.
[127] *Id.*
[128] 441 U.S. 322 (1979).
[129] 436 U.S. 371 (1978).

hand, was not a "fundamental" right, and therefore not protected by the privileges and immunities clause.[130]

In determining whether recreational hunting was a "fundamental right" and thus one that a state could not abridge for nonresidents, the Court considered *Geer* and the other early cases concerning the right of states to restrict wildlife resources to their own borders or for their own citizens. The plaintiffs argued that these cases had no remaining vitality. The Court disagreed. While acknowledging the numerous recent cases that had chipped away at the old state ownership cases, the Court was unwilling to "completely reject" those decisions because "[t]he fact that the State's control over wildlife is not exclusive and absolute in the face of federal regulation and certain federally protected interests does not compel the conclusion that it is meaningless in their absence."[131]

The relationship of the *Baldwin* court's discussion of the ownership cases to its conclusion that recreational sport hunting is not protected by the privileges and immunities clause is obscure. The Court seemed to be saying that the early cases that recognize states' authority "to preserve this bounty [*i.e.*, wildlife] for their citizens alone"[132] reflect an understanding early in our nation's history that access to that bounty was not considered a fundamental right. Notwithstanding that the "ownership" theory upon which those cases rested has since been discredited, they are still instructive in determining whether a "federally protected interest," *i.e.*, a constitutionally protected privilege or immunity, is at stake.[133]

Much clearer is the relationship of the discussion of the ownership cases in the opinion of the three dissenting Justices to the opinion's rationale. In the dissenting Justices' view, characterization of sport hunting as a fundamental or nonfundamental right is irrelevant. Rather, the appropriate inquiry is whether the state has a proper justification for discriminating against citizens of other states.[134] In the dissenters' view, there were only three possible justifications: conservation, cost, or the right of a state to do what it wants with that which it owns. The dissenters concluded that the record before them failed to reveal that the discrimination was necessary for any conservation purpose.[135] As to cost, the dissenters accepted the district court's finding that the fee differential was not justified on the

[130]436 U.S. at 386, 388. The three dissenting Justices argued that the fact of discrimination against nonresidents, not the nature of the right restricted, was the determining factor, but they did not dispute the majority's characterization of recreational sport hunting as a nonfundamental right. Thus, it appears that no member of the Supreme Court regards recreational sport hunting as a "fundamental" right.

[131]436 U.S. at 386–87.

[132]*Id.* at 384.

[133]*Id.* at 386.

[134]*Id.* at 402 (Brennan, J., dissenting).

[135]*Id.* at 403–4.

basis of cost allocation.[136] The third possible justification was dismissed in the following vivid terms:

> The lingering death of the [state ownership] doctrine as applied to a State's wildlife, begun with the thrust of Mr. Justice Holmes' blade in *Missouri v. Holland* . . . and aided by increasingly deep twists of the knife in [*Toomer* and other cases], finally became a reality in *Douglas v. Seacoast Products, Inc.*[137]

The Chief Justice, in a separate, concurring opinion, implicitly agreed with the dissenters that characterization of the right at stake was not the relevant inquiry. Thus, in his view, it was "the special interest of Montana citizens in its elk," rather than the nonfundamental character of the right to hunt those elk, "that permits Montana to charge nonresident hunters higher license fees without offending the Privileges and Immunities Clause."[138] Admitting that characterization of the state's special interest in wildlife as ownership was a "legal anachronism of sorts," the Chief Justice maintained that "[w]hether we describe this interest as proprietary or otherwise is not significant."[139] He then recounted the many recent decisions limiting state authority over wildlife and concluded that "[n]one of those cases hold that the Privileges and Immunities Clause prevents a State from preferring its own citizens in allocating access to wildlife within that State."[140] Had the Chief Justice not omitted *Toomer* from that catalog of cases, he would not have been able to make that patently wrong assertion.[141] Had he recognized the error, he would have been compelled either to join the dissenters or disagree with them on the basis of a conservation or cost justification for the discrimination. His only alternative was to embrace the anachronistic "fundamental rights" formulation of the majority's privileges and immunities test.[142]

The breath of air given the state ownership doctrine by *Baldwin* proved to be its last gasp. Less than a year later, concluding "that time has revealed the error" of *Geer*, the Supreme Court expressly overruled it in

[136]*Id.* at 404.
[137]*Id.* at 405.
[138]*Id.* at 393–94 (Burger, J., concurring).
[139]*Id.* at 392.
[140]*Id.* at 393.
[141]The Chief Justice may have felt that it was appropriate to exclude *Toomer* from his list of cases restricting state authority over wildlife because elsewhere in his opinion he noted that Toomer held "that the doctrine [of state ownership] does not apply to migratory shrimp located in the three-mile belt of the marginal sea." He distinguished those shrimp from "the elk involved in this case [which] are found within Montana and remain primarily within the State. As such they are natural resources of the State." 436 U.S. at 392. The shrimp, however, cannot be distinguished on that basis from the menhaden involved in Douglas, a case the Chief Justice included in his list.
[142]*See* note 130, *supra*.

Hughes v. Oklahoma.[143] The facts in *Hughes* were nearly identical to those in *Geer.* An Oklahoma law prohibited shipping out of the state, for purposes of sale, minnows seined or otherwise procured from the state's waters.[144] The defendant was a Texas minnow dealer arrested for transporting out of the state minnows purchased from a licensed Oklahoma dealer. The defendant challenged the constitutionality of the Oklahoma statute on the grounds that it unreasonably interfered with interstate commerce, even though no federal law preempted the state from regulating in this area. Seven members of the Supreme Court agreed and thus erected what may be the most formidable barrier yet to discriminatory state regulation of wildlife.

In the Court's view, the *Geer* ownership analysis had "been eroded to the point of virtual extinction" in subsequent decisions.[145] While recognizing "the legitimate state concerns for conservation and protection of wild animals underlying the 19th-century legal fiction of state ownership,"[146] the Court was unwilling to continue to regard wildlife as conceptually different from other natural resources in a state. The Court expressly overruled *Geer* by concluding that "challenges under the Commerce Clause to state regulations of wild animals should be considered according to the same general rule applied to state regulations of other natural resources."[147] Applying this rule, the Court found that, although conservation of its minnows may be a legitimate local purpose justifying some discrimination, less discriminatory means of achieving that purpose were available. Thus, the state's choice of means was unconstitutional.[148]

The most serious question to follow from *Hughes* is whether the *Baldwin* result would have been the same had the plaintiffs there based their challenge on the commerce clause rather than the privileges and immunities

[143]441 U.S. 322, 326 (1979).

[144]The prohibition did not apply to hatchery minnows or to persons leaving the state with no more than three dozen minnows in their possession. *See* 441 U.S. at 323–24 n. 1.

[145]*Id.* at 331.

[146]*Id.* at 336.

[147]*Id.* at 335.

[148]*Id.* at 337–38. Seven years later, the Supreme Court upheld a Maine law that prohibited the importation of bait fish in Maine v. Taylor, 477 U.S. 131 (1986). The action arose out of a prosecution of a bait fish importer by the United States for an alleged violation of the Lacey Act. The United States chose not to appeal the Court of Appeals determination that the Maine law was unconstitutional under Hughes. However, the state of Maine, which had intervened in the case, appealed to the Supreme Court. The Court reversed, holding that Maine's interest in preventing introduction of fish parasites and exotic species was a valid local purpose that could not be achieved through means less restrictive of interstate commerce because of the lack of an adequate sampling and inspection process for baitfish. The Court rejected Taylor's contention that the fear of introduced parasites and exotic species was speculative and likely exaggerated, noting that "Maine has a legitimate interest in guarding against imperfectly understood environmental risks, despite the possibility that they may ultimately prove to be negligible." 477 U.S. at 148.

clause. Several commentators have opined that it might not.[149] This view has considerable force. The Oklahoma law struck down in *Hughes* could hardly have affected interstate commerce more substantially than the Montana law upheld in *Baldwin.* As the *Hughes* dissent pointed out, Oklahoma permitted the export of as many minnows as anyone would want, so long as those minnows came from hatcheries and not from natural streams.[150] Though *Baldwin* held that recreation is not a fundamental right, there is no shortage of precedent for the proposition that recreation is commerce.[151] Indeed, Mr. Baldwin was an outfitter and licensed hunting guide whose livelihood depended in substantial part on out-of-state big game hunters.[152]

Had the *Hughes* commerce test been applied to the Montana differential license fee scheme, the Court would have had to examine whether the degree of burden placed upon interstate commerce was justified by a legitimate local public interest and whether that interest could be promoted with a lesser impact on interstate activities.[153] That, in effect, would have obliged the Court to make the same inquiry into the relationship between the fee differential and the difference in cost to the state of servicing residents and nonresidents, as it made in *Toomer.* Though the *Baldwin* majority refrained from expressing any view on that question, it is clear that the three dissenters believed the Montana statute failed the test.

The Fifth Amendment "Takings Clause"

One other constitutional provision, the Fifth Amendment's "takings clause," offers a potential restraint on both state and federal regulation

[149] *See, e.g.,* George C. Coggins, *Wildlife and the Constitution: The Walls Come Tumbling Down,* 55 Wash. L. Rev. 295, 318 (1980) ("Oddly enough, the Baldwin plaintiffs seem not to have raised objections under the commerce clause. Perhaps bedazzled by the similarities with Toomer, they may have forfeited a better argument."); Note, *Hughes v. Oklahoma and Baldwin v. Fish and Game Commission: The Commerce Clause and State Control of Natural Resources,* 66 Va. L. Rev. 1145, 1155 (1980) ("[A] commerce clause challenge might better protect nonresidents' individual rights than the privileges and immunities clause.").

[150] 441 U.S. at 345 (Rehnquist, J., dissenting).

[151] *See, e.g.,* United States v. International Boxing Club, 348 U.S. 236, 241 (1955). *Cf.* Flood v. Kuhn, 407 U.S. 258 (1972).

[152] 436 U.S. at 372. Other evidence in the case indicated that as many as half of the nonresident elk hunters used outfitters, that certain outfitters were dependent almost entirely upon business from nonresidents, and that a typical nonresident elk hunter spent $1,250 plus license fee and outfitter's fee for a week-long hunt. *Id.* at 374 n. 9 and 376.

[153] 441 U.S. at 331. This test derives from Pike v. Bruce Church, Inc., 397 U.S. 137 (1970), and was described by the Court as applying to challenges to state regulation of exports of natural resources. If hunting license fees are analogized to state severance taxes on natural resources, the test applied in Commonwealth Edison Co. v. Montana, 453 U.S. 609 (1981), might be applicable. This test requires that the tax be "fairly related to services provided by the State" as well as not discriminate against interstate commerce. *Id.* at 617, quoting Complete Auto Transit v. Brady, 430 U.S. 277 (1977) at 279.

to protect wildlife. The Fifth Amendment contains a series of proscriptions, the last of which is "nor shall private property be taken for public use, without just compensation."[154] Although the clearest examples of proscribed takings occur when the state or federal government physically occupies land to build a highway, dam, or other structure, a taking of private property can also occur when government regulation restricts property use. The Supreme Court held in 1922 that some degree of government regulation of private property is indispensable, but that compensation is owed when regulation goes "too far."[155] Many later cases have wrestled with the question of how far is "too far."

Whether governmental restrictions designed to protect wildlife can so interfere with private property as to give rise to a duty to compensate the property owner has been litigated frequently in both state and federal courts. With only a few exceptions, the clear weight of authority is that restrictions on killing wildlife do not result in a compensable taking of private property, even when restrictions impair a property owner's ability to conduct a profitable business or when the wildlife cause economic harm to the property owner.

In *Bailey v. Holland*,[156] for example, federal closure of the plaintiff's private land to waterfowl hunting was held not to constitute a taking of property despite the landowner's claim that his marsh property and $10,000 in improvements were rendered worthless as a result of the closure. Similarly, in *Lansden v. Hart*,[157] the court found no taking from a hunting closure, even though the plaintiff's hunting business was destroyed. State courts have generally reached the same result,[158] although two courts have ordered compensation in similar circumstances.[159] The Supreme Court in *Andrus v. Allard*[160] upheld a ban on the sale of lawfully acquired eagle feathers. The Court's holding was based, not on the qualified nature of an individual's property interest in wildlife, but on the fact that the owner was not deprived of its entire use.

There have also been many Fifth Amendment challenges to protective wildlife regulations based on damage to private property by the wildlife. These, too, have been almost universally rejected. For example, in *Moun-*

[154]U.S. Const. amend. V.

[155]Pennsylvania Coal Co. v. Mahon, 260 U.S. 393, 415 (1922).

[156]126 F.2d 317 (4th Cir. 1942).

[157]168 F.2d 409 (7th Cir.), *cert. denied*, 335 U.S. 858 (1948).

[158]*See* Collopy v. Wildlife Comm'n, 625 P.2d 994 (Colo. 1981); State v. McKinnon, 133 A.2d 885 (Me. 1957); Bauer v. State Game, Forestation and Parks Comm'n, 138 Neb. 436, 293 N.W. 282 (1940); Platt v. Philbrick, 47 P.2d 302 (Cal. 1935); and Maitland v. People, 23 P.2d 116 (Colo. 1933). *See also* Thomson v. Dana, 52 F.2d 759 (Or. 1931).

[159]Alford v. Finch, 155 So. 2d 790 (Fla. 1963) (suspicion of unfair dealing in that plaintiff's lands were closed so other lands could be opened to hunting); Allen v. McClellan, 405 P.2d 405 (N. Mex. 1965).

[160]444 U.S. 51 (1979).

tain States Legal Foundation v. Hodel, plaintiffs challenged restrictions imposed by the Wild Free-Roaming Horses and Burros Act because protected animals were allegedly consuming valuable forage on private land.[161] The court rejected this claim, reasoning that wild animals are a common property whose control and regulation are to be exercised "as a trust for the benefit of the people," quoting *Geer v. Connecticut* seven years after its narrow holding had been overruled.

In *Christy v. Hodel*, a sheep rancher had been assessed a civil penalty for killing a grizzly bear, listed as "threatened" under the Endangered Species Act, after bears had killed several of his sheep.[162] Christy claimed that the law, which prohibits killing a listed species, prevented him from protecting his sheep and thus effected a taking of his property. The court rejected this claim. It reasoned that the regulations themselves "do not purport to take, or even regulate the use of, plaintiffs' property. The regulations leave the plaintiffs in full possession of the complete 'bundle' of property rights to their sheep."[163] The plaintiffs' losses were simply the "incidental . . . result of reasonable regulation in the public interest."[164] Further, the *Christy* court found no constitutional right to defend property, a finding with which Justice White disagreed when the Supreme Court later declined to review the case.[165]

Most state court decisions to consider similar claims have come to the same result.[166] Only two state courts have required the state to compensate landowners for damage done to property by protected wildlife.[167] In both of these cases, it may be significant that the state had first offered to lease the land in question, but the landowner had declined.

It is thus apparent that courts generally find no duty to compensate landowners for economic losses that result from state or federal restrictions on the killing of wildlife. Laws like the Endangered Species Act, however, broadly prohibit not just killing protected species, but also activities that significantly damage their habitat.[168] Whether these restric-

[161]799 F.2d 1423 (10th Cir. 1986).

[162]857 F.2d 1324 (9th Cir. 1988).

[163]*Id.* at 1334.

[164]*Id.* at 1335.

[165]490 U.S. 1114, 1115 (1989) (White, J., dissenting). This aspect of the Christy decision was also criticized in Lauri Alsup, *The Right to Protect Property*, 21 Envt'l Law 209 (1991).

[166]*See* Cook v. State, 74 P.2d 199 (Wash. 1937); Barrett v. State, 220 N.Y. 243, 116 N.E. 99 (N.Y. Ct. App. 1917) (damage to timber by beavers not compensable because government has absolute authority to regulate taking of wildlife); Jordan v. State of Alaska, 681 P.2d 346 (Alaska Ct. App. 1984) (state regulation prohibiting the shooting of bears, even in defense of property, did not result in taking).

[167]State v. Herwig, 17 Wis. 2d 442, 117 N.W. 2d 335 (1962); Shellnut v. Ark. State Game & Fish Comm'n, 222 Ark. 25, 258 S.W. 2d 570 (1953).

[168]See discussion of Endangered Species Act prohibition of habitat modification in Chapter 7 at text accompanying notes 105–128.

tions will be treated differently under the Fifth Amendment from land use restrictions for other purposes has yet to be tested.

The reasoning in a recent Supreme Court takings decision, *Lucas v. South Carolina Coastal Council*,[169] suggests the possibility that land use restrictions may be imposed without offending the Fifth Amendment if they are aimed at protecting wildlife. In that case the Court held that where a land use regulation deprives a landowner of all economic value of the land, the government generally must pay compensation However, if the land use restriction inheres in the landowner's title, no compensation is owed. This qualification leaves open the possibility that because a landowner's property right has never been construed to extend to wildlife, and because under old English law the rights of private landowners were constrained by obligations to protect wildlife and its habitat,[170] restrictions to protect wildlife will not require compensation. Justice Scalia's majority opinion in *Lucas* also notes that regulations not "aimed at land" may enjoy a more liberal test.[171] Since wildlife protection regulations are directed at wildlife, rather than at land use *per se*, the door is left open to argue that the Fifth Amendment acts as less of a constraint on the government's regulatory powers when it acts to protect wildlife than when it acts for other purposes.[172]

[169]505 U.S. 1003 (1992).

[170]*See* note 3 *supra*.

[171]505 U.S. 1027, n. 14.

[172]For further discussion of the Fifth Amendment and regulation of habitat modification under the Endangered Species Act, see A. Kimberly Rockwell, *The Fifth Amendment Implications of Including Habitat Modification in the Definition of Harm to Endangered Species*, 11 J. Land Use & Envt'l L. 573 (1996); Patrick A. Parenteau, *Who's Taking What? Property Rights, Endangered Species, and the Constitution*, 6 Fordham Envt'l L.J. 619 (1995); Albert Gidari, *The Economy of Nature, Private Property, and the Endangered Species Act*, 6 Fordham Envt'l L.J. 661, 681–87 (1995); Oliver A. Houck, *Why Do We Protect Endangered Species, and What Does That Say About Whether Restrictions on Private Property to Protect Them Constitute "Takings"?* 80 Iowa L. Rev. 297 (1995).

Chapter 3

REGULATING COMMERCE IN WILDLIFE

INTRODUCTION

Most federal wildlife conservation laws regulate both taking of wildlife and commerce in wildlife and wildlife products. Restrictions on commerce serve in part to buttress federal prohibitions on taking. While federal authority for taking prohibitions has frequently been challenged, there has not been any serious doubt concerning the constitutional authority for federal regulation of interstate and foreign commerce in wildlife and wildlife products.

In fact, the first major federal wildlife statute, the Lacey Act of 1900,[1] was aimed solely at regulating interstate commerce in wildlife and refrained from asserting any federal authority over its taking.[2] That Act, supplemented until 1981 by the Black Bass Act of 1926,[3] is the cornerstone of federal efforts to conserve wildlife through regulating commerce. Other statutes, including the Migratory Bird Treaty Act,[4] Bald and Golden

[1]Act of May 25, 1900, Ch. 553, 31 Stat. 187–189 (current version at 16 U.S.C. §§ 701 and 3371–3378 and 18 U.S.C. § 42).

[2]In the debate that preceded passage of that statute, Congressman Lacey explained that his bill would not itself prohibit the taking of any wildlife because "to do that it would become necessary to enact a national game law, which . . . would be unconstitutional." 33 Cong. Rec. 4873 (1900). As described in Chapter 2, Lacey's comments reflected the thinking of the day that the federal role in wildlife regulation was extremely narrow because states "owned" wildlife within their borders.

[3]Act of May 20, 1926, Ch. 346, 44 Stat. 576 (codified at 16 U.S.C. §§ 851–856 (1976)) (repealed 1981).

[4]The Migratory Bird Treaty Act makes it unlawful for any person to "possess, offer for sale, sell, offer to barter, barter, offer to purchase, purchase, deliver for shipment, ship, export,

Eagle Protection Act,[5] Wild Free-Roaming Horses and Burros Act,[6] Marine Mammal Protection Act,[7] and the Endangered Species Act[8] also restrict commerce in the wildlife they seek to conserve.

A second purpose of the Lacey Act was to prohibit importation of wildlife injurious to agricultural interests. Other federal laws, such as Executive Order 11987,[9] the Federal Noxious Weed Act,[10] and the Nonindigenous Aquatic Nuisance Prevention and Control Act of 1990,[11] also aim to control the intentional or unintentional spread of injurious animals and plants. More recently, these laws have been prompted by recognition of the potential for harm to natural ecosystems from introduction of non-native species, as well as concern for agricultural and horticultural interests. This chapter first examines restrictions on commerce in unlawfully taken wildlife and wildlife parts and then turns to federal efforts to control injurious and non-native plants and animals.

PROHIBITION OF COMMERCE IN UNLAWFULLY TAKEN WILDLIFE

The Original Lacey Act

According to the House committee report that accompanied the Lacey Act, its "most important purpose" was "to supplement the State laws for

---

import, cause to be shipped, exported, or imported, deliver for transportation, transport or cause to be transported, carry or cause to be carried, or receive for shipment, transportation, carriage, or export" any migratory bird except as permitted under regulations of the Secretary of the Interior. 16 U.S.C. § 703. These prohibitions are considerably more expansive than the limited proscriptions contained in the 1916 Convention with Great Britain against the "shipment or export" of migratory birds from any state during its closed season and against the "international traffic" in unlawfully taken migratory birds. For a discussion of the Migratory Bird Treaty Act see Chapter 4.

[5]The Bald and Golden Eagle Protection Act prohibits not only the taking of bald and golden eagles, but also their possession, sale, purchase, barter, transportation, exportation, or importation. *See* 16 U.S.C. § 668. See Chapter 4 for a more detailed discussion of this law.

[6]The Wild Free-Roaming Horses and Burros Act prohibits the processing of wild horse or burro remains into commercial products and the sale of any wild horse or burro maintained on private land. *See* 16 U.S.C. § 1338(a). For a more detailed discussion of this law, see Chapter 10.

[7]The Marine Mammal Protection Act makes it unlawful to "transport, purchase, sell, export, or offer" to do any of the foregoing with respect to any illegally taken marine mammal or any product made from such a mammal. 16 U.S.C. § 1372(a)(4). For a more detailed discussion of the Marine Mammal Protection Act, see Chapter 5.

[8]The Endangered Species Act generally prohibits commercial acivities involving endangered species. *See* 16 U.S.C. § 1538(a). For a more detailed discussion of this law, see Chapter 7.

[9]Exec. Order No. 11987, § 2(a), 3 C.F.R. at 116 (1977 Comp.), *reprinted following* 42 U.S.C.A. § 4321.

[10]7 U.S.C. §§ 2801 *et seq.*

[11]16 U.S.C. § 4701 *et seq.*

the protection of game and birds,"[12] by making it unlawful for any person to deliver to a common carrier, or for any common carrier to transport from one state or territory to another, wild animals or birds killed in violation of state or territorial law.[13] To aid enforcement, the Act also required that all packages containing dead animals, birds, or parts thereof be clearly marked when shipped in interstate commerce.[14] The Lacey Act prescribed maximum fines of $200 for shippers and for consignees "knowingly receiving such articles," and $500 for carriers "knowingly carrying or transporting the same."[15]

To bolster state wildlife laws further, the Lacey Act also sought to prevent game from being shipped into a state in order to circumvent prohibitions on the sale of local game killed in violation of that state's laws. Section 5 of the Act, repealed in 1981, provided that

> the dead bodies, or parts thereof, of any wild game animals, or game or song birds transported into any State or Territory . . . shall . . . be subject to the operation and effect of the laws of such State or Territory . . . as though such animals or birds had been produced in such State or Territory.[16]

In this way, Congress sought to clarify an aspect of state power over commerce in wildlife that had been rife with uncertainty.

The issue of the Lacey Act's constitutionality has never reached the Supreme Court, but the few lower courts that considered the Act's constitutionality during its initial decades uniformly upheld it.[17] Most of the early Lacey Act litigation addressed a different question: How did this new federal law affect state power to regulate possession or sale of wildlife imported from other states?

Illustrative of the cases that presented this question is *People v. Bootman*, involving New York's Forest, Fish and Game Law of 1900.[18] The law prohibited possessing certain birds during the state's closed season. In a case decided shortly before the Lacey Act's passage, a New York court held that the state's law could not be applied to persons who had lawfully acquired birds in another state, brought them into New York during its

[12]H.R. Rep. No. 474, 56th Cong., 1st Sess. 2 (1900).

[13]Act of May 25, 1900, ch. 553, § 3 (current version at 16 U.S.C. § 3372(a)).

[14]*Id.* (current version at 16 U.S.C. § 3372(b)).

[15]*Id.* § 4 (current version at 16 U.S.C. § 3373).

[16]*Id.*. § 5 (codified at 16 U.S.C. § 667e (1976) (repealed 1981)). The provision's unusual wording came almost verbatim from the Wilson Original Package Act of 1890, 27 U.S.C. § 121, which permitted states to regulate or prohibit importation of alcoholic beverages. *See* H.R. Rep. No. 474, 56th Cong., 1st Sess. 3 (1900).

[17]*See, e.g.*, Rupert v. United States, 181 F. 87 (8th Cir. 1910), and Eager v. Jonesboro, Lake City and Eastern Express Co., 103 Ark. 288, 147 S.W. 60 (1912). For more recent cases, *see* note 71 *infra.*

[18]180 N.Y. 1, 72 N.E. 505 (1904).

open season, and continued to possess them after the season closed.[19] Four years later, the *Bootman* court again reached the same result, but this time the court commented on the Lacey Act's effect:

> If the federal statute had been passed first it would not be unreasonable to believe that the Legislature intended to so expand the meaning of our game laws as to forbid the possession of imported game during the close season. It was not passed, however, until after the enactment of the state law, and hence can have no effect upon its meaning.[20]

State v. Shattuck was a similar case in which a Minnesota court held that the Lacey Act removed all doubt as to a state's authority to prohibit possession of game lawfully acquired in another state.[21]

Shortly after the *Bootman* and *Shattuck* cases, the United States Supreme Court in *New York ex rel. Silz v. Hesterberg* upheld a state's power to apply its game laws to wildlife brought in from outside the state, wholly without regard to the Lacey Act.[22] *Silz* illustrated the extreme reach of state law: The court applied New York's law, which prohibited possession of grouse and plover during the closed season, to birds not found anywhere in the United States but imported from England and Russia.

The Black Bass Act

The original Lacey Act's language included all "wild animals or birds" without limitation. In practice, though, it apparently was construed to apply only to game birds and fur-bearing mammals.[23] In 1926 Congress enacted new legislation to provide essentially identical protection for two species of fish known commonly as black bass. The Black Bass Act, like the Lacey Act, prohibited interstate shipment of such fish taken contrary to state law, required the clear marking of all packages containing the fish, and authorized states to treat the fish, when brought in from other states, as though they had been produced in the receiving state.[24] Maximum penalties for violating the Act were $200, or three months' imprisonment, or both.[25]

Congress amended both the Lacey and Black Bass Acts several times. A

[19]People v. Buffalo Fish Co., 164 N.Y. 93, 58 N.E. 34 (1900).

[20]180 N.Y. at 7, 72 N.E. at 506–07.

[21]96 Minn. 45, 104 N.W. 719 (1905).

[22]211 U.S. 31 (1908). As a result of this holding, the repeal of section 5 of the original Lacey Act in 1981 had no practical consequence.

[23]*See* 67 Cong. Rec. 9385 (1926) (Remarks of Congressman Hawes).

[24]The Black Bass Act, which was codified at 16 U.S.C. §§ 851–856 (1976), was repealed and its provisions consolidated with those of the Lacey Act as a result of the Lacey Act Amendments of 1981, Pub. L. No. 97–79 § 9(b)(1), 95 Stat. 1079.

[25]16 U.S.C. § 853 (1976) (repealed 1981).

1935 amendment extended the former's prohibition on interstate commerce to include wild animals, birds, and parts or eggs thereof, captured or killed contrary to federal law or the laws of any foreign country.[26] Similarly, the Black Bass Act was amended in 1969 to apply to any fish taken contrary to the laws of a foreign country.[27]

The Tariff Act of 1930

Congress took a significant step to aid enforcement of foreign conservation laws even before the 1935 Lacey Act amendment, by enacting the Tariff Act of 1930. Section 527 of the Tariff Act provides that if the laws of any foreign country or political subdivision thereof restrict the taking or exporting of any wild mammal, bird, or part or product thereof, no such bird or mammal may be imported, directly or indirectly, into the United States without a certification from the United States consul at the place of export that it was not acquired or exported in violation of those laws.[28] Animals or products imported in violation of this provision may be seized and forfeited, whether or not the importer is culpable.[29]

The Tariff Act was a progressive measure in 1930, substantially enhancing the Lacey Act's effectiveness. The Tariff Act's narrow scope, however, has reduced its importance today. It covers only birds and mammals, whereas the Lacey Act now prohibits importation of, and interstate commerce in, any type of unlawfully acquired foreign fish or wildlife.[30] For

[26]Act of June 15, 1935, ch. 261, tit. II, § 201, 49 Stat. 380 (current version at 16 U.S.C. § 3372(a)(2)(A)).

[27]Endangered Species Conservation Act of 1969, Pub. L. No. 91–135, § 9(a), 83 Stat. 281 (repealed 1981). Earlier, the Black Bass Act had been amended to encompass all "game" fish in 1947, Act of July 30, 1947, ch. 348, 61 Stat. 517, and then to all United States fish in 1952, Act of July 16, 1952, ch. 911, § 2, 66 Stat. 736.

[28]19 U.S.C. § 1527(a).

[29]*Id.* § 1527(b). *See* United States v. Fifty-three (53) Eclectus Parrots, 685 F.2d 1131, 1133–34 (9th Cir. 1982). Until amended in 1981, the Lacey Act authorized forfeitures of illegally traded wildlife only in connection with "knowing and willful" violations of that Act. *Compare* 16 U.S.C. § 3374(a) with 18 U.S.C. § 43(e) (1976) (repealed 1981).

[30]In addition to the Lacey Act's prohibition on importing unlawfully taken foreign wildlife, the Tariff Act of 1962 prohibits importing the feathers or skin of any bird, subject to several exceptions. *See* 19 U.S.C.A. § 1202 and Additional U.S. Note to Chapter 5, Harmonized Tariff Schedule of the United States. Among the exceptions, skins bearing feathers of eight named species of foreign birds may be imported, subject to specified quotas and pursuant to permits from the Secretary of the Interior, for purposes of making artificial flies for fishing. In 1993, however, the U.S. Fish and Wildlife Service rescinded regulations that implemented these feather import quotas. *See* 58 Fed. Reg. 60524, 60525 (1993), rescinding 50 C.F.R. § 15.11 (1993). The Service described the rescinded regulations as "unnecessary and wasteful of government and private resources." It is unclear whether the Service's action rescinded the feather quotas themselves. Customs Service regulations continue to require Fish and Wildlife Service permits to import the species in question. *See* 19 C.F.R. § 12.29 (1995).

the categories of wildlife it covers, however, the Tariff Act is easier to enforce. The Act requires consular certification for certain imported birds and mammals. For wildlife not subject to the Tariff Act, no consular certification is required, and thus effective enforcement depends heavily upon customs agents' knowledge of foreign wildlife laws.[31]

The Tariff Act's exceptions present some anomalies. For example, animals imported for scientific or educational purposes are exempt.[32] This exemption may have reflected a policy judgment in 1930 that the United States would not assist in enforcing foreign wildlife laws if the purpose of importation were scientific or educational. Nevertheless, Congress reversed that policy when it amended the Lacey Act in 1935 to prohibit importation of *all* wildlife taken contrary to foreign law. The exemption in the Tariff Act thus is an anachronism serving no current policy.

It was unclear until 1982 whether the Tariff Act protects against the practice of exporting to the United States from one country wildlife unlawfully taken in a third country. The Act on its face prohibits importing, "directly or indirectly," wildlife taken contrary to the laws of the country of origin. The only certification it requires, however, is that of the United States consul "for the consular district in which is located the port or place from which such [animal] was exported." The Ninth Circuit interpreted the latter provision "to require proper documentation from the animal's country of origin, whether or not the United States importer was involved in the initial export from that country."[33]

The 1949 Lacey Act Amendments

In 1949, Congress significantly amended the Lacey Act a second time. The amendments prohibited importing "wild animals or birds" under conditions known to be "inhumane or unhealthful" and authorized the

[31]Fish and Wildlife Service regulations provide that, to obtain customs clearance for imported wildlife, the importer must provide "[a]ll permits, licenses or other documents required by the laws or regulations of the United States." 50 C.F.R. § 14.52(c)(2) (1995). Required documents will vary, depending upon what kind of animal is involved and whether it is protected by the Convention on International Trade in Endangered Species of Wild Fauna and Flora (hereinafter CITES). Under CITES, official export permits *must* accompany all specimens listed on any of its three appendices. See Chapter 14 at text accompanying notes 144–154. Although any country that is a party to CITES may list on Appendix III those species within its jurisdiction whose protection requires the cooperation of other parties, failure to list any protected species does not diminish the Lacey Act's requirement that only lawfully taken specimens may be imported.

[32]19 U.S.C. § 1527(c)(2).

[33]United States v. Fifty-three (53) Eclectus Parrots, 685 F.2d 1131, 1134 (9th Cir. 1982). Similarly, CITES requires "re-export" certificates whenever importation of listed species is from a country other than its country of origin. *See* Chapter 14 at text accompanying note 144.

Secretary of the Treasury to prescribe requirements for transporting such animals under humane and healthful conditions.[34] The amendment further provided that "the presence . . . of a substantial ratio of dead, crippled, diseased, or starving wild animals or birds shall be deemed prima facie evidence" of a violation. The ambiguity inherent in the statute's terms (including the very term "wild animals or birds") perhaps explains why there have been virtually no reported cases under this provision since it was enacted nearly five decades ago.[35]

Congress originally vested responsibility for administering the Lacey and Black Bass Acts in the Secretaries of Agriculture and Commerce, respectively. Responsibility for both laws was later transferred to the Secretary of the Interior, and after three decades of consolidated responsibility, administration was once again divided between the Secretaries of Commerce and the Interior in 1970.[36]

The 1981 Lacey Act Amendments

The first truly comprehensive amendments to the Lacey Act came in 1981. These amendments repealed the Black Bass Act and merged its provisions into the Lacey Act.[37] Congress also substantially expanded the latter's scope, revised the duties it imposes, and strengthened the enforcement handles it makes available. The only significant part of the Lacey Act left untouched was the provision governing importation of injurious wildlife.[38]

The major changes affected provisions prohibiting certain dealings in specimens taken, transported, or sold in violation of underlying state, federal, or foreign laws. These provisions formerly covered only wild vertebrates other than migratory birds, mollusks, and crustaceans;[39] they now apply to all wild animals, including those bred in captivity, and to some wild plants.[40] The underlying laws, violation of which can lead to a Lacey

[34]Act of May 24, 1949, ch. 139, § 2, 63 Stat. 89. Amendments in 1981 transferred the Secretary of the Treasury's responsibilities to the Secretary of the Interior. *See* 18 U.S.C. § 42(c).

[35]In United States v. States Marine Lines, Inc., 334 F. Supp. 84, 89 (S.D.N.Y. 1971), the court declined to rule that 15 percent mortality among shipped animals at the time of arrival or 30 percent mortality within a few days thereafter constituted the required "substantial ratio" for a *prima facie* violation of the statute.

[36]Reorg. Plan No. 4 of 1970, 35 Fed. Reg. 15627, 84 Stat. 2090.

[37]Pub. L. No. 97-79, 95 Stat. 1073 (1981).

[38]18 U.S.C. § 42(a) and (b). *See* text accompanying notes 87–94 *infra.*

[39]18 U.S.C. § 43 (1976) (repealed 1981).

[40]16 U.S.C. § 3372(a). Plants now subject to the Lacey Act include only those that are indigenous to any state and that are either listed on any appendix of CITES or pursuant to any state law providing for conservation of species threatened with extinction. *Id.* § 3371(f). Plant conservation laws of foreign countries are not among the underlying laws, violation of which can lead to a Lacey Act violation. *Cf. id.* §§ 3372 (a)(2)(B) and (a)(3)(B).

Act violation, now include Indian tribal laws and federal treaties, as well as state, federal, and foreign laws.[41] Finally, while formerly the law prohibited most interstate or foreign trafficking in specimens "taken, transported, or sold" contrary to underlying laws, it now also prohibits trafficking in specimens "possessed" contrary to any underlying state or foreign law.[42]

The 1981 amendments also substantially increased the penalties for Lacey Act violations. The maximum criminal fine was doubled to $20,000; the maximum jail sentence was increased from one to five years. Penalties of this magnitude can be assessed against importers, exporters, or those engaged in the purchase or sale of protected specimens if they knew that the specimens were taken, possessed, transported, or sold in violation of an underlying law.[43] With the exception of importers and exporters, however, these severe criminal penalties apply only when the market value of the specimens exceeds $350.[44] Under prior law, only those who "knowingly and willfully" violated the Lacey Act were subject to criminal penalties.[45] The law no longer requires the government to prove that a defendant knew of the Lacey Act's prohibitions and intended to violate them.

Offenses that fit within the above categories are considered felonies, by virtue of the maximum allowable penalties.[46] Misdemeanor offenses include any violation of the Act in which the violator knowingly engaged in conduct the Act prohibits and, in the exercise of due care, should have known that the specimens involved were taken, possessed, transported, or

[41]*Id.* § 3372(a)(1). 18 U.S.C. § 1165 prohibits trespass upon Indian reservations for the purpose of hunting, trapping, or fishing. The court in United States v. Sanford, 547 F.2d 1085, 1088 (9th Cir. 1976), held that transporting wildlife taken after an unlawful trespass is not a violation of the Lacey Act, since it is not the taking, but rather the trespass, that federal law prohibits. The 1981 amendments eliminated the significance of that distinction, since any taking after trespass violates Indian tribal law.

[42]Treaties and Indian tribal laws are treated differently from underlying state and foreign laws in that specimens "taken or possessed" contrary to the former are subject to the Lacey Act, whereas specimens "taken, possessed, transported, or sold" contrary to state laws are subject to the Act. *Compare* 16 U.S.C. § 3372(a)(1) *with id.* § 3372(a)(2)(A).

[43]*Id.* § 3373(d)(1). Criminal penalties do not apply to violations of the Act's container-marking requirements. See text accompanying note 53 *infra.*

[44]16 U.S.C. § 3373(d)(1)(B). For cases discussing the market value requirement, see text accompanying notes 82–83 *infra.*

[45]18 U.S.C. § 43(d)(1976) (repealed 1981).

[46]Even if the underlying state law violation constitutes only a misdemeanor, so long as the market value of the wildlife exceeds $350 and the guilty party is knowingly involved in the sale, purchase, import, or export of the illegally taken wildlife, the Lacey Act offense is punishable as a felony. *See* Sen. Rep. No. 97–123, 97th Cong., 1st Sess. (1981) at 15, *reprinted in* 1981 U.S.C.C.A.N. 1748, 1762 *See also* United States v. Thomas, 887 F.2d 1341, 1349 (9th Cir. 1989).

sold contrary to an underlying law.[47] The maximum criminal penalties for misdemeanor offenses are a $10,000 fine and a year in prison. Civil penalties also may be imposed where the defendant should have known of an underlying state law violation. Maximum civil penalties are $10,000 per violation.[48]

Congress also amended several other enforcement provisions. Specimens involved in any violation now may be forfeited under a strict liability standard.[49] Equipment used to violate the law may be forfeited if the defendant is convicted of a felony.[50] The Secretary of the Treasury may pay rewards to persons furnishing information that leads to conviction or other enforcement actions.[51] Finally, criminal violators may lose selected federal licenses or permits.[52]

The 1981 amendments significantly relaxed one part of the Lacey Act, the provision requiring marking of packages containing fish or wildlife. The law now merely requires the Secretaries of Commerce and the Interior to promulgate jointly package-marking regulations that are "in accordance with existing commercial practices."[53] The maximum penalty for violating these regulations is a $250 civil penalty.

Judicial Interpretation of the Lacey Act

Prior to the 1981 amendments, the modest penalties and the high standard of culpability for a Lacey Act violation made it an ineffective deterrent against lucrative wildlife crime. Consequently, federal prosecutors sometimes creatively interpreted other laws to get at illegal commercial wildlife activities. In at least one case, *United States v. Plott*,[54] prosecutors successfully charged wildlife poachers with a violation of the National Stolen Property Act, which makes it a crime to "transport[s] . . . in interstate or foreign commerce any goods . . . the value of $5,000, or more, knowing the same to have been stolen, converted, or taken by fraud."[55]

[47] 16 U.S.C. § 3373(d)(2). The court may also assess civil penalties in these situations.

[48] *Id.* § 3373(a)(1). The civil penalty may not exceed the maximum that may be imposed under the underlying law when the violation pertains to specimens valued at less than $350 and does not involve importation, exportation, or sale of such specimens. *Id.* In the case of violations of the Act's container-marking requirements, the maximum civil penalty is $250. *Id.* § 3373(a)(2).

[49] *Id.* § 3374(a)(1).

[50] *Id.* § 3374(a)(2).

[51] *Id.* § 3375(d).

[52] *Id.* § 3373(e).

[53] *Id.* § 3376(a). The regulations, not promulgated until 1987, are codified at 50 C.F.R. § 14.81–82 (1995).

[54] 345 F. Supp. 1229 (S.D.N.Y. 1972).

[55] 18 U.S.C. § 2314.

The defendant contended that poached alligator hides could not be considered "stolen" goods because the peculiar nature of a state's "ownership" of alligators was not the type of ownership contemplated by the term "stolen." The court disagreed on the ground that "taking from an owner in trust constitutes stealing or converting within the meaning of [the Stolen Property Act] just as much as taking from any other owner."[56] The court found the defendant guilty of a felony potentially punishable by up to ten years' imprisonment, while a Lacey Act violation would have been a only a misdemeanor.[57]

The major changes to the Lacey Act that Congress made in 1981—broadening the range of species and the categories of offenses to which it applies, dramatically increasing maximum penalties, and lowering the standard of proof that must be met to establish a criminal violation—have made the Lacey Act one of the most important and frequently used tools of federal wildlife law enforcement officials. While there were few reported Lacey Act decisions prior to 1981, now they are numerous. This section discusses some of the more important issues addressed in those decisions.

Prosecution Based on Violation of Foreign Law

Several cases have addressed the scope of the Lacey Act's prohibitions with respect to foreign law. Prior to the 1981 amendments, a Philadelphia reptile dealer argued successfully that violation of a general customs law of a foreign nation, applicable to all exports, could not be the basis for a Lacey Act violation.[58] The court reasoned that the purpose of customs laws is simply to raise revenue, whereas the foreign laws encompassed within the scope of the Lacey Act are limited to those "designed and intended for the protection of wildlife."[59]

The 1981 amendments' legislative history indicated that Congress was dissatisfied with this reasoning. The House report concluded that this view was "too restrictive," even though it agreed that the Lacey Act is not intended to subsume foreign laws "that are plainly and solely revenue laws with no specific reference to wildlife."[60] The Ninth Circuit relied upon this legislative history in *United States v. Lee,* a case upholding a Lacey

[56]345 F. Supp. at 1232.

[57]In another case, however, the Second Circuit Court of Appeals was unpersuaded by the Plott reasoning and refused to apply the Stolen Property Act to one who had transported illegally harvested clams in interstate commerce. United States v. Long Cove Seafood, Inc., 582 F.2d 159 (2d Cir. 1978). *But see* United States v. Tomlinson, 574 F. Supp. 1531 (D. Wyo. 1983) (applying the Stolen Property Act to defendants who transported wild horses taken in violation of the Wild Free-Roaming Horses and Burros Act).

[58]United States v. Molt, 599 F.2d 1217 (3rd Cir. 1979).

[59]*Id.* at 1218–19.

[60]H.R. Rep. No. 97–276, 97th Cong., 1st Sess. 14 (1981).

Act violation premised on a violation of a Taiwanese regulation prohibiting squid fishing vessels from taking salmon.[61] The accused argued that the Taiwanese regulation was really an economic rather than a conservation measure. The court refused to probe the underlying motive for the regulation and held that because it clearly related to wildlife, it was within the scope of the Lacey Act.[62]

The *Lee* court also concluded that foreign regulations are within the scope of the Lacey Act's reference to "foreign law." When Congress amended the Lacey Act in 1981, it created an ambiguity by prohibiting trade in wildlife taken "in violation of any law or regulation of any State or in violation of any foreign law."[63] The specific reference to "law or regulation" in the case of a state, and the narrower reference only to "law" in the case of a foreign nation, allowed the *Lee* defendants to argue that the Lacey Act "encompasses only foreign statutes, not foreign regulations."[64] The court rejected this argument, finding that Congress's clear intent to expand the Act's scope required the court to interpret the phrase "foreign law" so as "not to limit the Act's applicability, but instead to encompass the wide range of laws passed by" the world's diverse legal systems.[65]

Because of the Lacey Act's broad scope, wildlife importers sometimes face a daunting task in learning the legal requirements of the nations from which they obtain their wildlife. *United States v. 2,507 Live Canary Winged Parakeets (Brotogeris versicolorus)* illustrates how complex that task can be.[66] An exotic pet dealer imported a shipment of parakeets from Peru, accompanied by a CITES permit signed by Peru's Director of Forest and Fauna.[67] When the birds were seized under the Lacey Act and the Endangered Species Act, the importer filed an action to reclaim them. Although the proper Peruvian official had signed the permit, the court

[61]937 F.2d 1388 (9th Cir. 1991).

[62]In United States v. One Afghan Urial (*Ovis orientalis blanfordi*) Fully Mounted Sheep, 964 F.2d 474 (5th Cir. 1992), the court rejected a similar claim that because Pakistan's Imports and Exports Act was enacted to regulate trade, and not specifically to protect wildlife, it was not subsumed within the Lacey Act.

[63]16 U.S.C. § 3372(a)(2)(A).

[64]937 F.2d at 1391.

[65]*Id.* In reaching this conclusion, the court followed its earlier decision in United States v. 594,464 Pounds of Salmon, 871 F.2d 824 (9th Cir. 1989). The earlier case was a forfeiture case, whereas the Lee case was a criminal prosecution. Courts usually resolve ambiguity in a criminal statute in a defendant's favor to ensure that the defendant has fair notice of what the law requires. The Lee court, however, noted that the Lacey Act allows for criminal penalties only if a violator knew or should have known that his underlying action was unlawful. For that reason, the court was unwilling to reach a different result from what it had decided in the 594,464 Pounds of Salmon case.

[66]689 F. Supp. 1106 (S.D. Fla. 1988).

[67]CITES refers to the Convention on International Trade in Endangered Species of Wild Fauna and Flora, discussed in detail in Chapter 14.

concluded (based in part on testimony by the official's predecessor) that Peruvian law prohibited export of the species in question, and therefore the official who signed the permit lacked the legal authority to do so.

The dealer asserted an "innocent owner" defense, arguing that he had relied upon a Peruvian shipper to ensure that export was lawful, and that there was no evidence suggesting any culpability on the dealer's part. The court rejected this contention, concluding that the forfeiture provisions of the Lacey Act imposed strict liability upon the importer of unlawfully imported wildlife, without regard to culpability. Moreover, the court concluded that even if this defense were available, the dealer was not entitled to assert it.

> [T]he court utterly rejects the [dealer's] contention that a major importer of exotic birds and reptiles, highly experienced in dealing with the requirements of foreign wildlife importation laws, could reasonably rely on a foreign businessman with no apparent legal training to "warranty" the legality of the shipment.[68]

The Lacey Act occasionally requires United States courts not merely to interpret a foreign law, but to determine its validity under a foreign constitution. In *United States v. Mitchell*, the defendant in a Lacey Act prosecution contended that the provincial law he was accused of violating exceeded the power of the provinces under Pakistan's constitution.[69] The district court agreed and dismissed one count of the indictment. The government appealed.

After considering evidence from the American embassy in Pakistan, the Pakistani embassy in the United States, a foreign law specialist from the Library of Congress, and various Pakistani government officials and lawyers, the court of appeals concluded that "whether the Pakistani federal government would consider such a [provincial] regulation an encroachment upon its federal powers is a matter almost impossible to determine."[70] Ultimately, the court ducked this task by finding that even if the challenged regulation were unconstitutional, it was unconstitutional with respect to some, but not all, of the alleged acts of the defendant. It reinstated the dismissed count and allowed the government to proceed to trial.

Prosecution Based on Violation of State Law

A more familiar role for United States courts is deciding the constitutionality of federal or state laws under our own constitution. Defendants in several cases have argued that the Lacey Act allows an unconstitutional

[68]689 F. Supp. at 1119.
[69]985 F.2d 1275 (4th Cir. 1993).
[70]*Id.* at 1284.

delegation of legislative powers to foreign governments or to the states, a claim that courts have uniformly rejected.[71]

A defendant charged with a Lacey Act violation can argue, however, that the underlying state law is unconstitutional. In *United States v. Taylor*, the defendant temporarily succeeded in winning dismissal of a Lacey Act indictment by arguing that the Maine law he was accused of violating (prohibiting the importation of bait fish) represented an unconstitutional restriction on interstate commerce.[72] The United States chose not to appeal from this ruling by the court of appeals. However, the state of Maine, which had intervened in the case when the defendant challenged the validity of its law, sought review in the Supreme Court. The case thus came to the Supreme Court in an unusual posture: an appeal by a state of the reversal of a federal conviction of a private citizen.

The Court nonetheless granted review and upheld the challenged law under the principles of *Hughes v. Oklahoma*.[73] The Court ruled that Maine's interest in preventing introduction of fish parasites and exotic species was a valid local purpose that could not be achieved through means less restrictive of interstate commerce.

As the above cases illustrate, usually it is the defendant's burden to prove invalid the underlying law upon which a Lacey Act prosecution is based. In at least one context, however, the government must establish the validity of the underlying law. According to the Ninth Circuit in *United States v. Sohappy*,[74] because Indian hunting and fishing treaty rights are subject to state regulation only to the extent reasonable and necessary for conservation purposes, the government must establish that the underlying state law or regulation meets this test in order to prove a Lacey Act offense by an Indian with treaty hunting or fishing rights.[75] The *Sohappy* court also held that Lacey Act offenses based on underlying violations of Indian tribal law are applicable to both Indians and non-Indians.

Prosecution for Providing Guide Services

The question of whether those who provide guide services for illegal hunting can be charged with the Lacey Act offense of "selling" illegally

[71]United States v. Lee, 937 F.2d 1388 (9th Cir. 1991) (no unlawful delegation to foreign governments); United States v. Rioseco, 845 F.2d 299, 302 (11th Cir. 1988) (no unlawful delegation to foreign governments); and United States v. Bryant, 716 F.2d 1091, 1094–95 (6th Cir. 1983), *cert. denied*, 465 U.S. 1009 (1984) (no unlawful delegation to states).

[72]752 F.2d 757 (1st Cir. 1985), *rev'd* 477 U.S. 131 (1986).

[73]441 U.S. 322 (1979).

[74]770 F.2d 816 (9th Cir. 1985).

[75]In United States v. Williams, 898 F.2d 727, 729 (9th Cir. 1990), the court held that "[t]ribal wildlife laws are *per se* valid against tribal members" and therefore "[t]here is no requirement of conservation necessity for establishing the validity of *tribal* wildlife laws incorporated through the Lacey Act." (Emphasis in original.)

taken wildlife gave rise to conflicting decisions in the appellate courts, and Congress stepped in to resolve the conflict. In *United States v. Todd*, the Fifth Circuit held that "[a] commercial arrangement whereby a professional guide offers his services to obtain wildlife illegally is an offer to sell wildlife."[76] The court relied on the Senate report explaining the 1981 amendments, which includes the above statement virtually verbatim.[77]

The Ninth Circuit, however, disagreed in *United States v. Stenberg*.[78] The court reasoned that because a criminal statute must be construed narrowly so as to give fair notice to those to whom it applies, and because "[w]e do not believe an ordinary person would consider the sale of guiding services . . . to be a sale of wildlife," those who merely provide guide services cannot be charged with a Lacey Act offense.[79] The Ninth Circuit did, however, later uphold a conspiracy conviction against a guide who should have known that his clients were hunting without valid licenses.[80] Ultimately, Congress resolved the split in the circuits in 1988 by amending the Lacey Act to make clear that the sale of wildlife includes providing guide services.[81]

Other Issues

The *Todd* and *Stenberg* cases also addressed the question of how to determine the "market value" of illegally taken wildlife. This issue is important because whether the value is greater or less than $350 may determine whether the offense is punishable as a felony or a misdemeanor. In the *Todd* case, where the defendants offered illegal airborne hunts at prices ranging from $1,000 to $5,000, the court held that "[t]he best indication of the value of the game 'sold' in this manner is the price of the hunt."[82]

In *Stenberg*, where one defendant was charged with selling wildlife and guide services to undercover agents for prices in excess of $350, the court expressed concern "that the government should not be given absolute discretion to convert any violation of the [Lacey] Act into a felony by offering a defendant an excessively high price for the contraband."[83] The court recommended instructing juries not to regard the price paid by a government agent as conclusive evidence of market value.

The government offered perhaps its most novel interpretation of the Lacey Act when it prosecuted a goldfish farmer in *United States v. Car-*

[76]735 F.2d 146, 152 (5th Cir. 1984).

[77]Sen. Rep. No. 97–123, 97th Cong., 1st Sess. at 12 (1981), *reprinted in* 1981 U.S.C.C.A.N. 1748, 1759.

[78]803 F.2d 422 (9th Cir. 1986).

[79]*Id.* at 436.

[80]United States v. Thomas, 887 F.2d 1341 (9th Cir. 1989).

[81]Pub. L. No. 100–653, § 101, 102 Stat. 3825, codified at 16 U.S.C. § 3372(c).

[82]735 F.2d at 152.

[83]803 F.2d at 433.

penter.[84] The evidence suggested that the farmer and his employees had killed thousands of fish-eating birds each year by shooting, poisoning, or trapping. The government charged the farmer with the Lacey Act offense of "acquiring" illegally taken wildlife, on the theory that the farmer retrieved the killed birds for the purpose of secretly disposing of them, thereby "acquiring" them. The court rejected the government's position, which it characterized as "collaps[ing] the two steps required by the statute into a single step—the very act of knowingly taking the bird in violation of laws is, in the government's view, the act of acquiring the bird."[85] The court's decision suggests that one can "acquire" illegally taken wildlife within the meaning of the Lacey Act, only from another person.

COMMERCE AND INJURIOUS WILDLIFE

The risks to crops, gardens, and natural ecosystems of introducing injurious, nonindigenous species of plants and animals are substantial.[86] Federal legislation responded first to threats posed to agriculture and horticulture from intentional importation of injurious species. More recently, laws have focused on both intentional and nonintentional introductions of non-native species and the broad range of harms they pose to natural ecosystems.

Injurious Wildlife and the Lacey Act

Beyond supplementing state wildlife conservation laws, the Lacey Act sought to protect domestic agricultural interests by barring importation of wildlife injurious to agriculture. The original Lacey Act prohibited, except for "natural history specimens for museums or scientific collections," importing mongooses, fruit bats, English sparrows, starlings, and "such other birds or animals as the Secretary of Agriculture may from time to time declare injurious to the interest of agriculture or horticulture" and authorized the Secretary of the Treasury to issue regulations for implementing the section.[87] For seventy years, however, the Secretary failed to issue any regulations.[88]

[84]933 F.2d 748 (9th Cir. 1991).

[85]*Id.* at 750.

[86]An excellent and comprehensive discussion of the issue is found in Office of Technology Assessment, *Harmful Non-Indigenous Species in the United States* (1993). Interactions triggered by introduced species are a significant contributor to species endangerment. Forest Service, U.S. Dep't of Agriculture, General Technical Rep. RM-241, *Species Endangerment Patterns in the United States* 9 (1994).

[87]Act of May 25, 1900, ch. 553, § 2 (current version at 18 U.S.C. § 42). The duty originally vested in the Secretary of Agriculture is now exercised by the Secretary of the Interior. The Secretary of the Treasury shares with the two Secretaries the duty of enforcing the injurious wildlife importation restrictions. 18 U.S.C. § 42(a)(5).

[88]Current injurious wildlife regulations are found at 50 C.F.R. Part 16 (1995). The Public

During that period, Congress significantly expanded the government's authority to prohibit importing injurious wildlife. In 1960, the Department of the Interior persuaded Congress to amend the Act to clarify its authority over importation of injurious "birds or animals" by changing those words to "wild mammals, wild birds, fish (including mollusks and crustacea), amphibians, reptiles, or the offspring or eggs of any of the foregoing."[89] In an important step for wildlife conservation, the same amendment expanded the interests to be protected from injurious wildlife from the original "agriculture or horticulture" to "human beings . . . agriculture, horticulture, forestry . . . wildlife or the wildlife resources of the United States."[90]

The Department of the Interior did not seek to implement its new authority with comprehensive regulations until 1973. Those regulations proposed to treat all foreign wildlife as injurious to one or more of the interests specified in the statute.[91] If importation of a particular species could be shown to present a "low risk" of injury, however, its importation would not be restricted. Interior's proposal included a relatively short list of wildlife already determined to present a low risk.

The original proposal caused a storm of protest. It threatened to prohibit certain interests (most notably the pet trade) from importing much foreign wildlife and to complicate the task of others (particularly the zoological and scientific communities) by requiring import permits. Congress held a hearing to review Interior's proposal.[92] The Department came

---

Health Service also regulates importation of certain foreign wildlife under the Public Health Service Act, 42 U.S.C. § 264, to prevent the spread of communicable diseases. Current Public Health Service regulations regarding importation of turtles, tortoises, terrapins, and primates are found at 42 C.F.R. §§ 71.52 and 71.53 (1995). The Department of Agriculture, in order to protect domestic livestock from foreign-borne disease, also regulates importation of wild ruminants and swine under section 306 of the Tariff Classification Act, 19 U.S.C. § 1306.

[89]Act of Sept. 2, 1960, Pub. L. No. 86–702, § 1, 74 Stat. 753 (current version at 18 U.S.C. § 42(a)(1)). Even this change, however, did not entirely accomplish the Department's stated objective to clarify its authority to regulate "without question, the importation of fish, including mollusks and crustacea, *or any other type of animal found to be injurious*" (emphasis added). Department of the Interior letter transmitting the proposed bill, *reprinted in* Sen. Rep. No. 1883, 86th Cong., 2d Sess. at 4 (1960), *reprinted in* 1960 U.S.C.C.A.N. 3310, 3312.

[90]*Id.* The 1960 amendment also added a definition of the terms "wildlife" and "wildlife resources," as used in the quoted passage, which included "wild mammals, wild birds, fish (including mollusks and crustacea), and all other classes of wild creatures whatsoever, and all types of aquatic and land vegetation upon which such wildlife resources are dependent." Thus the range of species to be protected by importation restrictions is substantially broader than the range of wildlife that may be kept out of the country.

[91]*See* 38 Fed. Reg. 34970 (1973).

[92]Hearings on Proposed Injurious Wildlife Regulations Before the Subcomm. on Fisheries

under heavy pressure to back off its proposal as affected interests claimed economic ruin and contended that Interior had no authority to do what it proposed.

Two months after the congressional hearings, Interior issued a revised proposal that, while expanding the predetermined list of low-risk wildlife and modifying slightly the proposed procedures for issuance of permits, adhered to the same basic approach.[93] Interior took no further action, however, until August 1976, more than two and one-half years after the original proposal. Interior announced at that time that it was abandoning altogether its proposed approach in favor of seeking a legislative "clarification" of its authority. The Department's stated justifications for this reversal were congressional failure to appropriate the extra funds needed to carry out the program and fear of a legal challenge if the Department adopted its earlier proposals.[94]

Executive Order 11987: A Promise Unfulfilled

Four months after taking office in 1976, President Jimmy Carter issued a far-reaching executive order directing federal agencies to restrict introduction of "exotic" species into natural ecosystems under their jurisdiction and to encourage the states and others to do the same.[95] Nearly twenty years later, regulations implementing the order have yet to appear. The order remains in effect, however, and it is worthy of discussion, if only for the potential it offers.

The order focuses, significantly, not on the importation of exotic species, but on their "introduction." This term means "the release, escape, or establishment of an exotic species into a natural ecosystem."[96] Although the term "exotic species" includes only those species not naturally occurring anywhere in the United States, the order extends to their movement within the United States once they have been brought into the country. The order directs federal land managing agencies to restrict introduction of exotic species into natural ecosystems on land they administer, to the extent permitted by law.

---

and Wildlife Conservation and the Environment of the House Comm. on Merchant Marine and Fisheries, 93d Cong., 2d Sess. (1974).

[93] *See* 40 Fed. Reg. 7935 (1975).

[94] Interior had earlier advised Congress that it saw no merit to the assertion that the Lacey Act's legislative history precluded adopting a low-risk approach. Hearings, *supra* note 92, at 137 (statement of John Oberheu). In fact, the legislative history is silent regarding whether the Secretary's determinations must be made on an individual species basis or whether they may be made in more general terms. Arguments that the Lacey Act requires species-by-species determinations are based on doubtful inference rather than clear legislative history.

[95] Exec. Order No. 11987, *supra* note 9.

[96] *Id.* § 1(b).

A separate provision directs that any federal agency, to the extent it is authorized to restrict the *importation* of exotic species, shall restrict their *introduction* into *any* natural ecosystem. For injurious animals under the Lacey Act and for noxious weeds under the Federal Noxious Weed Act, the Departments of the Interior and Agriculture, respectively, are directed to restrict the introduction anywhere of these species. Export of native species for the purpose of introducing them into natural ecosystems of foreign countries is also to be restricted to the extent permitted by law.[97] Exceptions to these restrictions may be made if the Secretary of Agriculture or the Interior finds no adverse effect on natural ecosystems.[98]

Finally, the order directs federal agencies to "encourage the States, local governments, and private citizens to prevent the introduction of exotic species into natural ecosystems of the United States."[99] None of the order's provisions, however, applies to an exotic species for which the Secretary of Agriculture or the Interior finds that its introduction will not have an adverse effect on natural ecosystems.

The Federal Noxious Weed Act

A significant feature of the Executive Order is that the "exotic species" whose introduction it seeks to restrict include not only animals, but also plants.[100] Existing authority to regulate importation of plants, vested in the Department of Agriculture under the Plant Pest Act[101] and the Plant Quarantine Act,[102] is designed primarily to protect against introduction of diseases or pests harmful to other plants. The Federal Noxious Weed Act of 1974 provides added authority for restricting not only importation, but also interstate movement, of potentially harmful plants.[103]

The Federal Noxious Weed Act gives the Secretary of Agriculture extensive authority over "noxious weeds." The Secretary's authority is similar to, although more extensive than, the authority over injurious wildlife the Lacey Act gives the Secretary of the Interior. A "noxious weed" is a living plant of foreign origin that is either new to, or not widely prevalent in, the United States and that "can directly or indirectly injure crops, other useful plants, livestock, or poultry or other interests of agriculture, including irrigation, or navigation or the fish or wildlife resources of the United States or the public health."[104] The Secretary has extensive au-

[97] *Id.* § 2(c).
[98] *Id.* § 2(d).
[99] *Id.* § 2(a).
[100] *Id.* § 1(c).
[101] 7 U.S.C. §§ 147a, 149, and 150aa–150jj.
[102] *Id.* §§ 151–167.
[103] *Id.* § 2802.
[104] *Id.* §2802(c). Not surprisingly, the "noxious weeds" designated in regulations at 7 C.F.R. § 360.200 (1996) reflect a concern with restricting plant pests directly harmful to crops.

thority over both entry into, and movement within the United States of, designated noxious plants. The Secretary also may quarantine or require inspection of other articles and products as a means of preventing dissemination of noxious weeds.

The Department of the Interior had virtually no role to play with respect to the Federal Noxious Weed Act until 1990. The Secretary of Agriculture was authorized to cooperate with other federal agencies in eradicating or controlling noxious weeds and could appoint employees of other federal agencies to assist in carrying out the Act pursuant to cooperative agreements between the two agencies. The Department of the Interior had no explicit statutory responsibilities of its own.

Congress amended the Act in 1990 to add a new section concerning "undesirable plants" on federal lands.[105] This term has a much broader meaning than the "noxious weeds" that the original Act covered. The term encompasses plants "that are classified as undesirable, noxious, harmful, exotic, injurious, or poisonous, pursuant to State or Federal law."[106] Endangered species and plants indigenous to an area where control action is to be taken are excluded.

The definition is somewhat puzzling. First, the term "injurious" appears in the Lacey Act, but it is limited to animals, not plants. Second, the term "exotic" does not appear in any federal statute, but it is used in Executive Order 11987, where it refers to any plant or animal not naturally occurring, either currently or historically, in any ecosystem in the United States. Thus, if the Executive Order is subsumed within the term "federal law," as used in the Noxious Weed Act, all plants not native to the United States would be considered "undesirable."

The 1990 amendments require each federal land management agency to establish and fund an undesirable plants management program for lands under its jurisdiction. The program is to be carried out through cooperative agreements with state agencies that target particular plants or groups of plants for control or containment. Control efforts must use "integrated management systems" that may include preventive measures, herbicides, biological agents, cultural methods, and general land management practices.

The significance of the 1990 amendments to the Noxious Weed Act is that they give explicit authority to the Fish and Wildlife Service, National Park Service, and Bureau of Land Management to take the initiative in developing programs for control of undesirable plants on their lands. Because the definition of undesirable plants ultimately may hinge on state law classifications, the key to the success of these efforts will likely be close cooperation with the states.

[105] *Id.* § 2814.
[106] *Id.* § 2814(e)(7).

The Nonindigenous Aquatic Nuisance Prevention and Control Act of 1990

Deliberate importation is only one means, and probably not the most important means, of introducing non-native species.[107] For example, the zebra mussel, a Eurasian species, apparently was introduced to North American waters when an ocean-going vessel discharged its ballast waters. The substantial economic and ecological disruption caused by the zebra mussel prompted Congress in 1990 to enact the Nonindigenous Aquatic Nuisance Prevention and Control Act.[108] Much of the law focuses on controlling ballast water discharges because they are a prime pathway for aquatic nuisance introductions, but it is not limited to that problem.

The 1990 law is broader than any of the legal authorities that preceded it. The term "nonindigenous species" includes any plant or animal, as well as "other viable biological material" (e.g., a virus), that enters an ecosystem beyond its historic range.[109] Unlike under the Lacey Act and the Executive Order, nonindigenous species need not be foreign. Species native to one part of the United States are nonindigenous in other parts.

To be nonindigenous, however, is not necessarily to be a nuisance. The Act defines an "aquatic nuisance species" as "a nonindigenous species that threatens the diversity or abundance of native species or the ecological stability of infested waters, or commercial, aquacultural or recreational activities dependent on such waters."[110] This definition, like "injurious" species under the Lacey Act, requires some exercise of judgment. While the Lacey Act specifies that the Secretary of the Interior is to determine through rulemaking what species are "injurious," the 1990 Act specifies neither who is to determine which aquatic species are "nuisances" nor how they are to do it.

Nor does the 1990 Act specify precisely what to do about aquatic nuisance species. Rather, it creates a task force co-chaired by the Director of the Fish and Wildlife Service and the Undersecretary of Commerce for Oceans and Atmosphere.[111] The task force is to develop and implement a program to prevent the introduction and dispersal of aquatic nuisance species, and to monitor, study, and control these species. As part of this program, the task force is to "establish and implement measures . . . to minimize the risk of introduction of aquatic nuisance species," including

[107]See Office of Technology Assessment, *Harmful Non-Indigenous Species in the United States,* supra note 86 (1993).

[108]16 U.S.C. § 4701 *et seq.* The mussel clogged municipal and industrial water intake structures and out-competed native mollusks.

[109]*Id.* § 4702(9).

[110]*Id.* § 4702(2).

[111]*Id.* § 4721.

measures for "assessment of the risk that an aquatic organism carried by an identified pathway may become an aquatic nuisance species."[112]

The 1990 Act reflects a much more sophisticated understanding of the gravity of the introduced species problem and the complexity of addressing that problem than did the Lacey Act. Whether it will have any greater impact, however, remains to be seen.

[112] *Id.* § 4722(c)(1).

Part II

SPECIES CONSERVATION

This part of the book looks at what many consider to be the core of federal wildlife law. It examines the major federal laws and programs created to conserve particular groups of organisms. These are migratory birds (Chapter 4), marine mammals (Chapter 5), ocean fish (Chapter 6), and endangered species (Chapter 7).

The impetus for enactment of federal laws to conserve each of these groups of species was quite similar. First, the wildlife was regarded as an important resource with significant value—usually economic—to humans. Second, the states, which historically had exercised management authority over these species, were perceived to be incapable of effectively conserving the species in question. For example, individual states do not have jurisdiction over migratory species in all of the areas through which the species move. The same is true for nonmigratory species whose ranges exceed state boundaries. Consequently, Congress enacted laws capable of addressing the threat to these species throughout all or most of their ranges.

None of the federal programs discussed in these chapters displaced state authority altogether. Each of these programs treats the often delicate issue of shared state and federal responsibility for living resources in its own unique way. Thus, each of these federal laws could be considered an experiment in natural resource federalism.

Three of the four laws examined here date from the 1970s. They represent part of the wave of federal laws enacted during a brief period of congressional activism, unrivalled in American environmental history. The fourth law, the Migratory Bird Treaty Act, was enacted in 1918, long before there was a body of environmental law into which it could be fit. In contrast to the more recent three laws, the Migratory Bird Treaty Act is a relatively short, simple, and straightforward law. It has seldom been changed in the eighty years since its enactment. The other three were more complex from the beginning and have become even more so as a result of frequent congressional alteration. Their histories have been shaped significantly by the back-and-forth of court decisions and legislative responses.

Each of the laws examined here uses regulation of commercial and recreational hunting or fishing as its primary conservation tool. While hunting pressures unquestionably once were the major threat to the survival of most species of concern, a host of other threats are now recognized as of equal or greater importance. For example, the status of many species of migratory waterfowl, which have received most of the attention of the federal migratory bird program, recently has depended more on habitat conditions than on hunting limits. The decline of many species of migratory songbirds due to forest changes in the United States and elsewhere is perhaps the best illustration of the futility of trying to conserve species solely through controls on hunting. Yet alone among the four statutes considered here, the Endangered Species Act has tried to grapple with the broader set of threats to species survival. The controversies that have attended its implementation are testimony to the practical and political difficulties of doing so.

The primary threat to species survival on both a national and global scale is loss or degradation of habitat. With the exception of the Endangered Species Act and, to a lesser extent, recent amendments to the Fishery Conservation and Management Act, the laws examined in this part do not address the seemingly limitless threats to wildlife habitat. Part III of the book examines the many laws that do.

Chapter 4

WILD BIRDS

The first category of wildlife Congress chose to protect with targeted legislation was migratory birds. The Migratory Bird Treaty Act was enacted in 1918, following the signing of a treaty with Great Britain that committed the United States to conserving the species that migrate between Canada and our nation.

Other birds have since received congressional protection. Our national symbol, the bald eagle, was the focus of legislation in 1940, and in 1992, Congress adopted a law regulating import of exotic birds. This chapter discusses these three laws and litigation under them.

PROTECTION OF MIGRATORY BIRDS

> But for the treaty and the statute there might soon be no birds for any powers to deal with.
>
> *Missouri v. Holland* (1920)

Early in the twentieth century, federal legislation to stop the decline in migratory bird populations was struck down for lack of a constitutional underpinning.[1] In 1916 the United States signed a treaty with Great Britain on behalf of Canada, the Convention for the Protection of Migratory Birds,[2] both to provide the constitutional foundation for federal regula-

[1]See Chapter 2.

[2]Aug. 16, 1916, 39 Stat. 1702, T.S. No. 628 (hereinafter Canadian Convention). The events leading to the signing of the 1916 Convention, passage in 1918 of the Migratory Bird Treaty Act, and the Supreme Court's landmark decision in Missouri v. Holland, 252 U.S. 416

tion (the treaty power) and to initiate international action to halt the decline of migratory species. Congress passed the Migratory Bird Treaty Act[3] (hereinafter the Treaty Act) in 1918 to implement the treaty. In *Missouri v. Holland,* quoted above,[4] the Supreme Court upheld the Treaty Act's constitutionality.

The primary threat to migratory bird populations when the Act was passed was unrestrained shooting for commerce and sport.[5] Consequently, the main thrust of the Act and the treaty is to regulate "taking" of migratory birds, especially hunting.

The Treaty Act, however, also implements treaties that were signed with Mexico in 1936,[6] Japan in 1972,[7] and the Soviet Union in 1976.[8] By the 1970s, when the latter two treaties were signed, additional major human-induced threats to migratory bird populations included chemical poisoning and habitat destruction, as well as more limited hazards such as electric power poles.[9]

Courts have interpreted the Treaty Act to prohibit "taking" by means other than hunting—for instance, by chemical poisoning.[10] However, courts have refused to apply the Act's strictures to habitat destruction by logging.[11] Still new hazards are emerging, such as large, wind-driven power generators. It remains to be seen how the Treaty Act will affect these more recent threats, and others as yet unknown.

While there is significant overlap among the treaties underlying the Treaty Act, they differ in the birds that are covered, the limitations on taking, and the affirmative protective measures they require or encourage. These differences are of more than academic interest. The Ninth Circuit

---

(1920), upholding the constitutionality of both the Convention and the Act, are described in Chapter 2.

[3]Act of July 3, 1918, Ch. 128, 40 Stat. 755 (1918) (current version at 16 U.S.C. §§ 703–712).

[4]252 U.S. 416, 435 (1920).

[5]*See* George Cameron Coggins & Sebastian T. Patti, *The Resurrection and Expansion of the Migratory Bird Treaty Act,* 50 U. Col. L. Rev. 165, 168 & 171 (1979).

[6]Convention for the Protection of Migratory Birds and Game Mammals, Feb. 7, 1936, United States-Mexico, 50 Stat. 1311, T.S. No. 912 (hereinafter Mexican Convention).

[7]Convention for the Protection of Migratory Birds and Birds in Danger of Extinction, and their Environment, March 4, 1972, United States–Japan, 25 U.S.T. 3329 (hereinafter Japanese Convention).

[8]Convention Concerning the Conservation of Migratory Birds and Their Environment, Nov. 19, 1976, United States–U.S.S.R., 29 U.S.T. 4647, T.I.A.S. No. 9073 (hereinafter Soviet Convention).

[9]Coggins & Patti, *supra* note 5 at 174; Donell R. Grubbs, *Of Spotted Owls and Bald Eagles: Raptor Conservation Soars into the '90s,* 19 Cap. U. L. Rev. 451, 461 (1990).

[10]United States v. FMC Corp., 572 F.2d 902 (2d Cir. 1978); United States v. Corbin Farm Service, 444 F. Supp. 510 (E.D. Cal.), *aff'd,* 578 F.2d 259 (9th Cir. 1978). *See also* United States v. Van Fossan, 899 F.2d 636 (7th Cir. 1990).

[11]*See, e.g.,* Seattle Audubon Society v. Evans, 952 F.2d 297 (9th Cir. 1991).

Court of Appeals has held that subsistence hunting of migratory game birds in Alaska may be permitted only to the extent it is in accord with all four treaties,[12] a requirement that may well apply to other activities covered by the Act.[13] Accordingly, this section begins with a discussion of the treaties.

THE TREATIES: RESTRICTIONS ON TAKING

The 1916 Canadian Convention establishes three groups of migratory birds: (1) migratory game birds, (2) migratory insectivorous (insect-eating) birds, and (3) other migratory nongame birds. For each group, article II establishes a closed season[14] during which "no hunting shall be done except for scientific or propagating purposes under permits." Except for migratory game birds, the closed season is year round—in other words, no taking is allowed at any time of the year.[15] The one exception is that certain types of birds in the third category may be taken by Eskimos and Indians for food and clothing but may not be sold or offered for sale.[16]

With respect to migratory game birds, article II provides that the closed season shall be between March 10 and September 1[17] and that the "season for hunting shall be further restricted to such period not exceeding three and one-half months as the High Contracting Powers may severally deem appropriate and define by law or regulation."[18] Article V prohibits taking nests or eggs of migratory birds except for scientific or propagating purposes. Article VII authorizes the issuance of permits to kill any migratory birds that, "under extraordinary conditions, may become seriously injurious to agricultural or other interests in any particular community."

Each of the subsequent treaties, although patterned after the original

[12]Alaska Fish and Wildlife Fed'n v. Dunkle, 829 F.2d 933, 941 (9th Cir. 1987), *cert. denied,* 485 U.S. 988 (1988), discussed *infra* at text accompanying notes 133–149.

[13]See Andrus V. Allard, 444 U.S. 51, 62 n. 18 (1979), in which the Court stated that it is "hazardous to look to any single Convention for definitive resolution of a statutory construction problem. . . . [I]nasmuch as the Conventions represent binding international commitments, they establish *minimum* protections for wildlife." (Emphasis in original.)

[14]The treaty calls this the "close season."

[15]*See* discussion of Alaska Fish and Wildlife Fed'n v. Dunkle, 829 F.2d 933 (9th Cir. 1987), *cert. denied,* 485 U.S. 988 (1988), *infra* at text accompanying notes 133–149.

[16]Canadian Convention, art. II, § 3.

[17]*Id.*, art. II, § 1. A protocol amending the Convention to permit some subsistence hunting of game birds during this season was signed by both countries on December 14, 1995, but it has not been ratified. See discussion of native subsistence hunting at text accompanying notes 131–160 *infra.* For Atlantic coast shorebirds, a special closed season between February 1 and August 15 is prescribed.

[18]For a discussion of litigation about the meaning of the quoted language, see text accompanying notes 161–164 *infra.*

convention, differs somewhat. The Mexican treaty, for example, limits the length of the hunting season for migratory birds to a maximum of four months and requires that hunting be conducted "under permits issued by the respective authorities in each case."[19] In addition, it calls for the establishment of "refuge zones in which the taking of such birds will be prohibited,"[20] and it prohibits hunting from aircraft.[21] The Japanese Convention specifies no dates or lengths for hunting seasons but requires that whatever seasons are set by each party "avoid their principal nesting seasons and . . . maintain their populations in optimum numbers."[22] The Soviet Convention prohibits not only taking migratory birds, but also their nests or eggs and the "disturbance of nesting colonies."[23] Without specifying any dates or overall durations, it authorizes the parties to establish hunting seasons, so long as they assure "the preservation and maintenance of stocks of migratory birds."[24]

The treaties authorize other exceptions to the basic prohibition against taking. For example, all four treaties authorize takings to control bird-caused problems, although the extent of the exception differs from treaty to treaty. Whereas the Canadian Convention authorizes, under "extraordinary conditions," taking birds that "may become seriously injurious to

[19]Mexican Convention, article II(C). The Mexican Convention is quite unartfully drafted insofar as the establishment of closed seasons is concerned. Article II(A) provides generally for their establishment but does not specify their length; article II(C) then limits the permissible hunting period to no more than four months in each year; article II(D) prescribes a closed season of from March 10 to September 1 for wild ducks; and article II(E) prescribes a year round closed season for "migratory insectivorous birds." The ambiguity arises from article IV, however, which lumps all migratory birds into only two categories: migratory game birds and migratory nongame birds. If the latter grouping was intended to include any birds other than "migratory insectivorous birds," then the Convention fails to indicate what closed or open seasons are to apply to such other birds. The matter was further confused when, by exchange of notes on March 10, 1972, the governments of Mexico and the United States supplemented the 1936 Treaty by agreeing to a lengthy list of additional birds to be protected. These birds are described as "additions . . . to the list of birds set forth in Article IV," without specifying whether the same are to be considered "migratory game birds" or "migratory nongame birds." By administrative regulation, however, they are all treated as nongame birds, and protected from hunting. *See* 50 C.F.R. §§ 10.13, 20.11 (1995).

[20]Mexican Convention, art. II(B).

[21]*Id.* art. II(F).

[22]Japanese Convention, art. III, § 2. For a discussion of litigation concerning the requirement to maintain optimum numbers, see text accompanying notes 164–165 *infra.*

[23]Soviet Convention, art. II, § 1. The prohibition against disturbing nesting colonies is a new feature not found in any of the earlier treaties or in the Migratory Bird Treaty Act, although it is arguably subsumed in the prohibition against "taking," a term undefined in the treaties and the Treaty Act. Implementing regulations of the Fish and Wildlife Service, however, restrict the definition of taking in a way that would not seem to include nesting colony disturbance. *See* 50 C.F.R. § 10.12 (1995).

[24]Soviet Convention, art. II, § 2.

the agricultural or other interests in any particular community,"[25] the Mexican Convention authorizes taking birds only "when they become injurious to agriculture and constitute plagues."[26] The Japanese and Soviet Conventions, on the other hand, broadly authorize takings to protect persons or property.[27] The treaties authorize other exceptions as well.[28]

The considerable overlap among species protected by the four treaties thus results in differing substantive standards applicable to protection of the same bird.[29] There has even been controversy over what species are protected by the various conventions. The Canadian Convention describes the birds it protects by the common names of certain general groups, such as "woodpeckers," "gulls," and so forth.[30] The Mexican Convention, on the other hand, identifies protected birds solely in terms of their scientific family names. The Japanese Convention gives the full scientific name for each species of bird that it protects and further defines the term "migratory bird" to include only those species "for which there is positive evidence of migration" between the United States and Japan or which have subspecies or populations "common to both countries."[31] The Soviet Convention takes a similar, though still different, approach.[32]

[25]Canadian Convention, art. VII.

[26]Mexican Convention, art. II(E).

[27]Japanese Convention, art. III, § 1(b); Soviet Convention, art. II, § 1(d). For a case in which the court rejected a "defense of property" defense to a conviction under the Treaty Act for taking great horned owls, see United States v. Darst, 726 F. Supp. 286 (D. Kan. 1989), discussed *infra* at text accompanying note 183.

[28]Like the Canadian Convention, the Mexican Convention creates an exception for the taking of migratory birds for scientific and propagating purposes and also adds a new exception for museums. Mexican Convention, art. II(A). The Japanese and Soviet Conventions add educational and "other specific purposes not inconsistent with" the objectives of the Conventions to the list of taking exceptions. Japanese Convention, art. III, § 1(a); Soviet Convention, art. II, § 1(a). Both the Mexican and Japanese Conventions also contain specific exemptions for "private game farms." Japanese Convention, art. III, § 1(d); Mexican Convention, art. II(A).

[29]For a list of the species protected under the Treaty Act, see 50 C.F.R. § 10.13 (1995).

[30]Migratory game birds are identified in the Canadian Convention not only by their common names, but also by their scientific family names. *See* Canadian Convention, art. I, § 1. Migratory insectivorous birds under that Convention include not only those specifically listed by common name, but also "all other perching birds which feed entirely or chiefly on insects." *Id.* art. I, § 2.

[31]Japanese Convention, art. II, § 1.

[32]The Soviet Convention applies to all species and subspecies that migrate between the two countries and to those with separate populations sharing common breeding, wintering, feeding, or molting areas. Soviet Convention, art. I, § 1. The list of migratory birds may in effect be expanded unilaterally by either party at least as to areas under, or persons subject to, its own jurisdiction, by virtue of authority conferred by article VIII. That article authorizes either party, in its discretion, to treat any species or subspecies of bird as though it were a protected migratory bird under the Convention, as long as it belongs to the same family as any bird that is so protected.

THE TREATIES: OTHER MEASURES

While the prohibition against taking migratory birds is the major focus of all the treaties, the more recent treaties with Japan and the Soviet Union contain important measures aimed at protecting migratory bird habitat. For example, the Japanese Convention directs the parties to "endeavor to establish sanctuaries and other facilities for the protection or management of migratory birds."[33] It also directs the parties to "endeavor to take appropriate measures to preserve and enhance the environment of birds protected under" the Convention, such as by preventing "damage resulting from the pollution of the seas" and by controlling the importation and introduction of live animals and plants that are "hazardous to the preservation of such birds" or that "could disturb the ecological balance of unique island environments."[34]

The Soviet Convention directs each party to establish preserves, protected areas, and facilities for the conservation of migratory birds and to "manage such areas so as to preserve and restore the natural ecosystems."[35] Article IV of the Soviet Convention includes a broadly worded exhortation to take measures necessary to "protect and enhance the environment and to prevent and abate the pollution or detrimental alteration of that environment,"[36] and a number of more specific directives aimed at accomplishing that goal.[37]

Of potentially great importance is the requirement in article IV that each party identify "areas of breeding, wintering, feeding, and molting which are of special importance in the conservation of migratory birds within the areas under its jurisdiction."[38] These areas are to be included in a list to be appended to the Convention, and the parties are to "undertake measures necessary to protect [their] ecosystems . . . against pollution, detrimental alteration and other environmental degradation."[39]

A closely related provision authorizes the parties, by mutual agreement, to designate areas not under the jurisdiction of either of them as areas of special importance to the conservation of migratory birds.[40] Each party shall "undertake measures necessary to ensure that any citizen or person subject to its jurisdiction will act in accordance with the principles of this Convention in relation to such areas." What acting "in accordance with the principles of" the Convention means is not altogether clear. Presum-

[33]Japanese Convention, art. III, § 3.
[34]*Id.* art. VI.
[35]Soviet Convention, art. VII.
[36]*Id.* art. IV, § 1.
[37]*Id.* § 2.
[38]*Id.* § 2(c).
[39]*Id.*
[40]*Id.* § 3.

ably, it is intended to give these areas the same or a similar degree of protection as the areas of special importance within the jurisdiction of the parties. To date, however, no areas of special importance have yet been designated by the United States.

Article IV of the Soviet Convention requires each party to establish procedures for warning the other of "substantial anticipated or existing damage to significant numbers of migratory birds or the pollution or destruction of their environment."[41] Once warned, the parties are to cooperate "in preventing, reducing or eliminating such damage" and in rehabilitating the environment. Apparently, this provision was intended to facilitate early detection of and cooperative action in combatting major disasters, such as oil spills.

Article V of the Soviet Convention provides for special protection measures for migratory birds in danger of extinction. Whenever either party decides that any species, subspecies, or "distinct segment of a population" is endangered and establishes special measures for its protection, it is to inform the other party of its action. The other party is then directed to "take into account such protective measures in the development of its management plans." This is a broader mandate than that of a similar provision in the Japanese treaty that merely directs each party to control the exportation or importation of any species or subspecies found by the other party to be endangered.[42]

THE MIGRATORY BIRD TREATY ACT: STRUCTURE

The Migratory Bird Treaty Act implements all four of the conventions just described. Despite the fact that three of the treaties were signed after the Act's passage, the law reads very much today as it did when originally passed in 1918.[43]

The Act's proscriptions are broad.[44] Section 2 makes it unlawful "at any

[41] *Id.* § 2(a).

[42] Japanese Convention, art. IV, § 3.

[43] Notwithstanding the many differences among them, the ratification of each new convention did not result in a major overhaul of the Act but only in technical amendments that merely added appropriate references to each subsequent convention. Thus, to the extent that the new features of the later conventions cannot be subsumed within the general language of the Act, those features remain unimplemented by domestic legislation. That concern, however, is alleviated by virtue of the broad and general language the Act employs, as discussed below. For a discussion of the extent to which international treaties are self-executing without the need for implementing legislation, see Comment, *The Migratory Bird Treaty: Another Feather in the Environmentalist's Cap,* 19 S. D. L. Rev. 307, 311–16 (1974).

[44] The Act amplified in a number of respects the narrow terms of the Canadian Convention. The Convention prohibited only "hunting" migratory birds during their closed seasons. The amplification was obviously intentional, for the Act's caption described it as being "An Act to give effect to the convention . . . and for other purposes." 40 Stat. 755. The fact that the

time, by any means or in any manner," inter alia, to hunt, take,[45] capture, kill, possess, purchase, sell, barter, or transport, any bird protected by the treaties, any part, nest, or egg of a protected bird, or any product composed of any part, nest, or egg of a protected bird,[46] except as permitted by regulation of the Secretary of the Interior.[47]

The Secretary has wide discretion in promulgating regulations, but all rules are expressly made "[s]ubject to the provisions . . . of the conventions." The Secretary must determine "when, to what extent, if at all, and by what means" to permit the above activities.[48] In doing so, the Secretary must give "due regard to the zones of temperature and to the distribution, abundance, economic value, breeding habits, and times and lines of migratory flight of such birds."[49] Finally, very important to its administration, the Act provides that nothing therein prevents the states from making or enforcing laws or regulations that are consistent with the conventions or the Act or that give further protection to migratory birds, their nests, or eggs.[50]

Act's proscriptions go beyond those of its underlying treaties presents no constitutional problem, however, since the Supreme Court has ruled that the Commerce Clause provides an independent basis for the Act. *See* Andrus v. Allard, 444 U.S. 51 (1979).

[45]Unlike the Endangered Species Act, neither the Act nor its regulations define "take." See text accompanying notes 107–111, *infra,* for a discussion of the meaning of the word "take" in the Treaty Act as distinguished from that in the Endangered Species Act.

[46]Migratory Bird Treaty Act, ch. 128, § 2, 40 Stat. 755 (1918), 16 U.S.C. § 703. These prohibitions are considerably more expansive than the limited proscription contained in the Canadian Convention against the "shipment or export" of migratory birds from any state during its closed season and against the "international traffic" in unlawfully taken migratory birds. Canadian Convention, art. VI. Furthermore, it is noteworthy that the Act does not require that these activities be conducted in interstate commerce to be unlawful. The Act does, however, prohibit the transportation in interstate commerce of "any bird, or any part, nest, or egg thereof, captured, killed, taken, shipped, transported, or carried . . . contrary to the laws of the State, Territory, or district in which it was captured, killed, or taken, or from which it was shipped, transported, or carried." *Id.* § 705. The latter prohibition is not entirely redundant of the former because the latter applies to *all* birds, not just migratory birds, protected under the treaties. *See* Bogle v. White, 61 F.2d 930 (5th Cir. 1932). It is, however, with the possible exception of its application to nests, redundant of the Lacey Act. *See* Chapter 3.

[47]16 U.S.C. § 703. Congress initially vested authority to promulgate regulations in the Secretary of Agriculture but transferred it to the Secretary of the Interior in 1939. The Fish and Wildlife Service in the Department of the Interior is the implementing agency. The term "Secretary" in this section refers either to the Secretary of the Interior or the Secretary of Agriculture, as appropriate.

[48]16 U.S.C. § 704.

[49]*Id.*

[50]*Id.* § 708. To implement the Treaty Act, two types of federal regulations are promulgated. The first is of a general and continuing nature and governs such subjects as hunting methods, tagging and identification requirements, scientific and other permit requirements, and other similar matters. These are found at 50 C.F.R. Part 20, subparts A-J and L, and Part 21 (1995). The second type fixes season lengths, shooting hours, bag limits, and so forth, and is revised

The Act provides criminal penalties for violations.[51] Most violations are misdemeanors, punishable by a fine of not more than $500, imprisonment for not more than six months, or both.[52] Taking with intent to sell, offering for sale, sale, and related acts with respect to protected birds are felonies, punishable by a fine of not more than $2,000, imprisonment for not more than two years, or both.[53] Misdemeanors are crimes of strict liability,[54] but one must "knowingly" violate the Act for a felony conviction.[55] The Act also provides for forfeiture of guns, traps, vehicles, and other equipment used in violating the Act.[56]

The Treaty Act does not create a private right of action against the Secretary or other federal officials to compel them to comply with the Act. Case law, however, has established that affected parties may proceed under the Administrative Procedure Act (APA) against the Secretary and other federal officials who violate the Treaty Act.[57]

---

annually on the basis of bird population data and the recommendations of affected states, "Flyway Councils," and various advisory committees. These are found at 50 C.F.R. Part 20, Subpart K (1995). The end-product of these formalized proceedings is the promulgation of so-called "framework regulations," which offer individual states a range of choices regarding season lengths, shooting hours, bag limits, and so forth. Individual states select from among the choices offered in the framework regulations, and their selections are then published as final federal regulations. The annual process of formulating these regulations is described in more detail in Department of the Interior, Fish and Wildlife Service, Final Supplemental Environmental Impact Statement: Issuance of Annual Regulations Permitting the Sport Hunting of Migratory Birds (1988).

[51]The Secretary, however, cannot be compelled to prosecute violations. Alaska Fish and Wildlife Fed'n v. Dunkle, 829 F.2d 933 (9th Cir. 1987), *cert. denied*, 485 U.S. 988 (1988).

[52]16 U.S.C. § 707(a).

[53]16 U.S.C. § 707(b).

[54] See text accompanying notes 63–102 *infra*.

[55]16 U.S.C. § 707(b). Prior to 1986, the Treaty Act did not require knowledge for a felony conviction. Courts balked at convicting a defendant of a felony with no finding of intent. The court in United States v. Wulff, 758 F.2d 1121 (6th Cir. 1985), found that a felony conviction for violation of the Treaty Act, without proof of scienter, violates due process. In response to the decision in Wulff, in 1986 Congress amended section 6(b) to add the word "knowingly." The legislative history of this amendment explains what the change means:

> [T]he amendment will require proof that the defendant knew (1) that his actions constituted a taking, sale, barter, or offer to sell or barter, as the case may be and (2) that the item so taken, sold, or bartered was a bird or portion thereof. It is not intended that proof be required that the defendant knew the taking, sale, barter or offer was a violation of the subchapter, nor that he knew the particular bird was [a protected bird].

Sen. Rep. No. 99–445, 99th Cong. at 16, *reprinted in* 1986 U.S.C.C.A.N. 6113, 6128.

[56]16 U.S.C. § 707(c).

[57]*See, e.g.*, Alaska Fish and Wildlife Fed'n v. Dunkle, 829 F.2d 933, 938 (9th Cir. 1987), *cert. denied*, 485 U.S. 988 (1988); Humane Society v. Watt, 551 F. Supp. 1310, *aff'd without opinion* 713 F.2d 865 (D.C. Cir. 1983); National Rifle Association v. Kleppe, 425 F. Supp. 1101 (D.D.C. 1976). In Defenders of Wildlife v. EPA, 688 F. Supp. 1334 (D. Minn. 1988),

CHALLENGES TO RESTRICTIONS ON THE MANNER OF TAKING BIRDS

There have been numerous challenges to the Secretary's authority under section 3 of the Treaty Act to regulate migratory bird hunting. The most fundamental of these asserts that, insofar as the Act permits the Secretary to determine both the extent and the means by which migratory birds may be taken, it exceeds the powers conferred by the conventions, which authorize only the determination of closed seasons.

In *Cochrane v. United States*[58] and *Cerritos Gun Club v. Hall*,[59] plaintiffs challenged hunting restrictions on this ground. The courts in these cases upheld the Act and the Secretary's implementing regulations as a valid exercise of power conferred by both the Convention and the commerce clause. The *Cochrane* court reasoned, with respect to the Convention, that "the authority to deprive the hunters of *any* open season carries with it the power to provide for a limited open season *for limited purposes only*"[60] and, with respect to the Act as an exercise of the commerce power, that "the asserted limitations of the treaty may well be ignored for they offer no bar to the legislation as an act whose object is to regulate interstate commerce."[61]

In 1979, in *Andrus v. Allard*, the Supreme Court rejected the conclusion of at least one early decision that congressional authority for the regulation of migratory birds must rest upon the treaty power rather than the federal commerce power.[62] Thus the Court effectively concluded that the Constitution's commerce clause confers ample authority for federal regulation of migratory wildlife.

The Question of Scienter: Restrictions on Baiting

The Secretary's ban on hunting "[b]y the aid of baiting, or on or over any baited area,"[63] has generated more controversy than has any other

---

modified, 882 F.2d 1294 (8th Cir. 1989), the district court held that the plaintiffs could proceed under the APA for the EPA's violation of the Treaty Act and the Bald Eagle Protection Act. The appellate court found that because another statute (the Federal Insecticide, Fungicide, and Rodenticide Act) provided a specific procedure to challenge the regulations at issue, the APA action would not be allowed. 882 F.2d 1294 (8th Cir. 1989). The court did not take issue with the premise that citizens may proceed under the Treaty Act and the APA in the absence of another avenue for review of an agency decision.

[58] 92 F.2d 623 (7th Cir. 1937).

[59] 96 F.2d 620 (9th Cir. 1938).

[60] 92 F.2d at 626 (emphasis in original). Courts have not always found similar reasoning persuasive. *See, e.g.* Foster-Fountain Packing Co. v. Haydel, discussed in Chapter 2 at text accompanying note 83.

[61] 92 F.2d at 627.

[62] The decision rejected was United States v. Marks, 4 F.2d 420 (S.D. Tex. 1925).

[63] 50 C.F.R. § 20.21(i) (1995).

limitation on the manner of taking migratory birds. The controversy has arisen from the fact that courts have not required scienter, or guilty knowledge that an area has been baited, to find a violation.

As early as 1939, the court in *United States v. Reese* refused to read a scienter requirement into the Act, relying heavily upon purely pragmatic considerations:

> There appears no sound basis here for an interpretation that the Congress intended to place upon the Government the extreme difficulty of proving guilty knowledge of bird baiting on the part of persons violating the express language of the applicable regulations . . . but it is more reasonable to presume that Congress intended to require that hunters shall investigate at their peril conditions surrounding the fields in which they seek their quarry.[64]

Almost without exception, courts have followed the *Reese* holding,[65] even in one case in which the defendant relied upon a written communication from an Assistant Secretary of the Interior that attempted to clarify the meaning of the baiting regulation.[66] Courts have rejected the argument that failing to require scienter renders the Act unconstitutionally vague or overbroad.[67]

In the recent case of *United States v. Boynton*, the Fourth Circuit examined at length the Treaty Act's purpose, legislative history, and interpretation by other courts to conclude that the regulatory scheme clearly does not require proof of scienter on the part of either the hunter or the baiter, if the two are different.[68] In *Boynton*, defendants argued that their actions fell within an exception to the baiting prohibition which permits hunting in an area where grain has been scattered solely "as the result of normal agricultural planting or harvesting" or "as the result of bona fide agricultural operations or procedures."[69] Defendants had hunted in an area where a farmer had scattered grain purportedly in order to grow wheat for erosion control. The court held the exception inapplicable, finding that scattering grain as it had been done in this case was not a bona fide

[64]27 F. Supp. 833, 835 (W.D. Tenn. 1939).

[65]*See, e.g.*, United States v. Boynton, 63 F.3d 337 (4th Cir. 1995); United States v. Smith, 29 F.3d 270 (7th Cir. 1994); United States v. Engler, 806 F.2d 425 (3d Cir. 1986); United States v. Manning, 787 F.2d 431 (8th Cir. 1986); United States v. Chandler, 753 F.2d 360 (4th Cir. 1985); United States v. Catlett, 747 F.2d 1102 (6th Cir. 1984), *cert. denied* 471 U.S. 1074 (1985); United States v. Brandt, 717 F.2d 955 (6th Cir. 1983); United States v. Green, 571 F.2d 1 (6th Cir. 1977); Rogers v. United States, 367 F.2d 998 (8th Cir. 1966); United States v. Angueira, 744 F. Supp. 36 (D. Puerto Rico 1990), *aff'd*, 951 F.2d 12 (1st Cir. 1991).

[66]Clemons v. United States, 245 F.2d 298 (6th Cir. 1957).

[67]*See, e.g.*, United States v. Smith, 29 F.3d 270 (7th Cir. 1994); United States v. Traxler, 847 F. Supp. 492 (S.D. Miss. 1994).

[68]63 F.3d 337 (4th Cir. 1995).

[69]50 C.F.R. § 20.21(i) (1995).

agricultural operation or procedure because plants could not be expected to grow, and if they had, they would not have prevented erosion. The farmer's subjective intent was irrelevant.[70]

Only two circuits—the Fifth and the Tenth—have required scienter.[71] The Fifth Circuit held in *United States v. Delahoussaye* that "a minimum form of scienter—the 'should have known' form—is a necessary element of the offense."[72] Under this test, if the presence of bait "could reasonably have been ascertained," liability may be imposed.[73]

Although elimination of the scienter requirement has substantially eased the burden of law enforcement against baiting violators, there has been a considerable clamor for change because of the arrest and conviction of many allegedly unwitting violators. Among the examples were the defendants in *United States v. Catlett,*[74] who were convicted of taking birds over a field that was baited with corn nine days earlier.[75] There were only traces of corn left, and there was no evidence that the defendants either baited the field or knew that it was baited. The court appeared sympathetic with the defendants' plight, stating:

> The unfortunate defendants were apparently unaware of, and had not participated in, the baiting. . . .
>
> . . .
>
> We concede that it is a harsh rule and trust that prosecution will take place in the exercise of sound discretion only.[76]

[70]63 F.3d at 344–46. The court rejected the argument that the exception's use of the words "normal" and "bona fide" introduces a subjective element. The court expressly declined to follow the Sixth Circuit's finding in United States v. Brandt, 717 F.2d 955, 958 (6th Cir. 1983), that the "crucial inquiry" in such a case is the grain scatterer's intent. *See also* United States v. Sylvester, 848 F.2d 520 (5th Cir.1988), in which the court rejected the "bona fide agricultural operations" defense but did not indicate whether intent was relevant.

[71]United States v. Sylvester, 848 F.2d 520 (5th Cir. 1988); United v. Delahoussaye, 573 F.2d 910 (5th Cir. 1978); Allen v. Merovka, 382 F.2d 589 (10th Cir. 1967); United States v. Bookout, 788 F. Supp. 933 (S.D. Tex. 1992). *See also* United States v. Bryson, 414 F. Supp. 1068 (D. Del. 1976). The *Bryson* Court found two separate duties in the regulation. It followed *Allen* in finding that scienter is required for the defendant to be guilty of taking "by the aid of bait" but held that strict liability could be imposed for taking "on or over any baited area." The significance of the latter holding is diminished, however, by the court's literal reading that the birds taken must be directly over the bait itself. *Id.* at 1073. The Bryson court's restriction of the so-called "zone of influence" of the bait is unique. Other courts have declined to impose a spatial limitation on the "zone of influence," finding that it is unlawful to hunt birds attracted by bait regardless of where they are hunted. *See, e.g.,* United States v. Chandler, 753 F.2d 360 (4th Cir. 1985).

[72]573 F.2d 910, 912 (5th Cir. 1978).

[73]*Id.*

[74]747 F.2d 1102 (6th Cir. 1984), *cert. denied,* 471 U.S. 1074 (1985).

[75]The regulation considers an area "baited" for up to ten days after all bait has been removed. 50 C.F.R. §20.21(i) (1995).

[76]747 F.2d at 1103, 1105. *Cf.* United States v. Rollins, 706 F. Supp. 742 (D. Idaho 1989), discussed at text accompanying notes 101–102 *infra.*

The court "reluctantly" affirmed the conviction, deferring to the Sixth Circuit's prior decision in *United States v. Green.*[77]

Catlett was appealed to the Supreme Court, presenting an opportunity for the Court to resolve the division among the circuits. The Court denied certiorari, with Justice White dissenting.[78]

Calipatria Land Co. v. Lujan[79] illustrates the tension between the states and the federal government with regard to the anti-baiting regulation. A California regulation allowed licensed grain feeding clubs to bait areas and hunt ducks from 250 yards away from the baited area,[80] purportedly to draw birds away from crops. For years, the Fish and Wildlife Service did not enforce the federal anti-baiting regulation in California. In 1975, the Service signed an agreement with the California Department of Fish and Game, in which the Service agreed not to enforce the federal rule pending a study. The study was completed in 1980. In 1988, the Service terminated the agreement and served notice that it would begin enforcing the federal regulation in the 1990–1991 hunting season. The clubs filed suit to enjoin its enforcement. The court held that the federal regulation takes precedence over less restrictive state law, by virtue of the Supremacy Clause.[81]

Those objecting to the majority rule have often introduced legislation to redefine the baiting offense so as to apply only to those who actually place the bait or who hunt over a baited area with knowledge that it is baited.[82] Proposed legislation has also typically included an express grant of authority to the Secretary to close hunting areas found to be baited, although it is likely that such authority already exists.[83]

[77]571 F.2d 1 (6th Cir. 1977).

[78]471 U.S. 1074 (1985).

[79]793 F. Supp. 241 (S.D. Cal. 1990).

[80]Cal. Admin. Code tit. 14, R.54 § 336(c).

[81]793 F. Supp. at 244. The court disagreed with an earlier ruling in the same district that the Secretary must be deemed to have adopted the state regulation where the Secretary had knowledge of the state regulation and did not enforce the federal regulation. United States v. Oleson, 196 F. Supp. 688 (S.D. Cal. 1961).

[82]As an alternative to legislation or a "sea change" among the circuits, one commentator has suggested that the Secretary should revise the regulation to require the Delahoussaye "should have known" standard. The regulation thus would prohibit taking migratory birds "by the aid of baiting, or negligently on or over a baited area." Comment, *The Anti-Baiting Regulation Pursuant to the Migratory Bird Treaty Act: Have the Federal Courts Flown the Coop, or is the Regulation for the Birds?* 14 Geo. Mason U. L. Rev. 407 (1991). This Comment examines the baiting issue in depth.

[83]It seems that the Secretary's statutory authority to determine the extent and means of taking migratory birds, when combined with the often exercised authority to establish refuge zones, is broad enough to encompass closure of baited areas. Because of the emergency nature of such closures, it seems there would be good cause for noncompliance with informal rulemaking procedures. Indeed, the Secretary established an emergency system for closure or temporary suspension of migratory bird hunting seasons, when continuation of the season

Scienter and the Meaning of "Take": The Poisoning and Logging Cases

The premise of the baiting cases that scienter is not required to establish a violation of the Treaty Act, together with the ambiguity of the undefined term "take" that appears in both the Canadian Convention and the Treaty Act, raises the question of whether the logic of those cases might be applied to other activities that result in the killing of migratory birds as an unintended consequence. Courts have answered that question in the affirmative with respect to poisoning,[84] and in the negative with respect to habitat modification by logging.[85]

Poisoning

As noted above, the Treaty Act makes it unlawful to "by any means or in any manner, . . . pursue, hunt, take, capture, [or] kill," a protected bird.[86] Though the term "take" is not defined in the Act, regulations define it to mean "pursue, hunt, shoot, wound, kill, trap, capture, or collect, or attempt" any of the foregoing.[87] Courts have found that this language is broad enough to include poisoning.[88]

In *United States v. FMC Corp.*, the Second Circuit upheld the conviction of a pesticide manufacturer whose waste discharges into a storage pond caused the deaths of a number of migratory birds attracted to the pond.[89] The court found that the discharge of the toxic wastes was unintentional and that the defendant undertook remedial measures to keep birds from the pond once it learned of the problem. Nonetheless, the court imposed liability by analogizing the situation to those in which strict liability has been imposed on those who engage in "extrahazardous activities."[90]

A slightly different rationale was employed in *United States v. Corbin Farm Service*.[91] There an employee of a pesticide distributor, the owner of an

---

could pose an imminent threat to endangered or threatened species of wildlife, in just such a fashion. *See* 50 C.F.R. § 20.26 (1995).

[84]Defenders of Wildlife v. EPA, 688 F. Supp. 1334 (D. Minn. 1988), *modified*, 882 F.2d 1294 (8th Cir. 1989); United States V. FMC Corp., 572 F.2d 902 (2d Cir. 1978); United States v. Corbin Farm Service, 444 F. Supp. 510 (E.D. Cal.), *aff'd*, 578 F.2d 259 (9th Cir. 1978). *See also* United States v. Van Fossan, 899 F.2d 636 (7th Cir. 1990).

[85]Seattle Audubon Society v. Evans, 952 F.2d 297 (9th Cir. 1991); Citizens Interested in Bull Run, Inc. v. Edrington, 781 F. Supp. 1502 (D. Or. 1991).

[86]16 U.S.C. § 703.

[87]50 C.F.R. § 10.12 (1995).

[88]*See* cases cited in note 84 *supra*. For a discussion of similar unreported, and ultimately inconclusive, cases, see Coggins & Patti, *supra* note 5, at 183–85. The seminal discussion of the issue addressed in the text is found in the Comment cited in note 43 *supra*.

[89]572 F.2d 902 (2d Cir. 1978).

[90]*Id.* at 907. The court suggested that the production of pesticides was an extrahazardous activity for both humans and wildlife.

[91]444 F. Supp. 510 (E.D. Cal.), *aff'd*, 578 F.2d 259 (9th Cir. 1978).

alfalfa field, and an aerial pesticide applicator, were charged with violating the Treaty Act as a result of the spraying of the field with a registered pesticide that caused the deaths of a number of birds. The defendants sought to dismiss the charges on the ground that it was undisputed that they had no intent to kill any birds.[92] The court denied their pretrial motions on the basis that the relevant inquiry was not whether the defendants intended to kill birds but whether they acted with "reasonable care under the circumstances."[93] This standard is essentially that adopted by the Fifth Circuit in the *Delahoussaye* baiting case.

The court in *Corbin Farm Service* rejected the defendant's argument that death by poisoning did not fall within the Treaty Act's prohibition of taking because the Bald and Golden Eagle Protection Act expressly defined "take" to include "poison," while the Treaty Act did not.[94] The court found that Congress's 1978 amendment of the Bald Eagle Protection Act to include poison was intended to clarify the meaning of "take" in both statutes, rather than to restrict its meaning in the Treaty Act.[95]

The courts in both of these cases labored with the question of how to avoid carrying the implications of their holdings to the point of logical absurdity. Both suggest that such hypothetical examples as bird deaths caused by collisions with automobiles, glass windows, or the like should not form the basis for prosecution under the Act. The *FMC* court said that a "construction that would bring every killing within the statute . . . would offend reason and common sense" and that "the sound discretion of prosecutors and the courts" was an adequate safeguard against such possibilities.[96] By resting this conclusion upon prosecutorial discretion, however, the court implicitly acknowledged the great difficulty of determining what, for birds at least, should be deemed an extrahazardous activity.[97]

[92]In contrast, the defendant in United States v. Van Fossan, 899 F.2d 636 (7th Cir. 1990), intended to poison birds, but he did not intend to poison protected birds. In an attempt to rid his property of pigeons after the city had told him to do so, the defendant had sprinkled bird seed laced with strychnine on the ground. However, not only pigeons ate the tainted seed; so did grackles and mourning doves, which are protected under the Treaty Act. The defendant's conviction was affirmed.

[93]444 F. Supp. at 536. Two other issues were decided in this case. Relying upon the initial text, the court concluded that "poisoning" of birds is encompassed within the Act's prohibition of killing. *Id.* at 532. The court also ruled that a single act resulting in bird deaths can support only a single count, regardless of the number of birds killed by that act. *Id.* at 526–31. The latter ruling was upheld on appeal.

[94]The definition of "take" that appeared in the Bald Eagle Protection Act prior to 1972 included "pursue, shoot, shoot at, wound, kill, capture, collect," etc. Despite the breadth of this latter definition, Congress amended it in 1972 so as expressly to include "poison." *See* 16 U.S.C. § 668c.

[95]444 F. Supp. at 532.

[96]572 F.2d at 905.

[97]See note 90 *supra*.

The *Corbin Farm Service* court dismissed the hypothetical extreme examples in a manner that was at least consistent with its rationale for imposing liability. "[T]he hypothetical car driver," the court reasoned, "is not reasonably in a position to prevent the bird's death whereas a person applying pesticide might be able to foresee the danger and prevent it."[98] The reason that the latter person is presumed to be able to foresee a danger is apparently that, in the court's view, "[w]hen dealing with pesticides, the public is put on notice that it should exercise care to prevent injury to the environment and to other persons."[99] The court seems to be saying that the public at large, when engaged in certain types of activities, will be deemed to owe a duty of care to the environment.[100]

The court in *United States v. Rollins* applied the "due care" test and reversed the conviction of a defendant who had sprayed an alfalfa field with pesticides and inadvertently killed geese that ate the alfalfa.[101] The court found the Treaty Act "unconstitutionally vague" as applied to the defendant because there was no way to know that "common farming practices carried on for many years in the community" that had never killed geese before would be criminal.[102]

In *Defenders of Wildlife v. Environmental Protection Agency*, conservation groups brought an action to enjoin the Environmental Protection Agency's (hereinafter EPA) registration of strychnine as an approved above-ground rodenticide because use of the chemical also resulted in deaths of migratory birds, including threatened and endangered species.[103] The groups claimed that the EPA's action violated the Treaty Act, Bald Eagle Protection Act, and Endangered Species Act (ESA).[104] The district court agreed, finding that registration of strychnine constituted "taking" protected species under all three statutes.[105]

[98]444 F. Supp. at 535.

[99]*Id.* at 536.

[100]Coggins & Patti, *supra* note 5, at 192, attempt to devise a test to guide future decisions of this character and include a requirement that there be some degree of "culpability" in the action. Though they say their test is consistent with the baiting and poisoning cases discussed above, culpability has not been a requirement either of most of the baiting cases or of the *FMC* case. The *FMC* and *Corbin Farm Service* cases are also analyzed in Margolin, *Liability under the Migratory Bird Treaty Act,* 7 Ecol. L. Q. 989 (1979); Comment, *The Courts Take Flight: Scienter and the Migratory Bird Treaty Act,* 36 Wash. & Lee L. Rev. 241 (1979); and Comment, *Courts Hold Scienter Not Required for Conviction Under Migratory Bird Treaty Act,* 8 Envt'l L. Rep. (Envt'l L. Inst.) 10092 (1978).

[101]706 F. Supp. 742 (D. Id. 1989).

[102]*Id.* at 745. The court disagreed with the *FMC* court, *supra* note 89, that since the violation was only a misdemeanor, prosecutorial discretion would avert harsh punishment for innocent technical violations. The court said that "a violation of due process cannot be cured by light punishment." *Id.* at 745.

[103]882 F.2d 1294 (8th Cir. 1989).

[104]16 U.S.C. §§ 1531–43.

[105]688 F. Supp. 1334 (D. Minn. 1988), *modified,* 882 F.2d 1294 (8th Cir. 1989).

The court of appeals affirmed the ruling as to the ESA but reversed the ruling as to the Treaty Act and the Bald Eagle Protection Act on jurisdictional grounds.[106] Consequently, the court of appeals did not reach the issue of whether the EPA's action constituted a "taking" under the Treaty Act or the Bald Eagle Protection Act.

Habitat Destruction by Logging

In *Seattle Audubon Society v. Evans,* the Ninth Circuit Court of Appeals declined to interpret the Treaty Act's prohibition against "taking" protected birds to include destruction of northern spotted owl[107] habitat by logging.[108] The court found that differences in the definitions of "take" in the Treaty Act and the Endangered Species Act supported a finding that Congress did not intend the term in the Treaty Act to include habitat destruction.

In contrast to the Treaty Act's definition of "take," the ESA defines "take" to include "harass" and "harm."[109] Regulations define "harm" to include "significant habitat modification or degradation where it actually kills or injures wildlife by significantly impairing essential behavior patterns."[110] The court of appeals in *SAS v. Evans* agreed with the district court that the differences in definition are "distinct and purposeful"; thus "[h]abitat destruction causes 'harm' to the owls under the ESA but does not 'take' them within the meaning of the [Treaty Act]."[111]

[106]The court found that the availability of review of the EPA's registration of strychnine under FIFRA precluded general review under the APA and either the Treaty Act or the Bald Eagle Protection Act.

[107]While the northern subspecies of the spotted owl (*Strix occidentalis caurina*) is not migratory, the species *Strix occidentalis* is nevertheless included on the protected list. 50 C.F.R. § 10.13 (1995).

[108]952 F.2d 297 (9th Cir. 1991) (hereinafter SAS v. Evans). SAS v. Evans was a consolidated appeal of two district court cases, Seattle Audubon Society v. Robertson, Order on Motion for Summary Judgment and for Dismissal, No. C89–99(T)WD, 1991 WL 180099 (W.D. Wa. March 7, 1991), and Portland Audubon Society v. Lujan, No. Civ. 87–1160–FR, 21 Envt'l L. Rep. 21,341 (D. Or. May 8, 1991) (hereinafter PAS v. Lujan). The Seattle case was against the Forest Service, while the Portland case was against the Bureau of Land Management. Plaintiffs alleged that the agencies' timber sale plans violated the Treaty Act because they permitted clearcut logging of spotted owl habitat.

[109]16 U.S.C. § 1532.

[110]50 C.F.R. § 17.3 (1995).

[111]952 F.2d at 303. The court of appeals observed that Congress amended the Treaty Act in 1974, the year after passage of the ESA, but that Congress did not modify the Treaty Act's prohibitions to include "harm." *Id.* However, the 1974 Treaty Act amendment was essentially a technical amendment, to implement the 1972 Japanese Convention. The only substantive change Congress made was to add a prohibition of transactions involving "products" made from protected birds or their parts, nests, or eggs, a provision that was in both the Mexican and Japanese Conventions. The definition of "take" was not at issue. *See* Pub. L. 93–300, 88 Stat. 190 (1974); Sen. Rep. 93–851, 93d Cong., 2d Sess., *reprinted in* 1974 U.S.C.C.A.N. 3250. Moreover, the Japanese Convention acknowledged the importance of protecting the "environ-

As noted above, however, the court in *Corbin Farm Service* rejected the defendant's argument that Congress's failure to similarly amend the Treaty Act when it amended the Bald Eagle Protection Act in 1972 to include "poison" within the "take" prohibition, precluded a finding that the Treaty Act prohibits poisoning. The court found that the amendment was intended to clarify the meaning of "take" in both statutes, rather than to restrict its meaning in the Treaty Act.[112] This reasoning is at odds with the court's conclusion in *SAS v. Evans* that Congress would have amended the Treaty Act to include "harm" after passage of the ESA if it intended taking by habitat modification to apply to the Treaty Act. The Treaty Act's proscription of taking or killing "by any means or in any manner" would appear on its face to include habitat modification unless such activities were expressly *excluded.* No congressional action, including passage of the ESA, evidences intent to exclude habitat modification.[113]

The court of appeals in *SAS v. Evans* found the reasoning of the poisoning cases "inapposite,"[114] although it did not analyze the perceived differences between poisoning and logging. The court said only that "[the poisoning] cases do not suggest that habitat destruction, leading indirectly to bird deaths, amounts to the 'taking' of migratory birds within the meaning of the Migratory Bird Treaty Act."[115] Thus, the court seemed to attach pivotal significance to the fact that the death of owls occurs only indirectly from logging, whereas poisoning is directly and immediately responsible for bird deaths.[116] This reasoning may be difficult to square with the Treaty Act's prohibition against "kill[ing]" protected birds "at any time, by any means or in any manner," except as permitted by regulations adopted by the Secretary.[117]

A major issue in both *FMC* and *Corbin Farm Service* was the defendants'

---

ment" of migratory birds. Japanese Convention, art. VI. See discussion at text accompanying notes 33–42, *supra.*

[112]444 F. Supp. at 532.

[113]Note also that prior to 1977 the Fish and Wildlife Service required a permit to "scare" or "herd" migratory birds for depredation control purposes. *See* 50 C.F.R. § 21.41 (1976). Query: How could a permit be required for such activities unless they also constitute "takings"?

[114]952 F.2d at 303.

[115]*Id.*

[116]*But see* Babbitt v. Sweet Home Chapter of Communities for a Great Oregon, 115 S. Ct. 2407 (1995), in which the Supreme Court rejected a similar distinction in determining whether "take" under the ESA includes habitat destruction that causes injury to a protected species. See Chapter 7 for a discussion of the Sweet Home case.

[117]16 U.S.C. § 703. *See* Victor M. Sher, *Travels with Strix: The Spotted Owl's Journey Through the Federal Courts,* 14 Pub. Land L. Rev. 41, 70 (1993). Mr. Sher, who represented the plaintiffs in the spotted owl cases, states that the evidence that logging kills owls was not seriously contested; the district court in Portland Audubon Society v. Lujan noted that "in its own documents, the BLM recognizes that the planned timber sales in the habitat of the northern spotted owls will result in the deaths of owls."

lack of intent to kill birds. Both courts held that defendants could be convicted of violating the Treaty Act regardless of intent.[118] Intent was not at issue in the *SAS v. Evans* appeals, since both cases were civil actions against the government. Rather, the basis for the holding apparently was that intended or not, the conduct simply did not fall within the Treaty Act's prohibitions. The decision thus would appear to preclude criminal prosecution of a logger or timber company under the Treaty Act even if the logger or timber company intended to kill birds by logging.

The court in *Citizens Interested in Bull Run, Inc. v. Edrington*[119] followed the Ninth Circuit's decision in *SAS v. Evans* and advanced a restrictive interpretation of the Treaty Act that is contrary to the poisoning cases. There, the court found not only that logging spotted owl habitat did not take birds within the meaning of the Treaty Act, but that the Act is directed only at hunting and the sale of protected birds and was intended to apply only "to individual hunters and poachers."[120]

Prior to the Ninth Circuit's decision in *SAS v. Evans,* commentators had concluded that habitat modification that results in bird deaths would fall within the Treaty Act's proscriptions.[121] One commentator noted:

> Wildlife biologists have long recognized that habitat alteration and destruction are at least as important causes of wildlife population declines as direct killing, and wildlife statutes reflect that realization. In particular, the MBTA and the migratory bird treaties evidence an intent to protect bird populations from more than just hunting, and the recent Conventions strongly emphasize habitat . . . considerations.[122]

The potential consequences of holding that the Treaty Act prohibits killing or "taking" protected birds by habitat modification from logging likely deterred the Ninth Circuit from embracing this conclusion.[123] Reg-

[118]See discussion of scienter, *supra* at text accompanying notes 63–83. The court in Corbin Farm Service seems to have adopted a "duty of care" standard that is beyond that required in the FMC case or the baiting cases. However, even applying this standard, a court could consider whether there is a duty to log (or in the case of the government, to permit logging) only in a manner that is reasonably calculated to avoid migratory bird deaths.

[119]781 F. Supp. 1502 (D. Or. 1991).

[120]*Id.* at 1510.

[121]*See, e.g.,* Grubbs, *supra* note 9 at 469 ("[T]he MBTA provides a broad proscription against any disturbance of migratory birds"); Comment, *Of Birds and Men: The Migratory Bird Treaty Act,* 26 Idaho L. Rev. 371, 379 (1989) ("looking at the language of the statute, it would be anomalous to deny the applicability of the MBTA to" birds affected by logging).

[122]Coggins & Patti, note 5 *supra,* at 196 (citations deleted). Moreover, as discussed *supra* at text accompanying notes 33–42, both the Japanese Convention and the Soviet Convention explicitly addressed the need to protect habitat from adverse modifications.

[123]The court may also have been influenced by the fact that the Forest Service's logging plan was invalid under the National Forest Management Act, thus eliminating any need to apply the Treaty Act to prohibit logging in spotted owl habitat. The Bureau of Land Man-

ulations under the Treaty Act list "nearly every native, North American insectivorous and non-game species," as well as numerous species of game birds, regardless of whether the species is threatened or endangered.[124] Applying the Treaty Act's prohibitions to habitat modification would clearly have enormous implications, far beyond those that have triggered great controversy under the Endangered Species Act.[125]

Regulations Requiring the Use of Nontoxic Shot

The Secretary's authority to require that migratory waterfowl be hunted in designated areas with nontoxic (usually steel), rather than lead, shot was challenged unsuccessfully in *National Rifle Association of America v. Kleppe*.[126] The court upheld the challenged regulations on the ground that they had an adequate factual basis, without focusing at all on the scope of the Secretary's authority to regulate activities that have an indirect and unintentional adverse effect on migratory birds.

Though unsuccessful in court, opponents of lead shot restrictions have succeeded for several years in delaying them. Riders to the annual Department of the Interior appropriations legislation prevented the enforcement of those restrictions except in states that agreed to them. The Secretary, however, persisted in efforts to mandate the use of nontoxic shot. In 1986, the government adopted a rule initiating a five-year program to phase in a nationwide ban on the use of lead shot for waterfowl and coots by the 1991–1992 hunting season.[127] A key part of the rule provided that if states did not agree to nontoxic shot zones, the Secretary would close the areas to hunting of waterfowl and coots under the Treaty Act, the Bald Eagle Protection Act, and the ESA.[128] The final rule implementing the nationwide ban on lead shot was adopted in 1991.[129]

CHALLENGES TO THE SECRETARY'S AUTHORITY TO PERMIT HUNTING

There have been challenges to the administration of the federal migratory bird program as too lax, as well as charges that it unduly restricts

---

agement's logging plan was enjoined under the National Environmental Policy Act on remand to the district court. See Chapters 9 and 10 for a discussion of these cases.

[124]Coggins & Patti, *supra* note 5, at 172.

[125]See Chapter 7 at text accompanying notes 103–144.

[126]425 F. Supp. 1101 (D.D.C. 1976). Current regulations regarding use of nontoxic shot are found at 50 C.F.R. § 20.140 *et seq.* (1995).

[127]51 Fed. Reg. 42103 (1986).

[128]*Id.* at 42107. The Secretary's authority to take this step was not challenged.

[129]56 Fed. Reg. 22100 (1991). For an exhaustive discussion of the "lead shot vs. steel shot" issue, see Bruce B. Weyrauch, *Waterfowl and Lead Shot*, 16 Envt'l L. 883 (1986).

taking of birds. Cases have challenged the allowance of subsistence hunting in Alaska, regulations establishing hunting seasons, and regulations permitting hunting of declining species.[130]

Subsistence Hunting

Subsistence hunting of migratory birds by Alaskan natives has long been a source of controversy and is "one of the most troublesome issues surrounding the implementation of this country's migratory bird treaties."[131] The conflict is a particular example of a more widespread problem faced by wildlife managers: how to balance the needs of people who depend on wildlife for their food and the need to protect declining wildlife. Recent case law has shown the Treaty Act to be a poor vehicle for achieving a workable solution to this dilemma. If ratified, however, a 1995 protocol amending the 1916 Convention with Great Britain (also called the Canadian Convention) to permit some subsistence hunting of game birds during the closed season would open the door to a more realistic solution under the Treaty Act.[132]

In *Alaska Fish and Wildlife Fed'n and Outdoor Council v. Dunkle*, the Ninth Circuit Court of Appeals determined that the Treaty Act does not permit any subsistence hunting by Alaskan natives in closed seasons.[133] The court also held, however, that the government may not be compelled to enforce the Act.[134] Consequently, after *Alaska Fish and Wildlife Fed'n* the Fish and Wildlife Service was free to continue a long-standing policy of not punishing subsistence hunting, but it was not free, in an effort to reduce illegal hunting, to enter into agreements with Alaskan natives that permit a limited amount of hunting or to issue regulations permitting limited hunting.

Alaska Fish and Wildlife Fed'n concerned the hunting of migratory game birds on the Yukon-Kuskokwim Delta in western Alaska. Several species of geese that nest and raise their young on the Delta are an important part

[130]In Fund for Animals, Inc. v. Morton, Civil No. 74–1581 (D.N.J. 1974), plaintiffs raised a procedural challenge. Plaintiffs sought an injunction against the 1974–75 migratory bird hunting regulations because the Secretary had failed to prepare an environmental impact statement for the regulations, as the National Environmental Policy Act requires. The case was settled out of court when the Secretary agreed to prepare a programmatic impact statement.

[131]Sen. Rep. No. 95–1175, 95th Cong., 2d Sess., *reprinted in* 1978 U.S.C.C.A.N. 7641, 7645.

[132]Protocol Between the Government of the United States of America and the Government of Canada Amending the 1916 Convention Between the United Kingdom and the United States of America for the Protection of Migratory Birds in Canada and the United States, signed December 14, 1995 (hereinafter Canadian Protocol). See discussion of this protocol *infra*.

[133]829 F.2d 933 (9th Cir. 1987), *cert. denied*, 485 U.S. 988 (1988) (hereinafter Alaska Fish and Wildlife Fed'n).

[134]*Id.* at 138.

of the local native diet.[135] The geese are in residence on the Delta only in the nesting season, from March to September, which coincides with the closed season for migratory game birds under the Canadian Convention.

The Treaty Act as adopted in 1918 made no mention of subsistence hunting by Alaskan natives. To implement a provision permitting subsistence hunting in the 1972 Soviet Convention, Congress amended the Act in 1978 to allow the Secretary to adopt regulations permitting such hunting, with an important proviso: the regulations must be "[i]n accordance with" the four treaties.[136] Apparently, Congress believed the other treaties would be amended to conform to the Soviet Convention soon after the Act was amended.[137]

The Secretary did not adopt regulations with regard to subsistence hunting pursuant to the 1978 amendment. Official Fish and Wildlife Service policy since 1975 had been that subsistence hunting in Alaska during the closed season would not be prosecuted.[138] In 1984 and 1985 the Fish and Wildlife Service entered into agreements with the Delta natives in an effort to reduce hunting of particular species of geese. The agreements limited, but did not prohibit, subsistence hunting during the closed season.[139]

Sporting groups sued the Fish and Wildlife Service, alleging that the agreements, and the government's failure to punish subsistence hunting, violated the Act. The district court granted summary judgment against the plaintiffs on both claims, finding that the 1925 Alaska Game Law[140] permitted subsistence hunting and superseded the Treaty Act in this regard.

The court of appeals reversed the district court's decision with regard to the agreements. The court found that the 1925 Alaska Game Law did not supersede the Treaty Act, and thus the latter governs hunting of migratory birds in Alaska.[141] The court held that under the Treaty Act, "the

[135]829 F.2d at 935. These include the cackling Canada goose, white fronted goose, Pacific black brant, and emperor goose. They are all considered game birds.

[136]Fish and Wildlife Improvement Act, Pub. L. 95–616, § 3(h)(2), 92 Stat. 3112 (codified at 16 U.S.C. § 712(1)). Although the court relied on the legislative history of the 1978 amendment, which dealt only with subsistence hunting, there is no suggestion that Congress believed subsistence hunting is unique in the need to be consistent with all four treaties. *See* 829 F.2d at 941, citing Sen. Rep. No. 95–1175, *supra* note 131, at 7645.

[137]829 F.2d at 941.

[138]*Id.* at 935.

[139]*Id.* at 936.

[140]Act of Jan. 13, 1925, 43 Stat. 739 (hereinafter 1925 AGL). The 1925 AGL prohibited adoption of regulations restricting subsistence hunting unless the supply of the hunted species was in danger of extermination. 1925 AGL § 10.

[141]The court found that "the 1925 AGL was considered repealed under a general repealing clause" of the Alaska Statehood Act, Pub. L. 85–508, § 8(d), 72 Stat. 339 (1958), 48 U.S.C. notes prec. § 21. Moreover, the 1925 AGL could be read to be consistent with the Treaty

Secretary of the Interior is authorized to issue regulations permitting subsistence hunting, but only to the extent that the regulations are in accord with all four treaties.''[142] Consequently, the regulations must be in accord with the most restrictive treaty—in this case, the Canadian Convention—and thus ''the Secretary may adopt regulations that permit subsistence hunting for up to three and one-half months between September 1 and March 10 of each year.''[143] On remand, the district court found that the agreements were invalid under the Treaty Act.[144]

The court of appeals also held, however, that it lacked jurisdiction to review the plaintiffs' claim that the Fish and Wildlife Service had ''abrogated its statutory duty to enforce the closed hunting season required by the MBTA [Treaty Act].''[145] Citing *Heckler v. Chaney*,[146] the court found that failure of an agency to prosecute is reviewable only ''where the substantive statute has provided guidelines . . . for its enforcement powers.''[147] The court found that the Treaty Act had not done so and instead ''explicitly delegates the authority to adopt regulations and discretionary enforcement powers to the Secretary of the Interior.''[148]

The Supreme Court declined to review the Ninth Circuit's decision. As one commentator observed, ''Thus ended the only period in the last six decades when the MBTA's validity was seriously questioned.''[149]

In 1988, the Fish and Wildlife Service adopted a new enforcement policy regarding subsistence hunting in Alaska.[150] The 1988 policy enunciated two broad goals: to provide biological information about the geese to the native population, and to enlist native cooperation in maintaining the species. The policy prohibits (1) taking cackling or emperor geese at any time; (2) taking white-fronted or brant's geese during nesting; (3) taking eggs from any of the four species; and (4) using private or chartered aircraft to assist in hunting. The policy does not provide for any specific enforcement effort.

Alaska Fish and Wildlife Fed'n was significant in clearly establishing that the treaties underlying the Treaty Act are not to be disregarded in gov-

---

Act. See 829 F.2d at 942–45 for a discussion of the 1925 AGL and its relationship to the Treaty Act.

[142]829 F.2d at 941.

[143]*Id.* In addition to the Canadian Convention, the treaties with Mexico and Japan would have to be modified to permit subsistence hunting during closed seasons. *Id.* See discussion of the Canadian Convention's taking restrictions *supra* at text accompanying notes 15–18.

[144]19 Envt'l L. Rep. 20081 (D. Alaska 1988).

[145]829 F.2d at 938.

[146]470 U.S. 821 (1985).

[147]*Id.* at 833.

[148]829 F.2d at 938.

[149]Gregory F. Cook, *Government and Geese in Alaska*, 5 J. Envt'l L. & Litig. 29, 42 (1990). Mr. Cook represented the plaintiffs in Alaska Fish and Wildlife Fed'n.

[150]53 Fed. Reg. 16877 (1988).

ernment policy. However, the decision unequivocally left enforcement of the Treaty Act and the terms of the treaties wholly within the government's discretion. Moreover, the decision may have been a step backward in efforts to fashion a workable solution to the subsistence hunting problem.

It has been apparent for some time that no solution that completely prohibits subsistence hunting is likely to work.[151] The Canadian Convention would have to be amended, however, if any level of subsistence hunting is to be permitted. A 1979 Protocol Amendment to the Canadian Convention permitting subsistence hunting met with vigorous opposition from North American conservationists and was never ratified.[152] In 1995, the governments of the United States and Canada signed a second protocol amending the Convention to permit some subsistence hunting of game birds during the closed season.[153] The amendment has not yet been ratified by either government. Prior to the 1995 protocol, one commentator suggested that if the treaty were amended the Treaty Act should be amended simultaneously to "articulate clear standards for the [Fish and Wildlife Service] to follow in enforcing the law."[154]

The decision in *Alaska Fish and Wildlife Fed'n* that Treaty Act regulations must be consistent with all four treaties also made it clear that longstanding rules permitting subsistence taking of particular *nongame* birds in Alaska violate the Treaty Act.[155] Although the Canadian, Japanese, and Soviet treaties permit such taking,[156] the Mexican Convention contains no exception for native peoples from its taking prohibition, perhaps because

[151]*See, e.g.*, Cook, *supra* note 149 at 53. ("A total prohibition on the hunting of nesting waterfowl in Alaska's [Yukon-Kuskokwim] Delta appears to be in the same class of socially unrealistic legislation as this Nation's unsuccessful experiment with a prohibition on the sale of alcohol.")

[152]Cook, *supra* note 149 at 57. Conservationists objected to the Protocol's "imprecise language which seems to authorize 'subsistence' harvest at any time of the year, including nesting and breeding seasons." *Id.*

[153]Canadian Protocol, *supra* note 132. If ratified, the amendment would allow the Fish and Wildlife Service to issue regulations governing subsistence hunting of geese during the closed season. (Sport hunting would not be permitted.) Hunting of wild ducks, however, is prohibited by the Mexican Convention. Negotiations were to begin in 1996 to amend the Mexican Convention to permit subsistence duck hunting.

[154]Cook, *supra* note 149 at 52.

[155]Regulations permit year round subsistence taking of auks, auklets, guillemots, murres, and puffins by Eskimos and Indians in Alaska as well as of snowy owls and cormorants by any person in Alaska. *See* 50 C.F.R. § 20.132 (1995). The addition of snowy owls and cormorants was effectuated in 1973 without ever having been the subject of a proposed rulemaking open for public comments. At the time it was done, the Fish and Wildlife Service explained that its action was authorized by article I of the Mexican Treaty. *See* 38 Fed. Reg. 17841 (1973). In fact, not only did article I not authorize the action taken, but article II specifically prohibited it.

[156]Canadian Convention, art. II, § 3; Japanese Convention, art. III, § 1(e); Soviet Convention, art. II, § 1(c).

indigenous peoples comprise such a large portion of the Mexican population. Nevertheless, the regulations remain in effect.

The scope of Indian taking rights outside Alaska may be considerably greater than the Treaty Act on its face suggests, in areas where Indian treaties permit such taking. The question of Indian treaty rights was not raised in *Alaska Fish and Wildlife Fed'n* because native Alaskans do not have treaty rights.[157] In *United States v. Bresette,* however, a district court reversed the conviction of two Chippewa Indians for selling artifacts containing feathers of protected birds, on the grounds that existing Indian treaty rights that permitted the taking were not affected by passage of the Migratory Bird Treaty Act.[158] The court in *Bresette* distinguished the holding in *United States v. Dion,*[159] in which the Supreme Court found that the Bald Eagle Protection Act abrogates treaty rights to take eagles. In contrast to the Bald Eagle Protection Act, the *Bresette* court found, the Treaty Act contains no clear evidence of intent to abrogate treaty rights.[160]

Hunting Seasons and "Optimum Numbers"

In *Fund for Animals, Inc. v. Frizzell,*[161] plaintiffs contended that the Canadian Convention required either a single three-and-a-half-month season for all birds, or separate three-and-a-half-month seasons for each of the five families of migratory game birds designated in the Convention, and that the Japanese Convention required a determination of "optimum numbers" for each of the protected species prior to permitting their being hunted. Because of the procedural posture of the case, neither the district court nor the court of appeals addressed these claims. Accordingly, their merits will be briefly considered here.

With respect to the claim that the Canadian Convention requires a single, three-and-a-half-month migratory bird hunting season, the contention is that the Secretary's practice of so limiting the hunting season within each state, while prescribing successively later seasons for more southern states, results in a hunting season that exceeds six months when all states are lumped together. Literally, the Convention refers only to a "season for hunting" migratory game birds. Despite this literal reading of the

[157]*See* United States v. Bresette, 761 F. Supp. 658, 663 (D. Minn. 1991), citing Cohen, Handbook of Federal Indian Law, 739 (1982). See Chapter 13 for a discussion of Indian treaty rights and wildlife conservation laws.

[158]761 F. Supp. 658 (1991). The court in United States v. Cutler, 37 F. Supp. 724 (D. Id. 1941), had reached the same conclusion 50 years earlier.

[159]476 U.S. 734 (1986).

[160]See discussion of Dion *infra* at text accompanying notes 232–234. The court in Bresette also found that the Treaty Act is not a "nondiscriminatory conservation measure" permissible under Puyallup Tribe v. Dep't of Game of Washington, 391 U.S. 392 (1968).

[161]530 F.2d 982 (D.C. Cir. 1975), *aff'g* 402 F. Supp. 35 (D.D.C. 1975).

Convention, it has apparently been the continuous practice of both the United States and Canada to prescribe separate seasons for each state or province. This consistent administrative practice, together with the absence of any relevant negotiating history, persuaded the district court to find that plaintiffs did not have a substantial chance of success on the merits of the claim in the *Frizzell* case and thus to deny a preliminary injunction.[162] The court was also persuaded by the Department of the Interior's assertion that its interpretation of the three-and-a-half-month provision was appropriate because it gave greater flexibility in the management of migratory birds than would a uniform three-and-a-half-month season.[163]

Even more problematic is the meaning of the requirement in the Japanese Convention that bird hunting seasons be set so as "to maintain their populations in optimum numbers." Just as the negotiating history of the Canadian Convention failed to illuminate the meaning of the three-and-a-half-month provision, so too the negotiating history of the Japanese Convention sheds no light on the meaning of the term "optimum numbers." Plaintiffs in the *Frizzell* case argued that the term should have the same meaning as the term "optimum sustainable population" as used in the Marine Mammal Protection Act, because the Convention and the Act were drafted at approximately the same time. The government, on the other hand, argued that the Convention's failure to include a definition of the term indicated an intent not to borrow the Marine Mammal Protection Act's definition. Instead, the government urged that the term imposed no different duty on the Secretary than what he had always done in promulgating hunting regulations under the Treaty Act. The district court again concluded that the plaintiffs had failed to demonstrate a substantial likelihood of disproving the government's interpretation.[164]

[162]While courts typically give great deference to a long-established administrative interpretation of a statute, they are not bound thereby.

[163]It is doubtless true that Interior's position offers more flexibility. On the other hand, it would be possible to achieve still greater flexibility by subdividing individual states into multiple zones and proscribing separate hunting seasons for each. In this manner, each state could have a migratory bird hunting season that exceeds six months in length. While zoning of individual states has not been used much in migratory bird management, its use necessarily sharpens the question of what meaning, if any, the Convention's reference to a 3 1/2-month hunting season has.

[164]In fact, neither party offered a convincing explanation of the relationship between the language of the Convention and that of the Marine Mammal Protection Act. At the time the Japanese Convention was signed, the Act had not yet been passed. Moreover, the term "optimum sustainable population" did not even come into being until the Senate Commerce Committee reported out a marked-up bill three months after the signing of the Convention. Sen. Rep. No. 92–863, 92d Cong., 2d Session (1972). At the time the Convention was signed, the House bill, H.R. 10420, 92d Cong., lst Sess. (1971), contained the term "optimum sustainable yield." Accordingly, if there is any nexus between the terms used in the Japanese Convention and the Marine Mammal Protection Act, the latter would seem to constitute a

In evaluating the meaning of the term "optimum numbers" in the Japanese Convention, neither the parties nor the court adequately addressed one apparently relevant consideration. Since the term is not found in either of the preceding migratory bird conventions, its meaning may be linked to the Japanese Convention. Examining the preambles to each of the conventions, one finds the following statement of purpose in the Canadian Convention: "Many of these species are of great value as a source of food or in destroying insects which are injurious to forests and forage plants on the public domain." In a similar vein, the Mexican Convention states its purpose as being to "permit a rational utilization of migratory birds for the purpose of sport as well as for food, commerce and industry." In a strikingly dissimilar manner, the preamble to the Japanese Convention notes that the birds subject to its protection "constitute a natural resource of great value for recreational, aesthetic, scientific, and economic purposes, and that this value can be increased with proper management." Because this articulation of values is broader and more diverse than comparable articulations in the earlier treaties, it seems reasonable to suggest that the term "optimum numbers" is really a direction to optimize the values served.

Because the Secretary's authority to issue hunting regulations under the Treaty Act is "[s]ubject to the provisions and in order to carry out the purposes of the conventions," the Act implicitly incorporates the diverse purposes of the Japanese Convention, at least insofar as the birds protected by it are concerned. Thus, to the extent that those purposes are broader than, or otherwise different from, the purposes to be served by the earlier treaties, the Secretary's regulations should reflect them.[165]

Hunting of Declining Species

Plaintiffs in *Humane Society of the United States v. Watt* challenged the Secretary's decision to continue sport hunting of black ducks in the face of at least some evidence suggesting a steady, long-term decline in the black duck population.[166] The case was factually complex and hinged ultimately on the adequacy of the evidence available to support the Secretary's decision. In the court's view, evidence of such a decline was

---

refinement and clarification of a term, the meaning of which was still uncertain at the time the Convention was signed. For a discussion of the concept of "optimum sustainable population" in the Marine Mammal Protection Act, see Chapter 5.

[165]It is interesting to note that in Frizzell the government emphasized as a key factor in the promulgation of the Secretary's regulations the economic value of the waterfowl resource. Significantly, economic value appears last in the listing of purposes to be served by the Japanese Convention, just as it does in the Marine Mammal Protection Act. *See* 16 U.S.C. § 1361(6).

[166]551 F. Supp. 1310 (D.D.C 1982), *aff'd*, 713 F.2d 865 (D.C. Cir. 1983).

contradictory as to its cause and its gravity. Under such circumstances, the court was unwilling to disturb the Secretary's judgment. One of the few clear propositions of law asserted in the opinion was that the Treaty Act "does not impose upon the [Fish and Wildlife] Service a mandatory duty to prohibit the hunting of any species whose population, for whatever reason, is declining."[167] The court suggested that such a mandatory duty would exist only if the duck declined to such an extent that it became eligible for protection under the Endangered Species Act.[168]

In only one instance has a court held that the standards of the Treaty Act or the treaties it implements circumscribe the Secretary's authority to permit migratory bird hunting—*Alaska Fish and Wildlife Fed'n and Outdoor Council v. Dunkle*, which invalidated agreements permitting subsistence hunting in closed seasons.[169] The Secretary's authority also may be limited by other laws when authorization of hunting affects interests protected by the other laws. At least one court has held that the Endangered Species Act limits the Secretary's authority under the Treaty Act.[170] That case and the related case of *Connor v. Andrus*[171] are discussed in detail in Chapter 7.

INFRINGEMENT OF PRIVATE PROPERTY RIGHTS

Despite the holding of *Geer v. Connecticut* that individual property rights in wildlife can be acquired only in such manner as the government may allow,[172] landowners and others have challenged hunting prohibitions under the Treaty Act on the grounds that the prohibitions constitute unlawful "takings" of private property. Courts have uniformly upheld the Treaty Act against these claims.

The first case to address the issue in detail was *Bailey v. Holland*, which concerned the designation of certain private lands adjacent to Back Bay Migratory Waterfowl Refuge as a closed area in which no hunting of migratory birds was to be permitted.[173] The plaintiff landowner alleged that

[167]*Id.* at 1319.

[168]*Id.*

[169]829 F.2d 933 (9th Cir. 1987), *cert. denied*, 485 U.S. 988 (1988). See discussion of Alaska Fish and Wildlife Fed'n *supra* at text accompanying notes 133–149.

[170]Defenders of Wildlife v. Andrus, 428 F. Supp. 167 (D.D.C. 1977).

[171]453 F. Supp. 1037 (W.D. Tex. 1978).

[172]161 U.S. 519 (1896).

[173]126 F.2d 317 (4th Cir. 1942). The Secretary has, since 1931, designated several areas adjacent to federal wildlife refuges as closed areas. *See id.* at 323. Although the authority to make these designations arguably derives from the power conferred by the Treaty Act to determine the "extent" and "manner" of migratory bird hunting, it was nowhere made express until the 1936 Convention with Mexico, which provided in article II for the establishment of "refuge zones in which the taking of such birds will be prohibited." The Treaty Act has never been amended to incorporate the "refuge zone" concept of the Mexican

the Secretary's designation had in effect extended the boundaries of the refuge so as to encompass his land and argued that the only lawful way for the Secretary to do that was to purchase the land pursuant to the Migratory Bird Conservation Act.[174] The court disposed of the plaintiff's contentions as follows:

> If the Government wishes to do more in the way of protecting migratory birds than prohibiting their slaughter, *e.g.*, erect improvements to lessen the dangers resulting from the drainage of marshy areas, it must acquire some proprietary interest in the areas suitable for such uses. It was to meet this need that Congress enacted the Migratory Bird Conservation Act. . . . The 1929 Act does not deprive the Secretary of his regulatory authority under the 1918 statute. Merely because the Government purchases certain land in order to do more than prohibit hunting, it does not follow that compensation must be made for all land closed to hunting.[175]

Despite the unequivocal nature of the *Bailey* holding, the same issue was litigated ad infinitum in connection with the Secretary's designation of a closed area adjacent to Horseshoe Lake Preserve in Illinois. Plaintiffs lost in two federal actions[176] and a suit in the court of claims.[177]

On the broader issue of whether hunting rights constitute property rights, a few state court decisions have reached results contrary to those of the decisions described above.[178] Although those decisions have been criticized,[179] the Department of the Interior has, on at least some occasions, treated hunting rights as property rights by bringing condemnation proceedings against them when acquiring lands pursuant to the Migratory Bird Conservation Act.[180]

The Supreme Court in *Andrus v. Allard*[181] rejected a more limited "takings" challenge to the Treaty Act. Plaintiffs claimed that prohibiting the

---

Convention. Nonetheless, the Secretary has continued to designate areas adjacent to national wildlife refuges as closed areas in which migratory bird hunting is prohibited.

[174]The Migratory Bird Conservation Act is discussed in Chapter 8.

[175]126 F.2d at 324.

[176]Lansden v. Hart, 168 F.2d 409 (7th Cir.), *cert. denied*, 335 U.S. 858 (1948); Lansden v. Hart, 180 F.2d 679, 684 (7th Cir.), *cert. denied*, 340 U.S. 824 (1950); and Sickman v. United States, 184 F.2d 616 (7th Cir. 1950), *cert. denied*, 341 U.S. 939 (1951).

[177]Bishop v. United States, 126 F. Supp. 449 (Ct. Cl. 1954).

[178]*See, e.g.*, Alford v. Finch, 155 So. 2d 790 (Fla. 1963), and Allen v. McClellan, 75 N.M.400, 405 P.2d 405 (1965). For decisions *contra*, see Platt v. Philbrick, 8 Cal. App. 2d 27, 47 P.2d 302 (1935), Maitland v. People, 93 Colo. 59, 23 P.2d 116 (1933), State v. McKinnon, 153 Me. 15, 133 A.2d 885 (1957), Bauer v. Game, Forestation & Parks Comm'n., 138 Neb. 436, 293 N.W. 282 (1940), and Cook v. State, 192 Wash. 602, 74 P.2d 199 (1937).

[179]*See, e.g.*, Plimpton, *Power of State to Designate Game Preserves*, 6 Nat. Res. J. 361 (1966).

[180]*See, e.g.*, Swan Lake Hunting Club v. United States, 381 F.2d 238 (5th Cir. 1967), in which the "owner" of the hunting rights was a party other than the owner of the land.

[181]444 U.S. 51 (1979).

sale of artifacts containing protected bird parts that were lawfully acquired before passage of the Treaty Act, was an unconstitutional taking of property. Though the Court found it "undeniable" that a prohibition against the sale of the artifacts would "prevent the most profitable use of appellees' property," it refused to treat that as an unconstitutional taking. [182]

In *U.S. v. Darst*, the defendant raised a private property claim of a different sort.[183] Darst was convicted of violating the Treaty Act by catching great horned owls in leg traps and then killing them. He said that the owls were killing his chickens, and he claimed that he had a constitutional right to protect his property. The court found that there is no constitutional right to defend one's property against federally protected wildlife. The court cited *Christy v. Hodel*, in which the Ninth Circuit declined to find a constitutional right to kill an endangered grizzly bear to defend livestock.[184]

APPLICABILITY OF THE TREATY ACT TO CAPTIVE-RAISED BIRDS

The Tenth Circuit has twice considered the question of the Act's applicability to captive-reared birds. In 1978, in *United States v. Richards,* the court concluded that captive-reared falcons are subject to the Act.[185] A year later, in *United States v. Conners,* the same circuit concluded that captive-reared ducks are not.[186] The principal asserted basis for this different result is that two of the original three treaties refer to "wild" ducks whereas none of those treaties makes a similarly limited reference to falcons. That explanation is not only unpersuasive but, in light of the Supreme Court's subsequent decision in *Andrus v. Allard,*[187] clearly wrong as a basis for the result in Conners.

Andrus v. Allard involved a challenge to the Secretary's authority to prohibit the sale of artifacts containing parts of birds lawfully acquired before the Treaty Act applied to them. The Court upheld that authority on the basis that since "a flat proscription on the sale of wildlife, without regard to the legality of the taking" is "a traditional legislative tool for enforcing conservation policy," an explicit statutory exception was necessary to overcome the presumption that the Act was intended to be all-encompassing.[188]

[182]*Id.* at 66.
[183]726 F. Supp. 286 (D. Kan. 1989).
[184]857 F.2d 1324, 1329–30 (9th Cir. 1988). See Chapter 7 for a discussion of this case.
[185]583 F.2d 491 (10th Cir. 1978).
[186]606 F.2d 269 (10th Cir. 1979). For a case discussing the applicability of the Endangered Species Act to captive-reared wildlife, see Cayman Turtle Farm, Ltd. v. Andrus, 478 F. Supp. 125 (D.D.C. 1979), *aff'd* without opinion (D.C. Cir. Dec. 12, 1980), discussed in Chapter 7.
[187]444 U.S. 51 (1979).
[188]*Id.* at 61.

While not explicitly disapproving of the result in *Conners*, the Court rejected arguments based upon a literal reading of one of the treaties.[189] Instead, the Court held that the various treaties "establish *minimum* protections for wildlife; Congress could and did go further in developing domestic conservation measures."[190] Further, the Court cited the *Richards* decision approvingly.[191] Thus it can with confidence be asserted on the basis of the *Allard* decision, from which no member of the Court dissented, that *Conners* was wrongly decided.[192]

PROTECTION OF BALD AND GOLDEN EAGLES: THE BALD EAGLE PROTECTION ACT

Structure

In 1940 Congress, believing that the bald eagle was threatened with extinction and desiring to preserve it as the nation's symbol, enacted legislation to protect the eagle. The Bald Eagle Protection Act (hereafter Bald Eagle Act) made it a criminal offense, with certain exceptions, for any person to take or possess any bald eagle or any part, egg, or nest thereof.[193] The original Act prescribed maximum penalties for violations that are identical to the Treaty Act's misdemeanor penalties, six months' imprisonment and $500 in fines, or both.[194]

The 1940 Act contained several important exceptions that continue in effect today.[195] Among these are that the Secretary of the Interior may permit taking and possession of bald eagles "for the scientific or exhibition purposes of public museums, scientific societies, and zoological parks," as well as "for the protection of wildlife or of agricultural or other interests in any particular locality."[196] The Secretary may permit these activities only "after investigation" and upon determining "that it is compatible with the preservation of the bald eagle" to do so.[197] Further, the

[189] *Id.* at 62–63, n.18.

[190] *Id.* (emphasis in original). The Court relied upon the discussion at pages 74–76 of the initial text in support of this proposition.

[191] *Id.* at 63 n. 19.

[192] However, most of the administrative restrictions on taking wild mallards do not apply to captive-reared mallards. Regulations governing the taking of captive-reared mallards are found at 50 C.F.R. § 21.13 (1995).

[193] 16 U.S.C. §§ 668–668d.

[194] Act of June 8, 1940, Ch. 278, § 1, 54 Stat. 250 (1940) (current version at 16 U.S.C. § 668a).

[195] One original exception, eliminated by amendment in 1959, provided that the terms of the Act did not apply in Alaska. Act of June 25, 1959, Pub. L. No. 86–70, § 14, 73 Stat. 143 (1959).

[196] 16 U.S.C. § 668a.

[197] *Id.* The Secretary's regulations pertaining to the issuance of permits for scientific or

Act does not prohibit the possession of any bald eagle, or part, nest, or egg thereof, which was lawfully taken prior to passage of the Act.[198]

Congress has substantially amended the Bald Eagle Act three times. The Act is now more stringent in some respects and less so in others than the 1940 legislation.

The 1962 Amendments

The first amendment, in 1962, was designed to both add protection for golden eagles[199] and enhance the protection of immature bald eagles, which are difficult to distinguish from golden eagles. The amendment extended to golden eagles the same protection that applied to bald eagles.[200] The 1962 amendment also created two new exceptions. First, it provided that the Secretary could exempt, by permit, takings of bald and golden eagles "for the religious purposes of Indian tribes."[201] Congress emphasized the permit requirement with respect to bald eagles: "bald eagles may not be taken for any purpose unless, prior to such taking, a permit to do so is procured from the Secretary of the Interior."[202]

Second, the amendment provided that the Secretary, on request of the governor of any state, "shall authorize" taking golden eagles "for the purpose of seasonally protecting domesticated flocks and herds in such State, . . . in such part or parts of such State and for such periods as the Secretary determines to be necessary to protect such interests."[203] Although a number of "blanket permits" for taking depredating golden eagles were issued prior to 1969, the Secretary established as a matter of administrative policy in 1970 that no further blanket permits would be issued.[204]

---

exhibition purposes, found at 50 C.F.R. § 22.21 (1995), purport to apply to parts, eggs, and nests of eagles, as well as to the eagles themselves, whereas the statutory exception extends only to "specimens" of eagles. 16 U.S.C. § 668a.

[198] 16 U.S.C. § 668(a).

[199] The House report on the bill to amend the Act stated that because golden eagle feathers are used in Indian religious ceremonies and for tourist souvenirs, and because they are actively hunted by bounty hunters in some states, "there is grave danger that the golden eagle will completely disappear." H. R. Rep. No. 1450, 87th Cong., 2d Sess., at 2 (1962).

[200] Act of Oct. 24, 1962, Pub. L. No. 87–884, 76 Stat. 1246.

[201] 16 U.S.C. § 668a.

[202] *Id.*

[203] *Id.* The statute is unclear as to whether, with respect to this exception, the Secretary must determine that any permitted taking "is compatible with the preservation of" the golden eagle.

[204] The 1970 policy is found at 41 Fed. Reg. 50355 (1976). Despite the change in policy, administrative regulations continue to provide for issuance of blanket depredation control orders. *See* 50 C.F.R. 22.31 (1995). These regulations provide for publication in the Federal Register and notice to the requesting governor, *after* the Director's decision. With respect to other types of permits for taking bald or golden eagles, no public notice is required at any time.

The 1972 Amendments

Congress substantially amended the Act once again in 1972,[205] driven by the widely publicized deaths of several dozen bald and golden eagles as a result of the indiscriminate use by certain Wyoming ranchers of thallium sulfate, apparently as a means of coyote control, and the equally well-publicized deliberate shooting of several hundred eagles from helicopters, likewise at the behest of ranchers. With respect to the poisonings, the Department of the Interior took the view that under the Act as it then existed, no prosecutions could be brought unless the government could show that the person who placed the poison actually intended to kill eagles.[206] Accordingly, to lessen the degree of intent required to establish a violation, Congress amended the Act to provide that whoever "shall knowingly, or with wanton disregard for the consequences of his act," take any eagle, shall be subject to penalties prescribed.[207] At the same time, through an abundance of caution, Congress amended the Act's definition of the term "take" to include poisoning, though it is likely that this amendment was unnecessary.[208]

The 1972 amendments also substantially increased the maximum penalties for violations of the Act, and the penalties have remained the same since 1972. For first violations, maximum penalties are at $5,000, one year's imprisonment, or both; for subsequent violations, the applicable maxima are $10,000 and two years' imprisonment.[209] Another important amendment was the addition of two types of civil sanctions. First, civil penalties of up to $5,000 per violation may be assessed against any person taking or possessing any eagle, part, nest, or egg thereof.[210] Second, the

[205]Act of Oct. 23, 1972, Pub. L. No. 92–535, § 1, 86 Stat. 1064.

[206]*See* Hearings Before the Subcommittee on Agriculture, Environmental, and Consumer Protection of the Senate Committee on Appropriations on Predator Control and Related Problems, 92d Cong., 1st Sess. 44 (1971).

[207]16 U.S.C. § 668(a).

[208]The Secretary apparently thought the definition of "take" that appeared in the Act prior to 1972 encompassed poisoning, for the eagle regulations promulgated in 1963 provided that takings authorized pursuant to permit or depredation control order must be "by firearms, traps, or other suitable means *except by poison.*" 50 C.F.R. § 11.4 (1964) (emphasis added). Further, the undefined term "take" that appears in the Migratory Bird Treaty Act apparently is thought to include poisoning. See text accompanying note 10 *supra*.

[209]Though the maximum fine prior to 1972 was $500, actual fines levied were substantially less. According to the Senate report that accompanied the 1972 amendments, the average fine imposed for the 32 convictions under the Act during the preceding five years was less than $50. Sen. Rep. No. 92–1159, 92d Cong., 2d Sess. at 4 (1972), *reprinted in* 1972 U.S.C.C.A.N. 4281, 4288.

[210]The civil penalty provision, 16 U.S.C. § 668(b), is essentially identical to the criminal sanction provision as it was before the 1972 amendment added the words "knowingly, or with wanton disregard for the consequences." Accordingly, since the criminal sanction was thought to apply only to "willful" violations prior to 1972, the civil penalty provision arguably

Secretary may immediately cancel the federal grazing privileges of any person convicted of a violation of the Act.[211] Finally, the 1972 amendments introduced a "citizen bounty" by providing that any person giving information that leads to a conviction under the Act may receive up to one-half of any fine, up to $2,500.[212]

Like the amendments a decade earlier, the 1972 amendments were not meant solely to strengthen the Act's prohibitions. They also created a new exemption for the taking and possession of golden eagles for the purposes of falconry, subject to a permit.[213] However, only golden eagles initially taken "because of depredations on livestock or wildlife" may be taken for falconry purposes.[214]

The 1978 Amendments

The Bald Eagle Act's prohibition on taking eagles extends to taking their nests. Perceiving a significant and increasing potential conflict between this prohibition and future coal-mining activities in the West, Congress in 1978 added yet another exception to the Act's prohibitions. That exception authorizes the Secretary of the Interior to permit, by regulation, "the taking of golden eagle nests which interfere with resource development or recovery operations."[215] Although the legislative history indicates that Western coal development was the exclusive focus of congressional concern,[216] the Act neither defines "resource development or recovery" nor limits the exception to any geographic region. Similarly, the Act is unclear as to what, if any, standards govern the exercise of Secretarial discretion, though the legislative history suggests that compatibility with preservation of the species was intended.[217]

---

applies only to "willful" violations now. If so, the rather unusual result is that the Act requires a lesser standard of culpability for imposition of criminal sanctions than it requires for imposition of civil sanctions.

[211] 16 U.S.C. § 668(c).

[212] *Id.* § 668(a).

[213] General falconry regulations under the Treaty Act are found in 50 C.F.R §§ 21.28 (1995). Regulations governing falconry permits for golden eagles are found at 50 C.F.R. § 22.24 (1995).

[214] 16 U.S.C. § 668a. 50 C.F.R. § 22.24(f)(1) (1995) provides that golden eagles may be trapped for falconry purposes only from "a specified depredation area."

[215] Fish and Wildlife Improvement Act of 1978, Pub. L. No. 95–616, § 9, 92 Stat. 3110, 3114–3115 (codified at 16 U.S.C. § 668a).

[216] Sen. Rep. No. 95–1175, *supra* note 131, at 7645.

[217] *Id.* at 7646. This legislative history suggests that the compatibility standard applies to all of the exceptions contained in 16 U.S.C. § 668a. If so, the provision is unartfully drafted.

Relationship of the Migratory Bird Treaty Act and the Bald Eagle Protection Act

Many of the legal issues that have arisen under the Bald Eagle Act have also arisen under the Treaty Act.[218] For example, in *Andrus v. Allard* the Supreme Court simultaneously addressed the question of the application of the restrictions on commerce in the Bald Eagle Act and the Treaty Act to eagle parts acquired prior to the enactment of either law.[219] The Court found restrictions of both Acts applicable. With respect to the Bald Eagle Act, the Court cited a limited exception in the Act for the possession or transportation (but not sale) of eagles taken prior to its enactment. From this narrow exception the Court drew the negative inference that Congress intended no exception for sale of eagles.

Not all issues common to the Bald Eagle and Treaty Acts have been resolved in a consistent way, however. The court in *United States v. Martinelli*,[220] for example, held that a defendant charged with violating the Bald Eagle Act had a right to a jury trial even though the maximum penalties prescribed make the offense a "petty" offense[221] and although the clear weight of authority under the Treaty Act is to the contrary.[222] This difference, however, is no longer important, for the 1972 amendment, by increasing the penalties under the Bald Eagle Act, removed violations of that Act from the category of petty offenses.

The Scienter Requirement

One significant difference in the interpretation of the Treaty Act and the Bald Eagle Act concerns the requirement of intent. Under the Treaty Act, scienter, or guilty knowledge, is not an element of a violation, except for a felony conviction related to the sale of protected birds.[223] In *United*

[218]Many of these issues have arisen in criminal cases, but civil actions under the Administrative Procedure Act are permitted under the Bald Eagle Act as well as under the Treaty Act. *See, e.g.* Humane Society v. Lujan, 768 F. Supp. 360 (D.D.C. 1991) (the court "assumed without deciding" that plaintiffs could proceed under the APA). However, because there is no private right of action under either the Treaty Act or the Bald Eagle Act, citizens have no standing to sue nonfederal governmental units. *See* Protect Our Eagles' Trees (POETS) v. Lawrence, 715 F. Supp. 996 (D. Kansas 1989) (plaintiffs could not bring action against a city under the Bald Eagle Act for permitting developer to cut down eagle perching trees).

[219]444 U.S. 51 (1979).

[220]240 F. Supp. 365 (N.D. Cal. 1965).

[221]*See* 18 U.S.C. § 1(3).

[222]The Martinelli decision was based on Smith v. United States, 128 F.2d 990 (5th Cir. 1942), a case arising under the Migratory Bird Treaty Act. The Smith decision has not been followed, however, in subsequent Treaty Act cases. *See, e.g.*, United States v. Ireland, 493 F.2d 1208 (4th Cir. 1973), United States v. Jarman, 491 F.2d 764 (4th Cir. 1974), and United States v. Cain, 454 F.2d 1285 (7th Cir. 1972).

[223]See text accompanying notes 53–55, *supra*.

States v. Hetzel, however, the court insisted on a showing of criminal intent, holding that "[r]ules of decision developed under the Migratory Bird Treaty Act may not automatically be applied to a prosecution under the Bald Eagle Protection Act."[224] The *Hetzel* decision is a good example of the axiom that bad cases make bad law. In that action, the defendant was charged with violating the Bald Eagle Act because he possessed a pair of bald eagle talons. He apparently removed them from a dead bald eagle that he found while hunting on a national wildlife refuge, intending to give them to a Boy Scout organization. The defendant was convicted before a United States Magistrate and fined one dollar. Despite the nominal fine, the defendant appealed.

On appeal, the court emphasized the change in the degree of intent required for a violation as a result of the 1972 amendments. The court reasoned that since prior to 1972 only "willful" violations were punishable, the defendant's act, which occurred prior to the 1972 amendments, could not be punished, lest "thousands of Boy Scouts who have innocently obtained and now possess eagle feathers would also be subject to criminal prosecution by the government."[225] Thus the court seemed to hold that the requirement of "willfulness" was met only if it could be shown that the defendant knew that what he was doing violated the law.[226]

The court reached a different, and more convincing, result a year later in *United States v. Allard,* in which an Indian was charged with selling a bonnet containing golden eagle feathers.[227] The defendant sought to show that he was unaware that the sale was unlawful by introducing evidence that selling eagle feather bonnets was commonplace among Indians. Rejecting this argument, the court drew a distinction that had eluded the *Hetzel* court:

> The effect of the word "knowingly" is to require that . . . the defendant [know] that the feathers were golden eagle feathers, and I think it clear that a conviction would not be had were a person to sell golden eagle feathers thinking them to be turkey feathers. . . . The Act does not, and no statute that I recall seeing, makes the defendant's knowledge of the law an element of the crime.[228]

In distinguishing between knowledge of the law and knowledge of facts, the court's holding seems consistent with the purpose of the 1972 amend-

[224] 385 F. Supp. 1311, 1314 (W.D. Mo. 1974).

[225] *Id.* at 1316, quoting defendant's counsel James Daleo.

[226] In United States v. Mackie, 681 F.2d 1121 (9th Cir. 1982), the defendant was convicted under the Migratory Bird Treaty Act of offering to sell eagles and eagle parts to undercover agents. The court rejected the defendant's claim that the government was required to bring the case under the more specific Bald Eagle Act. The court found no indication that Congress intended to pre-empt the Treaty Act when it passed the Bald Eagle Act. Query whether, in a jurisdiction in which the court requires scienter for violations of the Bald Eagle Act, the government could prosecute under the strict liability standard of the Treaty Act.

[227] 397 F. Supp. 429 (D. Mont. 1975).

[228] *Id.* at 432.

ments, which were directed at practices that were not done with the objective of violating the law yet resulted in the deaths of large numbers of eagles.[229]

Indian Treaty Rights

The conflict between Indian treaty rights to hunt on their reservations and wildlife conservation laws has been a continuing source of controversy, like the conflict between subsistence hunting and the Treaty Act, discussed above. Prior to 1986, courts were split on the issue of whether the Bald Eagle Act abrogates treaty rights. In *United States v. White,* the Eighth Circuit Court of Appeals held that the Bald Eagle Act did not abrogate treaty rights.[230] Judge Lay wrote a strong dissent, a position with which the Ninth Circuit later agreed in *United States V. Fryberg.*[231]

In 1986, the Supreme Court resolved the conflict in *United States v. Dion.*[232] The Court held that the Bald Eagle Act abrogated Indian treaty rights to take bald and golden eagles.[233] The Court found that the 1962 amendment of the Act "reflected an unmistakable and explicit legislative policy choice that Indian hunting of the bald or golden eagle, except pursuant to permit, is inconsistent with the need to preserve those species."[234]

As noted above, the court in *United States v. Bresette* reached a different conclusion under the Treaty Act.[235] The court *in Bresette* distinguished the *Dion* decision, finding that in contrast to the Bald Eagle Act, the Treaty Act contains no clear evidence of intent to abrogate treaty rights.[236]

Infringement of Religious Freedom

Many Indians use eagle feathers, particularly golden eagle feathers, in religious ceremonies.[297] Consequently, the Bald Eagle Act has an impact

[229]Both the Hetzel and Allard decisions rely upon Morissette v. United States, 342 U.S. 246 (1952), in support of their opposite conclusions. The court in Morissette considered difference in criminal intent between crimes "malum in se" and crimes "malum prohibitum." The Allard court concluded that the distinction drawn in Morissette supported its findings that no criminal intent was required because wildlife laws are designed to achieve "social betterment" and because "the taking of wildlife is traditionally regulated by laws not requiring specific intent." 397 F. Supp. at 432–33.

[230]508 F.2d 453 (8th Cir. 1974).

[231]622 F.2d 1010 (9th Cir. 1980).

[232]476 U.S. 734 (1986).

[233]*Id.* at 745.

[234]*Id.* See further discussion of Dion in Chapter 13 at text accompanying notes 62–70.

[235]761 F. Supp. 658 (D. Minn. 1991).

[236]See discussion of Bresette at text accompanying notes 158–160, *infra. See* Chapter 13 for a discussion of Indian treaty rights and wildlife conservation laws generally.

[237]See text accompanying note 201 *supra.*

on Indians' rights to free exercise of their religion, as well as on Indian treaty rights to hunt. The issues are distinct, and the Supreme Court in *Dion* did not decide whether a ban on taking bald and golden eagles except by permit is an unconstitutional infringement of free exercise of religion because the defendant did not pursue this claim in the Court.[238]

In 1986, the year the Supreme Court decided *Dion*, two courts reached opposite conclusions with respect to whether the Bald Eagle Act unlawfully constrains the free exercise of religion. In *United States v. Abeyta*, a district court held that the Act does impermissibly constrain Indian religious freedom and dismissed the charge against the defendant for possessing golden eagle parts.[239] In contrast, the district court in *United States v. Thirty-Eight Golden Eagles* rejected a defense of religious freedom in a forfeiture case, a decision that was affirmed by the Ninth Circuit without opinion.[240] The district court found that the government has a compelling interest in protecting bald and golden eagles, since both are "species whose existence is threatened."[241]

The owner of the eagle carcasses also raised a defense under the American Indian Religious Freedom Act (AIRFA),[242] which provides that

> it shall be the policy of the United States to protect and preserve for American Indians their inherent right of freedom to believe, express, and exercise the traditional religions of the American Indian . . . including but not limited to access to sites, use and possession of sacred objects, and the freedom to worship through ceremonials and traditional rites.[243]

The court rejected the statutory claim, agreeing with the magistrate that AIRFA "was meant to insure that American Indians were given the protection that they are guaranteed under the First Amendment; it was not meant to in any way grant them rights in excess of those guarantees."[244]

In a different sort of religious freedom challenge, the pastor of an "all-

[238]However, an Eighth Circuit panel had rejected Dion's religious freedom claim. 762 F.2d 674, 680 (8th Cir. 1985).

[239]632 F. Supp. 1301 (D.N.M. 1986).

[240]649 F. Supp. 269 (D. Nev. 1986), *aff'd* 829 F.2d 41 (9th Cir. 1987). The forfeiture provision is 16 U.S.C. § 668b(b). Ten years earlier, the Ninth Circuit had rejected a religious freedom defense to a conviction for *selling* eagle parts. United States v. Top Sky, 547 F.2d 486 (9th Cir. 1976). The court in Top Sky found that selling eagle parts was not part of the Indian religion, and in fact was deplored by Indian religion. *Id.* at 488.

[241]649 F. Supp. at 276. *See also* United States v. Jim, 888 F. Supp. 1058 (D. Or. 1995) (prosecution of Native American under the Bald Eagle Act does not violate the Religious Freedom Restoration Act of 1993; government's interest in protecting declining species is compelling). See further discussion of United States v. Abeyta and United States v. Thirty-Eight Golden Eagles in Chapter 13 at text accompanying notes 88–94.

[242]42 U.S.C. § 1996.

[243]*Id.*

[244]649 F. Supp. at 278.

race'' church following Indian religious customs, who was denied a permit under the Bald Eagle Act to use eagle feathers in religious ceremonies, contended that the law was an impermissible establishment of religion because it discriminated in favor of Native Americans.[245] The court rejected the challenge, finding that the government did not violate the First Amendment by denying exemptions from the taking prohibition to non–Native Americans while allowing the exemption for Native Americans. The court used an equal protection analysis, finding that the special treatment of Indians was warranted by Congress's ''historical obligation to respect Native American sovereignty and to protect Native American culture'' and by the legislative history of AIRFA.[246]

Applicability of the Bald Eagle Protection Act to Federal Agencies

An as yet unlitigated question under the Bald Eagle Act is its application to federal agencies. Prior to 1982, the practice of the Department of the Interior had consistently been to require of federal agencies the same permits required of private persons when taking eagles for scientific, exhibition, or other authorized purposes. However, in 1982, Secretary of the Interior Watt's Solicitor issued an opinion concluding that the Act is inapplicable to the federal government.[247] Remarkably, however, that opinion never mentions *Andrus v. Allard,* the only case in which the Supreme Court had interpreted the Act at that time. Whereas that decision is premised upon the view that conservation legislation is to be broadly construed and its exceptions narrowly limited, an interpretation implicitly reaffirmed in *Dion,* the Solicitor's opinion proceeds from precisely the opposite assumption, arguing that unless the Act contains specific mention of its application to federal agencies, an exception for such agencies should be implied. The Solicitor's opinion is poorly reasoned and, in the authors' opinion, wrong.

Constitutionality of the Bald Eagle Protection Act

One final legal uncertainty regarding the Bald Eagle Act, which has never been raised in any reported litigation, is its very constitutionality. As was pointed out at the outset of this chapter, the purpose behind the original legislation was the protection of a symbol of the nation. Some

[245]Rupert v. Fish and Wildlife Service, 957 F.2d 32 (1st Cir. 1992).

[246]*Id.* at 35. *See also* United States v. Billie, 667 F. Supp. 1485 (S.D. Fla. 1987), in which the court rejected the defendant's contention that the Endangered Species Act impermissibly infringed upon his religious freedom because it failed to provide a permit system for Indians like the Bald Eagle Act does.

[247]Memorandum from William H. Coldiron, Solicitor of the Department of the Interior, to the Director of the U.S. Fish and Wildlife Service, June 30, 1982.

commentators have argued that there is no apparent treaty or commerce power justification for such legislation.[248] Nonetheless, it will probably be more difficult to have the Bald Eagle Act declared unconstitutional than it was to sustain the conviction of a Boy Scout leader in the *Hetzel* case.

Conclusion

Perhaps the best measure of the effectiveness of the Bald Eagle Protection Act was the decision of the Secretary in 1978 to list the bald eagle as an endangered species under the Endangered Species Act in all but five of the forty-eight coterminous states and to list it as a threatened species in the remaining five states.[249] That step indicates that, despite some thirty-eight years in which the taking of bald eagles had been prohibited,[250] their survival remained very much in jeopardy and could not be assured without the habitat protection measures afforded by the Endangered Species Act.[251]

PROTECTION OF EXOTIC BIRDS: THE WILD BIRD CONSERVATION ACT

Overview

In contrast to the Migratory Bird Treaty Act and the Bald Eagle Protection Act, which protect birds naturally found in the United States, the

[248] *See* Guilbert, *Wildlife Preservation Under Federal Law,* in *Federal Environmental Law* 588 n. 267 (E. Dolgin & T. Guilbert, eds. 1974). But see the discussion of the federal commerce power in Chapter 2. It also has been argued that the federal government has an inherent power to protect wildlife having "symbolic" value to the nation, wholly without regard to whether there exists any express constitutional basis for such protection. *See* Coggins & Hensley, *Constitutional Limits on Federal Power to Protect and Manage Wildlife: Is the Endangered Species Act Endangered?* 61 Iowa L. Rev. 1099, 1139–43 (1976).

[249] 43 Fed. Reg. 6233 (1978). A subspecies, commonly known as the Southern bald eagle, had been listed as an endangered species since 1967. The 1978 listing eliminated any distinction between the two subspecies.

[250] In fairness, it must be pointed out that the prohibition against taking has not in fact stopped the intentional killing of bald eagles. The 1971 massacre of several hundred eagles from helicopters, described previously, is a shocking example. More than three decades after enactment of the Bald Eagle Protection Act, shooting remained the major cause of bald eagle mortality. *See* 43 Fed. Reg. 6232 (1978).

[251] Some interesting, but probably only theoretical, problems arise from the listing of the bald eagle. For example, permits for otherwise unauthorized activities are available for a much narrower range of activities under the Endangered Species Act than under the Bald Eagle Act. Presumably the more restrictive provisions of the later act take precedence over the earlier act, but the matter is not entirely free from doubt. One must also question whether the Endangered Species Act's limited exception for subsistence taking by Alaskan natives can be read as eliminating the special Indian exemption in the Bald Eagle Act.

Wild Bird Conservation Act of 1992 (hereafter Wild Bird Act)[252] is directed at birds that are not naturally found in this country. The purpose of the Wild Bird Act is to "promote the conservation of exotic birds,"[253] which are defined as birds that are "not indigenous to the 50 States or the District of Columbia."[254]

The impetus for the Act was a recognition that populations of exotic wild birds "have declined dramatically due to habitat loss and the public's demand for pet birds."[255] In addition, "the mortality associated with the [pet] trade remains unacceptably high."[256] Congress found that as much as 30 percent to 50 percent of wild birds die before they leave the country of origin, and another 15 percent die in transit.[257] The United States is the largest importer of wild birds, importing hundreds of thousands of birds every year,[258] and it should "play a substantial role in finding effective solutions to these problems."[259]

Congress also found that the Convention on International Trade in Endangered Species of Fauna and Flora (CITES),[260] an international treaty regulating trade in wildlife to which the United States is a party, "has been ineffective in stemming the decline in wild bird populations."[261] Many exporting countries "lack the means to develop or effectively implement scientifically based management plans."[262] The Convention permits parties to "adopt stricter domestic measures for the regulation of trade in all species," and the Wild Bird Act provides a series

[252]Act of Oct. 23, 1992, Pub. L. No. 102–440, 106 Stat. 2224, codified at 16 U.S.C. §§ 4901–4916.

[253]16 U.S.C. § 4902.

[254]*Id.* § 4903(2)(A). Regulations define "indigenous" as "a species that is naturally occurring, not introduced as a result of human activity, and that currently regularly inhabits or breeds in the 50 states or the District of Columbia." 50 C.F.R. § 15.3 (1995).

[255]H. R. Rep. No. 102–749(II), 102d Cong., 2d Sess. (1992) at 7, *reprinted in* 1992 U.S.C.C.A.N. 1610.

[256]16 U.S.C. § 4901(1).

[257]H. R. Rep. No. 102–749(II), 102d Cong., 2d Sess. at 7 (1992), *reprinted in* 1992 U.S.C.C.A.N. 1610.

[258]*Id.* About half of the birds are finches, and half are parrots, including parakeets, macaws, cockatiels, cockatoos, and lovebirds. *Id.* A 1992 New York Zoological Society/Wildlife Conservation Society report estimates that at least 8.5 million birds were imported, legally or illegally, to the United States during the last decade, at least 85 percent of which were wild caught. Because so many birds die before leaving the country of origin or in transit, the report estimates that at least 16 million birds may have been trapped in the wild and shipped to the United States.

[259]16 U.S.C. § 4901(2).

[260]March 3, 1973, 27 U.S.T. 1087, T.I.A.S. No. 8249. See discussion of CITES in Chapter 14.

[261]H. R. Rep. No. 102–749(II), 102d Cong., 2d Sess. at 8 (1992), *reprinted in* 1992 U.S.C.C.A.N. 1610.

[262]16 U.S.C. § 4901(7).

of "nondiscriminatory measures that are necessary for the conservation of exotic birds."[263]

The Wild Bird Act aims to promote conservation of wild birds by

1. assisting wild bird conservation and management programs in the countries of origin of wild birds;
2. ensuring that all trade in species of exotic birds involving the United States is biologically sustainable and is not detrimental to the species; [and]
3. limiting or prohibiting imports of exotic birds when necessary to ensure that
   (A) wild exotic bird populations are not harmed by removal of exotic birds from the wild for trade; or
   (B) exotic birds in trade are not subject to inhumane treatment.[264]

Moratoria

The focus of the Wild Bird Act is regulating trade in birds, rather than regulating the "taking" of birds. The Act establishes moratoria on the import of species[265] of exotic birds listed on any CITES appendix[266] and permits the Secretary of the Interior to establish moratoria or quotas on species not listed under CITES.[267] The import bans on all species listed under CITES became effective on October 23, 1993, one year after the Act was passed.[268]

Approved Species List

The moratorium on import of a CITES-listed species may be lifted by the Secretary of the Interior. The Secretary is required to publish lists of "approved species" that are not subject to the import ban.[269] For non-captive bred species, in order to place the species on the approved list

[263] *Id.* §§ 4901(12) & (14). These last two findings are important to assure that the Act's restrictions are not struck down as unilateral measures in violation of international trade agreements.

[264] *Id.* § 4902.

[265] "Species" is defined to include subspecies and distinct population segments, as well as hybrids. *Id.* § 4903(7).

[266] *Id.* § 4904. Species protected by CITES are listed on various appendices to the Convention, depending on the degree of threat to the species. See Chapter 14 for a discussion of CITES.

[267] *Id.* § 4907.

[268] *Id.* § 4904.

[269] *Id.* § 4905. In January 1996, the Secretary published a final rule to implement procedures for establishing an approved list of wild birds listed in CITES appendices that can be imported. 61 Fed. Reg. 2084 (1996), to be codified at 50 C.F.R. § 15.3. See 59 Fed. Reg. 62255 (1994) for the rule to implement procedures for establishing an approved list of captive-bred birds that can be imported.

the Secretary must determine that (1) each country of origin for which the species is listed is effectively implementing CITES, and

> (2) A scientifically-based management plan for the species has been developed which—(A) provides for the conservation of the species and its habitat and includes incentives for conservation; (B) ensures that the use of the species is biologically sustainable . . . ; and (C) addresses factors relevant to the conservation of the species.[270]

The Secretary also must determine that "[t]he management plan is implemented and enforced," and that "[t]he methods of capture, transport, and maintenance of the species minimizes the risk of injury or damage to health, including inhumane treatment."[271] In making these determinations, the Secretary must "use the best scientific information available" and must "consider the adequacy of regulatory and enforcement mechanisms in all countries of origin."[272] Even when a species is on the approved list, the Secretary retains emergency authority to suspend imports of the species.[273]

For captive-bred species, to place the species on the approved list the Secretary must determine that "(1) the species is regularly bred in captivity and no wild-caught birds of the species are in trade; or (2) the species is bred in a qualifying facility."[274] The Act establishes requirements for qualifying facilities.[275]

Species Not Protected by CITES

As noted above, the Secretary may establish a moratorium or quota on species not protected by CITES, or on all species from a particular country.[276] To establish a moratorium or quota on a species, the Secretary must determine that the findings that must be made for a CITES-listed species to be on the "approved" list under section 4905(c) cannot be made, and that the moratorium or quota "is necessary for the conservation of the species."[277] To establish a moratorium or quota on all species of exotic birds from a country, the Secretary must determine that "the country has not developed and implemented a management program for exotic birds . . . that ensures both the conservation and the humane treatment of ex

[270] 16 U.S.C. § 4905(c).
[271] *Id.*
[272] *Id.* § 4905 (a)(3).
[273] *Id.* § 4904(b).
[274] *Id.* § 4905(b).
[275] *See id.* § 4906.
[276] *Id.* § 4907(a).
[277] *Id.*

otic birds.''[278] The Secretary shall terminate a moratorium or quota if he or she finds that the reasons for establishing it no longer exist.[279]

Petitions

The Wild Bird Act provides a procedure for petitioning the Secretary to take any of the actions authorized in the Act.[280] Any person may petition the Secretary, and within ninety days of submission of the petition, the Secretary must issue a preliminary ruling on whether the requested action might be warranted. There must be an opportunity for public comment on the petition, and the Secretary must publish a final ruling within ninety days of the close of the comment period.[281]

Prohibitions and Penalties

The Act prohibits importation of ''any exotic bird in violation of any prohibition, suspension, or quota on importation,''[282] and it provides for both civil and criminal penalties for violations.[283] The standard of scienter (knowledge) for civil penalties differs for persons in the business of importing exotic birds and other people: a commercial importer is subject to a maximum $25,000 penalty for violating the Act, while others are subject to that maximum only if they ''knowingly'' violate the Act.[284] The maximum penalty for persons other than commercial importers who unknowingly violate the Act is $500.[285] Any person who knowingly violates the Act is also subject to criminal penalties, including a fine, imprisonment for up to two years, or both.[286]

Exemptions

The Act provides several exemptions from its prohibitions, subject to a permit system.[287] A permit to import a bird not on the approved list may be authorized if the Secretary determines that ''importation is not detrimental to the survival of the species'' and the bird is being imported

[278] *Id.*
[279] *Id.*
[280] *Id.* § 4909.
[281] *Id.*
[282] *Id.* § 4910.
[283] *Id.* § 4912.
[284] *Id.* § 4912(a)(1)(A).
[285] *Id.* § 4912(a)(1)(C).
[286] *Id.* § 4912(a)(2). Note that the scienter standard for criminal penalties is the same for everyone, including commercial importers.
[287] *Id.* § 4911.

exclusively for scientific research, a cooperative breeding program, a zoological breeding or display program, or as a personally owned pet of a person returning to the United States after an absence of at least a year. The latter exemption is limited to no more than two birds per individual in one year.[288]

Exotic Bird Conservation Fund

In addition to regulating the importation of exotic birds, the Wild Bird Act establishes the Exotic Bird Conservation Fund "to provide financial and technical assistance for projects to conserve exotic birds in their native countries."[289] The fund consists of money received from penalties, fines, or forfeiture of property for violation of the Act, donations, and any amounts appropriated by the Secretary.

Appropriation

Finally, the act authorizes an appropriation of $5 million per year to implement the Act.[290]

Litigation under the Act

There has been only one reported case under the Wild Bird Act. In *Humane Society of the United States v. Babbitt,*[291] plaintiffs challenged a Fish and Wildlife Service rule that exempted from the moratorium on importing species listed in any CITES appendix a species listed in Appendix III that originated in a country that has not listed the species in Appendix III.[292] The court declared the regulation invalid, holding that the rule's "country of origin" provision violated the Act's plain language.[293] The court said that its reading of the Act not only comported with the language, but also was necessary to effectuate the Act's purposes, since birds

[288]The proposed regulations regarding import of pets provided that a "household" was limited to two per year. Because of objections from commenters and because the statute allows two per year per individual, the final regulation was changed. *See* 50 C.F.R. § 15.25(a) (1995).

[289]16 U.S.C. § 4913.

[290]*Id.* § 4915. The Congressional Budget Office estimated that about $700,000 would be needed each year to carry out the regulatory program, leaving approximately $4.3 million for grants for conservation programs. H. R. Rep. No. 749(II), 102d Cong., 2d Sess. at 18 (1992), *reprinted in* 1992 U.S.C.C.A.N. 1610, 1620.

[291]849 F. Supp. 814 (D.D.C. 1994).

[292]58 Fed. Reg. 60524 (1993).

[293]*See* 16 U.S.C. § 4904(c).

cross national boundaries and can be pulled across boundaries by trappers. In response to the court's ruling, the Fish and Wildlife Service announced that the moratorium covers all birds listed in any CITES appendix, regardless of country of origin.[294]

[294] 59 Fed. Reg. 26810 (1994).

Chapter 5

MARINE MAMMALS

BACKGROUND OF THE MARINE MAMMAL PROTECTION ACT

Until the late 1960s, federal wildlife legislation made no attempt to establish comprehensive programs to conserve any type of wildlife other than migratory birds. Enactment of the Endangered Species Preservation Act of 1966 and the Endangered Species Conservation Act of 1969 signalled a renewal of interest in establishing a more prominent federal presence in wildlife conservation. The programs these laws initiated, though, were relatively modest.[1]

By the early 1970s, pressure from many diverse interests had grown for a comprehensive and coordinated program to conserve the world's marine mammals. The pressures arose out of diverse perspectives. Commercial interests, some scientists, and traditionalists in the conservation community believed that marine mammals were an important commercial and food resource that, with proper management, could be used indefinitely through sustained harvests.[2] Other members of the scientific and conservation communities believed that because marine mammals played an important ecological role in marine systems, the first priority of federal policy ought to be to maintain their populations for ecological reasons. A third group comprised those who believed that marine mammals, because of their apparent intelligence and highly developed social systems,

[1]The Endangered Species Preservation Act of 1966 and the Endangered Species Conservation Act of 1969 are discussed in detail in Chapter 7 *infra* at text accompanying notes 3–27.

[2]The principal bill reflecting these interests was the Anderson-Pelly bill, H.R. 10420, 92d Cong., 1st Sess. (1971).

ought to be left undisturbed and made off-limits to human use.[3] This group was motivated primarily by their concern for the welfare of individual animals.

Whatever their motivation, nearly all were in agreement that the existing patchwork of laws and regulations pertaining to marine mammals was inadequate. Although the International Whaling Commission had been charged with safeguarding the world's whale stocks for future generations, its regulations were so ineffective that in 1970 eight species of whales were listed as threatened with worldwide extinction under the Endangered Species Conservation Act of 1969.[4] A few other species of marine mammals were subject to some form of federal authority, including the West Indian and Amazonian manatees,[5] the Mediterranean monk seal,[6] the sea otter,[7] and the North Pacific fur seal,[8] but the remainder were protected only by state law or not at all.

Because of the widely divergent interests of those seeking change, their common desire for a new management system for marine mammals did not produce unanimity as to the nature of that system. Instead, Congress had to forge a compromise between the seemingly irreconcilable views of the traditional "managers," the newer environmentalists, and the "protectionists."[9] The legislation that finally resulted was the Marine Mammal Protection Act of 1972 (MMPA).[10] Precisely because this compromise was necessary in order to pass any legislation at all, the Act articulated only broad, general policy goals and implemented them with specific directions

[3]The principal bill reflecting this view was the Harris-Pryor bill, H.R. 6554, 92d Cong., 1st Sess. (1971).

[4]35 Fed. Reg. 18320 (1970).

[5]The West Indian manatee was determined to be threatened with extinction under the Endangered Species Preservation Act of 1966. *See* 32 Fed. Reg. 4001 (1967). The Amazonian manatee was listed pursuant to the 1969 Act. *See* 35 Fed. Reg. 8495 (1970).

[6]The Mediterranean monk seal was determined to be threatened with worldwide extinction under the Endangered Species Conservation Act of 1969. *See* 35 Fed. Reg. 8495 (1970).

[7]Taking of sea otters on the high seas by persons subject to the jurisdiction of the United States was prohibited by section 301 of the Fur Seal Act of 1966, Pub. L. 89–702, 80 Stat. 1096 (formerly codified at 16 U.S.C. § 1171; repealed in 1983).

[8]North Pacific fur seals were at the time managed according to sustained-yield principles pursuant to the Interim Convention on the Conservation of North Pacific Fur Seals. *See* Department of Commerce, National Marine Fisheries Service, Final Environmental Impact Statement, Renegotiation of Interim Convention on Conservation of North Pacific Fur Seals, app. E (1976). For further discussion of fur seal conservation, see Chapter 14 at text accompanying notes 10–55.

[9]The legislative history is described well in Gaines and Schmidt, *Wildlife Population Management Under the Marine Mammal Protection Act of 1972*, 6 Envt'l L. Rep. (Envt'l L. Inst.) 50096, 50103–08 (1976).

[10]The MMPA was enacted as Pub. L. 92–522, 86 Stat. 1027. It has since been amended several times and is codified at 16 U.S.C. §§ 1361–1421h.

that were neither purely protectionist nor purely exploitive but almost always complex.

AN OVERVIEW OF THE MMPA

Whatever its complexities and ambiguities, the MMPA clearly was a major departure from the regulatory scheme it replaced. Most fundamentally, it completely preempted the states from any authority over marine mammals and substituted for the many state programs a single, comprehensive federal program. The central feature of this program was a moratorium of indefinite length, during which no marine mammals could be imported into the United States or taken by any person subject to United States jurisdiction. Limited exceptions were carved out for scientific and public display purposes, for taking by natives of the North Pacific and Arctic coasts, and for taking incidental to commercial fishing operations or pursuant to international treaty. Many of these initial exceptions were substantially revised and still other exceptions were added in later amendments.

Although the MMPA immediately preempted state authority over marine mammals, it also established a mechanism whereby states could regain their authority, together with federal financial assistance to carry out approved state programs. Similarly, although the MMPA established a moratorium on taking and importing marine mammals, it also established a detailed and highly formal mechanism whereby the moratorium could be waived and taking and importation permitted for any species or population stock of marine mammal found to meet the MMPA's novel and complex population criteria.[11] Later amendments to the MMPA have eroded the force of the moratorium and the law's novel population goals.

Finally, the MMPA attempted to establish an ambitious international program for marine mammal protection. As a component of that pro-

[11]One of the MMPA's major innovations was its protection for individual population stocks, as well as species, of marine mammals. The MMPA defines a "population stock" as "a group of marine mammals of the same species or smaller taxa in a common spatial arrangement, that interbreed when mature." 16 U.S.C. § 1362(11). The protection of stocks represented a significant departure from the Endangered Species Conservation Act of 1969, which applied only to species and subspecies. Pub. L. No. 91–135, § 3(a), 83 Stat. 275 (repealed 1973). This broadening of the scope of protection was in response to the recognition that the health of species and subspecies often depends upon the continuing viability of their various stocks. *See, e.g.*, the following statement of Dr. Kenneth S. Norris: "It is important that we preserve not only the species but the population structure as well, since this [is] part of the way animals have evolved in the world and have managed to meet the changes in the environment that assail [them]." Marine Mammals: Hearings on H.R. 690 and Related Bills Before the Subcomm. on Fisheries and Wildlife Conservation of the House Comm. on Merchant Marine and Fisheries, 92d Cong., 1st Sess. 410 (1971).

gram, the MMPA directed that its policies be the official policies of the United States in the negotiation and renegotiation of international agreements concerning marine mammals. Further, the MMPA provided authority for restricting the importation of certain products from foreign countries whose fishing or other practices impeded the attainment of the Act's goals.

FEDERAL AGENCY RESPONSIBILITIES UNDER THE MMPA

Primary responsibility for implementing the MMPA is shared by the Secretaries of Commerce and the Interior. The former, through the National Marine Fisheries Service, has authority with regard to all members of the order Cetacea (whales and porpoises) and all members, except walruses, of the order Pinnipedia (seals).[12] The Secretary of Commerce also implements, with respect to all marine mammals, the provisions of the MMPA governing incidental take of marine mammals in the course of commercial fishing operations.[13] The Secretary of the Interior, through the United States Fish and Wildlife Service, administers the MMPA with respect to all other marine mammals (manatees, dugongs, polar bears, sea otters, and walruses).[14]

In carrying out these duties under the MMPA, the Secretary is required to consult with the Marine Mammal Commission, an independent advisory body created by the MMPA.[15] The Commission was intended to serve as an impartial and nonpolitical source of expert scientific advice to the Secretary. To ensure the scientific integrity of the Commission, the MMPA requires that its three members be appointed by the President from a "list of individuals knowledgeable in the fields of marine ecology and resource management" unanimously agreed to by the Chairman of the Council on Environmental Quality, the Secretary of the Smithsonian Institution, the

[12]16 U.S.C. § 1362(12)(A)(i). This provision does not refer explicitly to the Secretary of Commerce, but rather to the "Secretary of the department in which the National Oceanic and Atmospheric Administration is operating." The latter agency was placed in the Department of Commerce by Executive Reorganization Plan No. 4 of 1970, 35 Fed. Reg. 15627 (1970), 84 Stat. 2090.

[13]16 U.S.C. § 1362(12)(B).

[14]*Id.* § 1362(12)(A)(ii). This sharing of jurisdiction, though a contentious issue at the time of the MMPA's enactment, was premised in part on the belief that there would soon be established a federal Department of Natural Resources in which all federal authority over wildlife would be consolidated. *See* George Cameron Coggins, *Legal Protection for Marine Mammals: An Overview of Innovative Resource Conservation Legislation*, 6 Envt'l L. 1, 26–27 (1975). This action never occurred. The Act imposes identical duties on the Secretaries of Commerce and the Interior for most purposes. Thus the term "Secretary" is used in this chapter to refer to either of them, except where noted otherwise.

[15]16 U.S.C. §§ 1401–07.

Director of the National Science Foundation, and the Chairman of the National Academy of Sciences.[16]

THE MMPA'S MANAGEMENT PRINCIPLES: MSY AND OSP

The management units under the MMPA include both species and "stocks." A stock is "a group of marine mammals of the same species or smaller taxa in a common spatial arrangement, that interbreed when mature."[17] Stocks were a familiar concept in traditional fisheries and marine mammal management. The management goals of the MMPA, however, were intended to be a sharp break from tradition.

Management of commercially valuable fish and other wildlife historically has been aimed at producing a "maximum sustainable yield" (or "MSY") of the exploited species or stock. Management for MSY, in its purest form, focuses solely on the effects of a given harvest level on the ability of the target species or stock to replenish itself.[18] This narrow focus may have been appropriate when the only recognized value of the animal being harvested was its commercial value. However, the MMPA declares that marine mammals have "esthetic and recreational as well as economic" value.[19] To further these and other values, the MMPA provides that the "primary objective" of marine mammal management "should be to maintain the health and stability of the marine ecosystem."[20] Thus, the MMPA sought to create a much broader vision in marine mammal management than had previously prevailed.

The MMPA further provides that "[w]henever consistent with" the primary objective of maintaining the health and stability of the marine ecosystem, "it should be the goal to obtain an optimum sustainable population keeping in mind the carrying capacity of the habitat."[21] The quoted language can only mean that there may be circumstances when maintaining an optimum sustainable population ("OSP") is inconsistent with the primary objective of marine ecosystem health and stability, and

[16] *Id.* § 1401(b)(1). The requirement that the individuals be unanimously agreed to by agency heads was added to the MMPA in 1984. Two years earlier, Congress had added a requirement that Commission members be appointed with the advice and consent of the Senate. These measures were enacted in response to the Reagan Administration's efforts to appoint to the Commission individuals lacking the prescribed qualifications who were recommended by the presidentially-controlled Council on Environmental Quality.

[17] *Id.* § 1362(10).

[18] For a discussion of MSY and related concepts, see F. Christy, Alternative Arrangements for Marine Fisheries: An Overview 23 (1973).

[19] 16 U.S.C. § 1361(6).

[20] *Id.*

[21] *Id.*

thus must give way.[22] What those circumstances might be are difficult to discern.[23] This lack of clarity is but one of several reasons the MMPA's population policy has been described as the Act's "most intricate and the most poorly articulated component."[24]

Although OSP has a major role to play in marine mammal management, the definition of this important term is far from clear. The Act provides:

> The term "optimum sustainable population" means, with respect to any population stock, the number of animals which will result in the maximum productivity of the population or the species, keeping in mind the carrying capacity of the habitat and the health of the ecosystem of which they form a constituent element.[25]

Two key variables in this definition are the "maximum productivity" of the species or stock and the "health of the ecosystem." The former is relatively quantifiable; the latter is much less so. The more one emphasizes the former, the more OSP resembles traditional living marine resource management standards, which were based solely on the annual "production" of new individuals to replace those taken the previous year.[26] To the extent one emphasizes the latter, OSP would seem to compel the minimum feasible disturbance to the ecosystem, especially where the effects of any action are largely uncertain.

Some writers have sought to reconcile the two components of the definition by suggesting that "productivity" need not mean only the narrow concept of biological productivity, but rather the broader concept of "eco-

[22]The relative status of the two objectives was clear in the bill passed by the House, which referred to obtaining an "optimum sustained yield" as a "secondary objective." H.R. 10420, 92d Cong., 1st Sess. § 2(6). The Senate amendment to the House-passed bill reversed these priorities, establishing the attainment of optimum sustainable populations as the primary objective of marine mammal management, and reducing ecosystem health and stability to the status of a factor to be kept in mind. The language quoted in the text makes it clear that the Conference Committee adopted the House priorities, despite deletion of the term "secondary" in the original House version.

[23]In explaining the Conference bill to the House, Congressman Dingell, then Chairman of the Subcommittee on Fisheries and Wildlife Conservation of the Committee on Merchant Marine and Fisheries, was likewise unable to foresee any incompatibility in the MMPA's two major goals, but he recognized that in the event of conflict, ecosystem health was to be paramount: "I will say that I cannot imagine a case in which the objectives of ecosystem stability and non-disadvantageous taking might conflict; but if they should, it is ecosystem protection which must prevail."

[24]Gaines and Schmidt, *supra* note 9, at 50101.

[25]16 U.S.C. § 1362(9). The original definition referred to "optimum carrying capacity" rather than "carrying capacity." The best explication of the original legislative history of the OSP standard can be found in Gaines and Schmidt, *supra* note 9, at 50103–08.

[26]*See* Gaines and Schmidt, *supra* note 9, at 50106.

logical productivity.''[27] That interpretation, however, is based more on the ''spirit'' of the MMPA than on anything expressed in the statute.[28] More convincing is the Marine Mammal Commission's conclusion that the OSP definition ''includes certain features which are potentially inconsistent and other features which call for subjective value judgments that are not amenable to quantification on the basis of available data.''[29]

The MMPA's definition of OSP, by referring to ''the number of animals,'' implies that for any given species or stock there is a single population size that represents its OSP. The implementing regulations since 1976, however, have defined OSP so as to refer to a range of population sizes rather than a single number of animals within a species or stock. The regulatory definition of OSP is

> a population size which falls within a range from the population level of a given species or stock which is the largest supportable within the ecosystem to the population level that results in maximum net productivity. Maximum net productivity is the greatest net annual increment in population numbers or biomass resulting from additions to the population due to reproduction and/or growth less losses due to natural mortality.[30]

Thus, the upper boundary of the OSP range represents the carrying capacity level, often equated with pre-exploitation abundance.[31] The lower boundary is close to, if not at, the level of traditional MSY-based manage-

[27] *Id.* at 50107.

[28] There is some support in the legislative history for this interpretation. Senator Stevens, principal author of the Senate bill, stated the following: ''It also requires a judgement, not only on the maximum population of the species, but on the maximum total productivity of the environment including all constituent elements.'' 118 Cong. Rec. 25258 (1972). Curiously, however, this explanation was offered in regard to the MMPA's original term ''optimum carrying capacity'' rather than ''optimum sustainable population.''

[29] Marine Mammal Commission, The Concept of Optimum Sustainable Populations 1 (undated).

[30] 50 C.F.R. § 216.3 (1994). This is the definition promulgated by the National Marine Fisheries Service. The U.S. Fish and Wildlife Service has endorsed this definition but has never included an OSP definition in its own MMPA implementing regulations. *See* 44 Fed. Reg. 2540, 2541–42 (1979).

[31] The possibility that a species or stock could exceed its OSP level might appear to be implicit in a provision of section 104 of the MMPA that refers to cases ''in which an application for a permit cites as a reason for the proposed taking the overpopulation of a particular species or population stock.'' 16 U.S.C. § 1374(b). A better interpretation, however, is that this provision refers to overpopulation in a particular area, rather than overpopulation more generally, because the provision goes on to require the Secretary to consider the alternative of translocating excess animals to a location formerly, but no longer, inhabited by the species or stock. The availability of unoccupied suitable habitat means that the population is not the ''largest supportable within the ecosystem'' (the upper boundary of OSP as defined in the regulations).

ment.[32] In practice, it has been necessary to determine (or at least estimate) the upper boundary, since the lower boundary has generally been set at 60 percent of the carrying capacity population.[33]

A marine mammal species or stock that is below its OSP level is considered to be a "depleted" species.[34] So too is any marine mammal that has been listed as threatened or endangered under the Endangered Species Act.[35] The rules governing what can and cannot be done to marine mammals often turn upon whether the species or stock involved is depleted.

THE MORATORIUM

The cornerstone of the MMPA as passed in 1972 was its "moratorium" on taking and importing marine mammals and marine mammal products. Although the Act defines the moratorium to mean a "complete cessation" of taking and a "complete ban" on importation,[36] there were several exceptions to these prohibitions in 1972 and the number has grown steadily in later decades.

The moratorium that exists today essentially applies only to taking and importation of marine mammals for recreational purposes or for commercial use of marine mammal products. Even for these purposes, the moratorium can be "waived," but obtaining a waiver is difficult. Taking that is merely incidental to commercial fishing and other activities can be authorized without a waiver, as can takings and importations for a wide variety of other purposes. A maze of detailed statutory provisions now governs these many activities. The discussion that follows provides a roadmap through the maze, starting with the Act's deceptively simple prohibition against "taking."

THE TAKING PROHIBITION

Section 102 of the MMPA prohibits any person or vessel subject to the jurisdiction of the United States from "taking" any marine mammal on the high seas or on the lands or waters of the United States.[37] The MMPA's

[32]In an early MMPA judicial decision, Judge Skelly Wright, while characterizing the statutory definition of OSP as "singularly unenlightening," was unable to conclude that "MSY is definitely inconsistent with OSP." Animal Welfare Institute v. Kreps, 561 F.2d 1002, 1014 (D.C. Cir. 1977).

[33]*See* 42 Fed. Reg. 64548 (1977); 45 Fed. Reg. 72178 (1980).

[34]16 U.S.C. § 1362(1)(A) and (B).

[35]*Id.* § 1362(1)(C).

[36]*Id.* § 1362(7).

[37]*Id.* § 1372(a)(1) and (2). National Marine Fisheries Service regulations prohibit any person subject to the jurisdiction of the United States from taking any marine mammal during the moratorium, without regard to the locus of the taking. *See* 50 C.F.R. § 216.11(c) (1995). However, in United States v. Mitchell, 553 F.2d 996 (5th Cir. 1977), the court held

moratorium on taking marine mammals from a particular species or stock could be waived, however, and taking allowed under such regulations as the Secretary deemed necessary to insure that the taking "will not be to the disadvantage" of the species or stock.[38] In order for taking not to disadvantage a species or stock, the Secretary had to determine that the species or stock was at its OSP level and would not be reduced below that level by the authorized taking. This determination could be made only through a rulemaking proceeding "on the record after opportunity for an agency hearing."[39] That is, a formal, adjudicatory hearing was required to waive the moratorium and authorize taking.

The procedure for authorizing taking, while complex, was at least straightforward. Less so was the scope of the taking prohibition itself. Section 3 of the MMPA defines "take" to mean "to harass, hunt, capture, or kill" any marine mammal or to attempt to do so.[40] Congress broke new ground by including harassment in the definition, but it did not define this new term. It did make clear that the term was to be interpreted expansively, encompassing even acts that are not intended to affect marine mammals adversely.[41]

Two recent cases explored the outer reaches of the take prohibition. Until 1991, the National Marine Fisheries Service's definition of "take" included "the negligent or intentional operation of an aircraft or vessel, or the doing of any other negligent or intentional act which results in disturbing or molesting a marine mammal."[42] The Service added the following to the take definition in 1991: "feeding or attempting to feed a marine mammal in the wild."[43] Operators of a commercial tour boat business that took tourists into the Gulf of Mexico for the purpose of feeding dolphins challenged the new prohibition in *Strong v. United States.*[44] The district court sided with the tour boat operators on the grounds that "to feed" is not among the dictionary definitions of "to harass." The court of appeals reversed the district court. It noted that

> the agency has been given substantial scientific evidence that feeding wild dolphins disturbs their normal behavior and may make them less able to search for food

---

that the MMPA did not restrict the taking of marine mammals by U.S. citizens in the territory of a foreign nation.

[38]16 U.S.C. § 1373(a).

[39]*Id.* § 1373(d).

[40]*Id.* § 1362(12).

[41]According to the House Committee report, "[t]he act of taking need not be intentional: the operation of motor boats in waters in which these animals are found can clearly constitute harassment." H.R. Rep. No. 707, 92d Cong., 1st Sess. at 23 (1971), *reprinted in* 1972 U.S.C.C.A.N. 4144, 4155.

[42]50 C.F.R. § 216.3 (1994).

[43]*See* 56 Fed. Reg. 11693, 11697 (1991).

[44]5 F.3d 905 (5th Cir. 1993).

on their own. It is therefore clearly reasonable to restrict or prohibit the feeding of dolphins as a potential hazard to them.[45]

Although feeding dolphins is therefore prohibited as a taking, firing a rifle at them to frighten them away from food is not, at least according to the court in *United States v. Hayashi.*[46] That case involved the prosecution of a commercial fisherman who fired two rifle shots near porpoises that he believed were eating the catch off his fishing line. Convicted in the trial court, Hayashi appealed. In overturning his conviction, the court of appeals reasoned that the meaning of the term "harass" in the statutory definition of "take" could only be derived by reference to the words around it: hunt, capture, and kill. According to the court, each of these "involve[s] direct and significant intrusions upon the normal, life-sustaining activities of a marine mammal," and so, therefore, must any act to be considered harassment.[47]

Had the court stopped there, its reasoning might have seemed defensible. Instead, however, it went on at some length to characterize the dolphins' interest in the food on Mr. Hayashi's line as "abnormal" behavior. Characterizing it in this way allowed the court to find an escape from the legal "Catch-22" situation that the court believed otherwise confronted Hayashi. Noting the regulatory prohibition against feeding marine mammals, the court argued that allowing marine mammals to feed on a baited line or a hooked fish might run afoul of this prohibition. "Guilty of 'harassment' by 'feeding' if he did nothing, and guilty of 'harassment' by 'disturbing' if he took steps to prevent the feeding, our hypothetical Hayashi would face possible criminal prosecution under the MMPA no matter which course he chose."[48] The solution to this dilemma, the court reasoned, lay in recognizing the fact that what the dolphins were doing was not normal and, in the words of the court, "[d]eterrence of abnormal marine mammal activity is not proscribed."[49]

The court's formulation creates as many problems as it solves. The familiar behavior of dolphins "riding" the bow wave of a boat is presumably no more "normal" than that of eating fish off a hook, since it, too, is made possible by human activities. Under the court's view, however, firing a gun into the water to watch dolphins scatter from the bow of a boat would not be harassment because it is not a "significant" intrusion on a "normal" behavior. Yet, it is hard to imagine that Congress did not intend to reach activities of that sort when it prohibited harassment of marine mammals.

[45] *Id.* at 906–7.
[46] 22 F.3d 859 (9th Cir. 1993).
[47] *Id.* at 864.
[48] *Id.* at 866.
[49] *Id.* at 864.

In 1994 Congress added a definition of "harassment" to the MMPA. The definition rejects the court's reasoning in *Hayashi* and proscribes "any act of pursuit, torment, or annoyance which (i) has the potential to injure a marine mammal or marine mammal stock in the wild; or (ii) has the potential to disturb a marine mammal or marine mammal stock in the wild by causing disruption of behavioral patterns, including, but not limited to, migration, breathing, nursing, breeding, feeding, or sheltering."[50] Conspicuously missing from this definition is any limitation to "normal" behavioral patterns or any of the key words (direct, sustained, and significant) that the *Hayashi* court used.[51]

WAIVER OF THE MORATORIUM

It is technically incorrect to refer to a waiver of "the moratorium," for what is in fact waived is only that part of the moratorium applicable to a particular species or stock of marine mammal, and then only to the extent provided in the waiver. In making these "partial waivers" the Secretary must act on the basis of the "best scientific evidence available and in consultation with the Marine Mammal Commission."[52] If satisfied that the proposed waiver will be "compatible with" the MMPA and if "assured that the taking of such marine mammal is in accord with sound principles of resource protection and conservation as provided in the purposes and policies" of the MMPA, the Secretary may waive the moratorium and permit the taking pursuant to "suitable regulations."[53]

The net effect of these requirements is that prior to any waiver, the Secretary must determine that the affected species or stock is within the range of OSP and that the taking authorized by the waiver will not reduce it below that level.[54] This conclusion is reinforced by the MMPA's requirements regarding the promulgation of regulations to implement the waiver. These regulations must ensure that the takings they authorize "will not be to the disadvantage of those species and population stocks and will be consistent with the purposes and policies" of the MMPA.[55]

Once having decided to propose regulations to implement a waiver decision, the Secretary must follow procedural requirements much more stringent than those ordinarily applicable to agency rulemaking. The reg-

[50]16 U.S.C. § 1362(18).

[51]Although Congress in 1994 clearly rejected the Hayashi court's reasoning, it nevertheless endorsed the result. *See* text accompanying notes 94–105, *infra*.

[52]16 U.S.C. § 1371(a)(3)(A).

[53]*Id.*

[54]*See* Gaines and Schmidt, *supra* note 9, at 50009.

[55]16 U.S.C. § 1373(a). Although the MMPA does not define the term "disadvantage," it is commonly understood as a shorthand way of incorporating the MMPA's prohibition against taking that would reduce a species or population stock below its OSP level.

ulations must "be made on the record after opportunity for an agency hearing."[56] This requirement brings into play the full panoply of procedures applicable to an adversarial administrative hearing, rather than the more typical "notice and comment" procedures of informal rulemaking. The most important of these is the right to cross-examine. Moreover, the hearing's scope is not limited solely to the proposed regulation but extends to the decision to waive the moratorium. Finally, when proposing regulations, the Secretary must publish and make available to the public a statement describing the existing population levels of the affected species or stock, the proposed regulations' anticipated impact on the OSP of the species or stock, and the evidence upon which the Secretary's proposed action is based.[57]

Once promulgated, the regulations do not by themselves authorize any taking or importation of marine mammals. The Secretary also must issue a permit, consistent with the regulations, specifying the number of animals to be taken, where they are to be taken, over what time period, and other similar conditions.[58] Where the permit is for importation, as distinct from taking, there are additional requirements. The Secretary may not permit importation of marine mammals that were taken inhumanely, that were pregnant at the time of taking, or that were "nursing at the time of taking, or less than eight months old, whichever occurs later."[59]

Before issuing a permit, the Secretary must publish notice of the permit application in the Federal Register and allow thirty days for public comment.[60] If, during that period, any person requests a hearing on the application, the Secretary may order a hearing.[61] The permit applicant has the burden of demonstrating that the proposed taking or importation is consistent with the regulations and with the policies of the MMPA.[62] The Secretary's decision to grant or deny a permit is subject to judicial review at the behest of the applicant or any party opposed to the application.[63]

[56]*Id.* § 1373(d).

[57]*Id.* It was the Secretary's failure to publish this statement (or even to make the required underlying determinations) that was the narrow basis for the court's holding in Committee for Humane Legislation, Inc. v. Richardson, *infra* note 68.

[58]16 U.S.C. § 1374(b). The Secretary also may issue "general permits." *Id.* § 1374(h). General permits have been issued to commercial fishing associations to authorize the incidental taking of marine mammals by individual members of those associations.

[59]*Id.* § 1372(b). For a discussion of these requirements, *see* Animal Welfare Institute v. Kreps, *supra* note 32.

[60]16 U.S.C. § 1374(d)(2).

[61]*Id.* § 1374(d)(4).

[62]*Id.* § 1374(d)(3).

[63]*Id.* § 1374(d)(6). In Animal Welfare Institute v. Kreps, supra note 32, at 1006, the court concluded that "Congress implicitly intended to confer standing to challenge waiver regulations on the same categories of people to whom it gave standing to challenge permits."

This is the only express authorization of judicial review of any administrative action under the MMPA.

When the MMPA was passed, Congress presumably intended that the complex procedures it crafted to accomplish a waiver of the moratorium would be the principal mechanism for allowing the taking of marine mammals. In fact, however, this process has rarely been used.[64] Instead, Congress has marginalized the significance of the waiver provisions by establishing an ever-broadening list of takings that are lawful without regard to the moratorium or the waiver provisions. Today, the waiver process remains available to those who might propose recreational or commercial hunting of marine mammals, or who wish to import marine mammal products for commercial purposes. For most others, however, there are simpler ways around the MMPA's supposed moratorium.

EXCEPTIONS TO THE MORATORIUM: TAKING INCIDENTAL TO COMMERCIAL FISHING OPERATIONS

The taking prohibition is subject to a long list of exceptions. The most complex of these pertain to taking that is "incidental" to fishing activities. The concept of "incidental taking" in this context encompasses a wide range of activities, including both some that are intended to "take" marine mammals and some that are not.

In the tuna fishery of the eastern tropical Pacific Ocean, fishers deliberately chase and attempt to capture (at least temporarily) schools of surface-swimming dolphins in order to catch the tuna that typically swim beneath them. In this instance, "taking" (i.e., chasing and encircling) the dolphins is quite deliberate, but since the ultimate purpose is to catch tuna, the deliberate taking of dolphins is considered "incidental" to the fishery. In other fisheries, marine mammals can become entangled or caught in nets intended for fish. The fishers do not intend to catch the marine mammals and, in fact, would prefer not to, but this incidental capture still occurs. Finally, marine mammals may try to "steal" fish from nets. To prevent them from doing so, fishers may try to frighten them away with small explosives or gunshots. This form of deliberate harassment is also generally considered to be incidental to the fishery.

The MMPA as originally enacted made few distinctions among the fisheries in which incidental taking of marine mammals occurred. Only the commercial tuna fishery of the eastern tropical Pacific Ocean was singled out for special treatment. Incidental taking in all other commercial fish-

[64]A moratorium waiver to allow the commercial importation of seal skins from South Africa in 1976 was overturned on the grounds that it violated the MMPA's strictures against importing animals that were nursing or less than eight months old when taken. *Id.*

eries was subject to a uniform set of requirements: Taking incidental to commercial fishing operations was prohibited unless specifically authorized by regulation.

Even when incidental taking was authorized, however, the MMPA somewhat ambiguously sought to minimize or eliminate it. The MMPA declared an "immediate goal that the incidental kill or incidental serious injury of marine mammals permitted in the course of commercial fishing operations be reduced to insignificant levels approaching a zero mortality and serious injury rate."[65] All commercial fisheries were to be regulated under this broad framework.

Over the years, however, Congress has repeatedly revised the statutory requirements applicable to commercial fishing. It has created a dazzling array of distinctions never contemplated by the original drafters of the law. One set of requirements applies to foreign vessels lawfully fishing in U.S. waters; another applies to U.S. vessels. California sea otters are treated differently from pinnipeds, and both are treated differently from all other marine mammals. The various commercial fisheries that incidentally take marine mammals are subject to differing requirements that depend upon the frequency of incidental taking within the fishery. The tuna fishery of the eastern tropical Pacific Ocean has its own set of detailed requirements, unlike those applicable to any other fishery. Underlying these many distinctions is a complex set of biological, economic, and political factors.

The Tuna-Dolphin Conflict

In the eastern tropical Pacific Ocean, in the waters off Central and northern South America, an unusual association of fish and marine mammals occurs. Schools of large yellowfin tuna regularly swim directly beneath surface-swimming schools of dolphins. This association is not known to occur anywhere else in the world. Precisely why it occurs there remains a mystery. The fact that schools of tuna also are found beneath other surface objects, such as floating logs, suggests that the tuna associate with the dolphins, and not vice versa. But even this is unclear, and it is possible that there is some mutual gain to both fish and dolphin.

Tuna fishers were quick to recognize the potential value of this unusual bond. Finding large schools of tuna in the open ocean is a daunting task. It is considerably easier with a guide to lead fishers to the fish. Schools of dolphins, active and visible at the ocean's surface, became the guides once the link between tuna and dolphins was known.

By the late 1950s, the technology of tuna fishing in the eastern tropical Pacific had adapted to exploit this link. Smaller "long-line" and "bait-

[65]16 U.S.C. § 1371(a)(2).

fishing'' boats, in which fishers literally hooked tuna on a long, multiple-hooked line or with pole and line, were replaced by massive ''purse seiners.'' The purse seiners used speed boats to encircle with huge nets both surface-swimming dolphin schools and the tuna beneath them. The big seiners could travel farther from shore and stay out longer than the smaller boats they displaced. Helicopters stationed aboard the seiners dramatically expanded the area of ocean fishers could search for signs of dolphins. The efficiency of this very capital-intensive method of fishing assured U.S. vessels of the lion's share of the catch. With the advent of purse seining, the tuna fishery of the eastern tropical Pacific Ocean became largely a U.S. fishery.

Purse seining was a very efficient method of catching tuna, but it took a staggering toll on dolphins. Attempting to swim through, rather than over, the nets, dolphins became entangled and drowned. When winds or currents caused the cork line of the net to roll over and cover the ocean surface with the net, hundreds of dolphins were killed. Estimates of annual dolphin mortality in the early years of purse seining were in the hundreds of thousands. By the time the MMPA was passed in 1972, an estimated five million dolphins had perished in the tuna fishery.[66]

For more than a decade, the toll on dolphins taken by U.S. tuna boats was largely unknown and unseen by all but the fishers themselves. The impact of purse seining on dolphins came to light as a result of the on-board observations of a young scientific researcher, William F. Perrin. An aroused public, influenced by a popular television program called ''Flipper'' that featured a dolphin by the same name, eventually demanded action to reduce or halt dolphin deaths in the tuna fishery. The tuna-dolphin issue was one of the major concerns at the time Congress debated and enacted the MMPA. The rest of this section examines how Congress has tried to resolve this controversy.

There have been three distinct phases in this effort. The first was a brief period in which Congress believed that a ''quick fix'' would be found. This optimism soon was followed by a decade of progressively more restrictive regulation to reduce killing of dolphins by the U.S. fleet. In the third phase, coincident with the passing of U.S. predominance in the eastern tropical Pacific fishery, the U.S. has sought to influence the fishing practices of nations that now ply the waters once dominated by U.S. boats.

1972–1976: Technological Optimism

Congress was well aware of the tuna-dolphin problem in 1972. Congress apparently believed, however, that a little ingenuity and research would produce a technological solution. Consequently, taking incidental to commercial fishing operations was exempted from the Act's general prohibi-

[66]*See* 59 Fed. Reg. 58285, 58288 (1993).

tion against taking marine mammals. The exemption was to last for only two years, while a $2 million research program was undertaken. The research was aimed at reducing incidental mortality "to the maximum extent practicable" by improving fishing methods and gear. If incidental taking continued after this two-year grace period, it would be subject to regulation, but it would not require a formal waiver of the Act's moratorium on taking marine mammals. The "immediate goal" of any regulation was to be that "the incidental kill or incidental serious injury of marine mammals permitted in the course of commercial fishing operations be reduced to insignificant levels approaching a zero mortality and serious injury rate."[67]

The technological optimism that led to the two-year exemption proved unfounded. When the grace period ended in 1974, elimination of incidental mortality in the commercial tuna fishery was as elusive as it had been two years earlier. The legal duty of the National Marine Fisheries Service became to regulate the incidental taking of dolphins to achieve the Act's somewhat nebulous "immediate goal."

The government's response to this mandate was to issue regulations generally requiring the fishing practices and gear that seemed most effective in reducing incidental mortality. The regulations, however, failed to set an upper limit on the number of dolphins that could be killed. Environmental groups challenged this failure in court. The government responded that it lacked the data to assess the impact of permitted incidental taking on the optimum sustainable populations of the affected dolphin stocks. In the opinion of federal district court judge Charles Richey, that was not good enough. Richey's 1976 decision in *Committee for Humane Legislation, Inc. v. Richardson*[68] ushered in a new and critical phase in resolution of the tuna-dolphin controversy.

1976–1984: "Ratcheting Down"

The most important impact of Judge Richey's ruling was to force the National Marine Fisheries Service to set a ceiling on the number of dolphins that could be incidentally killed by the U.S. tuna fleet. On June 11, 1976, an annual quota of 78,000 dolphins was set. This quota was reached before the end of October. The government eventually terminated the permit allowing the fleet to take dolphins, temporarily closing down the U.S. tuna fishery.

New regulations promulgated in 1977 sought to reduce incidental dolphin mortality by progressively "ratcheting down" the annual quota each year. The quota dropped from 59,000 in 1977 to 31,500 in 1980. In fact, the fleet's estimated mortality already had dropped to fewer than 20,000

[67] 16 U.S.C. § 1371(a)(2).

[68] 414 F. Supp. 297 (D.D.C, 1976), *aff'd*, 540 F.2d 1141 (D.C. Cir. 1976).

by 1978. This apparently dramatic decline had been achieved in part by shrinkage in the fleet's size.

In 1980 the government issued new regulations for 1981 to 1985. These reduced the quota still further, to 20,500 dolphins annually.[69] Unlike the regulations they replaced, however, these maintained a constant quota throughout the five-year period. The government concluded that, absent further shrinkage in the U.S. fleet's size or new technological developments, significantly lower mortality could not be achieved.

Despite its progress in reducing dolphin mortality, the U.S. fleet feared that the "immediate goal" provision of the Act would be used to force further reductions that were neither technologically nor economically achievable. Responding to this concern, Congress in 1981 clarified the "immediate goal" provision by declaring that for the tuna fleet it required "a continuation of the application of the best marine mammal safety techniques and equipment that are economically and technologically practicable."[70] At the same time, however, the 1981 amendments to the Act directed the Secretary of Commerce to undertake "research into new methods of locating and catching yellowfin tuna without the incidental taking of marine mammals."[71] If that research bore fruit, it might be technologically and economically practicable to reduce quotas still further and possibly to eliminate incidental taking altogether.

Compliance with quotas and other regulations was monitored through a program of on-board observers. The observers provided the only means of enforcing the Act. But fishing captains objected to being required to carry a federal agent whose purpose was to collect data for use in enforcing the Act. They contended that the Constitution's Fourth Amendment prohibition against warrantless searches and seizures prohibited the use of on-board observers as enforcement agents. This issue came to a head in *Balelo v. Baldridge.*[72] The district court ruled that using for enforcement purposes data collected by on-board observers violated the Fourth Amendment. The Ninth Circuit Court of Appeals, reviewing the case *en banc,* reversed this decision.

Congress next addressed the tuna-dolphin issue in the 1984 amendments to the MMPA. The amendments appeared to extend the annual quota of 20,500 indefinitely. In fact, the quota was reduced, because it was now to include all incidentally taken coastal spotted and eastern spinner dolphins. These two depleted stocks had not previously been included in the overall quota. Prior to the 1984 amendments, no incidental taking of depleted stocks could be authorized, but under certain circumstances

[69] 45 Fed. Reg. 72178 (1980).
[70] 16 U.S.C. § 1371(a)(2).
[71] *Id.* § 1380(a).
[72] 724 F.2d 753 (9th Cir.), *cert. denied,* 467 U.S. 1252 (1984).

the government had treated incidental take of individuals from these stocks as "accidental" take and did not count them toward the quota.[73] The 1984 amendments legitimized this policy but limited the number that could be taken from each stock and required that the take be counted toward the overall quota. The 1984 amendments also gave the government authority to lower the quota if it appeared that incidental taking had a significant adverse effect on a marine mammal population. The effect of the 1984 amendments was thus to ratchet down still further the dolphin quota for the U.S. fleet.

At the time it appeared that remarkable progress had been made in solving a challenging environmental problem. The U.S. tuna fleet's incidental mortality had declined from several hundred thousand annually to fewer than 20,000. This dramatic decline, however, was largely attributable to the fact that fewer U.S. boats fished the waters of the eastern tropical Pacific Ocean. In the mid-1970s, more than 150 U.S. purse seine vessels plied these waters. They accounted for nearly 70 percent of the fishing capacity in the region.[74] By 1987, however, the size of the U.S. fleet in the eastern tropical Pacific was sharply reduced. Many boats had shifted their effort to the western Pacific while others now operated under foreign flags. The foreign fleet, with seventy vessels, was roughly twice the size of the U.S. fleet.[75] A fishery that had long been dominated by the U.S. fleet had been transformed to one dominated by foreign boats. Eventually, nearly all U.S. boats disappeared from the region.

With foreign boats outside the direct regulatory control of the MMPA, the gains in reducing dolphin mortality by U.S. boats were likely to be cancelled out by increased mortality by foreign boats. Congress began searching for ways to influence foreign fishing practices.

1984–Present: Addressing the International Problem

Congress recognized from the outset the possibility that stringent regulation of the U.S. tuna fleet would put it at a competitive disadvantage with foreign fleets. This was not merely an economic concern, but also a conservation concern. Any shift from U.S. to foreign fishing meant that less fishing effort would be subject to direct control under the MMPA. To address this problem, Congress in the original MMPA directed the Secretaries of Commerce and State to begin negotiations within the Inter-

[73]The reason for this policy was purely practical. A few dolphins from depleted stocks occasionally would occur in large schools comprised primarily of dolphins of other stocks. The presence of dolphins from depleted stocks was not always apparent until after the net had already been deployed. The resulting incidental take of dolphins of the depleted stock was considered accidental and not treated as a violation of the prohibition against incidentally taking individuals of a depleted stock.

[74]Marine Mammal Commission, Annual Report to Congress, 1994, at 116 (1995).

[75]Sen. Rep. No. 100–592, 100th Cong., 2d Sess., at 7 (1988).

American Tropical Tuna Commission to reduce incidental taking of marine mammals in the tuna fishery. Congress also authorized an embargo against importing fish products from countries using fishing methods prohibited for U.S. fishers or that resulted in marine mammal mortality in excess of U.S. standards.

Significant progress eventually was made in the Inter-American Tropical Tuna Commission, but it took far longer than Congress likely anticipated in 1972. The Commission had been established by international treaty in 1949 to regulate tuna harvest in the eastern tropical Pacific. Protecting dolphins from incidental mortality in the fishery was a concern wholly outside its original mission. In 1977 the Commission nevertheless established a program to station observers aboard purse seiners. It did not, however, provide for any regulation of the vessels' fishing practices.

The embargo provisions of the original MMPA proved to be largely an empty threat. Compliance with the Act's requirements was certified on the basis of "on paper" representations about a foreign country's requirements. Whether the representations reflected what the country's vessels actually did at sea was doubtful. There were few observers aboard foreign vessels until the late 1980s, and reliable information about the vessels' actual practices and dolphin kill rates was quite limited.

Congress sought in 1984 to ensure not only that countries exporting tuna to the United States had programs to protect dolphins during tuna fishing, but also that the programs were achieving results like those of the U.S. fleet. Congress added a requirement that the average rate of dolphin mortality by a foreign fleet had to be comparable to that of the U.S. fleet.[76] If not, an embargo against importing tuna from that country would be imposed. Information from on-board observers clearly would be needed to enforce this new requirement.

The National Marine Fisheries Service was slow to implement the new requirements. It did not publish regulations to implement them until 1988,[77] a delay caused mostly by the government's reluctance to precipitate a major trade dispute with Mexico and other nations. It was the looming prospect of congressional hearings on MMPA reauthorization in 1988 that prompted the agency to issue regulations required by amendments four years earlier.

With the 1988 reauthorization, Congress added still more requirements to the MMPA's provisions governing tuna imports from foreign nations.[78] These included observer requirements, kill rates no more than twice the kill rate of the U.S. fleet (eventually declining to no more than 1.25 times the U.S. rate), and limits on the percentages of total mortality of coastal

[76]Pub. L. No. 98–364, 98 Stat. 440, § 101(a)(2) (codified at 16 U.S.C. § 1371(a)(2)).
[77]53 Fed. Reg. 8,910 (1988).
[78]*See* Pub. L. No. 100–711, § 4(a)(2) (codified at 16 U.S.C. § 1371(a)(2) *et seq.*).

spotted and eastern spinner dolphins. New requirements applicable to the U.S. fleet, including a prohibition against so-called "sundown sets" (encircling a school of dolphins shortly before sundown, a practice that often resulted in high dolphin mortality because the operation was completed in darkness), also limited foreign boats through the MMPA's comparability requirements. Even more significantly, Congress extended the embargo authority to intermediary countries not themselves engaged in fishing.[79] Yellowfin tuna imports from an intermediary nation were to be embargoed if the country did not embargo tuna from fishing nations that failed to meet the MMPA's standards.

In April 1990, Starkist Seafood Company announced that it would no longer purchase any tuna caught in association with dolphins and that it would begin labelling its canned tuna as "dolphin safe." Two other major tuna processing companies followed suit almost immediately.[80] This action was a response to intense pressure from dolphin conservation and animal rights groups. It dramatically changed the business climate for both U.S. and foreign tuna fleets. Although MMPA regulations allowed U.S. boats to catch tuna in association with dolphins, the processing companies' actions effectively deprived the boats of access to the U.S. market for their tuna. The processors' actions also made meeting the MMPA's comparability requirements superfluous for foreign fleets.

Congressional action followed swiftly on the heels of the processors' announcements. The Dolphin Protection Consumer Information Act[81] defined the term "dolphin safe" and made it illegal to sell any tuna with this label unless it met the Act's definition. To be dolphin safe, it was not sufficient if no dolphins were killed in the process of catching the tuna, or even if the labelled tuna was not caught by encircling a school of dolphins. Rather, *all* of the tuna caught on the vessel's trip must have been caught by means other than encircling dolphins. The result was that if a vessel only once set its nets around a school of dolphins, with or without killing any dolphins, none of the tuna caught on that entire fishing trip could be labelled as "dolphin safe."

These actions forced the hand of the foreign fishing fleets. Mexico had become the dominant player in the eastern tropical Pacific fishery. In August 1990 its tuna products were embargoed under the MMPA's comparability requirements. Mexico promptly challenged the MMPA's requirements on the basis that they violated the General Agreement on Tariffs and Trade (GATT), the international agreement governing trade among nations that seeks to reduce barriers to trade.

Under then applicable GATT procedures, complaints by one nation

[79] 16 U.S.C. § 1371(a)(2)(C).

[80] H. Rep. No. 101–579, 101st Cong., 2d Sess. 6–7 (1990).

[81] Pub. L. 101–627, § 901, 104 Stat. 4465 (1990), codified at 16 U.S.C. § 1385.

against another were heard initially by a special dispute panel. The panel that heard Mexico's complaint ruled in its favor in September 1991.[82] The ruling had to be ratified by the full GATT Council of Representatives to become effective. Moreover, a GATT determination that a nation has engaged in a prohibited trade practice does not require that the practice cease. Rather, it authorizes the complaining country to engage in retaliatory measures. Thus, the GATT panel ruling neither invalidated the MMPA's requirements nor obliged the United States to change them. As a practical matter, however, it put the United States in the difficult position of trying to resolve an important conservation issue at home without impairing a broader agenda to liberalize trade with Mexico and the rest of the world.[83]

Action by the GATT Council of Representatives was postponed while Mexico and the United States engaged in diplomatic efforts to resolve the dispute. These efforts eventually led to further MMPA amendments in 1992. The 1992 amendments encouraged negotiation of an international agreement to establish a "global moratorium" for at least five years on tuna fishing by means of purse seining on schools of dolphins. The amendments spell out in considerable detail the required elements of a possible agreement, which was to be secured by March 1, 1994. Until that date, the MMPA's embargo requirements would not apply to any nation that transmitted a formal communication committing itself to the planned global moratorium. The few boats remaining in the U.S. fleet could continue fishing in association with dolphins under sharply reduced quotas until 1994, unless no major purse seining nation had entered into a moratorium agreement by that date. In that case, the U.S. fleet could continue fishing in association with dolphins under progressively declining quotas until December 31, 1999, after which time its authority to do so expires.

No nation agreed to the proposed "global moratorium." Even without a moratorium, the International Dolphin Conservation Act of 1992[84] effectively closed the U.S. market to tuna caught in association with dolphins after 1994. This closure occurred despite the dramatic progress made by foreign vessels fishing in the eastern tropical Pacific Ocean. Dolphin mortality caused by foreign vessels declined from more than 80,000 in 1989, to fewer than 4,000 in 1993. Most of the decline resulted from dramatic reductions in the average number of dolphins killed each time a net was "set" around a school of dolphins. Some of the decline also

[82]GATT, United States—Restrictions on Imports of Tuna, B.I.S.D., 39th Supp. 155, para. 5.22 (1993), 30 I.L.M. 1594 (1991).

[83]The European Community filed a separate GATT challenge to the MMPA's embargo provisions with respect to intermediary nations in 1992. The dispute resolution panel found in 1994 that the embargo provisions violated GATT. *See* United States-Restrictions on Imports of Tuna, GATT Doc. DS29/R (1994), 33 I.L.M. 839 (1994).

[84]Pub. L. No. 102–523, 106 Stat. 3425.

resulted from changed fishing practices. Less fishing was targeted on dolphins, and more by encircling floating logs or other flotsam, or directly on tuna schools not associated with dolphins.[85] Meanwhile, the U.S. fleet has all but disappeared from the region where it once reigned supreme. In 1995, only five U.S. vessels still fished in the eastern tropical Pacific.[86]

An Assessment of the Tuna-Dolphin Issue

It has taken more than two decades to achieve the MMPA's original goal of reducing the incidental mortality of dolphins in the tuna fishery to a rate "approaching zero." The decline in dolphin mortality has been accompanied by the virtual elimination of the U.S. fleet as a significant component of the tuna fishery. The extent to which the MMPA's regulation of the U.S. fleet contributed to its near demise is open to debate.

Much of the dramatic reduction in incidental mortality has resulted from improved efficiency in purse seining associated with dolphins. It has not, however, been possible to eliminate dolphin mortality altogether. With existing technology, eliminating all dolphin deaths can be accomplished only by shifting fishing effort to floating logs and tuna schools not associated with dolphins. Such a shift in fishing practices, while beneficial for dolphins, may have other undesirable consequences. The yellowfin tuna caught by encircling floating logs or schools of tuna not associated with dolphins are typically much smaller than those caught by setting on dolphins. Consequently, many more fish are discarded as too small to be used. Moreover, the bycatch of other animals, including sharks, billfish, and endangered sea turtles, is higher in these log and school sets than in dolphin sets.[87] As in many areas of environmental policy, one worthy goal may have been achieved at the expense of others.

Tuna-fishing nations have continued their efforts to persuade the United States to lift its embargo on tuna caught in association with dolphins. Twelve nations, including the United States, signed an international agreement in 1995 that would allow the embargo to be lifted but would carefully regulate tuna-fishing in the eastern tropical Pacific.[88] There currently is disagreement in the United States as to the wisdom of

[85]In 1992, member nations of the Inter-American Tropical Tuna Commission and some non-member nations agreed to a set of measures to reduce incidental dolphin mortality in the tuna fishery, including both overall quotas and individual vessel quotas. This agreement aided the foreign fleet's significant improvement. *See* Marine Mammal Commission, Compendium of Selected Treaties, International Agreements, and Other Relevant Documents on Marine Resources, Wildlife, and the Environment, vol. II at 1174, 1369.

[86]Marine Mammal Commission, Annual Report to Congress, 1995, at 102 (1996).

[87]*See id.* at 104.

[88]This agreement is the Declaration of Panama, signed on October 4, 1995. *See* Marine Mammal Commission Annual Report to Congress, 1995, at 107 (1996).

lifting the embargo, and consequently the MMPA has not been amended to implement the agreement.[89]

Other Fisheries

The Regulatory Framework from 1972 to 1988

The considerable public attention given to the tuna-dolphin issue meant that perceptions of this problem dominated in 1972 when Congress created the general framework to address incidental taking in all commercial fisheries. Important factors in the tuna-dolphin "model" were the incidental death of literally tens of thousands of marine mammals each year, a presumed significant adverse impact upon the affected mammal stocks, and the expectation that technological innovations could dramatically lower mortality. Most other commercial fisheries that incidentally took marine mammals did not fit this model. Most of the other fisheries took many fewer marine mammals, were much less likely to have significant effects upon affected marine mammal stocks, and had little prospect for reduced mortality through technological innovation.

Despite these important differences, other fisheries could obtain relief from the MMPA's prohibition against incidental taking only through the arduous process of securing a permit after a formal adjudicatory hearing to determine whether the expected level of incidental taking would adversely affect any marine mammal stock. A number of practical obstacles made the grant of a permit unlikely, including the paucity of information about current and historic marine mammal stock abundance upon which to estimate optimum sustainable populations for the various stocks. Unlike the large purse seiners of the tuna fishery, which could easily accommodate on-board observers, many of the boats in other fisheries were too small to do so. In addition, the relative infrequency with which a given vessel took marine mammals made it more expensive and difficult to obtain useful information from on-board observers. Thus, despite the MMPA's new requirements, boats in many different commercial fisheries undoubtedly took marine mammals incidental to their operations, just as they had before the law was passed.

Almost no one was satisfied with this situation. Although many fishers continued fishing as they had always done, if they incidentally took a marine mammal they were violating a federal law. There was little likelihood that law enforcement agents would ever observe the violation, but a disgruntled employee could report it. From the government's point of view,

[89]For a detailed discussion of developments in this controversy, see Marine Mammal Commission Annual Report to Congress, 1995, at 109–110 (1996).

not only did it lack a means of enforcing the law, but the fishers' reluctance to acknowledge incidental takings meant it could not assess the biological significance of those takings.

Congress amended the MMPA in 1981 to address these problems. For any American fishery that incidentally took only "small numbers" of non-depleted marine mammals and had only a "negligible impact" on any marine mammal species or stock, the Secretary could allow the taking free of the permitting requirements that were applicable to other fisheries.[90] The new provision was limited to U.S. citizens in order to exclude the Japanese high seas salmon gillnet fishery in the U.S. fishery conservation zone off Alaska. This fishery took a thousand or more dolphins each year.

The Kokechik *Decision*

At the time Congress acted in 1981, a general permit allowing incidental taking of marine mammals in the Japanese high seas salmon gillnet fishery was scheduled to expire soon. In 1982, however, Congress effectively extended the permit until 1987 by amending the North Pacific Fisheries Act of 1954.[91] This action set the stage for a series of events that eventually led to a complete rewriting of the MMPA's requirements with respect to commercial fisheries. The immediate catalyst for change was not some new biological discovery or conservation principle. Rather, it was the desire of American salmon fishing interests to keep Japanese salmon vessels out of U.S. waters.

The Secretary of Commerce in 1987 issued a new general permit to the Federation of Japan Salmon Fisheries Cooperative Association, authorizing incidental take of 6,039 Dall's porpoises over a three-year period. The Japanese Federation also sought permission to take northern fur seals and sea lions. The Secretary denied this latter request because the fur seal was a depleted species and because there was insufficient information to determine the sea lion's status.

The Kokechik Fishermen's Association, an Alaskan subsistence fishing organization, challenged the permit. Several environmental organizations did likewise. The result was a court decision, *Kokechik Fishermen's Ass'n v. Secretary of Commerce,*[92] that a permit could not be issued for the incidental taking of one species of marine mammal if other species of marine mammals were sure to be taken in the course of the same fishing activities and the other mammals were either depleted or of unknown population status. The decision raised the possibility that other fisheries, operating either

[90]Pub. L. No. 97–58, § 2(1)(C), 95 Stat. 980 (*repealed,* 1988).

[91]*See* the Commercial Fisheries Research and Development Act of 1982, Pub. L. No. 97–389, § 201, 96 Stat. 1949 (amending 16 U.S.C. § 1034).

[92]839 F.2d 795 (D.C. Cir. 1988), *cert. denied sub nom.,* Verity v. Center for Envt'l Educ., 488 U.S. 1004 (1989).

under general permits or "small take" exceptions, might be similarly vulnerable to legal challenge. Consequently, although the plaintiffs accomplished their immediate goal of invalidating the Japanese permit, commercial fishing interests were so alarmed that they sought a comprehensive rewrite of the MMPA's provisions on taking marine mammals incidental to fishing.

There was general agreement that the existing system had not worked well and needed to be changed. There was no immediate agreement, however, on what the long-term changes should be. Congress enacted an interim solution that effectively waived the prohibitions against incidental taking for vessels that voluntarily participated in a program to gather information about the nature and magnitude of fishery interactions with marine mammals through observers, reports by the fishers themselves, and other sources.[93] The interim regime was to be replaced after three years by a permanent system. After extending the interim regime twice, in 1994 Congress enacted a comprehensive set of rules for incidental taking of marine mammals in commercial fisheries.

The 1994 Amendments

The 1994 amendments introduced important new concepts and terms into the MMPA lexicon. Among these were "potential biological removal levels," "strategic stocks," and "take reduction plans." The foundation for implementing these concepts was a requirement that the Secretary prepare a "stock assessment" for each marine mammal stock occurring in U.S. waters.[94] A stock assessment summarizes information about the size and distribution of the stock, and factors affecting it. This information is used to estimate the "potential biological removal level" for the stock. The potential biological removal level is a conservative estimate of the number of animals that can be removed from a stock without impairing the stock's ability to reach or maintain an optimum sustainable population level.[95] Any stock for which the level of direct human-caused mortality exceeds the potential biological removal level is considered to be a "stra-

[93]The interim exemption was codified as section 114 of the MMPA, 16 U.S.C. § 1383a.

[94]16 U.S.C. § 1386.

[95]The potential biological removal level for any stock is the product of three factors: (1) the minimum population estimate, (2) half the maximum theoretical or estimated net productivity rate of that stock at a small population size, and (3) a discretionary recovery factor that can be set as low as 0.1 or as high as 1.0. *See id.* § 1362(20). In theory, in a stable marine mammal population at the carrying capacity of its habitat, births will, on average, be exactly offset by deaths. Such a population does not grow and therefore has a net productivity rate of zero. If the population is reduced, however, to a level below what its habitat can support, births will exceed deaths and the population will grow. Theory predicts that the rate of growth will increase as the population is reduced in size until, at a small size, it reaches some theoretical maximum rate of growth that is determined by the species' inherent biological limits (*e.g.*, age at sexual maturity, gestation period, life span, etc.).

tegic stock." So too is any stock designated as depleted under the MMPA, listed as threatened or endangered under the Endangered Species Act, or declining and likely to be listed within the foreseeable future.[96]

The significance of designating any marine mammal stock as a strategic stock depends upon whether it is incidentally taken in a commercial fishery and, if so, the type of fishery in which it is taken. Section 118 of the MMPA requires the Secretary to classify commercial fisheries into one of three categories: (1) those with frequent incidental mortality and serious injury of marine mammals, (2) those with occasional mortality and injury, and (3) those with only a remote likelihood of these impacts.[97] If a strategic stock of marine mammals interacts with a fishery in either of the first two categories, the Secretary must prepare a "take reduction plan" for that stock.[98]

Each take reduction plan is to have an immediate goal, to be achieved within six months, and a long-term goal, to be achieved within five years. The immediate goal is to reduce incidental mortality and serious injury in commercial fisheries below the potential biological removal level. The long-term goal is to reduce these impacts to "insignificant levels approaching a zero mortality and serious injury rate."[99] The initial draft of a take reduction plan is to be drafted by a "take reduction team" established by the Secretary. The Secretary may modify the team's recommendations and must publish regulations to implement any final plan. A final plan can encompass a wide variety of measures, including fishery-specific limits on mortality, gear requirements, and voluntary actions.[100]

For a fishery subject to a take reduction plan, compliance with the plan satisfies the MMPA requirements with respect to incidental taking of marine mammals. For other fisheries, if vessel owners register their vessels with the Secretary, submit periodic required reports, and comply with the observer requirements the Secretary imposes, any incidental taking of marine mammals that occurs during fishing operations satisfies the MMPA requirements.[101]

The 1994 amendments are a significant departure from the principles established in the original MMPA. The 1972 law imposed a blanket moratorium on taking marine mammals and a virtual bar against waiving the moratorium for depleted stocks. The 1994 amendments grew out of a recognition that these requirements could not be enforced and probably

[96] *Id.* § 1362(19).

[97] *Id.* § 1387(c)(1). The current list of fisheries assigned to the three categories specified in section 118 was published at 60 Fed. Reg. 67063 (1995).

[98] 16 U.S.C. § 1387(f). The Secretary also may develop take reduction plans for non-strategic stocks that interact with "category 1" fisheries.

[99] *Id.* § 1387(f)(2).

[100] *Id.* § 1387(f)(9).

[101] *Id.* § 1387(f)(3)(D).

were unnecessary in any event to achieve the law's conservation objectives. Although somewhat ambiguous, the amendments reflect the ascendancy of the view that rather than seeking to eliminate incidental deaths of marine mammals, it is sufficient to keep mortality within acceptable limits.

The tension between the goals of eliminating incidental mortality and simply controlling it was implicit in the original formulation of the MMPA's "approaching . . . zero mortality" goal. The qualifying word "approaching" raised two obvious questions: how close to zero must one come to achieve this goal, and how quickly must one get there? The answer to the latter question seemed implicit in characterization of the goal as "immediate," although Congress specified no precise target date. The practical result was that the goal, though "immediate," was always perceived to be some time in the future.

By 1994, Congress was finally clear about the deadline: April 1, 2001, nearly thirty years after the MMPA's enactment.[102] Ironically, the 1994 Congress specified that "immediate" meant within six months when describing the much more modest "immediate goal" of take reduction plans (to reduce mortality and serious injury below potential biological removal levels). Conversely, the five-year goal was characterized as a "long-term" goal.[103] The meaning of "immediate" either has changed over time or varies depending upon the context in which it is used.

While Congress in 1994 finally clarified the timetable, it left unanswered the other important question: how close to zero mortality must one come? Incidental mortality and serious injury must be reduced to "insignificant levels," a standard that would appear to be satisfied so long as the affected stocks are capable of achieving or maintaining optimum sustainable population levels.[104] Had Congress stopped there, its intent would have been reasonably clear. However, it went on to state that mortality and serious injury must be "approaching a zero" rate.

The requirement that serious injury and mortality rates be approaching zero implies continuous improvement. Simply achieving and maintaining a very low rate would not appear to suffice, since a static rate, no matter how low, cannot be said to be "approaching zero." On the other hand, Congress in 1994 added a provision that fisheries "shall not be required to further reduce their mortality and serious injury rates" if they maintain insignificant serious injury and mortality levels approaching a zero rate."[105] This provision, though apparently intended to assure fishing interests that they would not be subject to continuous demands for further reductions, contains an inherent self-contradiction: injury and mortality

[102] *See id.* § 1387(b)(1).
[103] *Id.* § 1387(f)(2).
[104] *See id.* § 1387(b)(1).
[105] *Id.* § 1387(b)(2).

levels won't be required to be reduced, provided they are being reduced (i.e., approaching zero). Rather than evidence of drafting ineptitude, the MMPA's incidental taking goals reflect the sort of purposeful ambiguity by which opposing interests often reconcile their differing objectives. Unable to reach agreement on all details, they embrace a verbal formulation that leaves each the freedom to argue that its interpretation is what Congress truly intended.

Special Treatment for Certain Species

The regulatory framework described above applies generally to incidental taking of marine mammals in commercial fisheries. For certain species, however, special provisions apply. For marine mammals that are listed as threatened or endangered under the Endangered Species Act, both the above requirements (found in Section 118) and the special requirements of section 101(a)(5)(E) apply.[106] The latter requirements authorize incidental taking of these species under permits issued by the Secretary if the Secretary determines that the taking will have a "negligible impact" on the species and a recovery plan for it has been or is being developed.[107] Where a recovery plan exists, any take reduction plan developed under Section 118 applicable to the same species must be consistent with it.[108]

Neither section 118 nor section 101(a)(5)(E) applies to the California sea otter, a threatened species under the Endangered Species Act. The sea otter cannot be incidentally taken in commercial fisheries except in the manner originally authorized by the MMPA (i.e., pursuant to a permit issued following an adjudicatory hearing and a waiver of the moratorium). The otter's special status is due partly to 1986 legislation that established terms for an experimental introduction of sea otters at California's San Nicholas Island,[109] and partly to the highly effective efforts of the nonprofit group, Friends of the Sea Otter, in championing the otter's interests.

Incidental taking of marine mammals in the course of commercial fishing operations is generally inadvertent and unintentional. This is not always so, however. In the tuna purse seine fishery, for example, fishing vessels intentionally encircle dolphin schools for the purpose of capturing

[106] *See id.* § 1387(a)(2). Section 101(a)(5)(E) of the MMPA is codified at *id.* § 1371 (a)(5)(E).

[107] On August 31, 1995, the National Marine Fisheries Service determined that these requirements were met with respect to three endangered marine mammal stocks (humpback whales of the central north Pacific, and both eastern and western Steller sea lion stocks). The agency issued a single interim permit to 24 category I and II fisheries authorizing incidental taking from these stocks. *See* 60 Fed. Reg. 45399.

[108] *Id.* § 1387(f)(11).

[109] Pub. L. No. 99-625, 100 Stat. 3500 (1986).

tuna beneath those schools. In other fisheries, fishers sometimes try to protect their catch or gear from marine mammals by harassing nearby animals. Even though this is deliberate, it is considered incidental to commercial fishing operations.

With regard to lethal take, the MMPA generally prohibits the intentional killing of marine mammals in the course of commercial fishing operations.[110] However, a long-standing dilemma concerning the depredation of salmon by sea lions at Ballard Locks in Seattle, Washington, led to congressional authorization of intentional lethal take in certain circumstances. Congress in 1994 authorized the Secretary, upon application from the state, to permit the state to kill identifiable individual pinnipeds that were negatively affecting salmon that are either threatened or endangered or approaching this status.[111] Recognizing the potential for a similar situation in the future involving aquaculture resources and pinnipeds in the Gulf of Maine, in 1994 Congress also authorized a Gulf of Maine "Pinniped-Fishery Interaction Task Force" and directed it to study the problem and make recommendations for addressing it.[112]

EXCEPTIONS TO THE MORATORIUM: TAKING INCIDENTAL TO OTHER ACTIVITIES

Commercial fisheries are not the only enterprises that kill or harass marine mammals unintentionally and inadvertently. Other activities in marine environments also "take" marine mammals incidentally. Collisions with boats are perhaps the leading cause of non-natural mortality among manatees. Manatees are also occasionally crushed in the gates of south Florida's many water control structures. Offshore oil and gas activities can disrupt marine mammal migrating, feeding, and breeding activities; seismic testing associated with oil and gas exploration can stun, panic, or even kill nearby marine mammals.

Despite these and other examples of incidental taking of marine mammals, until 1981 there was no simple way to authorize these takings. The law prohibited all takings without a waiver of the moratorium. The possibility of criminal or civil penalties hung like the sword of Damocles over the heads of those engaged in activities that could take marine mammals. In reality, both they and the government knew that the sword was unlikely to fall. The knowledge that their activities might be illegal and potentially punishable still created a strong sense of discomfort for many.

Congress resolved this dilemma in 1981. It authorized the Secretary to allow taking of marine mammals incidental to nonfishing activities, pur-

[110] 16 U.S.C. § 1387(a)(5).
[111] *Id.* § 1389(b).
[112] *Id.* § 1389(h).

suant to regulations minimizing the activities' impact and requiring monitoring and reporting of any takings.[113] A theoretically absolute, but almost never enforced, prohibition was replaced by a more readily enforceable, limited authorization.

The Secretary's new authority to allow incidental takings could be exercised only for specified activities by U.S. citizens in specified regions, and then only if the Secretary determined that only "small numbers" of marine mammals would be taken and that the total taking within any five-year period would have a "negligible impact" on affected stocks or species.[114] In addition, the Secretary must find that the authorized taking "will not have an unmitigable adverse impact on the availability of such species or stock for taking for subsistence uses" by Alaskan natives.[115] Among the activities for which regulations authorizing incidental taking have been issued are on-ice seismic exploratory activities in the Beaufort Sea and space shuttle activities at Vandenberg Air Force Base in California. Sonic booms from the space shuttle were thought to have the potential to take seals and sea lions on California's Channel Islands.[116]

Since 1992, incidental harassment of marine mammals during nonfishing activities can be authorized by an even simpler means.[117] For these activities, the Secretary can issue authorizations, after notice and an opportunity for public comment, without the added step of issuing regulations. Essentially, the same findings must be made by the Secretary as described in the preceding paragraph. However, these incidental harassment authorizations are limited to one year, unlike the five-year maximum that applies to authorizations pursuant to regulations.

PERMITS FOR SCIENTIFIC RESEARCH, PUBLIC DISPLAY, AND OTHER ACTIVITIES

The MMPA's restrictions on taking marine mammals incidental to fishing and other activities have been steadily eased over time. Similarly, the activities for which permits may authorize deliberate taking and importation have been steadily expanded. The original MMPA authorized permits to take and import marine mammals, without a waiver of the moratorium, for purposes of scientific research and for public display. Scientific research was clearly the more favored of these activities, in that

[113]The 1981 amendments are codified at *id.* § 1371(a)(5)(A)–(C).

[114]The NMFS implementing regulations effectively treat the "small numbers" and "negligible impact" limitations as synonymous. They define "small numbers" as "a portion of a marine mammal species or stock whose taking would have a negligible impact on that species or stock." 50 C.F.R. § 228.3 (1995).

[115]16 U.S.C. § 1371(a)(5)(A)(i).

[116]*See* 50 C.F.R. § 228.21 (1995).

[117]*See* 16 U.S.C. § 1371(a)(5)(D).

research permits could be issued for any marine mammal.[118] The MMPA's restrictions against importing marine mammals that were pregnant, nursing, or less than eight months old at the time of taking also do not apply to scientific research permits.[119] Public display permits, on the other hand, could only be issued for mammals that were not depleted.

To avoid the loss of unique research opportunities, the Secretary can issue a scientific research permit in advance of the normal thirty-day public review period that follows a permit application.[120] In 1994, responding to complaints about the burden of permit requirements, Congress authorized the Secretary to issue "general authorization and implementing regulations" allowing, without the need for an individual permit, scientific research that results only in disturbance, rather than injury, to marine mammals.[121] The primacy of scientific research among the purposes for which permits may be granted thus has been assured.[122]

The need for authority to permit a broader set of activities became apparent during efforts to establish an experimental population of California sea otters, a marine mammal also protected by the Endangered Species Act (ESA). Capturing and relocating otters to establish a new population was essential to accomplishing the otter's recovery. Under the ESA, these activities could be permitted under a provision authorizing the taking of protected animals to "enhance the propagation or survival" of threatened or endangered species.[123] The MMPA, however, had no counterpart, and the government was concerned that it might be vulnerable

[118]In 1994 Congress imposed new limitations on the issuance of scientific research permits for depleted marine mammals. A scientific research permit may not authorize lethal taking from a species or stock that is depleted "unless the Secretary determines that the results of such research will directly benefit that species or stock, or that such research fulfills a critically important research need." 16 U.S.C. § 1374(c)(3)(B). The second half of this standard broke new ground in that it placed upon the Secretary the burden of establishing the relative importance of various research needs.

[119]*See* 16 U.S.C. § 1372(b)(1)–(4). The 1988 amendments to the MMPA lifted these latter restrictions with respect to the importation of marine mammals for public display when the Secretary determines that importation is "necessary for the protection or welfare of the animal." *Id.* at § 1372(b). This change was prompted by the desire to make it possible to import for public display purposes polar bear cubs orphaned when Canadian officials killed nuisance female bears. *See* H. Rep. No. 100–970, 100th Cong., 2d Sess. 1988 at 33, *reprinted in* 1988 U.S.C.C.A.N. 6154, 6174.

[120]16 U.S.C. § 1374(c)(3)(A). Early issuance of a research permit can also occur if delay "could result in injury to a species, stock, or individual."

[121]*Id.* § 1374(c)(3)(C).

[122]The National Marine Fisheries Service has categorically excluded both scientific research and public display permits from the requirement to prepare either an environmental assessment (EA) or an environmental impact statement (EIS) under the National Environmental Policy Act. However, two cases have raised the possibility that an EA or EIS may be required under some circumstances. *See* Jones v. Gordon, 792 F.2d 821 (9th Cir. 1986), and Greenpeace U.S.A. v. Evans, 688 F. Supp. 579 (W.D. Wash. 1987).

[123]*See* 16 U.S.C. § 1539(a)(1)(A).

to legal challenge if it stretched the MMPA's permit authority for scientific research to authorize the relocation. To remove any doubt, Congress enacted special legislation authorizing the otter relocation effort.[124]

The general problem brought to light by the sea otter example was rectified in 1988 by MMPA amendments authorizing the taking or importation of marine mammals for the purpose of "enhancing the survival or recovery of a species or stock."[125] This provision was adapted from the ESA's similar provision. The ESA provision has been widely used to allow taking or importing protected species for captive breeding, without any necessary connection between those captive breeding efforts and other efforts aimed at conserving the species in the wild. The MMPA provision, however, sought to establish this connection, at least in the case of depleted species. It provides that a permit to enhance the survival or recovery of a depleted marine mammal species or stock may allow captive maintenance only if the Secretary "requires that the marine mammal or its progeny be returned to the natural habitat of the species or stock as soon as feasible."[126] This more restrictive MMPA requirement takes precedence over the analogous ESA permitting requirements.[127]

The 1988 amendments to the MMPA also added detail to the requirements applicable to public display permits. The National Marine Fisheries Service, through its implementing regulations, previously had assumed responsibility for deciding whether the proposed public display would confer a "substantial public benefit" and for assessing the adequacy of the display facilities.[128] The 1988 amendments narrowed the agency's discretion. A public display permittee must offer an education or conservation program, but the adequacy of the program is judged by the "professionally recognized standards of the public display community" itself.[129] Permittees also must either be registered or hold a license under the Animal Welfare Act, and open their facilities to the general public on a regularly scheduled basis.[130]

It was not clear prior to 1988 whether a second permit is necessary to transfer an animal lawfully taken for display purposes to another person engaged in the same or another permissible purpose. The 1988 amendments resolved this uncertainty by allowing sale or transfer of any animal for which a public display permit has been granted to another person for public display, scientific research, or enhancing survival purposes.[131] This

[124]Pub. L. No. 99–625, 100 Stat. 3500 (1986).

[125]16 U.S.C. § 1374(c)(4)(A).

[126]*Id.* § 1374(c)(4)(B)(iii).

[127]*See id.* § 1543.

[128]*See* 50 C.F.R. § 216.31(c) (1995).

[129]16 U.S.C. § 1374(c)(2)(A)(i).

[130]*Id.* § 1374(c)(2)(A)(ii) and (iii). The Animal Welfare Act is codified at 7 U.S.C. § 2131–2159.

[131]16 U.S.C. § 1374(c)(2)(B)(ii). Prior to the 1994 amendments, NMFS had approved

amendment reduces the number of circumstances in which permits are required.

The 1994 amendments added two other categories of activities for which importation or taking permits may be issued. One allows importation of trophies from legal polar bear hunts in Canada.[132] The other authorizes the Secretary to issue permits "for photography for educational or commercial purposes involving marine mammals in the wild."[133] The latter change was needed despite the fact that there is no prohibition against photographing marine mammals or any requirement to obtain a permit to photograph marine mammals. Photographers may disturb marine mammals' feeding or otherwise disrupt their behavior. The disturbance constitutes a taking of the animal through harassment and is technically illegal. By authorizing photography permits, the MMPA eliminates a technical violation and enables regulation of activities that may hinder marine mammal conservation.

Even without a permit, public employees and persons designated in a cooperative agreement with the Secretary may take marine mammals for the animal's protection or welfare, to protect public health and welfare, or to remove nuisance animals.[134] Whales, dolphins, and other marine mammals occasionally strand themselves on beaches or in shallow coastal waters. When strandings occur, humans almost always make an effort to return the animals to safety. Through the use of cooperative agreements with qualified local organizations or individuals, the Secretary can authorize appropriate people to carry out rescue efforts without fear of violating the MMPA. In addition, Congress in 1992 enacted the Marine Mammal Health and Stranding Response Act[135] to develop contingency plans for mass strandings and to ensure collection of standardized tissue data and other information.

ALASKAN NATIVES

Native Alaskans living along the coastline have for many centuries depended upon the bounty of the sea for their survival. Marine mammals have been an important source of dietary protein and clothing and an important aspect of their traditional culture. Had the MMPA fallen with

---

these transfers by informal "letters of agreement," avoiding the public notification requirements applicable to permits. This practice was challenged in Citizens to End Animal Suffering and Exploitation, Inc. v. The New England Aquarium 836 F. Supp. 45 (D. Mass. 1993). The challenge was dismissed for lack of standing.

[132]16 U.S.C. § 1374(c)(5). For a discussion of international agreements regarding polar bear conservation, see Chapter 14 at text accompanying notes 99–133.

[133]16 U.S.C. § 1374(c)(6).

[134]*Id.* § 1379(h). Authority to enter into cooperative agreements is provided by 16 U.S.C. § 1382(c).

[135]*Id.* § 1421–1421h (subchapter V of the MMPA).

full force on Alaska's native coastal communities, the impact would have been much greater than for other communities.

Congress recognized the unique dependence upon marine mammals of many native Alaskan communities and accommodated their needs. The MMPA's prohibition against taking marine mammals generally does not apply to Indians, Aleuts, or Eskimos who dwell on the Alaskan coast, if the taking is done either for subsistence or to create and sell "authentic native articles of handicrafts and clothing."[136] However, to be permissible the taking must not be "accomplished in a wasteful manner."[137] These purposes are, in effect, statutorily preferred uses of marine mammals, for they require no permit or other authorization from the Secretary. The Secretary may, however, regulate taking of a member of any depleted species or stock by Alaskan natives, regardless of the taking's purpose. In addition, the MMPA's provisions authorizing transfer of management authority to the states provide a separate mechanism for regulating native Alaskan taking.

OTHER PROHIBITED ACTS AND PENALTIES

The taking prohibition is the most basic of the MMPA's restrictions and provides the basis for many of the MMPA's other restrictions. For example, it is unlawful to possess, transport, sell, offer to sell or purchase, or export any illegally taken marine mammal or any product of an illegally taken marine mammal.[138] It is also unlawful to import any marine mammal taken in violation of the MMPA or the laws of another country.[139] A separate prohibition makes it an offense "for any person to use any port, harbor, or other place under the jurisdiction of the United States to take or import marine mammals or marine mammal products."[140] Finally, except pursuant to a permit for scientific research or to enhance the survival or recovery of a species or stock, it is unlawful to import any marine

[136] *Id.* § 1371(b). An extensive discussion of this provision's legislative history can be found in Katelnikoff v. U.S. Dep't of Interior, 657 F. Supp. 659 (D. Alas. 1986). The court upheld an Interior Department regulation limiting the native handicraft exemption to items commonly produced on or before the date the MMPA was enacted. A later revision of the regulation that excluded items made from sea otters from the handicraft exemption was invalidated in Didrickson v. U.S. Dep't of Interior, 982 F.2d 1332 (9th Cir. 1992).

[137] 16 U.S.C. § 1371(b)(3). In United States v. Clark, 912 F.2d 1087, 1089 (9th Cir. 1990), the court rejected the claim that "custom has now changed to the point that taking walrus for a few parts alone (head, oosik and flippers) should be considered non-wasteful, even if the remainder of the animal is left behind and even if much greater use of the animal was made in the past."

[138] 16 U.S.C. § 1372(a)(3) and (4).

[139] *Id.* § 1372(c)(1).

[140] *Id.* § 1372(a)(2)(B).

mammal that is pregnant, nursing, or less than eight months old at the time of taking, or taken in a manner the Secretary deems inhumane.[141]

Both civil and criminal penalties may be imposed for violations of these prohibitions, as well as for violation of a permit or regulation issued under the MMPA. Criminal penalties may only be imposed against one who "knowingly violates" any of these requirements.[142] The maximum penalties are the same today as they were when the MMPA was enacted in 1972: $20,000 and a year in prison for criminal offenses, and $10,000 in civil penalties.[143] One category of offense is singled out for significantly lower penalties: If a fishing vessel with a valid authorization to engage in a fishery that causes frequent or occasional mortality or serious injury of marine mammals fails to display evidence of the authorization, a fine of not more than $100 may be imposed.[144] On the other hand, any vessel that is employed in the unlawful taking of any marine mammal may have its entire cargo seized and forfeited and be fined a civil penalty of not more than $25,000.[145]

FEDERALISM UNDER THE MMPA: THE TRANSFER OF MANAGEMENT AUTHORITY TO THE STATES

The relationship of state and federal authorities under the MMPA has been a contentious issue. Subject to certain exceptions, section 109 of the MMPA provides that "[n]o State may enforce, or attempt to enforce, any State law or regulation relating to the taking of any species . . . of marine mammal within the State."[146] Although this provision expressly refers only to "taking," in *Fouke Company v. Mandel* the court interpreted the MMPA to have implicitly preempted any state law respecting importation of marine mammals as well.[147]

Much like its moratorium, however, the MMPA's preemption of state regulatory authority may be overcome. States can regain their authority by meeting certain conditions. The simplest and easiest form of state participation is found in the MMPA's provision for "cooperative arrangements" between the Secretary and a state. The Secretary may delegate to a state the "administration and enforcement" of the MMPA.[148] The

141 *Id.* § 1372(b)

142 *Id.* § 1375(b).

143 *Id.* § 1375.

144 *Id.* § 1387(c)(3)(C).

145 *Id.* § 1376.

146 *Id.* § 1379(a).

147 386 F. Supp. 1341 (D. Md. 1974). For criticism of this decision, see Note, *Federal Preemption: A New Method for Invalidating State Laws Designed to Protect Endangered Species*, 47 U. Colo. Rev. 261 (1976).

148 16 U.S.C. § 1379(k).

MMPA does not define the term "administration," which arguably could include any of the Secretary's duties. It seems certain that a much narrower meaning was intended, however, because elsewhere the MMPA imposes stringent procedural requirements on the return of management authority to the states.

Section 109 of the MMPA as originally enacted allowed the Secretary to waive the moratorium for any species or stock of marine mammal, and return management authority for it to a state, if the Secretary found that the state had laws and regulations consistent with the terms of the waiver and other MMPA requirements. Alaska sought almost immediately to regain management authority over ten marine mammal species, and in 1975 the Secretary of the Interior returned management of the Pacific walrus to the state.[149] The Secretaries of the Interior and Commerce later conditionally approved Alaska's request for a waiver and return of management with respect to the remaining species.[150] Before the conditions to effectuate this return were met, however, events brought the Alaskan transfer request to an abrupt halt.

Shortly after Alaska resumed management of Pacific walrus, and while administrative proceedings continued regarding its request to manage nine other marine mammals, Alaskan natives filed a suit that had a major impact not only on these proceedings but on the MMPA itself. The question presented in *People of Togiak v. United States*[151] was whether the statutory exemption from the moratorium for subsistence and handicraft taking by Alaskan natives lapsed when Alaska resumed management. At the time walrus management was returned to Alaska, the Department of the Interior decreed that the native exemption was thereby rescinded and superseded by Alaska's laws and regulations,[152] which significantly restricted native taking.[153] Interior reasoned that the MMPA provision exempting native taking from the moratorium also exempted native taking from the provision preempting state management of marine mammals. Therefore, Alaska was free to regulate native taking *with or without* a return of management. The initial edition of this work expressed doubt that such reasoning would prevail.[154]

In an opinion denying its motion to dismiss, the court rejected Interior's statutory interpretation and found that the native exemption survived both the waiver and return of management to Alaska. Noting the

[149]40 Fed. Reg. 59459 (1975).

[150]44 Fed. Reg. 2540 and 2547 (1979).

[151]470 F. Supp. 423 (D.D.C. 1979).

[152]41 Fed. Reg. 14372, 14373 (1976).

[153]Alaska Administrative Code, section 81.340(4) (Reg. 50, Oct. 1976) (*repealed* July 4, 1980).

[154]First edition at 367.

native exemptions of the Endangered Species Act,[155] the Fur Seal Act,[156] and the Walrus Protection Act,[157] the court held that MMPA's native exemption was a deliberate balance Congress struck between two pervasive federal responsibilities: protection of wildlife and trust protection of Indians.[158] In a later order, the court enjoined the federal government from approving any state laws that restrict native taking permitted under the MMPA.[159]

In response to *Togiak*, Alaska elected to return management authority over Pacific walrus to the federal government and stated its intention not to pursue further the return of management for the other nine species.[160] Alaska believed that the equal protection clause of its constitution barred it from carrying out a management program containing a racially or ethnically based exemption. Accordingly, the Department of the Interior repealed its regulations conditionally approving Alaska's request and suspended the waiver for walrus.[161]

The *Togiak* decision and Alaska's response forced Congress to reconsider the MMPA's provisions regarding state and federal relations. The result was a major overhaul of section 109 and related provisions in 1981.[162] The amendments apply to all states seeking return of marine mammal management authority.

The new provisions relating to resumption of state management spell out in copious detail both the procedures and standards that govern transfers of authority. There are now two major stages in the process. The first is approval by the Secretary of a state program for conservation and management.[163] At this stage, the Secretary is not required or authorized to determine the size or OSP range of marine mammal species or stocks or to waive the moratorium against taking. The Secretary must, however, find that the state program includes an adequate process for making these determinations. Management authority may be transferred to the state if the Secretary finds that this and other requirements are met. The transfer does not, however, authorize the state to permit taking of the species or stock to which the transfer applies. The state then must carry out the second stage in the transfer process.

[155]16 U.S.C. § 1539(e).

[156]The Fur Seal Act was comprehensively revised in 1983. The native exemption in the revised law is found at 16 U.S.C. § 1153.

[157]Act of Aug. 18, 1941, ch. 368, 55 Stat. 632 (subsequently omitted from 16 U.S.C. as obsolete).

[158]470 F. Supp. at 428.

[159]No. 77–0264 (D.D.C. Jan. 21, 1980).

[160]44 Fed. Reg. 45565 (1979).

[161]*Id.*

[162]Act of Oct. 9, 1981, Pub. L. No. 97–58, § 4, 95 Stat. 982–86 (codified at 16 U.S.C. § 1379).

[163]16 U.S.C. § 1379(b)(1).

In the second stage the state must determine, prior to permitting the taking of any species or stock, that (1) the species or stock is at its OSP, and (2) the number of animals that may be taken without reducing the species or stock below its OSP.[164] The process by which the state must make these determinations is similar to the adjudicatory hearing process that the Secretary must follow when promulgating regulations to implement a waiver of the moratorium. The state must present its evidence in support of the proposed determinations and subject the evidence to cross-examination.[165] Judicial review comparable to the review available in federal courts to challenge federal agency action must be available as part of the state process.[166] In the case of Alaska, the state also must adopt a statute and regulations ensuring that taking for subsistence purposes is the priority consumptive use of the species.[167]

Once the determinations made during the state process are final, the state may resume regulating taking of the species of stocks that were the subject of the process, but not for all purposes. For example, the Secretary retains authority to regulate, subject to state approval and except when carried out by or for the state itself, taking for scientific, public display, or survival enhancement purposes.[168] The Secretary retains similar authority to regulate taking of marine mammals incidental to commercial fishing and other activities in the U.S. exclusive economic zone (EEZ).[169] Where the range of a species or stock extends beyond the waters of the state into the EEZ, the Secretary and the state must enter into an agreement allocating the allowable take among the various permissible purposes. Highest priority must be given to subsistence uses, in the case of Alaska, and to taking within the EEZ.[170]

The 1981 amendments also authorized the Secretary to make grants to states to help them develop programs on which to base a request for return of management.[171] Prior to these amendments, the Secretary could make grants only to help states implement programs. This change and the general revision of section 109 in 1981 were intended to induce states to seek return of management authority. None has yet done so, however.

[164] *Id.* § 1379(b)(1)(C)(i).

[165] *Id.* § 1379(c)(2).

[166] *Id.* § 1379(c)(4).

[167] *Id.* § 1379(f)(1)(A)(ii). Section 109 contains its own special definition of "subsistence uses," taken nearly verbatim from the Alaska National Interest Lands Conservation Act. *Compare* 16 U.S.C. § 1379(f)(2) *with* 16 U.S.C. § 3113.

[168] 16 U.S.C. § 1379(b)(3)(B)(ii).

[169] *Id.* § 1379(b)(3)(B)(i).

[170] *Id.* § 1379(d)(1). The Secretary's regulation of taking for subsistence purposes, or hunting, in the EEZ must be consistent with state regulation, if a state so requests.

[171] *Id.* § 1379(j).

THE MMPA AND INTERNATIONAL RELATIONS

The 1972 MMPA included an ambitious international agenda. It called for "the convening of an international ministerial meeting on marine mammals before July 1, 1973, [to negotiate] a binding international convention for the protection and conservation of all marine mammals."[172] More than two decades later, no convention exists. In fact, no meeting has been held.

Less ambitious was a directive to "initiate negotiations as soon as possible for the development of bilateral or multilateral agreements with other nations for the protection and conservation of all marine mammals."[173] The envisioned agreements have not been reached. There has been progress, however, with respect to some marine mammals, most notably dolphins of the eastern tropical Pacific Ocean and commercial whaling throughout the world.

The MMPA that exists today is significantly different from the law Congress enacted in 1972. Those changes reflect the practical problems in implementing the MMPA's original objectives and the need to accommodate competing interests. These realities are evident in domestic implementation of the MMPA and no less so in the international context. As with the MMPA's elusive "immediate goal" of reducing marine mammal incidental mortality in commercial fisheries to rates approaching zero, perhaps it is best to regard the ambitious international objectives of the MMPA as goals toward which one can make progress but never fully achieve.

[172] *Id.* § 1378(a)(5).
[173] *Id.* § 1378(a)(1).

Chapter **6**

OCEAN FISH

Ocean fisheries are managed under the Fishery Conservation and Management Act of 1976 (also called the Magnuson Act).[1] The Act introduced approaches to planning not previously found in federal wildlife law.

This chapter first describes the Act's background and then examines the statutory framework and litigation regarding the Act's conservation measures. The chapter will focus on those issues most important to conservation of fisheries.[2]

BACKGROUND TO THE FISHERY CONSERVATION AND MANAGEMENT ACT

The right of all nations to fish on the high seas was a fundamental principle of international law by the beginning of the seventeenth century.[3] Disputes were frequent, however, as to whether specific areas constituted the "high seas" or the "territorial sea" of a particular coastal

[1]16 U.S.C. §§ 1801–1882. The Act is named for its prime sponsor, the late Senator Warren Magnuson (D-WA).

[2]An in-depth analysis of the Act's management aspects, procedural requirements, and enforcement provisions is beyond the scope of this book. A detailed and thorough source of information on the Act is Jon Jacobson *et al.*, Federal Fisheries Management: A Guidebook to the Magnuson Fishery Conservation and Management Act (1985 rev. ed. & Supps. 1987, 1989, & 1991). The 1996 amendments to the Act (see discussion *infra*) will likely lead to a revised edition of Jacobson's work. Another excellent overview can be found in Eldon V.C. Greenberg's chapter on Ocean Fisheries in Sustainable Environmental Law at 371–441 (Campbell-Mohn *et al.* eds., Envt'l L. Inst. 1993).

[3]*See* Sen. Rep. No. 94–416, 94th Cong., 1st Sess. at 5 (1975) (hereinafter Sen. Rep. No. 94–416).

nation.[4] Most disputes were localized matters involving neighboring states until the twentieth century. With the development of modern fishing vessels capable of traveling great distances and capturing enormous quantities of fish, the problem assumed greater international significance. Adding to the complexity of this problem was President Truman's declaration in 1945 that the United States would regard it as proper

> to establish conservation zones in those areas of the high seas contiguous to the coast of the United States wherein fishing activities have been or in the future may be developed and maintained on a substantial scale . . . and all fishing activities in such zones shall be subject to regulations and control.[5]

Truman apparently intended only to proclaim a policy of seeking negotiated international agreements with other nations fishing in nearby waters. Some interpreted his remarks, however, to mean that the United States recognized the validity of a coastal nation unilaterally establishing a special "conservation zone" between its territorial waters and the high seas, where it alone would regulate all fishing. Chile promptly responded by declaring for itself a conservation zone extending out 200 miles from its coast.[6]

The Truman proclamation had become an accepted tenet of international law by 1958. In that year, the International Conference on the Law of the Sea concluded a Convention on Fishing and Conservation of the Living Resources of the High Seas, which proclaimed that coastal nations have a "special interest in the maintenance of the productivity of the living resources in any area of the high seas adjacent to [their] territorial sea[s]."[7] The Convention further provided that, if efforts to reach a negotiated agreement with nations fishing in these adjacent areas are unsuccessful, the coastal nation may "adopt unilateral measures of conservation appropriate to any stock of fish or other marine resources."[8] The Convention did not specify the size of the adjacent area in which unilateral actions could be taken, nor did the simultaneously concluded Convention on the Territorial Sea and Contiguous Zone specify the limits of the territorial sea.[9]

Most nations claimed territorial seas of three miles and conservation zones of twelve miles. The United States did so with passage of the Bartlett Act, which prohibited any foreign vessel from fishing in the United States'

[4]*See* Chapman, *The Theory and Practice of International Fishery Development-Management,* 7 San Diego L. Rev. 408 (1970).

[5]Presidential Proclamation No. 2667, 59 Stat. 884, 10 Fed. Reg. 12303 (1945).

[6]Sen. Rep. No. 94–416, *supra* note 3, at 7.

[7]April 29, 1958, [1966] 17 U.S.T. 138, T.I.A.S. No. 5969, art. 6, § 1.

[8]*Id.* art. 7, § 1.

[9]April 29, 1958, [1964] 15 U.S.T. 1606, T.I.A.S. No. 5639.

territorial sea or contiguous fisheries zone, or from taking any sedentary living resources from the continental shelf unless specifically permitted by the Secretary of the Treasury or pursuant to international agreement.[10]

Reliance upon international agreements to prevent the overexploitation of the oceans' living resources was established as a basic tenet of United States policy well before the Truman proclamation. The 1911 agreement pertaining to North Pacific fur seals and the 1931 convention on whaling were among the earliest expressions of this policy.[11] The United States concluded bilateral agreements with Canada pertaining to the Pacific halibut fishery and sockeye salmon of the Fraser River system in 1923 and 1930 respectively.[12] The principal regulatory measure of the former is the establishment of an overall quota on the annual take of halibut. The latter established a quota and attempted to allocate it equally between the two nations.

By the mid-1970s, the United States was a party to some twenty international fishing agreements.[13] The regulatory measures of these agreements varied greatly. Some, like the Convention for the Establishment of an Inter-American Tropical Tuna Commission, merely authorized an international commission to make nonbinding recommendations to member governments.[14] Others, like the International Convention for the High Seas Fisheries of the North Pacific Ocean, obligated particular nations to abstain from taking certain types of fish in particular areas.[15] Still others, like the International Convention for the Northwest Atlantic Fisheries, attempted to allocate the aggregate annual take of multiple species among more than a dozen participating nations.[16]

It was clear by 1970 that at least some of the international fishery agreements were failing in their principal aim of conserving fish. A variety of factors contributed to this failure. Agreements usually left enforcement to each signatory nation, with the result that agreements seldom were vig-

[10]16 U.S.C. §§ 1081–1086 and 1091–1094 (1976) (repealed as of March 1, 1977).

[11]See Chapter 14 for a discussion of these treaties.

[12]The current version of the halibut agreement is known as the Convention for the Preservation of the Halibut Fishery of the Northern Pacific Ocean and Bering Sea, March 2, 1953, United States-Canada, 5 U.S.T. 5, T.I.A.S. No. 2900. It is implemented through the North Pacific Halibut Act of 1982, 16 U.S.C. §§ 773–773k. The agreement pertaining to salmon of the Fraser River was replaced by the Pacific Salmon Treaty, implemented by the Pacific Salmon Treaty Act of 1985, Pub. L. No. 99–5, 99 Stat. 7.

[13]For a listing, see Sen. Rep. No. 94–416, *supra* note 3, app. I.

[14]May 31, 1949, 1 U.S.T. 230, T.I.A.S. No.2044. The Convention is implemented by the Tuna Convention Act of 1950, 16 U.S.C. §§ 951–961.

[15]May 9, 1952, United States-Canada-Japan, 4 U.S.T. 380, T.I.A.S. No. 2786. In 1992, the United States signed a new treaty governing North Pacific fisheries with Canada, Japan, and the Russian Federation. Convention for the Conservation of Anadromous Stocks in the North Pacific Ocean, Feb. 11, 1992, S. Treaty Doc. No. 102–30, 102d Cong., 2d sess. at art. V–VII.

[16]Feb. 8, 1949, [1950] 1 U.S.T. 477, T.I.A.S. No. 2089.

orously enforced.[17] There was no means of enforcement whatever against nations that chose not to join the agreement. Few international agreements provided a mechanism for limiting entry into the fishery, encouraging overcapitalization and thus political pressure for larger harvests than could be scientifically justified. Finally, often the scientific data base was inadequate to assure that a given level of taking would not deplete the resource. This last problem was particularly acute when the fishery involved several interrelated stocks or species of fish.[18]

Many nations began to recognize that international agreements had failed to conserve fish. A growing number of nations, particularly those with valuable fishing areas in nearby waters, began to assert ever broader claims to exclusive fishery zones. This trend, together with other emerging disputes concerning development of the ocean's resources, prompted the United Nations to convene the Law of the Sea Conference in 1974 in an effort to resolve the many issues left unsettled by the 1958 Conventions.

At the 1974 Conference and its subsequent session in 1975, a consensus appeared to emerge in favor of a uniform 12-mile territorial sea and 200-mile resource conservation zone for all coastal nations. The United States somewhat reluctantly joined in this view, provided that it be coupled with agreement on a number of other vigorously disputed issues.[19] As the Conference headed for its third session in 1976, it seemed doubtful that an overall agreement could be reached. Moreover, by that time thirty-six coastal nations had declared exclusive fishing zones beyond twelve miles.[20] Meanwhile, the foreign fishing fleets just outside the twelve-mile fisheries zone of the United States grew larger, and so did their harvest of fish.[21]

Weary of the slow pace of international negotiation and alarmed at the impact on the domestic fishing industry of the growing foreign fleets, Congress passed the Fishery Conservation and Management Act in the spring of 1976. Opponents of the Act had reservations about the wisdom of unilateral action by the United States and preferred to wait for possible agreement at the third session of the United Nations Law of the Sea Conference, which was just opening.[22] It would take another six years, however, for the Conference finally to adopt a treaty providing for coastal nation management of resources in the 200-mile zone.[23]

[17]Jacobson, *supra* note 2, at 7.

[18]Each of these problems is discussed in Chapman, *supra* note 4.

[19]*See* Sen. Rep. No. 94–416, *supra* note 3, at 8–9.

[20]*Id.*, app. II.

[21]*See* H.R. Rep. No. 94–445, 94th Cong., 1st Sess. at 34–41, *reprinted in* 1976 U.S.C.C.A.N. at 606–609 (hereinafter House Report). The House Report noted that between 1950 and 1969 "the world production of fish multiplied about threefold, from 20 million metric tons to about 63 million metric tons; *yet the U.S. share of the catch . . . remained at a relatively fixed level of between 2.0 and 2.2 million tons.*" *Id.* at 35 (emphasis in original).

[22]Jacobson, *supra* note 2, at 7–8.

[23]United Nations Convention on the Law of the Sea, Oct. 21, 1982, U.N. Doc. A/

ESTABLISHMENT OF UNITED STATES MANAGEMENT AUTHORITY

The foundation of the Fishery Conservation and Management Act is its establishment of a 197–mile-wide zone contiguous to the 3–mile-wide territorial sea of the United States, originally called the "fishery conservation zone" and now called the "exclusive economic zone" or "EEZ."[24] Within that zone, the Act asserts for the United States exclusive management authority, not merely over all fish, but also over "mollusks, crustaceans, and all other forms of marine animal and plant life other than marine mammals and birds."[25] In addition, at all places outside the zone, except the recognized territorial sea or fishery conservation zone/EEZ of a foreign nation, the Act claims for the United States the same exclusive authority over anadromous species of fish that spawn in its fresh or estuarine waters.[26] Finally, wherever the continental shelf extends beyond the EEZ, the Act claims for the United States exclusive management authority over certain sedentary species found there.[27]

Prior to the 1990 amendments to the Act, the law excluded "highly migratory species" of fish from United States management authority in the fishery conservation zone/EEZ.[28] "Highly migratory species" were defined as those "species of *tuna* which, in the course of their life cycle,

---

CONF.62/122 (1981), 21 I.L.M. 1261 (1982) (*opened for signature* Dec. 10, 1982, entered into force, 1994). For over a decade the United States Senate did not even consider ratification of the treaty, primarily because of objections by the executive branch to its provisions about mining the deep seabed. In October 1994, President Bill Clinton sent the Convention to the United States Senate requesting its advice and consent to U.S. accession. For a discussion of the relationship between the Magnuson Act and the Convention, should the United States become a party, see William T. Burke, *Implications for Fisheries Management of U.S. Acceptance of the 1982 Convention on the Law of the Sea*, 89 Am. J. Int'l L. 792 (1995). *See also* Christopher C. Joyner, *Ocean Fisheries, U.S. Interests, and the 1982 Law of the Sea Convention*, 7 Geo. Int'l Envt'l L. Rev. 749 (1995).

[24]16 U.S.C §§ 1802(11), 1811. Amendments to the Act in 1986 redesignated the fishery conservation zone as the EEZ. Act of Nov. 14, 1986, Pub. L. No. 99–659, § 101(a), 100 Stat. 3706.

[25]16 U.S.C. §§ 1802(12), 1811.

[26]*Id.* § 1811(b)(1). Anadromous fish hatch in fresh water, migrate to the ocean to mature, and return as adults to spawn in their natal waters. The rationale that supports special treatment for anadromous fish is that the "host state" in which they spawn expends substantial sums to enhance their populations, thus giving it paramount rights to the fish wherever they may be found. *See* Sen. Rep. No. 94–416, *supra* note 3, at 24. Anadromous fish in the North Pacific are governed by the Convention for the Conservation of Anadromous Stocks in the North Pacific Ocean, *supra* note 15.

[27]16 U.S.C. §§ 1802(7), 1811(b)(2). The Act expressly identifies certain corals, crabs, mollusks, and sponges as "Continental Shelf fishery resources" and authorizes the Secretary of Commerce to add other species to that list if they are sedentary at their harvestable stage. *Id.* § 1802(7).

[28]*Id.* § 1813 (1985) (amended 1990).

spawn and migrate over great distances in waters of the ocean.''[29] Thus, the United States did not claim management authority over tuna and did not recognize management authority of other coastal nations over tuna beyond their territorial seas.[30] The tuna exception was carved out for the domestic tuna industry, which depends in large part upon access to waters within 200 miles of several Latin American nations.[31]

Both foreign fishing for tuna, and the value of tuna caught by United States vessels in their own EEZ, had increased considerably by 1990. It had become clear that more active management of tuna was necessary.[32] The Fishery Conservation Amendments of 1990[33] brought tuna in the EEZ within the jurisdiction of the United States, as of January 1, 1992.[34] The amendments changed the definition of ''highly migratory species'' to include other pelagic species as well as tuna,[35] and provided for international cooperation in managing highly migratory species.[36]

THE DEVELOPMENT AND IMPLEMENTATION OF FISHERY MANAGEMENT PLANS

Probably more important than the fact of claiming exclusive management authority over species within and beyond the EEZ is the Act's requirement that they be managed in accordance with comprehensive plans to be drawn up by ''Regional Fishery Management Councils.'' The Councils are composed of both state and federal officials. In prescribing mechanisms for developing these plans and the substantive standards the plans must meet, the Act broke significant new ground in federal wildlife law.

[29] *Id.* § 1802(14) (1985) (amended 1990) (emphasis added).

[30] *See* Sen. Rep. No. 101–414, 101st Cong., 2d Sess. at 4 (1990), *reprinted in* 1990 U.S.C.C.A.N. 6279. The tuna exception did not apply to other pelagic species, including marlin, oceanic sharks, sailfishes, and swordfish. The United States asserted jurisdiction over these species in the EEZ.

[31] *See* H.R. Rep. No. 94–455, *supra* note 21, at 42–43. *See also* Comment, *The 200-Mile Exclusive Economic Zone: Death Knell for the American Tuna Industry*, 13 San Diego L. Rev. 707 (1976).

[32] *See* Sen. Rep. No. 101–414, *supra* note 30, at 3–5, and H.R. Rep. No. 101–393, 101st Cong., 1st Sess. (1989).

[33] Pub. L. No. 101–627, 104 Stat. 4436 (1990).

[34] *See* 16 U.S.C. §§ 1811, 1812.

[35] *Id.* § 1802(20). This change conforms the definition to the broader one in the 1982 United Nations Convention on the Law of the Sea, *supra* note 23. The change did not alter previously-asserted United States jurisdiction over non-tuna pelagic species.

[36] Congress recognized that ''[d]ue to the nomadic nature of the fish, efforts to conserve and manage highly migratory species require a high degree of international cooperation and coordination if they are to be effective.'' Sen. Rep. No. 101–414, *supra* note 30, at 3. Section 103(a) of the amendments provides that the United States ''shall cooperate'' with other nations involved in fishing for these species ''with a view to ensuring conservation'' of such species. 16 U.S.C. § 1812. For further discussion of these amendments, see Jacobson, *supra* note 2, at 13–17 (1991 Supp.).

Primary responsibility for developing management plans rests with the eight Regional Councils.[37] Each Council has authority over the fisheries seaward of the states comprising it.[38] The voting members of each Council include the principal official with marine fishery management responsibility from each state in the region, the regional director of the National Marine Fisheries Service for the geographic area concerned, and from four to twelve persons appointed by the Secretary of Commerce from lists of qualified individuals submitted by governors of the region's states.[39]

Each Council is responsible for preparing and submitting to the Secretary a fishery management plan with respect to each fishery "under its authority that requires conservation and management."[40] In preparing plans, the Councils must conduct public hearings in the area concerned "so as to allow all interested persons an opportunity to be heard."[41]

[37]16 U.S.C. § 1852.

[38]Certain coastal states are represented on more than one Council. For example, Florida is included in both the South Atlantic and the Gulf Councils, which have jurisdiction over fisheries in the Atlantic Ocean and the Gulf of Mexico, respectively. The "states" represented include the Virgin Islands, Puerto Rico, Guam, and American Samoa.

[39]The Act defines qualified individuals as those who "by reason of their occupational or other experience, scientific expertise, or training, are knowledgeable regarding the conservation and management, or the commercial or recreational harvest, of the fishery resources of the geographical area concerned." *Id.* § 1852(b)(2)(A). Amendments to the Act in 1996 added to the Pacific Council "one representative of an Indian tribe with Federally recognized fishing rights. . . ." Sustainable Fisheries Act, Pub. L. No. 104–297, § 107(b)(3) (1996) (to be codified at 16 U.S.C. § 1852(b)(5)(A)) (hereinafter Pub. L. No. 104–297). Each Council also includes, as nonvoting members, one representative each from the Fish and Wildlife Service, the Coast Guard, the Department of State, and the appropriate Marine Fisheries Commission, if any. 16 U.S.C. § 1852(c)(1). The Pacific Council has an additional nonvoting member appointed by the Governor of Alaska, *id.* § 1852(c)(2).

[40]*Id.* § 1852(h)(1). The Act defines a "fishery" as "one or more stocks of fish which can be treated as a unit for purposes of conservation and management and which are identified on the basis of geographical, scientific, technical, recreational, and economic characteristics." *Id.* § 1802(13); *see also* 50 C.F.R. § 602.13 (1995) (National Standard 3—Management Units). The fishery can be a single species or several, and it may be limited to certain ranges or harvest methods. For example, the Pacific Council manages five species of salmon under a single management plan. Jacobson, *supra* note 2, at 154 n.7. If any fishery overlaps the geographical areas of authority of two or more Councils, the Secretary may designate one Council to prepare the plan or require that it be prepared jointly. 16 U.S.C. § 1854(f)(1). Regulations implementing plans are found in 50 C.F.R., Parts 625 *et seq.* (1995).

[41]16 U.S.C. § 1852(h)(3). It is unclear whether procedural irregularities affecting public participation in plan development are a sufficient basis for overturning regulations implementing the plan. In Washington Trollers Association v. Kreps, 645 F.2d 684 (9th Cir. 1981), a divided court of appeals held that although the Act requires only that a summary of information relied upon appear in the plan, the provisions of the Act encouraging public participation in the development of plans require that information relied upon "must be reasonably available to the interested public." 645 F.2d at 686. Since there was disagreement as to whether the information was available to the public, the court concluded that there was an issue of fact to be resolved at trial, rendering the district court's award of summary judgment inappropriate. *But see* Alaska Factory Trawlers Ass'n v. Baldridge, 831 F.2d 1456,

The Secretary has final authority to approve or disapprove a management plan.[42] The Secretary must review the plan to determine if it is consistent with the national standards, the Act's other provisions, and other applicable laws.[43] If a Council fails altogether to submit a plan "after a reasonable period of time" for any fishery that "requires conservation and management," or to modify a draft plan disapproved by the Secretary, the Secretary may prepare a plan.[44] When the Secretary prepares a plan, it must be submitted to the appropriate Council for consideration and comment.[45]

A Council must submit to the Secretary, simultaneously with a management plan, proposed regulations to implement the plan.[46] The Secretary is responsible for promulgating final regulations.[47]

In an emergency, the Secretary may promulgate regulations necessary to address the emergency, "without regard to whether a fishery manage-

---

1460 (9th Cir. 1987), a later case in the same circuit, in which plaintiffs challenged the Secretary's rules implementing the Gulf of Alaska groundfish management plan because during development of the plan, the Council discussed it at closed mealtime meetings. The court stated that it could overturn the Secretary's decision only if plaintiffs showed that the alleged irregularities affected the Secretary's decision, a showing not required in Washington Trollers.

[42]16 U.S.C. § 1854(a). This power has been administratively delegated to the appropriate Regional Director of the National Marine Fisheries Service. Jacobson, *supra* note 2, at 92. The Act establishes a complex process for review, comment, and approval of plans. For procedures and deadlines, see 16 U.S.C. § 1854(a).

[43]16 U.S.C. § 1854(a)(1)(A). Applicable laws include the Coastal Zone Management Act, Endangered Species Act, Marine Mammal Protection Act, National Environmental Policy Act, Marine Protection, Research and Sanctuaries Act, and Administrative Procedure Act, among others. Courts have held that applicable laws also include Indian treaties. *See, e.g.*, Parravano v. Babbitt, 70 F.3d 539 (9th Cir. 1995); Washington State Charterboat Ass'n v. Baldridge, 702 F.2d 820, 823 (9th Cir. 1983), *cert. denied*, 464 U.S. 1053 (1984). Courts have deferred to the Secretary's discretion in determining whether to prepare an environmental impact statement pursuant to the National Environmental Policy Act. *See* Greenpeace Action v. Franklin, 982 F.2d 1342 (9th Cir. 1992); Northwest Environmental Defense Center v. Brennen, 958 F.2d 930 (9th Cir. 1992). *See* Jacobson, *supra* note 2, at 53 (1985) and 12 (1991 Supp.) for a discussion of consistency between fishery management plans and the Coastal Zone Management Act.

[44]16 U.S.C. § 1854(c)(1). In Conservation Law Foundation v. Franklin, 989 F.2d 54 (1st Cir. 1992), fishing groups challenged a consent decree establishing a timetable for amending a management plan that had been negotiated by conservation groups and the Secretary, on the ground that the Secretary had failed to wait for the Council to propose an amendment. The court rejected the challenge, finding that to require the Secretary to wait until the Council acted would "hold the Secretary hostage to the Councils." *Id.* at 60. The court held that "what constitutes a 'reasonable time' under the statute is solely within the Secretary's discretion," and that "the Secretary may generate her own revisions to Council-generated plans, if the council fails to revise after a reasonable time." *Id.* at 61.

[45]16 U.S.C. § 1854(c)(4)(A).

[46]*Id.* § 1853(c).

[47]*Id.* § 1854(b).

ment plan exists for such fishery."[48] The Secretary *must* promulgate emergency regulations if a Council unanimously requests this action.[49] Emergency regulations are subject to judicial review on the same basis as other regulations.[50]

The 1996 amendments to the Act expanded the Secretary's ability to respond rapidly to overfishing. The Sustainable Fisheries Act authorized the Secretary to promulgate "interim measures" to address overfishing using the same procedures that are established for promulgating emergency regulations.[51] The amendments defined "overfishing" as "a rate or level of fishing mortality that jeopardizes the capacity of a fishery to produce the maximum sustainable yield on a continuing basis."[52]

CONTENT OF FISHERY MANAGEMENT PLANS: REQUIRED PROVISIONS

All management plans, whether prepared by the Secretary or a Council, must incorporate conservation and management measures "to prevent overfishing and rebuild overfished stocks, and to protect, restore and promote the long-term health and stability of the fishery."[53] The measures must comply with ten national standards set forth in the Act. The Act directs the Secretary to establish advisory guidelines based on these standards to assist in plan development.[54]

[48] *Id.* § 1855(c)(1). An emergency regulation which changes any existing management plan may remain in effect for not more than 180 days. This period may be extended for no more than another 180 days after publication of a notice in the Federal Register if the public has had an opportunity to comment on the emergency regulation. *Id.* § 1855(c)(3)(B), as amended by Pub. L. No. 104–297, § 110(b).

[49] 16 U.S.C. § 1855(c)(2)(A). If the Council's request is not by unanimous vote, the Secretary *may* promulgate emergency regulations.

[50] *See id.* § 1855(f)(1). *See also* S. Conf. Rep. No. 94–711, 94th Cong., 2d Sess. at 55 (1976), *reprinted in* 1976 U.S.C.C.A.N. at 678. In Southeastern Fisheries Association v. Mosbacher, 773 F. Supp. 435 (D.D.C. 1991), fishing interests challenged the Secretary's emergency closure of the redfish fishery in the Gulf of Mexico pending a 20% escapement of juvenile fish from inshore waters. The fishery had grown dramatically in the 1980s, when cajun cooking became popular and diners sought "blackened redfish"; in 1980, less than 50,000 pounds of redfish were harvested in the EEZ, while in 1986, the EEZ catch was 8.1 million pounds. *Id.* at 437. The court upheld the Secretary's action.

[51] Pub. L. 104–297, § 110(b) (to be codified at 16 U.S.C. § 1855(c)).

[52] Pub. L. No. 104–297, § 102(8) (to be codified at 16 U.S.C. § 1802(29)). The amendments also set forth planning and reporting requirements to address overfishing. See further discussion of overfishing and the planning and reporting requirements *infra* at text accompanying notes 57–73.

[53] 16 U.S.C. § 1853(a)(1)(A). The Act defines "conservation and management" to include all measures "which are required to rebuild, restore, or maintain, and which are useful in rebuilding, restoring, or maintaining, any fishery resource and the marine environment." *Id.* § 1802(5).

[54] *Id.* § 1851(b). This section expressly provides that the guidelines do not have the force and effect of law. Guidelines are found at 50 C.F.R. §§ 602.10–602.17 (1995).

There were seven national standards prior to the 1996 amendments. They required that conservation and management measures: (1) prevent overfishing and assure an optimum yield from each fishery, (2) be based on the best scientific information available, (3) provide for the management of individual or interrelated stocks as a unit, (4) not discriminate between residents of different states, (5) consider efficiency, (6) allow for contingencies, and (7) minimize costs.[55]

Congress added three new national standards in 1996. Conservation and management measures now must also (8) minimize adverse economic impacts on fishing communities, (9) minimize bycatch, and (10) promote the safety of human life at sea.[56] In 1996, Congress also elaborated further on plan requirements for preventing overfishing and minimizing bycatch, as well as conserving essential fish habitat. These requirements are discussed in the following sections.

Measures to Prevent Overfishing

The most important of the national standards is the first. The Senate report accompanying the original Act described the overfishing prohibition as "the most basic objective of fishery management."[57] Twenty years later, Congress was still struggling with how best to prevent overfishing and rebuild depleted stocks.

As discussed above, in 1996 Congress defined "overfishing" as "a rate or level of fishing mortality that jeopardizes the capacity of a fishery to produce the maximum sustainable yield on a continuing basis."[58] Each management plan must "assess and specify the present and probable future condition of, and the maximum sustainable yield and optimum yield from, the fishery."[59] "Maximum sustainable yield," or MSY, is "the largest average annual catch or yield that can be taken over a significant period of time from each stock under prevailing ecological and environmental conditions."[60]

The second element of Standard 1, "optimum yield," was broadly defined in the statute prior to 1996:

The term "optimum," with respect to the yield from a fishery, means the amount of fish

[55] *Id.* §§ 1851(a)(1)-(7).

[56] Pub. L. No. 104–297, § 106(b) (to be codified at 16 U.S.C. §§ 1851(a)(8),(9), and (10)).

[57] Sen. Rep. No. 94–416, *supra* note 3, at 31.

[58] Pub. L. No. 104–297, § 102(8) (to be codified at 16 U.S.C. § 1802(29)). The new statutory definition is similar to the previous administrative definition. *See* 50 C.F.R. § 602.11(c)(1) (1995).

[59] 16 U.S.C. § 1853(a)(3).

[60] 50 C.F.R. § 602.11(d)(1) (1995).

(A) which will provide the greatest overall benefit to the Nation, with particular reference to food production and recreational opportunities; [and]
(B) which is prescribed as such on the basis of the maximum sustainable yield from such fishery, as modified by any relevant economic, social or ecological factor.[61]

This definition vested the Councils and the Secretary with a great deal of discretion in making this crucial determination. The traditional fisheries management standard of "maximum sustainable yield" (MSY) was intended to serve only as the point of departure for a determination of optimum yield.[62] How far the Councils or the Secretary chose to depart from that standard was probably beyond any effective judicial review.[63]

The determination of optimum yield, however, is critical. It establishes not only the maximum amount of harvest on a continuing basis,[64] but also

[61]16 U.S.C. § 1802(21) (1994) (amended 1996). See 50 C.F.R. § 602.11 (1995) for guidelines on determination of optimum yield.

[62]50 C.F.R. § 602.11(d)(3) (1995). The House Report, *supra* note 21, at 47–48, contains the following useful discussion of "optimum sustainable yield," a term that was defined in essentially the same manner as the term "optimum yield" in the Act prior to the 1996 amendments:

> Once the MSY of the fisheries or stock has been determined with reasonable scientific accuracy, and the same determination made with respect to the total biomass of an ocean area where many different, but inter-related fisheries occur, the developer of a management plan can begin to think in terms of the optimum sustainable yield (OSY). Thus while biologists in the past have tended to regard any unused surplus of a fishery as a waste, the resource manager may well determine that a surplus harvest below MSY will ultimately enhance not only the specific stock under management, but also the entire biomass. Conversely, the fisheries manager may determine that the surplus harvest of the entire biomass must be reduced substantially below MSY, in order to restore a valuable depleted stock which is taken incidentally to the harvesting of other species in this biomass. . . .
>
> The preceding concepts relate to the biological well-being of the fishery. The concept of optimum sustainable yield is, however, broader than the consideration of the fish stocks and takes into account the economic well-being of the commercial fishermen, the interests of recreational fishermen, and the welfare of the nation and its consumers. The optimum sustainable yield of any given fishery or region will be a carefully defined deviation from MSY in order to respond to the unique problems of that fishery or region. It cannot be defined absolutely for all stocks of fish or groups of fishermen, and will require careful monitoring by the Regional Marine Fisheries Councils and the Secretary of Commerce.

[63]*See, e.g.,* Maine v. Kreps, 563 F.2d 1043 (1st Cir. 1977), and Northwest Environmental Defense Center v. Brennen, 958 F.2d 930 (9th Cir. 1992), discussed at text accompanying notes 155–169 *infra.*

[64]Quotas may exceed optimum yield in any given year. *See* 50 C.F.R. § 602.11(g)(1)(1995) ("The specification of [optimum yield] in an FMP [fishery management plan] is not automatically a quota or ceiling, although quotas may be derived from the [optimum yield] where appropriate.") and 50 C.F.R. § 602.11(g)(2) (1995):

> Exceeding [optimum yield] does not necessarily constitute overfishing, although they might coincide. Even if no overfishing resulted, continual harvest at a level above a

the permissible level of foreign participation in the fishery. Each management plan must specify "the capacity and the extent to which fishing vessels of the United States, on an annual basis, will harvest the optimum yield."[65] Only that portion of the optimum yield that fishing vessels of the United States will not harvest can be made available to foreign fishers.[66] In addition, Councils may limit participation of United States fishing vessels, based on optimum yield, in order to prevent overfishing.[67]

In the 1996 amendments Congress elaborated upon the definition of optimum yield. Subparagraph A was revised to add "taking into account the protection of marine ecosystems."[68] A new subparagraph (C) provides that "in the case of an overfished fishery," the optimum yield is one that "provides for rebuilding to a level consistent with producing the maximum sustainable yield."[69] Perhaps most significant, subparagraph B was changed to provide that optimum yield is the amount of fish prescribed on the basis of the maximum sustainable yield "as reduced" by any relevant social, economic, or ecological factor, rather than "as modified" by any of the three factors.[70] This last change places a new ceiling on optimum yield in the case of an overfished fishery: it must be no higher than MSY. This change clearly puts a new limit on the Councils' and Secretary's discretion. The practical significance of the limit depends upon how often optimum yield previously had been set above MSY for fisheries that would be identified as "overfished" under the new standards.

These changes regarding overfishing suggest that Councils and the Secretary are to pay more attention to ecological factors in determining optimum yield and are to lower optimum yield more readily to prevent overfishing and to rebuild overfished stocks. Combined with other new requirements regarding overfishing, the revised definition may increase the Act's effectiveness in addressing this core issue. In particular, Councils are required in their management plans to

> specify objective and measurable criteria for identifying when the fishery to which the plan applies is overfished (with an analysis of how the criteria were determined

---

fixed value [optimum yield] would violate national standard 1 because [optimum yield] was exceeded (not achieved) on a continuing basis.

[65]16 U.S.C. § 1853(a)(4)(A).

[66]*Id.* § 1853(a)(4)(B).

[67]*Id.* § 1853(b)(6). The 1996 amendments imposed a moratorium on new limited access systems. See discussion of measures limiting access to a fishery *infra* at text accompanying notes 115–131.

[68]Pub. L. No. 104–297, § 102(7) (to be codified at 16 U.S.C. § 1802(28)). One commentator characterized this change as a "welcome but hard to implement criterion." David Fluharty, *Magnuson Fishery Management and Conservation Act Reauthorization and Fishery Management Needs in the North Pacific Region,* 9 Tul. Envt'l L.J. 301, 322 (1996).

[69]Pub. L. No. 104–297, § 102(7) (to be codified at 16 U.S.C. § 1802(28)).

[70]*Id.*

and the relationship of the criteria to the reproductive potential of stocks of fish in that fishery).[71]

Another new section of the Act requires the Secretary to report annually to Congress and to each Council on the status of fisheries and to identify fisheries that are overfished or approaching being overfished, using the Council's criteria.[72] For fisheries identified as overfished or approaching overfishing, the law now sets a timetable and establishes procedures for revising management plans and regulations to incorporate measures designed to prevent overfishing and to rebuild overfished fisheries.[73]

Measures to Minimize Bycatch

Concern about the incidental catch of nontarget species, or "bycatch," in domestic and international fisheries "has grown dramatically in recent years."[74] More than 27 million metric tons of nontargeted fish are discarded as waste each year.[75] Bycatch is a particular problem with certain fishing methods, such as trawling.[76] Incidental catch is not limited to fish;

[71] *Id.* § 108(a)(7) (to be codified at 16 U.S.C. § 1853(a)(10)).

[72] *Id.* § 109(e) (to be codified at 16 U.S.C. § 1854(e)). The statute provides that a fishery "shall be classified as approaching a condition of being overfished if . . . the Secretary estimates that [it] will become overfished within two years." *Id.* For a fishery that is overfished, a fishery management plan must

> (A) specify a time period for ending overfishing and rebuilding the fishery that shall—
> (i) be as short as possible, taking into account the status and biology of any overfished stocks of fish, the needs of fishing communities, recommendations by international organizations in which the United States participates, and the interaction of the overfished stock within the marine ecosystem; and
> (ii) not exceed 10 years, except in cases where the biology of the stock of fish, other environmental conditions, or management measures under an international agreement in which the United States participates dictate otherwise.

Id. (to be codified at 16 U.S.C § 1854(e)(4)).

[73] *Id.* Again, the practical impact of this requirement depends upon the degree to which Councils already were adopting these measures. For example, prior to the amendments the North Pacific Council had defined overfishing levels and had adopted rebuilding plans for overfished stocks. *See* Fluharty, *supra* note 68, at 320.

[74] U.S. Dept. of Commerce, *Our Living Oceans: Report on the Status of U.S. Living Marine Resources* 29 (1992) (hereinafter *Our Living Oceans*). Bycatch are fish and other animals that are accidentally caught by fishing vessels that are targeting other species. *Id.* For a discussion of bycatch in high seas driftnet fishing, see discussion at text accompanying notes 242–258, *infra.*

[75] Food and Agriculture Organization of the United Nations, A Global Assessment of Fisheries Bycatch and Discards 19 (1995).

[76] *See, e.g.,* Peter Weber, *Abandoned Seas: Reversing the Decline of the Oceans* 35 (Worldwatch Paper No. 116, 1993); Eugene C. Bricklemyer, Jr. *et al., Discarded Catch in U.S. Commercial Marine Fisheries,* in Audubon Wildlife Report 1989/90 at 260–61 (William J. Chandler ed., 1989).

large numbers of marine mammals, seabirds, and sea turtles also are caught in fishing nets.[77]

The Fishery Conservation and Management Act contained no provisions specifically addressing bycatch in the U.S. EEZ until 1990. The 1990 fishery amendments declared it congressional policy to assure that the fishery conservation and management program "considers the effects of fishing on immature fish and encourages development of practical measures that avoid unnecessary waste of fish."[78] The amendments stopped far short of requiring management plans to adopt measures to reduce bycatch.

In fact, most of the bycatch provisions in the 1990 amendments were in response to requirements that shrimpers use devices to reduce the catch of threatened and endangered sea turtles.[79] Rather than require use of similar devices to reduce bycatch of fish, the amendments provided for a three-year research program to assess the impact on fishery resources of incidental harvest by the shrimp trawl industry in the Gulf of Mexico and South Atlantic.[80] Congress also directed the Secretary to design a program to "evaluate the efficacy of technological devices and other changes in fishing technology" in reducing bycatch in the shrimp trawl industry.[81] During the three-year research period, the Secretary could not restrict shrimp harvesting or require the use of any technological device to reduce incidental catch.[82] This prohibition was to expire on January 1, 1994.[83]

Conservation groups and others strongly urged Congress to address the bycatch issue in the 1996 reauthorization:

> The Marine Fish Conservation Network advocates that the Magnuson Act . . . should include a definition of undesirable bycatch, establish a new national standard to minimize the negative impact of bycatch on fish populations and the marine ecosystem, and require managers to include provisions to reduce the incidental capture of nontargeted species in fishery management plans.[84]

[77]*See, e.g.*, Bricklemyer, *supra* note 76, at 261.

[78]16 U.S.C. § 1801(c)(3) (1994) (amended 1996). The 1990 amendments made no reference to incidental catch of non-fish species.

[79]See Chapter 7 at text accompanying notes 163–164 for a discussion of measures to protect sea turtles under the Endangered Species Act.

[80]16 U.S.C. § 1854(g) (1994) (amended 1996).

[81]*Id.* § 1854(g)(4) (1994) (amended 1996).

[82]*Id.* § 1854(g)(6)(A) (1990) (amended 1996).

[83]*Id.* § 1854(g)(6)(B) (1994) (amended 1996). Despite this Congressional restriction, the management plan for the Gulf of Mexico shrimp fishery prescribes measures such as the regulation of trawler mesh size to prevent accidental catch of shrimp below a certain size, and the use of new trawler gear to reduce the incidental catch of finfish. Jacobson, *supra* note 2, at 32 (1991 Supp.) (citing Gulf of Mexico Fishery Management Council, Fishery Management Plan for the Shrimp Fishery of the Gulf of Mexico, United States Waters 8–19 (1981)).

[84]Suzanne Iudicello, Scott Burns, and Andrea Oliver, *Putting Conservation into the Fishery Conservation and Management Act: The Public Interest in Magnuson Reauthorization*, 9 Tul. Envt'l L.J. 339, 343 (1996).

Congress adopted these proposals in large part.

First, Congress set forth the following definition of "bycatch":

> [F]ish which are harvested in a fishery, but which are not sold or kept for personal use, and includes economic discards and regulatory discards. Such term does not include fish released alive under a recreational catch and release fishery management program.[85]

Congress then adopted a new national standard relating to bycatch:

> (9) Conservation and management measures shall, to the extent practicable,
> (A) minimize bycatch and
> (B) to the extent bycatch cannot be avoided, minimize the mortality of such bycatch.[86]

To meet the new national standard, management plans must

> establish a standardized reporting methodology to assess the amount and type of bycatch occurring in the fishery, and include conservation and management measures that, to the extent practicable and in the following priority—
> (A) minimize bycatch; and
> (B) minimize the mortality of bycatch which cannot be avoided.[87]

Management plans may, but are not required to, include harvest incentives for fishers to "employ fishing practices that result in lower levels of bycatch or in lower levels of the mortality of bycatch."[88]

Additional provisions are directed at specific fisheries. The North Pacific Council is directed to reduce the total amount of bycatch occurring in the North Pacific and authorized to use fees and nontransferable an-

[85]Pub. L. No. 104–297, § 102(1) (to be codified at 16 U.S.C. § 1802(2)). The Act defines "economic discards" as "fish which are the target of a fishery, but which are not retained because they are an undesirable size, sex, or quality, or for other economic reasons." *Id.* § 102(3) (to be codified at 16 U.S.C. § 1802(9)). "Regulatory discards" are "fish harvested in a fishery which fishermen are required by regulation to discard whenever caught, or are required by regulation to retain but not sell." *Id.* § 102(9) (to be codified at 16 U.S.C. § 1802(33)).

[86]*Id.* § 106(b) (to be codified at 16 U.S.C. § 1851(a)(9)).

[87]*Id.* § 108(a)(7) (to be codified at 16 U.S.C. § 1853(a)(11)). Plans also must assess the type and amount of fish caught and released alive in recreational catch and release programs and include measures to reduce the mortality of these fish. *Id.* (to be codified at 16 U.S.C. § 1853(a)(12)). Plans must be amended to meet these new requirements within 24 months after the date of enactment of the law. *Id.* § 108(b).

[88]*Id.* § 108(c)(7) to be codified at 16 U.S.C. § 1853(b)(10)(1).

nual allocations of regulatory discards as incentives to reduce bycatch.[89] With regard to the shrimp trawl fisheries in the Gulf of Mexico and the South Atlantic, the Secretary must conclude within nine months of enactment of the law data collection to assess the impact on fishery resources of incidental harvest by the trawlers.[90] Within twelve months the Secretary must complete a program to minimize to the extent practicable the incidental mortality of bycatch in shrimp trawl operations through technology or changes in fishing operations.[91]

One obvious limitation of the Act's new bycatch provisions is that the definition excludes all nonfish incidental catch. As noted above, bycatch includes marine mammals, seabirds, and sea turtles. While some Councils use discretionary provisions to address nonfish bycatch problems in their management plans,[92] they are not required to do so. The amendments preserve the status quo in this respect. Citizens who wish to use the law to compel protection for nonfish species against incidental take must continue to rely on laws other than the Magnuson Act.[93]

Further, the bycatch provisions leave a great deal of discretion to the Councils. Measures must only "minimize" bycatch, "to the extent practicable." The Councils are not directed to use any particular method for determining to what extent it is practicable to minimize bycatch. Nor are they required to establish goals for bycatch reduction, require fishing methods or gear that is likely to reduce bycatch, or define objective ways to measure whether bycatch has been minimized. In fact, the bycatch provisions do not require any reduction at all in bycatch. With one exception, any Council is free to determine that bycatch has already been minimized and that no further reduction is practicable.[94] Councils may, of course, choose to establish objective goals or other specific requirements to reduce bycatch as part of their conservation and management measures, but the law does not direct them to do so. It is also possible

[89] *Id.* § 117(a)(3) (to be codified at 16 U.S.C. §§ 1862(f) and (g)). The North Pacific Council is given a specific directive to "lower, on an annual basis for a period of not less than four years, the total amount of economic discards occurring in the fisheries in its jurisdiction." *Id.* (to be codified at 16 U.S.C. § 1862(f)).

[90] *Id.* § 206.

[91] *Id.*

[92] See discussion of discretionary provisions *infra* at text accompanying notes 111–131. Measures used to reduce non-fish bycatch might include gear restrictions, area closures, and the like.

[93] Other laws that provide some protection for these species are the Endangered Species Act (see Chapter 7), Marine Mammal Protection Act (see Chapter 5), and Migratory Bird Treaty Act (see Chapter 4). The Migratory Bird Treaty Act has been used to limit seabird kills by marine fishers in at least one instance in state waters. *See* Bricklemyer, *supra* note 76, at 273. For discussion of bycatch reduction in high seas driftnet fishing, see discussion *infra* at text accompanying notes 242–258.

[94] *See* note 89, *supra*.

that the Secretary will add precision to the somewhat vague statutory language by promulgating guidelines for Councils to use in meeting their new obligation to minimize bycatch.

As noted above, management plans prior to 1996 were permitted but not required to include measures to reduce bycatch, and some Councils already had chosen to do so.[95] It is likely that most Councils inclined to adopt measures to reduce bycatch have already done so. Those that are not so inclined probably can find ways to avoid meaningful bycatch reduction measures even under the new requirements, but the public process engendered by the new provisions could prompt other Councils to adopt or improve bycatch reduction measures.

The 1996 amendments' most significant impediment to reducing bycatch might be found not in any weakness of the bycatch amendments themselves. Rather, amendments curbing Council discretion to limit excess fishing capacity may be the more important obstacles. Overcapitalization and the consequent "race to fish" result in fishers having a short-term perspective. This perspective can be a major contributor to high bycatch rates, and a limited access system can "begin to change the time horizons of fishers from short to long term."[96] This issue is discussed below in the section on limitation of access to a fishery.

Identification of Essential Fish Habitat

Although nonfishing activities that affect fish habitat may adversely affect a fishery, Councils have no authority to regulate any activities other than fishing.[97] Conservation groups and others nevertheless have tried to

[95] *See, e.g.*, C & W Fish Co., Inc. v. Fox, 931 F.2d 1556, 1561 (D.C. Cir. 1991) (drift gillnets banned in the Atlantic king mackerel fishery on the basis that the ban would reduce bycatch), and National Fisheries Institute v. Mosbacher, 732 F. Supp. 210 (D.D.C. 1990) (Atlantic billfish management plan prohibited commercial fishers from possessing billfish in the EEZ, based in part on the desire to reduce incidental catch of billfish by commercial fishers).

[96] *See, e.g.*, Dayton L. Alverson, Mark H. Freeberg, Steven A. Murawski, and J.G. Pope, A global assessment of fisheries bycatch and discards 175 (FAO 1994). *See also* Dean J. Adams, *Bycatch and the IFQ System in Alaska: A Fisherman's Perspective*, in University of Alaska Sea Grant Program, Solving Bycatch: Considerations for Today and Tomorrow 211 (1995).

[97] The explanatory notes to the Guidelines for Fishery Management Plans state:

> [The National Oceanic and Atmospheric Administration] recognizes that a decline in stock size or abundance may occur independent of fishing pressure and that adverse changes in essential habitat may increase the risk that fishing effort will contribute to a stock collapse. Regardless of the cause of a decline, however, the Act limits a Council's authority in addressing the situation. The only direct control available under the Act is to adjust fishing mortality. . . . If man-made environmental changes are contributing to the downward trends, in addition to controlling effort Councils should recommend restoration of habitat and other ameliorative programs, to the extent possible, and consider whether to take action under section 302(i) of the Act.

draw attention to the importance of protecting fish habitat if the Act's conservation goals are to be achieved.[98] Amendments in 1990 and 1996 placed new emphasis on habitat protection. Many of the 1996 reforms were supported by a coalition of conservation groups that, for the first time, played a significant role in Magnuson Act reauthorization.[99]

The Act as amended in 1996 requires management plans to "describe and identify essential fish habitat for the fishery, . . . minimize to the extent practicable adverse effects on such habitat caused by fishing, and identify other actions to encourage the conservation and enhancement of such habitat."[100] "Essential fish habitat" is defined as "those waters and substrate necessary to fish for spawning, breeding, feeding or growth to maturity."[101] Within six months of the enactment of the 1996 amendments, the Secretary is to "establish by regulation guidelines to assist the Councils in the description and identification of essential fish habitat in fishery management plans."[102] Within twenty-four months the Councils are to submit to the Secretary plan amendments that identify essential fish habitat.[103]

The only required provisions in a management plan directly relating to fish habitat are those noted above. However, the Act encourages and, in some instances, requires, communication between the Secretary or the Councils and federal or state agencies whose actions may adversely affect fish habitat. Examples of these actions are an offshore oil and gas lease, approval of a permit for dredge or fill disposal in a wetland or other waters

---

50 C.F.R. Pt. 602, Subpt. B, App. A at 30 (1995). *See also* 50 C.F.R. § 602.11(c)(7)(i) (1995) ("Whether [downward] trends are caused by environmental changes or by fishing effort, the only direct control provided by the Act is to reduce fishing mortality") and § 602.11(c)(7)(ii) ("Unless the Council asserts . . . that reduced fishing effort would not alleviate the problem, the [fishery management plan] *must* include measures to reduce fishing mortality regardless of the cause of the low population level") (emphasis added). The Councils' and Secretary's inability to regulate habitat has been criticized as a serious shortcoming in the Act. *See* Global Marine Biological Diversity 264 (Elliott Norse ed., 1993).

[98] *See* Iudicello *et al., supra* note 84, at 343. The authors observe, for example, that while 98% of commercially exploited fish species are dependent on estuarine areas, "Texas estuaries have lost almost 90% of their historic freshwater inflows due to upstream diversions, and since 1900, Louisiana has lost 1.1 million acres of coastal wetlands due to oil and gas production activities." *Id.*

[99] *See id.* at 341, 344.

[100] Pub. L. No. 104–297, § 108(a)(3) (to be codified at 16 U.S.C. § 1853(a)(7)). The amendments also added several references to the importance of fish habitat in the "Findings and Purposes" section. *See id.* §§ 101(1)-(3).

[101] *Id.* § 102(3) (to be codified at 16 U.S.C. § 1802(10)). This broad definition allows inclusion of inland, freshwater habitat used by anadromous fish for spawning, rearing, and feeding.

[102] *Id.* § 110(a)(3) (to be codified at 16 U.S.C. § 1855(b)(1)(A)).

[103] *Id.* § 108(b).

under section 404 of the Clean Water Act, a timber sale in a national forest that may cause erosion and sediment deposit on spawning habitat, or a grazing lease that may result in stream degradation.

The 1990 amendments provided that a Council "shall comment on and make recommendations concerning" any state or federal action that "is likely to substantially affect the habitat of an anadromous fishery resource under its jurisdiction."[104] A Council was permitted to comment on proposed state or federal actions that "may affect the habitat of a fishery resource under its jurisdiction."[105]

The 1996 amendments retained the Council comment and recommendations provisions, but they also strengthened the Secretary's and Councils' roles with respect to federal actions that could adversely affect essential fish habitat. Most important is the requirement that federal agencies "consult" with the Secretary on proposed actions that "may adversely affect any essential fish habitat identified under this Act."[106] If the Secretary determines that a proposed agency action would adversely affect essential fish habitat, the Secretary "shall recommend to such agency measures that can be taken by such agency to conserve such habitat."[107] Within thirty days of receiving recommended measures, the proposing agency must respond to the recommendations. The response must include detailed measures the agency proposes for avoiding or mitigating impacts to fish habitat, and if the proposed measures are not consistent with the Secretary's recommendations, the agency "shall explain its reasons for not following the recommendations."[108]

The 1996 amendments also added a requirement that the Secretary review all Commerce Department programs and "ensure" that they "further the conservation and enhancement of essential fish habitat."[109] Finally, the Secretary must provide information to other federal agencies to further habitat conservation.[110]

Congress apparently has done what it believes it can to address the habitat issue, short of giving the Secretary authority to regulate actions by other agencies. The amendments unquestionably will increase the attention given to fish habitat, both in agency decisionmaking and in research efforts, and this by itself is likely to have a salutary effect. The only require-

[104]16 U.S.C. § 1852(i)(1)(B) (1994) (amended by Pub. L. 104–297, § 110(a)(3), to be codified at 16 U.S.C. § 1855(b)(3)(B)). The 1990 amendment was a direct response to concerns of California salmon fishers and "is intended to increase the Council's participation and influence in decisions affecting habitat critical to the survival of anadromous species." H.R. Rep. No. 101–393, 101st Cong., 1st Sess. (1989).

[105]16 U.S.C. § 1852(i)(1)(A) (1994) (revised by Pub. L. 104–297, § 110(a)(3), to be codified at 16 U.S.C. § 1855(b)(3)(A)).

[106]Pub. L. No. 104–297, § 110(a)(3) (to be codified at 16 U.S.C. § 1855(b)(2)).

[107]*Id.* (to be codified at 16 U.S.C. § 1855(b)(4)(A)).

[108]*Id.* (to be codified at 16 U.S.C. § 1855(b)(4)(B)).

[109]*Id.* (to be codified at 16 U.S.C. § 1855(b)(1)(C)).

[110]*Id.* (to be codified at 16 U.S.C. § 1855(b)(1)(D)).

ments that apply to other federal agencies and that appear to be enforceable, however, are procedural. The comment and consultation provisions with respect to actions of other government agencies require no substantive results. One nevertheless may hope that these provisions will result in more informed decisionmaking and, consequently, reduced impact on fish habitat.

The only substantive habitat provisions are that (1) management plans must "minimize to the extent practicable adverse effects on such habitat caused by fishing," and (2) the Secretary must "ensure" that Commerce Department programs further habitat conservation. The fact that little is known about the habitat impacts of many fishing practices, along with the ubiquitous qualifying language "to the extent practicable," are likely to limit the significance of the former. Both provisions allow for substantial Council and Secretarial discretion. Councils may determine what is "practicable," and the Secretary may determine what furthers habitat conservation and what does not.

CONTENT OF FISHERY MANAGEMENT PLANS: DISCRETIONARY PROVISIONS

In General

In addition to specifying provisions that must appear in management plans, the Act permits the Councils and Secretary to include others at their discretion. These are, for example, gear restrictions, area closures, limitations on the catch, sale, and transshipment of fish, and others.[111] A plan may "require that one or more observers be carried on board a vessel of the United States engaged in fishing for species that are subject to the plan, for the purpose of collecting data necessary for the conservation and management of the fishery."[112] A plan also may "assess and specify the effect which the conservation and management measures of the plan will have on the stocks of naturally spawning anadromous fish in the region."[113] Finally, a plan may require United States fishing vessels to obtain a permit to participate in a particular fishery.[114]

Special Treatment for Limitation of Access to a Fishery

The Act prior to 1996 allowed management plans to include "a system for limiting access to the fishery in order to achieve optimum yield."[115]

[111] 16 U.S.C. § 1853(b).

[112] *Id.* § 1853(b)(8).

[113] *Id.* § 1853(b)(9).

[114] *Id.* § 1853(b)(1).

[115] *Id.* § 1853(b)(6). If the Secretary prepared a plan, it could include a system for limiting access only if first approved by a majority of the voting members of the appropriate Council. *Id.* § 1854(c)(3) (amended 1996).

Congress permitted the Councils and the Secretary to take this action because overcapitalization of the fishing industry is a significant problem for fish conservation. In the words of a recent Senate report,

> [i]n many fisheries, both in the United States and throughout the world, too many fishermen are chasing too few fish. Such excess harvesting capacity may result in social and economic pressure to raise the allowable catch levels in the fishery, increase competition among fishermen, and reduce economic stability of fishery participants.[116]

The law required the Council or Secretary to take into account the following factors when developing such a system:

> (A) present participation in the fishery;
> (B) historical fishing practices in, and dependence on, the fishery;
> (C) the economics of the fishery;
> (D) the capability of fishing vessels used in the fishery to engage in other fisheries;
> (E) the cultural and social framework relevant to the fishery; and
> (F) any other relevant considerations.[117]

In addition, under the national standards, any allocation of fishing privileges among United States fishers had to be:

> (A) fair and equitable to all such fishermen;
> (B) reasonably calculated to promote conservation; and
> (C) carried out in such manner that no particular individual, corporation, or other entity acquires an excessive share of such privileges.[118]

[116]Sen. Rep. No. 104–276, 104th Cong., 2d Sess. (1996). *See also* Iudicello *et al.*, *supra* note 84, at 346:

> Many major U.S. fisheries are seriously overcapitalized. In New England, thousands of fishing vessels compete for dwindling stocks of cod and haddock. In the Gulf of Mexico, the annual commercial quota for red snapper is exhausted in a matter of weeks. Here and elsewhere a race for fish has developed as the size and power of fishing fleets have overmatched sustainable catch levels. The oft-cited phenomenon of too many boats chasing too few fish is among the most important threats to fish populations, marine biodiversity in general, and the economic vitality of fishing communities.

[117]*Id.* § 1853(b)(6) (1994) (amended 1996). (The 1996 amendments left this provision substantially unchanged. They simply added "and any affected fishing communities" at the end of subparagraph (E). Pub. L. No. 104–297, § 108(c)(3)). By requiring consideration of these diverse factors, the Act illuminates the meaning of National Standard 5, which, while requiring consideration of "efficiency in the utilization of fishery resources," specifies that no conservation and management measure "shall have economic allocation as its sole purpose." 16 U.S.C. § 1851(a)(5).

[118]*Id.* § 1851(a)(4).

One form of limited access system was described in *Sea Watch International v. Mosbacher*, where the court upheld an individual transferable quota (ITQ) system for the Mid-Atlantic ocean quahog clam fishery.[119] The Mid-Atlantic Council previously had limited access to the surf clam fishery by imposing a moratorium on the entry of new vessels, coupled with a permit system in which permits were restricted to 184 vessels with a history of surf clam fishing in the region. Each vessel received a permit to fish for a fixed percentage of the year's quota, based on vessel fishing history. The court explained the system as follows:

> The permits were tied to the individual vessels for which they were issued, and could only be transferred together with those vessels. The vessels could not be replaced unless they sank, were destroyed by fire or otherwise left the fishery involuntarily. Thus, only vessels originally awarded permits could fish in the Mid-Atlantic surf clam fishery.[120]

Concerned about vessels shifting from the surf clam fishery to the less-regulated quahog fishery, the Council applied the same quota system to the quahog fishery in 1988. The court upheld this action, even though the annual quahog catch quota had never been reached. The court found that "[b]oth regulators and fishermen have an interest in having ground rules established before any problem matures. . . . [T]he Act does not mandate any finding of necessity before fishery access can be limited."[121]

The court in *Sea Watch International* thus interpreted the Act to authorize the Secretary to implement a limited access system to prevent overfishing, even without a finding that the fishery was already overfished. In stark contrast, the 1996 amendments suspend the Councils' and Secretary's authority to implement a limited access system, even if a fishery already is overfished.

When the House passed the Sustainable Fisheries Act in early 1996 with a five-year moratorium on the adoption of any new system of individual fishing quotas (IFQs),[122] critics loudly denounced it and urged the Senate

[119] 762 F. Supp. 370 (D.D.C. 1991).

[120] *Id.* at 372–73.

[121] *Id.* at 379 (citing 50 C.F.R. § 602.15(c) (1995): "In an unutilized or underutilized fishery, [limited access] may be used to reduce the chance that [overfishing or overcapitalization] will adversely affect the fishery in the future."). For a discussion of ITQ's and the surf clam fishery, see Comment, *Harnessing Market Forces in Natural Resource Management: Lessons from the Surf Clam Fishery*, 21 B.C. Envt'l Aff. L. Rev. 335 (1994). See references cited in Jacobson, *supra* note 2, at 46–47, n.10, for further discussion of limiting access to fisheries. *See also* William J. Milliken, *Individual Transferable Fishing Quotas and Antitrust Law*, 1 Ocean & Coastal L.J. 35 (1994).

[122] The amendments define "individual fishing quota" as

> a Federal permit under a limited access system to harvest a quantity of fish, expressed by a unit or units representing a percentage of the total allowable catch of a fishery that may be received or held for exclusive use by a person.

to delete it from the final law. They believed that systems for limiting access to a fishery are essential management tools that should be available to Councils in appropriate circumstances.[123] Proponents of the moratorium contended that the moratorium was needed to allow for a nationwide study of IFQ programs.[124]

The version of the bill that passed the Senate included the moratorium but shortened it from five years to four. The Senate version was enacted.[125] The Act provides that Councils may develop IFQ systems but may not submit them to the Secretary before October 1, 2000, and the Secretary may not implement them before that date.[126] The National Academy of Sciences is directed to conduct a study of IFQs and submit a final report to Congress by October 1, 1998.[127] The Councils and the Secretary are required to consider the Academy's report when submitting and approving IFQ proposals after the moratorium ends.[128] During the moratorium, three existing IFQ systems may continue and may be amended.[129]

Critics of the moratorium wanted Councils to retain the ability to use IFQs not only to address overfishing problems and improve fishing safety, but also to reduce bycatch. They argued that the current system of allowing all fishers to fish only during short, specified periods causes them to use fishing gear and methods that are dangerous and wasteful but catch the most fish as fast as possible.[130] The "race for fish" thus contributes to

Pub. L. No. 104–297, § 102(5) (to be codified at 16 U.S.C. § 1802(21)). An ITQ system, at issue in Sea Watch International, is one form of IFQ. It refers to a system in which an individual quota may be transferred to another.

[123]*See, e.g.*, Fluharty, *supra* note 68, at 325–26:

> The most important adjustment the Senate could make would be to permit the [North Pacific] Council to keep the IFQ mechanism in its management toolbox and available for use in appropriate circumstances. . . .
>
> . . . Delaying consideration of IFQs means using tools identified by the Council as less effective at solving management problems.

[124]*Id.* at 326.

[125]Pub. L. No. 104–297, § 108(e) (to be codified at 16 U.S.C. § 1853(d)(1)(A)).

[126]*Id.*

[127]Pub. L. No. 104–297, § 108(f). The report is to include recommendations to implement a national policy with respect to IFQ's. It is to analyze IFQ systems in effect in the United States and elsewhere, as well as alternative measures that accomplish the same objectives as IFQ's. The law contains numerous other directives as to how the study is to be conducted. *Id.*

[128]Pub. L. No. 104–297, § 108(e) (to be codified at 16 U.S.C. § 1853(d)(5)). The amendments also specify procedures that must be established in any IFQ program submitted after the moratorium. *Id.*

[129]These are the North Pacific halibut and sablefish, South Atlantic wreckfish, and Mid-Atlantic surf clam and quahog IFQ programs. Pub. L. No. 104–297, § 108(e) (to be codified at 16 U.S.C. § 1853(d)(2)(B)).

[130]*See* text accompanying note 96, *supra*.

bycatch problems, and critics contend that the moratorium will make it harder for Councils to effectively address these problems.[131]

JUDICIAL REVIEW OF MANAGEMENT PLANS

Within 30 days after final regulations implementing a management plan are promulgated, any person may file a petition for judicial review in any federal district court.[132] The Secretary must respond to a petition for judicial review within forty-five days of service unless the court extends the period for good cause.[133] Although a reviewing court may not enjoin implementation of challenged regulations pending completion of its review,[134] it may declare them invalid if they are found to be "arbitrary and capricious" or if they were otherwise unlawfully promulgated.[135] A court will not review a plan directly, but it will review the Secretary's determination that a plan is in conformity with the Act's requirements.[136] Most important, the Secretary's decision is presumed to be valid, and it will be upheld if it is supported by the record.[137]

[131]Proponents of IFQ systems point to the North Pacific Council's Community Development Quota program for groundfish fishers in Alaskan Native villages in the Bering Sea, which is a form of IFQ program. It has resulted in "better utilization of fish . . . [and] decrease in bycatch rates." Fluharty, *supra* note 68, at 310–11.

[132]16 U.S.C. § 1855(f)(1) (as recodified by Pub. L. No. 104–297, § 110(a)(2), which changed § 1855(b) to § 1855(f)). Courts have held that judicial review of implementing regulations may be obtained more than 30 days after their promulgation when the challenge is asserted as a defense in a forfeiture proceeding, Jensen v. United States, 743 F. Supp. 1091 (D.N.J. 1990), or in an enforcement proceeding, United States v. Seafoam II, 528 F. Supp. 1133, 1138 (D. Alaska 1982). The 30-day limit does, however, bar a later direct challenge to regulations. Hanson v. Klutznick, 506 F. Supp. 582, 585–86 (D. Alaska 1981). *See also* Northwest Environmental Defense Center v. Brennen, 958 F.2d 930, 933–34 (9th Cir. 1992) (30-day period begins to run on date of publication in Federal Register, not date regulations were filed with Office of the Federal Register); Sea Watch International v. Mosbacher, 762 F. Supp. 370, 374 (D.D.C. 1991) (same).

Actions taken to implement management plans, including closing a fishery to commercial or recreational fishing, are also subject to judicial review. 16 U.S.C. § 1855(f)(1) & (2) (previously § 1855 (b)(1) & (2)). This provision was added in 1990, apparently in response to the decision in Kramer v. Mosbacher, 878 F.2d 134 (4th Cir. 1989) (30-day period began to run when regulations were promulgated, rather than later date when Secretary issued notice of intent to close fishery).

[133]16 U.S.C. § 1855(f)(3)(A) (as recodified by Pub. L. No. 104–297 § 110(a)(2)).

[134]*Id.* § 1855(f)(1)(A); Kramer v. Mosbacher, 878 F.2d 134 (4th Cir. 1989). Upon motion of the petitioner, however, the court shall "expedite the matter in every possible way." *Id.* § 1855(f)(4).

[135]16 U.S.C. § 1855(f)(1) (as recodified by Pub. L. No. 104–297, § 110(a)(2)).

[136]Washington Trollers Ass'n v. Kreps, 466 F. Supp. 309 (W.D. Wash. 1979), *rev'd on other grounds*, 645 F.2d 684 (9th Cir. 1981). *Accord*, Louisiana v. Baldridge, 538 F. Supp. 625, 628 (E.D. La. 1982); Pacific Coast Federation of Fishermen's Ass'n v. Secretary of Commerce, 494 F. Supp. 626, 627 (N.D. Cal. 1980).

[137]*See, e.g.*, J.H. Miles & Co., Inc. v. Brown, 910 F. Supp. 1138 (E.D. Va. 1995); (C & W

Determination of Conformity with National Standard 1: Overfishing and Optimum Yield

Courts have had several opportunities to review whether the Secretary's determinations of optimum yield were consistent with the overfishing standard. In all cases, they have deferred to the Secretary's discretion and upheld the determination. It would appear that in only one case, *Northwest Environmental Defense Center v. Brennan,*[138] would the 1996 amendments regarding overfishing have changed the result.

The first major case involving review of the Secretary's determination that a plan conformed with Standard 1 was *Maine v. Kreps,* in which plaintiffs challenged the determination of optimum yield in the preliminary fishery management plan for the herring fishery of the Georges Bank area.[139] That fishery had been governed by the International Commission for the Northwest Atlantic Fisheries (ICNAF) prior to enactment of the Magnuson Act.[140] The United States withdrew from ICNAF in late 1976, after ICNAF had set a herring quota of 33,000 metric tons for 1977, of which 12,000 tons was to be allocated to United States fishers. In her preliminary plan, the Secretary of Commerce fixed 33,000 metric tons as the optimum yield of the fishery and allocated 21,000 metric tons to foreign fishers, the precise allocation developed by ICNAF.

The plaintiffs alleged that 33,000 metric tons was an impermissibly high optimum yield in view of the status of the herring stock. Under the statute, the optimum yield of a fishery was to be based on the fishery's MSY, "as modified by any relevant economic, social, or ecological factor."[141] The evidence before the Secretary was that the stock was at less than half the size that would support an MSY slightly below the level at which recruitment failure was feared.[142] The plaintiffs argued that under such circum-

---

Fish Co. v. Fox, 931 F.2d 1556 (D.C. Cir. 1991); Kramer v. Mosbacher, 878 F.2d 134 (4th Cir. 1989); Alaska Factory Trawler Ass'n v. Baldridge, 831 F.2d 1456 (9th Cir. 1987). One commentator asserts that "regulations under the Magnuson Act are essentially impervious to judicial review." Robert J. McManus, *America's Saltwater Fisheries: So Few Fish, So Many Fishermen,* 9 Nat. Resources & Envt'l 13, 15 (Spring, 1995).

[138] 958 F.2d 930 (9th Cir. 1992).

[139] 563 F.2d 1043 (1st Cir. 1977). See 16 U.S.C. § 1821(g) (§ 1821(h) before the 1996 amendments) for provisions pertaining to preliminary management plans. These plans were intended to be a transition measure when the Act became effective.

[140] ICNAF was created by the International Convention for the Northwest Atlantic Fisheries, Feb. 8, 1949, 1 U.S.T. 477, T.I.A.S. No. 2089.

[141] 16 U.S.C. § 1802(21). This section was amended to read "as reduced by any relevant economic, social, or ecological factor" in 1996 and recodified as 16 U.S.C. § 1802(28). See discussion *supra* at text accompanying note 70.

[142] Evidence introduced at trial indicated a somewhat less dire situation. *See* 563 F.2d at 1048 nn. 7 & 8.

stances the Secretary had a duty to fix the optimum yield at a level that would assure the quickest recovery of the stock by excluding all foreign fishing.[143]

Without any effort to explore the meaning of the overfishing standard, the court found that the plan did not violate the standard because the proposed harvest was expected to allow an annual increase of 10 percent in the stock.[144] The court concluded that there was "nothing in the Act which prescribes a particular annual rate at which a below-par stock need be rebuilt."[145] Rather, the rate of stock growth was to reflect the Secretary's judgment of what would "provide the greatest overall benefit to the Nation."[146]

Though the court's cryptic discussion of "overfishing" makes analysis speculative, the opinion does allow some inferences about the scope of discretion allowed the Secretary. There were two benchmark population levels to which the court attached importance. One is the level at which MSY can be realized; the other is the level at which "recruitment failure" occurs. The former is explicitly tied to the statute; the latter is not. When a stock is below the level at which MSY can be realized, which the court characterized as "depleted,"[147] the court seemed to suggest that the overfishing standard requires that harvest levels must allow at least some growth toward the MSY level. Though the court never said so explicitly, the required rate of stock growth probably will be greater (or, if not, it will at least require a greater justification) the closer the stock size is to the level of potential recruitment failure. Finally, the overfishing standard probably would prohibit setting a harvest level that kept or drove a stock below the latter level.

Although the court did not find that the Act prohibits the Secretary from fixing an optimum yield figure for a depleted stock at a level that allows foreign fishing, it did find that the Act requires that the record of the Secretary's decision clearly "reflect a rational weighing" of the considerations that led to the decision.[148] Because the record before it did

[143]The plaintiffs clearly did not think the duty to assure prompt stock recovery should be met at the expense of United States fishers. They challenged the allocation of 12,000 metric tons for United States fishers as too low, even though the fishers had never taken more than 4,600 tons in any year since 1960. *Id.* at 1048.

[144]563 F.2d at 1049.

[145]*Id.* at 1048–49.

[146]16 U.S.C. § 1802(21)(A) (currently found at § 1802(28)(A)).

[147]563 F.2d at 1052. Though the Act nowhere uses the term "depleted," the National Marine Fisheries Service has used it in the same sense as the court did. *See* H.R. Rep. No. 445, 94th Cong., 1st Sess. 95 (1975). For an alternative definition, see Beaver, *Herring, Sardines, and Foreign Affairs: Determination of Optimum Yield Under the Fishery Conservation and Management Act of 1976*, 53 Wash. L. Rev. 729, 747 n. 101 (1978).

[148]563 F.2d at 1049.

not do so, the court remanded the case to the district court and ordered the Secretary to explain the basis for her optimum yield decision.[149]

Following remand, the Secretary supplemented the record with affidavits emphasizing the adverse impact that an abrupt termination of foreign fishing would have on international fisheries relations. The district court held that the affidavits demonstrated an adequate basis for the 33,000 metric-ton optimum yield figure, and the court of appeals affirmed.[150] The latter court concluded that, although historic fishing patterns and past scientific cooperation were factors explicitly required by the statute to be considered in allocating permissible foreign harvest among nations, they could also be taken into account in the initial setting of the optimum yield figure.[151] The potential significance of that decision, however, was diminished by the emphasis the court gave to the fact that "[t]his is a transitional year. What is reasonable now may be less so later."[152]

It is unclear whether the 1996 amendments to the Act would change the result in *Maine v. Kreps.* The changes in the "optimum yield" definition would have an effect only if optimum yield had been set above MSY.[153] However, in the case of a fishery the Secretary has identified as "overfished" under the new standards, a management plan must specify a timetable for rebuilding the fishery that is "as short as possible," taking various factors into account.[154] If the fishery in *Maine v. Kreps* had been governed by the new standards and had been identified as overfished, the optimum yield level would have to conform with the Council's plan for rebuilding the fishery.

In *Northwest Environmental Defense Center v. Brennen,* the court again upheld the Secretary's determination that a plan's optimum yield did not violate the overfishing standard.[155] In *Brennen,* plaintiff environmental organizations challenged the harvest quota for Oregon coastal, naturally occurring (OCN) salmon. The court rejected the challenge, emphasizing the Secretary's "broad discretion to define optimum yield and overfishing."[156]

Unlike the court in *Maine v. Kreps,* the *Brennen* court carefully analyzed the overfishing/optimum yield standard, and the opinion is instructive. The disputed harvest quota affected the 1986 run of coho salmon. Coho are anadromous fish; that is, they hatch in fresh water, migrate to the

[149] *Id.* at 1051–52.

[150] State of Maine v. Kreps, 563 F.2d 1052 (1st Cir. 1977).

[151] *Id.* at 1056.

[152] *Id.* For a discussion of foreign policy considerations as factors in the determination of optimum yield, see Beaver, *supra* note 147, at 741–48.

[153] See discussion at text accompanying note 70, *supra.*

[154] See note 72, *supra,* and accompanying text.

[155] 958 F.2d 930 (9th Cir. 1992).

[156] *Id.* at 935.

ocean to mature, and return as adults to spawn in the stream in which they were born. Most coho return to spawn three years after hatching. In 1983, adverse climatic conditions resulted in a crash of the coho stock, and only 57,000 OCN coho returned to spawn.[157] This particular stock would return to spawn again in 1986, and because of the low numbers in 1983, plaintiffs were concerned that harvest levels for the 1986 season set by the Council were too high for the stock's long-term survival.[158]

The plan had established an escapement goal (the number of fish that should be allowed to escape ocean harvest to spawn) of 170,000 for 1986 and 200,000 for 1989.[159] However, harvest quotas for 1986 were expected to result in escapement of only 142,000 coho. The plaintiffs filed their first suit, *Northwest Environmental Defense Center v. Gordon*, challenging the 1986 harvest quota as overfishing.[160]

Before this suit could be resolved, the plan was amended in 1987 to adopt an "abundance-dependent" method for setting escapement goals for OCN coho salmon.[161] Using this method, escapement goals vary with the estimated stock size for the year. The plan established a minimum escapement goal of 135,000 for stock sizes under 270,000, a goal of one-half the stock for stock sizes between 270,000 and 400,000, and a goal of 200,000 for stock sizes exceeding 400,000 fish.[162] The plaintiffs filed a second suit, challenging the 1987 amendment.[163] The district court granted summary judgment for the defendants in both suits. The cases were consolidated on appeal.[164]

The basis for the second suit was that spawning escapement of 200,000 OCN coho was necessary for maximum sustainable yield, a stipulation to which the Secretary agreed, yet the abundance-dependence method of determining escapement levels allowed a level lower than 200,000 whenever the run fell below 400,000 fish.[165] Plaintiffs contended that overfishing means "any fishing which depletes stocks, even if it does not threaten extinction or irreversible damage. Thus, according to [plaintiffs], overfishing 'means all fishing which exceeds' maximum sustainable yield."[166] The court rejected this contention, finding that the Act "con-

[157]*Id.* at 933.

[158]*Id.*

[159]*Id.*

[160]849 F.2d 1241 (9th Cir. 1988).

[161]958 F.2d at 933.

[162]*Id.*

[163]958 F.2d 930 (9th Cir. 1992).

[164]The opinion cited in note 160, *supra*, is an earlier decision on the issue of whether the plaintiffs' first suit was moot. The court of appeals held that the case was not moot, even though the 1986 harvest season had ended. The court of appeals remanded the case to the district court, which then granted summary judgment for the defendants.

[165]958 F.2d at 935.

[166]*Id.* at 934.

templates maximum utilization of fishery resources consistent with the long-term health of the fishery. . . . Harvest levels above maximum sustainable yield do not necessarily constitute overfishing within the meaning of National Standard 1.''[167]

The court deferred to the Secretary's interpretation of "overfishing" and his determination that optimum yield allows harvest levels higher than MSY. The court cited the House report accompanying the Act's passage to support the Secretary's decision.[168] The court also emphasized that the Secretary had set a minimum escapement level of 135,000 fish. Plaintiffs could not demonstrate that a spawning escapement level of 135,000, "whether in one year or in several, is inconsistent with the long-term health of the OCN coho stock.''[169]

The 1996 amendments, as discussed above, would prevent the Secretary from setting an optimum yield level above MSY. It appears the Secretary's determination in *Northwest Environmental Defense Center* would be inconsistent with the new requirements.

Courts also have deferred to the Secretary when an optimum yield that was *below* MSY was established.[170] In *C & W Fish Co. v. Fox*, the plaintiffs challenged a ban on drift gillnets in the Atlantic king mackerel fishery.[171] They argued that without the use of drift gillnets, the maximum harvest of mackerel could not be realized, and that Standard 1 requires management plans to achieve "full utilization of the fishery resources.''[172] In other words, the court said, the plaintiffs "read Standard 1 to mandate that all [management plans] promote the maximum sustainable yield.''[173] The court rejected this reading of the Act, finding that a fishery management plan "can comply with Standard 1 if there are social, economic, or ecological factors that justify the pursuit of a yield less than the maximum

[167] *Id.* at 935.

[168] *Id.* at 935 n. 3. The court found that "[t]he language of the statute is unambiguous and vests broad discretion in the Secretary," and that therefore the court "need not resort to legislative history." *Id.* However, the court cited the House Report despite this conclusion, to counter the plaintiff's reliance on Sen. Rep. No. 94–416, *supra* note 3. The plaintiffs argued that the Senate Report contemplated that harvest above maximum yield would be "rare and should be only temporary." *Id.*, citing Sen. Rep. No. 94–416 at 3.

[169] *Id.* at 935. The court further noted that the plaintiffs' argument that a low escapement level in one year will lead to depressed populations in future years was not borne out by the record: the record low escapement level of 57,000 in the El Nino year of 1983 produced an estimated stock abundance of 286,000 in 1986. *Id.* at 935–36.

[170] Such a determination clearly would be permitted under the 1996 amendments.

[171] 931 F.2d 1556 (D.C. Cir. 1991).

[172] *Id.* at 1562.

[173] *Id.* *See* J. H. Miles & Co., Inc. v. Brown, 910 F. Supp. 1138, 1148 (E.D. Va. 1995), in which the court said that optimum yield is not the same as maximum yield; rather it is maximum sustainable yield. Management measures must aim to achieve optimum yield from each fishery on a continuing basis, not optimum yield in a single year.

sustainable yield."[174] The court appeared to accept the Secretary's contention that the "overfishing" half of Standard 1 takes precedence over the "optimum yield" half of the standard,[175] potentially a very significant holding and one that implicitly was endorsed by Congress in the 1996 amendments.

The Secretary's seemingly contradictory rationales for the ban on drift gillnets, however, made the court's finding merely dictum. The Secretary argued first that a reduction in optimum yield was justified because of the threat of overfishing posed by drift gillnets, and second, that the ban would not reduce the likelihood that optimum yield would be harvested.[176] The court rejected the first rationale as unsupported in the record, but it accepted the second rationale, citing evidence that the introduction of gillnets in 1986 had not increased the mackerel catch. Thus, the court upheld the ban not because of any threat that drift gillnets posed to the long-term health of the fishery, but in deference to the Secretary's determination of optimum yield and because the plaintiffs were unable to show that the ban would prevent optimum yield from being reached.

One case in which conservation groups challenged a management plan for allowing overfishing resulted in a settlement that provided for development of a "groundfish rebuilding program" for several stocks of groundfish in New England waters.[177] The agreement defined a "groundfish rebuilding program" as a management plan "which can be expected to eliminate the overfished condition of cod and yellowtail flounder stocks in five years after implementation and to eliminate the overfished condition of haddock stocks in ten years after implementation."[178] The agreement provided that the New England Fishery Management Council would be given the initial opportunity to develop a rebuilding program, but that if the Council did not do so by the prescribed deadline, the Secretary would do so, also according to an agreed-upon schedule.[179] It is unclear

[174]931 F.2d at 1562. The court cited 50 C.F.R. § 602.11(d) (1995) (maximum sustainable yield is only a starting point for determination of optimum yield) and 50 C.F.R. § 602.11(f)(3) (1995) (factors relevant to determination of optimum yield).

[175]931 F.2d at 1562.

[176]*Id.*

[177]Conservation Law Foundation of New England v. Mosbacher, Civil Action No. 91 11759–MA, 1991 WL 501640 (D. Mass. Aug. 28, 1991).

[178]*Id.*

[179]During consent negotiations in this case, fishers' groups sought to intervene. *See* Conservation Law Foundation v. Mosbacher, 966 F.2d 39 (1st Cir. 1992). The district court denied intervention, but the court of appeals reversed, finding that the groups had demonstrated an adverse effect from the settlement negotiations and that they were not adequately represented by the Secretary. *Id.* at 44. While the appeal seeking intervention was pending, however, the district court entered the consent decree. The fishing groups then challenged the consent decree, primarily on procedural grounds. *See* Conservation Law Foundation v. Franklin, 989 F.2d 54 (1st Cir. 1992). The court rejected the challenge. With respect to

whether the Secretary's final regulations implementing the rebuilding program could be challenged as a violation of the consent decree if they were not likely to result in rebuilding the fishery within the prescribed five- or ten-year periods. The court stated that "the rebuilding targets in the consent decree are not rules, but rather periods that *may* be incorporated into a final rebuilding program contemplated by the consent decree."[180] Under the 1996 amendments, rebuilding standards must be part of a management plan for an overfished fishery and as such would be determinative.[181]

Determination of Conformity with National Standard 2: Best Available Scientific Information

Whether or not the Secretary in a particular case relied on the best available scientific information in determining optimum yield is a critical question, since the Secretary's determination of optimum yield will be overturned only if it is arbitrary and capricious.[182] Presumably, if the information the Secretary relied on was not the best available, the determination of optimum yield could be seen as arbitrary. Several courts have explored the meaning of this second national standard.

Louisiana v. Baldridge[183] involved a challenge to the Gulf of Mexico Fishery Management Council's shrimp management plan on the ground that one of its provisions violated the Act's requirement that management measures be based upon the best scientific information available.[184] The court rejected that contention because the scientific information proffered by the plaintiffs had previously been submitted to the Council and thus was part of the administrative record before the Secretary.[185] In effect, the court treated the requirement that management measures be "based upon" the best scientific information as requiring only that the Council and the Secretary "consider" this information. The court's decision effectively requires plaintiffs challenging plans for violating the second national standard to find new and better scientific information outside the

---

the substance of the decree, the court found that the Secretary has discretion to establish the five- and ten-year rebuilding goals, since Section 1853(b)(10) provides that the Secretary may include "such other measures, requirements, or conditions and restrictions as are determined to be necessary and appropriate for the conservation and management of the fishery." "The Secretary, thus, has broad discretion concerning the contents of a [fishery management plan]." *Id.* at 61.

[180] 989 F.2d at 61 (emphasis added).

[181] *See* note 72, *supra.*

[182] See text accompanying notes 134–135, *supra.*

[183] 538 F. Supp. 625 (E.D. La. 1982).

[184] 16 U.S.C. § 1851(a)(2).

[185] 538 F. Supp. at 629.

administrative record. In the authors' view, that is a dubious reading of the second standard.

National Fisheries Institute v. Mosbacher involved a challenge to regulations governing the commercial and recreational harvest of Atlantic Ocean billfish.[186] The Council had determined that to prevent overfishing of billfish, the management plan would prohibit commercial fishers from possessing billfish within the management unit. The plaintiffs challenged this measure and the scientific basis for the finding that such a prohibition was necessary, since the evidence concerning the status of billfish was inconclusive.[187]

The court rejected the plaintiffs' claim that the decision violated Standard 2, stating that the Act "does not force the Secretary and Councils to sit idly by, powerless to conserve and manage a fishery resource, simply because they are somewhat uncertain about the accuracy of relevant information."[188] The court cited the applicable guideline, which states, "The fact that scientific information concerning a fishery is incomplete does not prevent the preparation and implementation of [a fishery management plan]."[189]

Similarly, the court in *Southeastern Fisheries Institute v. Mosbacher* upheld the closure of the directed commercial red drum fishery pending a 20 percent escapement of juvenile fish from inshore waters, despite the fact that the decision was based on "imperfect information."[190] The court noted that commercial harvesting of redfish had developed so rapidly that the Council and the Secretary had to make rapid decisions.[191]

The court in *Northwest Environmental Defense Center v. Brennen*[192] rejected the plaintiffs' challenge to the Secretary's consideration of socioeconomic factors, as well as biological factors, as a violation of Standard 2. The court found that the Act's definition of optimum yield as MSY "modified by any relevant economic, social, or ecological factor," and as the harvest level that will "provide the greatest overall benefit to the Nation,"[193] supports the Secretary's contention that the Act requires consideration of

[186]732 F. Supp. 210 (D.D.C. 1990).

[187]The court noted that because billfish are wide-ranging pelagic species, it is difficult to get accurate information about their movements and status. *Id.* at 212.

[188]*Id.* at 220.

[189]50 C.F.R. § 602.12(b) (1995). On another issue, the court determined that a measure in the plan prohibiting purchase, barter, trade, or sale of fish caught in the management unit did not exceed the Secretary's authority. The Secretary had determined that one way to conserve the billfish resource was to prevent development of a commercial market for billfish. 732 F. Supp. at 220.

[190]773 F. Supp. 435, 442 (D.D.C. 1991).

[191]*Id.*

[192]958 F.2d 930 (9th Cir. 1992).

[193]16 U.S.C. § 1802(21) (amended and recodified to § 1802(28) in 1996).

economic and social, as well as ecological, factors.[194] As discussed above, the 1996 amendment of the optimum yield definition continues to allow consideration of relevant economic, social, and ecological factors, but those factors now may serve only to *reduce* optimum yield from the MSY level.

Courts have deferred to the Secretary's determination of how much economic analysis is necessary in developing a plan. In *Pacific Coast Federation of Fishermen's Ass'n v. Secretary of Commerce*,[195] the plaintiffs had argued that "the Plan must include a detailed analysis of the fishing industry before any regulations can be adopted."[196] The court concluded that, if it could have appeared to the Secretary that the Regional Council "considered the economic impact of its Plan on the salmon fishing industry," the court would not disturb the Secretary's finding that the plan was adequate.[197] Neither "in-depth forecasting" nor "a rigorous exercise in microeconomic analysis" is required, in the court's view, because these tasks are "not fitted for an agency whose job is to weigh broad environmental and economic elements."[198] Specifically, courts have held that the Secretary need not conduct a cost-benefit analysis before approving a plan; nor must the Secretary demonstrate that a chosen alternative is the least restrictive.[199]

STATE AND FEDERAL RELATIONS UNDER THE ACT

As previously described, the Magnuson Act's unique Regional Councils have primary responsibility for developing fishery management plans. These hybrid state-federal agencies represent a significant innovation having no clear counterpart elsewhere in federal wildlife law.[200] The Council system is "an imaginative combination of local and federal expertise."[201]

The Act is also unique in its scheme for accommodating state and federal regulatory roles. The Fisheries Management and Conservation Act does not include any automatic preemption of state regulatory authority

[194] 958 F.2d at 936.

[195] 494 F. Supp. 626 (N.D. Cal. 1980).

[196] *Id.* at 629.

[197] *Id.* at 631.

[198] *Id.*

[199] Alaska Factory Trawler Ass'n v. Baldridge, 831 F.2d 1456, 1460 (9th Cir. 1987); National Fisheries Institute v. Mosbacher, 732 F. Supp. 210, 222 (D.D.C. 1990).

[200] The hybrid character of the Regional Councils created uncertainties as to whether they should be regarded as "federal agencies" for purposes of the National Environmental Policy Act, section 7 of the Endangered Species Act, and other statutes. Federal regulations treat the Councils like federal agencies and have done so consistently. *See* 50 C.F.R., Part 604 (1994), which was deleted as unnecessary in 60 Fed. Reg. 39271 (1995).

[201] Jacobson, *supra* note 2, at 44.

over fishing, in contrast to the Marine Mammal Protection Act and the Endangered Species Act of 1973. With the one exception discussed below, the Act preserves the right of the states to regulate all fishing within their boundaries.[202] It even provides that management plans established for fisheries within the EEZ may incorporate "the relevant fishery conservation and management measures of the coastal States nearest to the fishery."[203]

The only time the federal government may preempt the state's authority to regulate fishing within its boundaries is when the Secretary of Commerce, after notice and an opportunity for a full adversarial hearing, makes two findings: (1) that the fishing in a fishery covered by a fishery management plan "is engaged in predominantly within the EEZ and beyond such zone," and (2) that a state has taken or omitted to take any action, "the results of which will substantially and adversely affect the carrying out of such fishery management plan."[204]

The Secretary thus has the burden of establishing the legitimacy of federal authority. State authority is protected by what is in effect a presumption of validity. That presumption can be overcome, but only if rigorous adversarial procedures are followed. After making the necessary findings, the Secretary must notify the state and the appropriate Council that the Secretary intends to regulate the fishery within the boundaries of the state pursuant to the plan and implementing regulations in effect.[205] At any time thereafter, the state may apply for reinstatement of its authority over the fishery, and, if the Secretary finds that the reasons for which authority was assumed no longer prevail, the authority shall promptly be terminated.[206]

Where state laws conflict with federal regulations under the Act, the Secretary must preempt state laws. In *Southeastern Fisheries Ass'n v. Mosbacher*, fishing interests challenged the Secretary's failure to preempt state landing laws that restricted or prohibited landing redfish from the indirect redfish fishery in the Gulf of Mexico.[207] While federal regulations set a 100,000-pound quota for the fishery, they also provided that commercial fishers landing red drum from an indirect fishery had to comply with state landing and possession laws. In effect, the court said, the Secretary told fishers that they may catch the fish, but they may not land them. The court found that this created an impermissible conflict under the Act.[208]

[202]16 U.S.C. § 1856(a).

[203]*Id.* § 1853(b)(5).

[204]*Id.* § 1856(b)(1).

[205]The Secretary's regulatory authority does not extend to the internal waters of any state. *Id.*

[206]*Id.* § 1856(b)(2).

[207]773 F. Supp. 435 (D.D.C. 1991).

[208]*Id.* at 440. *See also* Vietnamese Fishermen Ass'n of America v. California Department of

Prior to the Act, a growing body of decisional law recognized the jurisdiction of coastal states to regulate fishing activities occurring beyond their traditional three-mile seaward boundaries.[209] By negative implication, the Act preserved this authority with respect to vessels "registered" under the laws of the state, concurrent with the exercise of federal authority.[210] The 1996 amendments explicitly provide that a state "may regulate a fishing vessel outside the boundaries of the State" if the vessel is registered in the state and either there is no fishery management plan for the fishery or the state's laws and regulations are consistent with the plan.[211] Because the Act fails to define the term "registered," however, the precise reach of the Act on state extraterritorial jurisdiction remains in doubt.[212]

In *People v. Weeren,* a California court enjoined state prosecution of a violation of state fishing law by owners of a federally documented vessel outside California's territorial waters.[213] The court held that the Act's provision prohibiting states from "directly or indirectly" regulating fishing in the fishery conservation zone precluded extraterritorial enforcement of state fishing laws despite the fact that defendants were California residents and had committed "preparatory acts" within the state. The court also held that the boating license California had issued the defendants did not constitute "registration" under California law. Similarly, in *United States v. Seafoam II,*[214] the federal district court for Alaska held that boats to which the state of Alaska had issued various fishing licenses were not

Fish & Game, 816 F. Supp. 1468 (N.D. Cal. 1993) (extra-territorial enforcement of a California ban on gillnetting for rockfish was preempted by federal groundfish regulations allowing gillnetting in those waters). For a discussion of state extra-territorial jurisdiction over fisheries under the Magnuson Act and a critique of the latter decision, see Casenote, *Vietnamese Fishermen Ass'n v. California Dep't of Fish & Game: Should Regional Fishery Councils Determine EEZ Preemption of State Laws?* 1 Ocean & Coastal L.J. 255 (1995).

[209]Skiriotes v. Florida, 313 U.S. 69 (1941) (state may assert jurisdiction over extraterritorial conduct of state citizens); Bayside Fish Flour Co. v. Gentry, 297 U.S. 422 (1936) (state may regulate fishing beyond its borders if the fish caught is later "landed" within the state); State v. Bundrant, 546 P.2d 530 (Alas.), *appeal dismissed sub nom.* Uri v. Alaska, 429 U.S. 806 (1976) (state may assert jurisdiction over fishery directly associated with the state).

[210]*See* 16 U.S.C. § 1856(a)(3) (1985) (amended 1996), which provided in part: "[A] State may not directly or indirectly regulate any fishing vessel outside its boundaries, unless the vessel is registered under the law of that State."

[211]Pub. L. No. 104–297, § 112(a) (to be codified at 16 U.S.C. § 1856(a)(3)(A)). The amendments also allow Alaska to regulate vessels that are not registered in Alaska but are operating in a fishery in the EEZ off Alaska for which there was no federal fishery management plan as of August 1, 1996. The authority under this section terminates when a fishery management plan is adopted for the fishery. Pub. L. No. 104–297, § 112(a) (to be codified at 18 U.S.C. § 1856(a)(3)(C)).

[212]For a discussion of congressional intent regarding abrogation of preexisting sources of state extraterritorial authority, see Comment, *The Fishery Conservation and Management Act of 1976: State Regulation of Fishing Beyond the Territorial Sea,* 31 Maine L. Rev. 303 (1980).

[213]93 Cal. App. 3d 541, 155 Cal. Rptr. 789 (1977).

[214]528 F. Supp. 1133 (D. Alaska 1982).

"registered" under the state's laws for purposes of making the boats "vessels of the United States" within the meaning of the Act.[215] Under the 1996 amendments, it probably would not matter whether the *Seafoam II* was registered in Alaska, but that is true only for Alaska. The cases leave fishers and the other states in confusion as to the scope of state extraterritorial jurisdiction.[216]

REGULATION OF FOREIGN FISHING

International Fishery Agreements

Even more important than the Act's provisions pertaining to state and federal relations are its international ramifications. As of March 1, 1977, no foreign fishing was permitted in any fishery over which the United States had exclusive management authority unless conducted pursuant to an international agreement in effect on April 13, 1976, or pursuant to a "governing international fishery agreement" (or "GIFA") negotiated under the Act.[217] In addition, all foreign fishing is required to be conducted in accordance with permits issued under the Act.[218]

The primary importance of these requirements is the degree to which they interject Congress into the realm of international fisheries negotiations. The governing international fishery agreements are not intended to be "treaties" in the constitutional sense, which are negotiated by the President and ratified by two-thirds of the Senate, but are instead "binding commitments" which become effective 120 days after being transmitted to Congress, unless during that period either house passes a resolution prohibiting their coming into force.[219]

[215]16 U.S.C. § 1802(31).

[216]Since the Act does not define registration, many states have sought to provide their own expansive definitions. After the Act's passage, Maine redefined registration to include any vessel "which is used to bring a marine organism into the State or its coastal waters." Me. Rev. Stat. Ann. Tit. 12 § 6001(36). Similarly, Oregon redefined registration to include the licensing of boats for fishing. Or. Rev. Stat. § 508.265. It is questionable whether Congress anticipated such broad concepts of registration. The Vessel Documentation Act of 1980, Pub. L. No. 96–594, 94 Stat. 3453, revised the laws relating to federal documentation of vessels. The revision does not clarify the Act's ambiguity.

[217]16 U.S.C. § 1821(a). The 1996 amendments added a fourth type of agreement, a Pacific Insular Area fishery agreement. Pub. L. No. 104–297, § 105(d) (to be codified at 16 U.S.C. § 1824(e)).

[218]16 U.S.C. § 1821(a)(3), as amended by Pub. L. No. 104–297, § 105(a).

[219]*Id.* §§ 1821(c), 1823, as amended by Pub. L. No. 104–297, §§ 105(a) and (c). When President Gerald Ford signed the Act into law, he expressed his concern about certain aspects of it, one of which was that it "purport[ed] to encroach upon the exclusive province of the Executive relative to matters under international negotiations." Statement by the President Upon Signing H.R. 200 into Law, *reprinted in* Senate Comm. on Commerce, 94th Cong., 2d Sess., A Legislative History of the Fishery Conservation and Management Act of 1976, at 35 (1976).

The central feature of a governing international fishing agreement is its acknowledgment of the exclusive fishery management authority of the United States. In addition, it obligates the signatory nation, and the owners and operators of the fishing vessels of that nation, to abide by all regulations issued by the Secretary under the Act, to cooperate in the Act's enforcement, to permit authorized United States observers on board its vessels and to reimburse the United States for the cost of the observers, to pay all applicable permit fees, and to take no more than the share of the total allowable level of foreign fishing allocated to such nation.[220] The total allowable foreign catch for any fishery (*i.e.*, the part of the optimum yield not harvested by United States vessels) is to be allocated among foreign nations according to criteria specified in the Act, by the Secretary of State in cooperation with the Secretary of Commerce.[221]

Congress has amended the Act several times to place additional restrictions on foreign fishing operations in the EEZ. For example, as of 1978, foreign vessels may receive at sea from United States vessels only the portion of the optimum yield of a particular fishery that exceeds the capacity of United States fish processors.[222] Amendments in 1979 required the Secretary of State to reduce, by at least 50 percent, the allocation of fish to any nation whose nationals, directly or indirectly, are conducting fishing operations or engaging in trade or taking that diminishes the effectiveness of the International Convention for the Regulation of Whaling.[223]

A package of 1980 amendments, embodied in the American Fisheries Promotion Act,[224] struck even more directly at foreign fishing in the EEZ. As its name implies, the purpose of the 1980 Act was to promote the American fishing industry, principally at the expense of the foreign industry. The legislation that originated in the House was designed to phase out foreign fishing in the EEZ altogether.[225] As finally passed, it provides a mechanism whereby the total allowable level of foreign fishing in a particular fishery may be reduced below what would otherwise be available to foreigners by subtracting the anticipated harvest by United States vessels from the optimum yield.[226] The 1980 Act also requires, subject to limited exceptions, that "a United States observer . . . be stationed aboard each foreign fishing vessel while that vessel is engaged in fishing within the

[220]16 U.S.C. § 1821(c).

[221]*Id.* §§ 1821(d),(e).

[222]*Id.* §§ 1824(b)(6), 1857(3).

[223]*Id.* § 1821(e)(2). This sanction is in addition to the discretionary sanctions that may be imposed under the Pelly Amendment to the Fishermen's Protective Act, discussed in Chapter 14.

[224]The Act is Part C of Title II of Pub. L. No. 96–561, 94 Stat. 3296.

[225]H.R. Rep. No. 1138, 96th Cong., 2d Sess. 45, *reprinted in* 1980 U.S.C.C.A.N. 6869, 6901.

[226]The complex formula by which the allowable foreign catch may be calculated under this alternative is set forth in 16 U.S.C. § 1821(e).

exclusive economic zone."[227] The full cost of this observer coverage is to be borne by the foreign fishers and is in addition to the permit fee.[228]

When the United States has entered into a governing international fishing agreement with any foreign nation, that nation must submit to the Secretary of State an application for a permit for each of its fishing vessels that wishes to engage in any fishery under United States management authority.[229] The Secretary of State must publish the application in the Federal Register and promptly transmit copies to the Secretary of Commerce, each appropriate Council, the Secretary of the department in which the Coast Guard is operating, and appropriate committees of Congress. Interested persons may submit comments to any Council to which the application has been transmitted. That Council in turn must consider those comments and submit its own comments to the Secretary of Commerce within forty-five days.[230]

After considering the Councils' views and consulting the Secretary of State and the Secretary of the department in which the Coast Guard is operating, the Secretary of Commerce may approve an application if it meets the requirements of the Act.[231] With respect to any approved application, the Secretary must establish conditions and restrictions that include the requirements of any applicable fishery management plan or preliminary fishery management plan.[232] If the foreign nation notifies the Secretary of State of its acceptance of the conditions and restrictions, it shall be issued permits for the vessels for which it had applied. The Secretary may charge "reasonable fees" for these permits.[233]

Although the Act permits foreign fishing to continue pursuant to any international agreement in effect on the date of enactment of the Act, it prohibits any agreement (other than a treaty) from being renewed, extended, or amended after June 1, 1976, unless it is in conformance with the requirements applicable to governing international fishery agreements.[234] As to treaties in effect on the date of enactment, the Act directs that they be renegotiated so as to make them consistent with the provi-

[227] *Id.* § 1821(i)(1) (redesignated § 1821(h)(1) by Pub. L. No. 104–297, § (a)(4)). This requirement applies to all foreign fishing activities, whether carried out pursuant to a governing international fishing agreement under section 201(c) of the Act, *id.* § 1821(c), or pursuant to a treaty such as the International Convention for the High Seas Fisheries of the North Pacific Ocean, *supra* note 15.

[228] 16 U.S.C. § 1821(i)(4) (redesignated as § 1821(h)(4) by Pub. L. No. 104–297, § 105(a)(4)). A 1980 amendment required full observer coverage of foreign vessels whose operations might result in incidental taking of billfish. *Id.* § 1827.

[229] *Id.* § 1824(b)(1).

[230] *Id.* § 1824(b)(4),(5).

[231] *Id.* § 1824(b)(6).

[232] *Id.* § 1824(b)(7).

[233] *Id.* § 1824(10).

[234] *Id.* § 1822(c).

sions of the Act. If they are not so renegotiated "within a reasonable period of time," the Act expresses the "sense of Congress" that the United States should withdraw from them.[235] Any fishing done by foreign vessels pursuant to existing international agreements after March 1, 1977, must be done pursuant to "registration permits."[236]

One other international aspect of the Act concerns access of United States fishers to fisheries over which a foreign nation asserts exclusive management authority. The Act directs the Secretary of State, upon the request of the Secretary of Commerce, to initiate negotiations for the purpose of securing for United States fishers "equitable access" to these fisheries.[237] If the Secretary of State is unable to conclude an access agreement "within a reasonable period of time," because of the foreign nation's refusal to negotiate in good faith or to commence negotiations, the Secretary must certify that fact to the Secretary of the Treasury.[238] The Secretary of the Treasury must take "necessary and appropriate" action immediately to prohibit the importation into the United States of all fish and fish products from the fishery involved.[239] The same importation restrictions may be imposed if a foreign nation seizes a United States vessel when the United States does not recognize the foreign nation's claim of jurisdiction.[240]

Finally, the 1996 amendments attempted to address the bycatch issue for foreign fishers as well as for United States fishers. The Act now requires the Secretary of State, in cooperation with the Secretary of Commerce, to "seek to secure an international agreement to establish standards and measures for bycatch reduction that are comparable to the standards and measures applicable to United States fishermen for such purposes," for any fishery regulated by the Act for which the Secretary determines that such an agreement is "necessary and appropriate."[241]

The Moratorium on High Seas Driftnet Fishing

Driftnet fisheries in the North Pacific have been a source of concern to the United States since Japan initiated a salmon driftnet fishery more than

[235] *Id.* § 1822(b).

[236] *Id.* § 1824(c).

[237] *Id.* § 1822(a)(4).

[238] *Id.* § 1825(a).

[239] *Id.* § 1825(b)(1). If the Secretary of State so recommends, the Secretary of the Treasury shall also prohibit importation of fish or fish products from other fisheries of the foreign nation.

[240] *Id.* § 1825(a)(4)(C). For the criteria according to which the United States will refuse to recognize the claim of a foreign nation to an exclusive economic zone or the equivalent fishery conservation zone beyond its territorial sea, see *id.* § 1822(f).

[241] Pub. L. No. 104–297, § 105(b)(2) (to be codified at 16 U.S.C. § 1822(h)).

thirty-five years ago.[242] In the 1980s, growing public concern about depletion of salmon and steelhead stocks, as well as incidental take of other fish, marine mammals, sea turtles, and seabirds, led to national and international calls for a ban on high seas large-scale driftnet fishing.[243]

Driftnets are panels of plastic webbing, suspended vertically in the water by floats at the top and weights at the bottom. The nets are set at sundown and allowed to drift with the current during the night. They create walls as long as thirty nautical miles, through which nothing larger than the openings in the mesh can pass.[244] Although driftnet fisheries target on particular species such as squid, there is a high incidental catch of non-target species.[245] An estimated 600 miles of nets (which are not biodegradable) are discarded annually. These nets continue to "ghost fish," or kill fish after they have been discarded, for years.[246] In 1990, about 1,000 driftnet vessels from Japan, Korea, and Taiwan were found in the North Pacific.[247]

Congress passed the Driftnet Impact Monitoring, Assessment and Control Act of 1987 (Driftnet Act) as a first step in reducing the effects of these fisheries.[248] The Driftnet Act required the United States to seek agreements with the driftnet fishing nations to provide reliable monitoring of fishing activities, and it authorized sanctions against fish products of nations that refused to enter into such agreements.[249] By 1989, the United States had entered into agreements with Japan, Korea, and Taiwan, providing for significantly increased observer coverage on fishing vessels, the placement of satellite tracking devices on fishing vessels, and boarding access for United States enforcement officials.[250] In December 1989, the United Nations General Assembly passed a resolution calling for an international moratorium on large-scale high seas driftnet fishing by June 30, 1992.[251] The resolution provided, however, that a moratorium

[242]Sen. Rep. No. 101–414, *supra* note 30, at 5. In recent years, there has been increased concern about the impact of the albacore tuna driftnet fleet in the South Pacific as well. *Id.* at 6.

[243]*See* U.N. General Assembly Resolution No. 44/225 (Dec. 22, 1989).

[244]Sen. Rep. No. 101–414, *supra* note 30, at 5.

[245]In addition to marine mammals and seabirds, an estimated 30–50 percent of the catch of fish is lost when fish die and drop out of the nets before fishers retrieve them. *Id.* at 5.

[246]*Id.* at 5.

[247]*Id. See also* Jacobson, *supra* note 2, at 3 (Supp. 1991), and references cited at 39, note 7.

[248]Pub. L. 100–220, Title IV, §§ 4001–4009 (1987), 101 Stat. 1477, codified as a note to 16 U.S.C.A. § 1822. The Driftnet Act applies to fishing with nets 1.5 miles long or longer. *Id.* § 1802(23).

[249]Pub. L. No. 100–220, Title IV, §§ 4004, 4006.

[250]Sen. Rep. No. 101–414, *supra* note 30, at 6.

[251]GA Res. 44/225, U.N.G.A. (1989). The resolution called for the moratorium to take effect in the South Pacific by July 1, 1991, a date set forth in the Convention for the Prohibition of Fishing with Long Driftnets in the South Pacific, Nov. 29, 1989, 29 I.L.M. 1449

would not be imposed in a region, or could be lifted, "should effective conservation measures be taken based upon statistically sound analysis to be jointly made" by participants in the fishery.

Congress amended and expanded upon the Driftnet Act in the Driftnet Act Amendments of 1990.[252] The amendments stated that it was United States policy to implement the moratorium called for by the 1989 United Nations resolution and to "secure a permanent ban on the use of destructive fishing practices, and in particular large-scale driftnets, . . . beyond the exclusive economic zone of any nation." [253] The amendments required the Secretary of Commerce, through the Secretary of State, to initiate international negotiations with all nations engaged in large-scale driftnet fishing, leading to agreements to implement the ban on this fishing.[254]

The scientific analysis called for in the 1989 United Nations resolution was undertaken in the North Pacific, and the scientific team presented its report in 1991.[255] Arguing that the report failed to show no substantial adverse impacts from driftnet fishing, the United States introduced a resolution into the United Nations General Assembly calling for termination of high seas pelagic driftnet fishing by December 31, 1992.[256] The resolution had a large number of co-sponsors and was adopted without a vote on December 20, 1991. Subsequently, Japan, Korea, and Taiwan terminated their high seas driftnet fisheries.[257] In 1992, Congress further amended the Fishery Conservation and Management Act to deny United States port privileges to vessels of nations that conduct large-scale driftnet fishing beyond their exclusive economic zone, and to prohibit imports of fish, fish products, and sport fishing equipment from these nations.[258]

(Wellington Convention). As a result of the Wellington Convention, virtually all large-scale driftnet fishing has ceased in the South Pacific. H.R. Rep. No. 262(II), 102d Cong., 2d Sess., *reprinted in* 1992 U.S.C.C.A.N. 4103.

[252]Pub. L. 101–627, § 107(a) (1990), 104 Stat. 4441 (codified at 16 U.S.C. § 1826).

[253]16 U.S.C. § 1826(c).

[254]*Id.* § 1826(d). Among other provisions, the agreements are to require that vessels be equipped with satellite transmitters, provide for statistically reliable monitoring of harvest through on-board observers, grant United States officials the right to board driftnet vessels on the high seas and inspect for violations, impose time and area restrictions to prevent interception of anadromous species, require that all large-scale driftnets are constructed "insofar as feasible," with biodegradable materials, and minimize the taking of nontarget fish species, marine mammals, sea turtles, seabirds, and endangered species.

[255]Scientific Review of North Pacific High Seas Driftnet Fisheries, Sidney, B.C., June 11–14 1991. *See* William T. Burke, *et al.*, The United Nations Resolutions on Driftnet Fishing: An Unsustainable Precedent for High Seas and Coastal Fisheries (1993) (School of Marine Affairs, University of Washington), for a summary of the results of the report.

[256]GA Res. 46/215, U.N.G.A. (1991), 31 I.L.M. 241 (1992).

[257]Burke, *supra* note 255, at 26. This paper examines the scientific studies and criticizes the process leading to the decision to ban large-scale driftnet fishing.

[258]High Seas Driftnet Fisheries Enforcement Act, Pub. L. 102–582, §§ 101, 102, 104 (1992), 106 Stat. 4901 (codified at 16 U.S.C. §§ 1826a-c). Before the United States imposes

PROHIBITIONS AND PENALTIES

The Magnuson Act prescribes a formidable array of both civil and criminal penalties for violations of the Act itself, the regulations and permits issued under it, and any applicable governing international fishing agreement. A civil penalty of $100,000 per day may be assessed for any violation.[259] The same civil penalty may be assessed against anyone who ships, transports, offers for sale, sells, purchases, imports, exports, or has custody, control, or possession of any fish taken or retained in violation of the Act, the regulations and permits issued under it, and any applicable governing international fishery agreement.[260] Among other acts, it is also illegal for any person to assault or interfere with an observer, or to engage in large-scale driftnet fishing that is subject to the jurisdiction of the United States.[261]

Criminal penalties may be imposed against United States fishers if they refuse to permit an authorized enforcement officer on board their vessels, interfere with the officer's search and inspection activities, resist a lawful arrest, interfere with the apprehension or arrest of another, or knowingly submit false information to an official. In addition, it is a criminal offense to "forcibly assault, resist, oppose, impede, intimidate, sexually harass, bribe, or interfere with any observer."[262] These offenses are punishable by a fine of not more than $100,000, or imprisonment for not more than six months, or both.[263] The maximum penalties are $200,000 and ten years' imprisonment when such offenses are committed with the use of a dangerous weapon or in a manner that injures an enforcement officer or an observer, or places either in fear of imminent bodily injury.[264]

Criminal penalties of the same amounts may be imposed against foreign fishers for the same offenses. In addition, it is a criminal offense for any foreign fisher to engage in any fishery under the exclusive management authority of the United States, except in accordance with an applicable permit.[265] The maximum fines and prison terms for that offense are likewise $100,000 and six months, respectively.

In addition to the foregoing, any fishing vessel, including all of its gear

sanctions on a nation identified as violating the ban, the President must attempt to obtain an agreement that the nation will immediately terminate large-scale driftnet fishing. 16 U.S.C. § 1826a(b)(2).

[259] *Id.* § 1858(a).

[260] *Id.* § 1857(1)(G).

[261] *See id.* § 1857 for a listing of prohibited acts.

[262] *Id.* § 1857(1)(L), as amended by Pub. L. No. 104–297, § 113(c).

[263] *Id.* § 1859(a)(1), (b).

[264] *Id.* § 1859(b). The penalties for injuring an observer, or placing an observer in fear of injury, were added in 1990. Pub. L. No. 101–627, § 115(b), 104 Stat. 4455 (1990).

[265] 16 U.S.C. § 1859(a)(2).

and cargo, is subject to forfeiture for any of the above described violations.[266] Also subject to forfeiture are all fish taken or retained in connection with or as a result of any violation of the Act. The Act establishes a rebuttable presumption that all fish found on board a seized vessel were taken or retained in violation of the Act.[267] Finally, the Secretary of Commerce may revoke, suspend, or modify a permit, or prohibit the issuance of a permit, for a violation of the Act or failure to pay a civil or criminal penalty imposed under the Act.[268]

Two cases in the federal district court for Alaska decided several important issues pertaining to the Act's enforcement. The first, *United States v. Tsuda Maru*,[269] held that section 311(b)(1)[270] authorizes warrantless inspections or searches of vessels fishing in the fishery conservation zone (now EEZ) even without any probable cause to believe that the vessel had violated the Act. The court held that this authorization was constitutionally permissible because the fishing industry was a "pervasively regulated" industry. The court emphasized the long history of regulation of the industry, extending back to the first federal licensing laws of 1793.[271] The second case, *United States v. Kaiyo Maru No. 53*, reached the same result, although it emphasized the comprehensive nature of federal regulation rather than the length of time the industry had been regulated.[272]

Other enforcement issues decided in *Kaiyo Maru* included the holding that under section 310(a)[273] the courts have discretion as to the extent of forfeiture to be applied against an offending vessel but are required to order total forfeiture of its unlawful catch.[274] The court also rejected the constitutional claims that the arrest of the vessel without prior notice and hearing was unlawful[275] and that the Act unlawfully discriminated against aliens.[276]

[266] *Id.* § 1860(a).

[267] *Id.* § 1860(e). The 1996 amendments added another presumption: "[T]hat any vessel that is shoreward of the outer boundary of the exclusive economic zone of the United States or beyond the exclusive economic zone of any nation, and that has gear on board that is capable of use for large-scale driftnet fishing, is engaged in such fishing." Pub. L. No. 104–297, § 114(d) (to be codified at 16 U.S.C. § 1860(e)).

[268] 16 U.S.C. § 1858(g).

[269] 470 F. Supp. 1223 (D. Alaska 1979).

[270] 16 U.S.C. § 1861(b)(1).

[271] 470 F. Supp. at 1229.

[272] 503 F. Supp. 1075 (D. Alaska 1980). The result in these two cases should be compared with the contrary holding under the Marine Mammal Protection Act in Balelo v. Baldridge, 724 F.2d 753 (9th Cir. 1983), *cert. denied*, 467 U.S. 1252 (1984). *See* Chapter 5 at text accompanying note 72.

[273] 16 U.S.C. § 1860(a).

[274] 503 F. Supp. at 1088–89.

[275] *Id.* at 1087–88.

[276] *Id.* at 1086. The court relied upon the reasoning of United States v. Tsuda Maru, 479 F. Supp. 519 (D. Alaska 1979).

CONCLUSION

Close to two decades of experience with the Fishery Conservation and Management Act have demonstrated its strengths and weaknesses. In 1990, Congress made the following observation:

> In the 14 years since the passage of the Magnuson Act, one of its goals, the "Americanization" of fisheries within the U.S. EEZ, has been largely achieved. Unfortunately, the elimination of foreign fishing has not always been accompanied by the expected improvement in our ability to conserve and manage our fishery resources. In several fisheries, stocks are at an all-time low, and the fishing effort is higher than at the peak of foreign fishing activities in the 1970s. Many U.S. fisheries appear to be overcapitalized, and the adverse impact of fishing on other living marine resources such as marine mammals, sea birds, sea turtles, and nontarget fish species is a recognized problem. In April 1990, the Commerce Committee received a letter signed by more than 200 well-known fishery scientists calling for an immediate moratorium on entry to all major fisheries of the United States.[277]

The recommended moratorium has not been adopted. In fact, the moratorium adopted in the 1996 amendments is exactly the opposite: a moratorium on implementation of limited access systems. Concern about the declining status of major fisheries in the United States and worldwide continues to mount.[278]

Fishery law expert Eldon Greenberg has observed that although "[t]he

[277]Sen. Rep. No. 101–414, *supra* note 30, at 3.

[278]*See* Food and Agriculture Organization of the United Nations, *Review of the State of World Fishery Resources: Marine Fisheries*, FAO Fisheries Circular No. 884 (1995) (FAO technical papers present a "sobering diagnosis of the state of world marine resources"); *Our Living Oceans* at 12 (for all U.S. regions combined, 45% of stocks of living marine resources are over-utilized); *The Tragedy of the Oceans*, The Economist, March 19th, 1994, at 21 ("Almost all the 200 fisheries monitored by the FAO are fully exploited. One in three is depleted or heavily over-exploited, almost all in the developed countries"); Peter Weber, *Abandoned Seas: Reversing the Decline of the Oceans* 32 (Worldwatch Paper No. 116, 1993) (United Nations Food and Agricultural Organization has concluded that "depletion of various stocks has occurred in virtually all coastal states throughout the world."); Jessica Mathews, *The last great catch—a global tragedy of the commons*, Seattle Times Mar. 20, 1994 at B5. Ms. Mathews noted:

> Thirteen of the 17 major global fisheries are depleted or in serious decline. The other four are over-exploited or fully exploited. . . . The global catch has been declining since 1989. Long before that, rising tonnages masked a shift from valuable species, such as flounder, haddock and swordfish, to much less edible ones, such as spiny dogfish, skate and shark—all that was left.
>
> . . . Since the United States took control of its 200–mile offshore zone, . . . [d]espite ever-greater effort—bigger boats, sonar, more days at sea—the catch of nine of the 12 Atlantic groundfish stocks has collapsed. The take of such species as cod, haddock and flounder is down by 70 percent to 85 percent. Clam and oyster harvests are down by half. Pacific salmon are nearing commercial or biological extinction. In the Gulf of Mexico it is the same story.

Magnuson Act, as envisioned by its sponsors, was primarily a conservation-oriented statute, focused upon the biological aspects of managing fish stocks, . . . the focus of managers has been on the social and economic interests of the users.''[279] In some fisheries, there has been ''a history of lack of willingness to impose significant, biological conservation measures.''[280] Mr. Greenberg questions whether the national standards, ''the heart of the statute,'' are ''so broad as to be virtually devoid of content, and . . . incapable of imposing significant constraints'' on the actions of the Councils or the Secretary of Commerce.''[281]

The Regional Council system, innovative as it was in 1976, has been criticized for ''institutionaliz[ing] special interests in fishery management,'' resulting in allowable catch quotas that are not biologically based but instead attempt to satisfy all those who want to fish.[282] While some Councils have adopted strong measures to deal with difficult issues such as bycatch and limiting access to fisheries, the broad discretion given the Councils and the Secretary by both the Act and the courts makes it virtually impossible to compel action to conserve fisheries if officials do not choose to do so.[283]

The 1996 amendments to the Act addressed overfishing, bycatch, and habitat conservation more vigorously than had previous amendments. Some of the amendments are likely to improve fishery conservation. The moratorium on programs to limit access to a fishery could, however, be a setback in reducing fishing capacity, which most knowledgeable observers regard as a fundamental problem for conserving fisheries. It remains to be seen whether the 1996 amendments will bring us closer to the goal of fishery conservation that the Act established over 20 years ago.

[279]Eldon V.C. Greenberg, *The Magnuson Act After Fifteen Years: Is It Working?* at B-9 (paper presented at the 1992 National Fishery Law Symposium, Washington, D.C., October 15–16, 1992).

[280]*Id.* at B-6.

[281]*Id.* at B-4.

[282]Norse, *supra* note 97, at 263. *See also* Greenberg, *supra* note 279, at B-6. (Author questions ''how to ensure that Council members will vote in the national interest rather than in their self-interest'' and ''whether the Council system, which involves competitors regulating themselves . . . is inherently unworkable.'') Challenges to Council composition have generally not been successful. *See, e.g.*, Northwest Environmental Defense Center v. Brennen, *supra* note 63, at 937 (plaintiffs lacked standing to challenge Council composition absent showing that composition of Council caused injury of which plaintiffs complained).

[283]One commentator observes that ''the case law construing the Magnuson Act grants what is probably the maximum deference to the Secretary, while the statute itself requires the Secretary, in turn, to grant maximum deference to the industry-dominated councils.'' McManus, *supra* note 137.

Chapter 7

ENDANGERED SPECIES

Concern about the extinction of wildlife species as a result of human activities was an important factor in the earliest federal efforts to regulate wildlife use. Debates that preceded passage of the Lacey Act in 1900 reveal that Congress was distressed about the virtual extermination of the passenger pigeon and the drastic depletion of many other bird species.[1] The Lacey Act, however, like most of the legislation that followed it, was quite limited as to the wildlife it protected and the range of protection it afforded. Only within the last three decades has the need to confront the problem of endangerment more directly and more comprehensively become apparent.

Since 1966, three successive federal statutes and one major international treaty have attempted to establish a coordinated program to forestall what appeared to be the inevitable destruction of numerous wildlife species. Today that program uses several regulatory tools to conserve wildlife: it restricts the taking of species in danger of extinction or likely to become so, regulates trade in them, provides authority to acquire habitat needed for their survival, and mandates that federal agencies consider the impacts of their activities on these species. In addition, the federal program uses an array of measures to address the global aspects of the species extinction problem.

This chapter examines the many facets of the federal program to conserve endangered species. It begins by tracing the program's development

[1] *See* remarks of Congressman John F. Lacey at 33 Cong. Rec. 4871 (1900). The House report that accompanied the Lacey bill noted that "[in] many of the States the native birds have been well-nigh exterminated." H.R. Rep. No. 56–474, 56th Cong., 1st Sess. 1 (1900).

from its inception in 1966. The bulk of the chapter is devoted to a detailed examination of the complex provisions of the Endangered Species Act of 1973,[2] the first truly comprehensive federal effort to conserve imperiled wildlife.

PREDECESSORS TO THE MODERN ENDANGERED SPECIES ACT

The Endangered Species Preservation Act of 1966

The Endangered Species Preservation Act of 1966[3] marked the formal beginning of the federal effort to protect endangered species. The 1966 Act directed the Secretary of the Interior to "carry out a program in the United States of conserving, protecting, restoring and propagating selected species of native fish and wildlife that are threatened with extinction."[4] Although Congress noted that a wide variety of causes, including loss of habitat, overexploitation, disease, and predation, contributed to wildlife endangerment, the only clearly discernable measure the 1966 Act sanctioned was habitat protection. The Act authorized the Secretary to use the land acquisition authority of existing laws to carry out the conservation program. It also created a new source of authority permitting the Secretary to use up to $15 million from the Land and Water Conservation Fund to acquire lands for the program.[5]

Beyond land acquisition authority, the contents of the new endangered species program were vague. For example, the Secretary was directed to review other programs in the Department of the Interior with a view to using them "to the extent practicable" for furthering the purposes of the endangered species program and to "encourage other Federal agencies to utilize, where practicable, their authorities in furtherance of" the pro-

[2]16 U.S.C. §§ 1531–1543.

[3]Pub. L. No. 89–669, §§ 1–3, 80 Stat. 926 (repealed 1973) (hereinafter "1966 Act"). Sections 4 and 5 of the 1966 Act, which consolidated various land units under the authority of the Department of the Interior into a single refuge system and established general standards for their administration, were subsequently designated as the National Wildlife Refuge System Administration Act of 1966. *See* Pub. L. No. 91–135, § 12(f), 83 Stat. 275, 283 (1969). That Act is discussed in Chapter 8.

[4]1966 Act § 2(a). Although the Department of the Interior was of the view that its authority under existing law was adequate to initiate an endangered species program, Congress refused to appropriate funds for that purpose until Interior's legislative authority was clarified. The principal purpose of the 1966 Act was to provide this clarification. See Sen. Rep. No. 89–1463, 89th Cong., 2d Sess. at 3 (1966), *reprinted in* 1966 U.S.C.C.A.N. 3342, 3344. For a useful discussion of some of the endangered species efforts of the Department of the Interior prior to the 1966 legislation, see S. L. Yaffee, Prohibitive Policy: Implementing the Federal Endangered Species Act 34–35, 190 n. 12, and 192 nn. 27–28 (1982).

[5]1966 Act §§ 2(b),(c). The Act provided that no more than $5 million could be appropriated annually from the Land and Water Conservation Fund for such purposes and that no more than $750,000 could be spent for the acquisition of any one area.

gram.[6] Closely related to these directives was the Act's declaration of congressional policy that the Secretaries of the Interior, Agriculture, and Defense, and the heads of all agencies within their departments,

> shall seek to protect species of native fish and wildlife, including migratory birds, that are threatened with extinction, and, insofar as is practicable and consistent with the primary purposes of such bureaus, agencies, and services, shall preserve the habitats of such threatened species on lands under their jurisdiction.[7]

The new program was directed at native wildlife "threatened with extinction." A species was determined to be threatened with extinction upon a finding by the Secretary of the Interior

> after consultation with the affected States, that its existence is endangered because its habitat is threatened with destruction, drastic modification, or severe curtailment, or because of overexploitation, disease, predation, or because of other factors, and that its survival requires assistance.[8]

The Secretary was directed to "seek the advice and recommendations of interested persons," including wildlife scientists, and to publish in the Federal Register the names of all species found to be threatened with extinction.[9] The Act did not specify any further procedures for the Secretary to follow in finding that a species was threatened with extinction. Finally, in addition to the obligation to consult with the states prior to listing endangered species, the Secretary was directed to "cooperate to the maximum extent practicable with the several States" in carrying out the program.[10]

While the 1966 Act marked a significant first step in the effort to protect endangered species, it had a number of serious limitations. The Act's most notable weakness was that it placed no restriction whatever on the taking of any species. That power remained solely with the states.[11] Nor did the

[6] 1966 Act § 2(d).
[7] *Id.* § 1(b).
[8] *Id.* § 1(c).
[9] *Id.*
[10] *Id.* § 3(a).
[11] Arguably, the declaration of policy that the Interior, Agriculture, and Defense Departments "seek to protect" endangered species obligated them to prohibit the taking of such species on lands under their jurisdiction. The legislative history seems to support this interpretation. *See* Letter from the Office of the Secretary of the Interior to the Senate Commerce Comm., Sept. 7, 1965, quoted in Conservation, Protection and Propagation of Endangered Species of Fish and Wildlife: Hearings on S.2217 Before the Subcomm. on Merchant Marine and Fisheries of the Senate Comm. on Commerce, 89th Cong., 1st Sess. 32 (1965). However, at that time the federal government's authority to regulate the taking of wildlife on federal land in the absence of any threatened harm to the land was much in doubt. See Chapter 2 at text accompanying notes 61–81.

Act restrict interstate commerce in endangered species. Moreover, the Act mandated very little in the way of habitat protection, apart from the limited acquisition authority conferred on the Secretary. The other major federal landholding agencies were directed to preserve habitat only "insofar as practicable and consistent with the primary purposes" of the agencies.[12] Finally, the 1966 Act applied only to "native" wildlife, offering no protection for foreign wildlife in danger of extinction.

The Endangered Species Conservation Act of 1969

Congress remedied some of the deficiencies of the 1966 Act three years later by enacting the Endangered Species Conservation Act of 1969.[13] The 1969 Act supplemented the 1966 Act and expanded somewhat the earlier Act's acquisition authority, defined the types of wildlife subject to protection under the 1966 Act, and expanded the scope of the Lacey and Black Bass Acts. Its major innovation, however, was that it authorized the Secretary to promulgate a list of wildlife "threatened with worldwide extinction" and to prohibit most importation of species on the list.

For domestic wildlife, the changes the 1969 Act introduced were relatively minor and, in one case, unnecessary if the earlier legislation had been given a literal interpretation. The 1969 Act amended the 1966 Act by defining "fish or wildlife" as "any wild mammal, fish, wild bird, amphibian, reptile, mollusk, or crustacean."[14] This definition was viewed as an expansion of the earlier Act's scope only because the Department of the Interior had limited the undefined term "fish and wildlife" in that Act to vertebrate animals.[15] Similarly, the 1969 Act amended section 3 of the Lacey Act to expand its prohibition on interstate and foreign commerce in unlawfully taken wild birds and mammals to include reptiles, amphibians, mollusks, and crustaceans.[16] Finally, the 1969 Act expanded somewhat the earlier Act's acquisition authority by permitting appropriation of up to $1 million annually for three years to acquire privately owned lands within the boundaries of areas already administered by the Secretary,[17] and by increasing from $750,000 to $2.5 million the maximum amount the Secretary could spend to acquire any one area with funds from the Land and Water Conservation Fund.[18]

[12] 1966 Act, § 1(b).

[13] Pub. L. No. 91–135, 83 Stat. 275 (1969) (hereinafter "1969 Act").

[14] *Id.* § 12(a).

[15] *See* Letter of the Assistant Secretary of the Interior to the Speaker of the House of Representatives, dated January 17, 1969, in H.R. Rep. No. 91–382, 91st Cong., 1st Sess. at 18 (1969).

[16] 1969 Act § 7(a). A major purpose behind this amendment was to protect the American alligator. *See* H.R. Rep. No. 91–382, 91st Cong., 1st Sess. at 9 (1969).

[17] 1969 Act § 12(c).

[18] *Id.* § 12(b). *See* note 5 *supra.*

In contrast to the 1969 Act's limited impact on domestic species, its international aspects were major and far-reaching. Most significantly, it authorized the Secretary of the Interior to promulgate a list of species or subspecies of fish or wildlife "threatened with worldwide extinction,"[19] and to prohibit their importation into the United States except for certain limited purposes.[20] The Secretary's determinations of what species or subspecies were threatened with worldwide extinction were to be based on essentially the same factors as those for native species under the 1966 Act. The determinations with regard to worldwide extinction were also to be "based on the best scientific and commercial data available to him."[21] Moreover, just as decisions to list native species required consultation with affected states, decisions to list foreign wildlife required "consultation, in cooperation with the Secretary of State, with the foreign country or countries in which such fish or wildlife are normally found."[22] However, unlike the earlier Act, the 1969 Act required the Secretary to list species using the rulemaking procedures of the Administrative Procedure Act.[23]

[19]1969 Act § 3(a).

[20]*Id.* § 2. One of the few issues ever litigated under the 1969 Act concerned its effect on a state's authority to restrict trade in wildlife determined by the state to be in danger of extinction. Two cases challenged New York laws that prohibited the sale of certain wildlife species and products derived from them, where the species protected were not among the list of federally protected species. In both cases, the claim that the 1969 act preempted the states from any regulatory authority over such matters was rejected. A.E. Nettleton Co. v. Diamond, 27 N.Y. 2d 182, 264 N.E. 2d 118, 315 N.Y.S. 2d 625 (1970) *appeal denied sub nom.* Reptile Prod. Ass'n v. Diamond, 401 U.S. 969 (1971); Palladio, Inc. v. Diamond, 321 F. Supp. 630 (S.D.N.Y. 1970), *aff'd*, 440 F.2d 1319 (2d Cir. 1971). For a resolution of the same question under the 1973 Act, see text accompanying notes 371–372 *infra*. For a related decision under the Marine Mammal Protection Act, *see* Fouke Company v. Mandel, 386 F. Supp. 1341 (D. Md. 1974).

[21]1969 Act §3(a). The standard of "best scientific and commercial data available" has been carried over into the Endangered Species Act of 1973, 16 U.S.C. §§ 1533(b)(1)(A) and 1536(a)(2), and, in slightly modified forms, the Marine Mammal Protection Act ("best scientific evidence available"), 16 U.S.C. § 1373(a), and the Fishery Conservation and Management Act ("best scientific information available"), 16 U.S.C. § 1851(a)(2). When the standard was formulated in 1969, the House Committee noted the concern of some witnesses "that the legislation does not contain meaningful, objective standards to guide the Secretary when making a determination whether certain species are threatened with worldwide extinction," but concluded that it did not "believe specific standards can be written into the legislation without harming the effect of the legislation. Existing species are so varied that a standard to fit all appears incredibly complex and cumbersome." H.R. Rep. No. 91–382, 91st Cong., 1st Sess. 6 (1969). It remains to be determined whether the standard of best scientific evidence implies an affirmative duty to undertake research designed to produce data not currently available before any action potentially harmful to wildlife may be taken. See the discussion at text accompanying notes 263–267, *infra.*

[22]1969 Act § 3(a).

[23]*Id.* § 3(d). Another innovation in the 1969 Act was its requirement that the Secretary review the list at least quinquennially and determine whether the listed species should continue to be treated as endangered. *Id.* § 3(a).

The 1969 Act carved out a few exceptions to its general prohibition against importing endangered wildlife. The Secretary was authorized to permit importation "for zoological, educational, and scientific purposes, and for . . . propagation . . . in captivity for preservation purposes."[24] In addition, "to minimize undue economic hardship," the Secretary could authorize a person who had entered into a contract to import an endangered species prior to its listing to import quantities of the species that the Secretary determined to be appropriate, for a period of up to one year.[25] In addition to the general ban on importing endangered wildlife, the 1969 Act amended the Black Bass Act to prohibit transporting in interstate or foreign commerce any fish taken contrary to the law of a foreign country, thus bringing the Black Bass Act into line with a similar amendment to the Lacey Act made in 1935.[26]

A significant international aspect of the 1969 Act was its direction to the Secretary of the Interior to take several affirmative steps to achieve a coordinated international effort to conserve wildlife. Most importantly, the Act directed the Secretary to "seek the convening of an international ministerial meeting" at which would be concluded "a binding international convention on the conservation of endangered species."[27] The result four years later was the Convention on International Trade in Endangered Species of Wild Fauna and Flora (CITES), discussed in detail in Chapter 14.

THE ENDANGERED SPECIES ACT OF 1973

Even as CITES was being signed in early 1973, it was apparent that the task of conserving endangered wildlife in the United States would require a more comprehensive effort than the 1966 and 1969 Acts had established. In particular, it was widely believed that the protection those Acts afforded was often too little and too late. President Nixon had expressed this sentiment a year earlier in his Environmental Message of February 8, 1972, in which he said that existing law "simply does not provide the kind of management tools needed to act early enough to save a vanishing species."[28] Congress also had already recognized the need for earlier action to conserve species when it provided in the Marine Mammal Protection Act that marine mammals might be considered "depleted," and therefore eligible for special protection, even before they became endangered.[29]

[24] *Id.* § 3(c).

[25] *Id.* § 3(b). *Compare* the similar exemption in the Marine Mammal Protection Act, discussed in Chapter 5.

[26] 1969 Act § 9(a). See Chapter 3 at text accompanying notes 27–28.

[27] *Id.* § 5(b).

[28] 8 Weekly Comp. Pres. Doc. 218, 223–224 (Feb. 8, 1972).

[29] For a discussion of the concept of "depletion" under the Marine Mammal Protection Act, see Chapter 5 at text accompanying notes 17–35.

The fact that the protections of the 1966 and 1969 Acts did not come into play early enough was not their only shortcoming. They also failed to give any protection to endangered populations of otherwise healthy species. Again, the Marine Mammal Protection Act, through its provision for protection of individual population stocks, offered a model for change.[30]

The existing federal endangered species program was hobbled in three other critical respects. The first was that it did not prohibit taking endangered species, instead leaving undisturbed the states' traditional authority to regulate taking of resident wildlife. Here too, by providing for an indefinite federal preemption of state authority, the Marine Mammal Protection Act demonstrated Congress's willingness to reconsider this traditional allocation of responsibility.[31] Second, while the 1966 and 1969 Acts obligated some federal agencies to avoid adverse impacts of proposed federal activities on endangered species and their habitats, the obligation was limited to a few designated agencies and was hedged by considerations of what was "practicable and consistent with the primary purposes" of those agencies. Finally, it was evident by 1973 that the problem of endangerment was not limited to vertebrates, mollusks, and crustaceans, but affected virtually all phyla of animals and plants.

To remedy these and other perceived inadequacies, Congress enacted the Endangered Species Act of 1973.[32] Both its statement of congressional findings and purposes and its definitions reflect the truly comprehensive sweep intended for the Act. Recognizing that endangered species of wildlife and plants "are of esthetic, ecological, educational, historical, recreational, and scientific value to the Nation and its people,"[33] the Act declares the bold purpose of providing "a means whereby the ecosystems upon which [they] depend may be conserved."[34] To accomplish this, it further declares a policy "that *all* Federal departments and agencies shall seek to conserve endangered species and threatened species and shall utilize their authorities in furtherance of the purposes of this [Act]."[35] The conservation measures it requires include "all methods and procedures which are necessary to bring any endangered species or threatened species to the point at which the measures provided pursuant to this [Act] are no longer necessary."[36] Finally, to eliminate any chance for a more restrictive interpretation, the Act defines the wildlife and plant species

[30]See Chapter 5 at text accompanying note 17.

[31]For a discussion of the respective state and federal roles under the Marine Mammal Protection Act, see Chapter 5 at text accompanying notes 146–171.

[32]Pub. L. No. 93–205, 87 Stat. 884 (1973) (current version at 16 U.S.C. §§ 1531–1543).

[33]16 U.S.C. § 1531(a)(3).

[34]*Id.* § 1531(b).

[35]*Id.* § 1531(c) (emphasis added).

[36]*Id.* § 1532(3).

eligible for its protection as including any member of the animal or plant kingdoms.[37]

The Act's Fundamental Units: Endangered Species, Threatened Species, and Critical Habitats

The 1973 Act builds its conservation program on three fundamental units. These include two classifications of species—those that are "endangered" and those that are "threatened"—and a third classification of geographic areas called "critical habitats." An understanding of these fundamental concepts is the first step in appreciating how the Act is intended to work.

The 1973 Act borrowed from CITES and the Marine Mammal Protection Act the principle that the "species" to be protected include not only true species and subspecies, but also distinct populations. For reasons not readily apparent, when the Act was originally passed it drew a distinction in this regard between plants and animals. Only for animals did the term "species" encompass distinct populations. A 1978 amendment further limited to vertebrates the authority to list and separately protect populations.[38] Although the rationale for this amendment is not evident, it almost certainly had less to do with any biological distinction between vertebrates and other life forms than with a political desire to limit the number of listed taxa.

The authority to list and separately protect individual populations provides the flexibility to apply the Act's conservation measures selectively to those populations of a species that are currently in trouble, while leaving unregulated healthy populations of the same species. Avoiding local extirpation of a species is desirable not only because a series of local extirpations frequently leads to endangerment of the species as a whole, but also because of the ecological, recreational, aesthetic, and other values populations provide in their localities.

The concern has sometimes been expressed, however, that the authority to list separate populations could be used to protect peripheral populations of otherwise abundant species. That is, since many common species are uncommon or rare at the edge of their range, the Act could protect these peripheral populations even though the species as a whole is not in peril. Whatever the theoretical merits of this concern, the authority to protect populations has been used sparingly in practice and often to protect United States populations of species that are more common and unlisted in other countries. Protecting these latter populations is sound

[37] *Id.* §§ 1532(8) and (14).

[38] Pub. L. No. 95–632, § 2(5), 92 Stat. 3751, 3752 (1978) (hereinafer "1978 Amendments") (codified at 16 U.S.C. § 1532(16)).

policy. Indifference to the loss of United States populations of species currently common in other countries would leave responsibility for insuring the species' ultimate survival entirely with other nations.

Just as the 1973 Act borrowed from CITES and the Marine Mammal Protection Act the principle of protecting geographic populations, so too did it borrow the concept of recognizing differing degrees of vulnerability and establishing protective measures appropriate to those differences. Thus the Act establishes two groups of protected species: "endangered" and "threatened." The former includes "any species which is in danger of extinction throughout all or a significant portion of its range."[39] The latter includes "any species which is likely to become an endangered species in the foreseeable future throughout all or a significant portion of its range."[40] The category of "threatened" was intended not only as a means of giving some protection to species before they became endangered, but also as a means of gradually reducing the level of protection for previously endangered species that had been successfully "restored" to the point at which the strong protective measures provided for that category were no longer necessary.[41]

Endangered species are protected by a number of prohibitions described later in this chapter. Threatened species are protected by "such regulations as . . . [are] necessary and advisable to provide for the conservation of such species," which may be as restrictive as the prohibitions applicable to endangered species.[42] Because the Act defines "species" to include distinct geographic populations (at least for vertebrate animals), it is possible that a species will be subject to stringent protection as an

[39]16 U.S.C. § 1532(6). For a discussion of some of the conceptual difficulties of determining what constitutes a "significant portion" of a species' range, see R.R. Lachenmeier, *The Endangered Species Act of 1973: Preservation or Pandemonium*, 5 Envt'l L. 29, 36–37 (1974). The Fish and Wildlife Service's implementing regulations give no guidance. The only species ineligible for protection as endangered species are insects "determined by the Secretary to constitute a pest whose protection under the provisions of this Act would present an overwhelming and overriding risk to man." 16 U.S.C. § 1532(6). No occasion has yet arisen to test this limited exception, and it seems unlikely that it ever will, for the closer any insect pest approaches to the brink of extinction, the "risk to man" that it poses will almost necessarily be reduced in significance.

[40]16 U.S.C. § 1532(20). When the Marine Mammal Protection Act was passed, its category of "depleted" species was in part the functional equivalent of the threatened status described in the text. When the Act was amended by the Endangered Species Act of 1973, the depleted category became, in effect, a third category—marine mammals likely to become threatened.

[41]Assistant Secretary of the Interior Nathaniel P. Reed in 1972 likened the procedure of moving a given species from the endangered list to the then-proposed threatened list "to that of a hospital where the patient is transferred from the intensive care unit to the general ward until he is ready to be discharged." Endangered Species Conservation Act of 1972: Hearings on S.249, S.2199 and S.3818 Before the Subcomm. on the Environment of the Senate Comm. on Commerce, 92d Cong., 2d Sess. 70 (1972).

[42]16 U.S.C. § 1533(d).

endangered species in one area, less stringent protection as a threatened species in another, and no protection elsewhere.[43]

The Act's flexibility is increased still further by the grant of authority to designate any species as endangered or threatened if it is so similar in appearance to any other listed species that effective protection of the latter species requires listing of the former.[44] This provision, derived from a similar provision in CITES,[45] is potentially of very broad scope, since it is often difficult to distinguish the products of one species from those of related species. To date, however, this authority has rarely been used.

The final building block of the Endangered Species Act is the concept of "critical habitat." As enacted in 1973, the Act neither defined this term nor specified a procedure for its designation, although it did impose a substantive duty, discussed later, that federal agency actions not modify or destroy critical habitat.

In 1978, Congress added a definition of critical habitat to the Act. Critical habitat may be a portion of the area occupied by a listed species, the entirety of the species' occupied area, or even areas outside the currently occupied area. The key in each instance is whether the area is "essential to the conservation of the species." Since the Act's broad definition of "conservation" encompasses measures to assure not only the survival of a listed species, but also its recovery,[46] critical habitat can include areas into which the future expansion of a listed species is essential to assure its survival or recovery.

While preserving broad authority to designate any area essential to the conservation of a listed species, the Act's definition was intended to discourage simply equating the existing range of a listed species with its critical habitat and to require instead a careful examination of the precise physical or biological features of most importance to conservation of the species. Once designated, the main functional significance of a critical habitat is to guide federal agencies in fulfilling their obligations under section 7 of the Act. These important obligations will be examined later in this chapter.

[43]The gray wolf is an example.
[44]16 U.S.C. § 1533(e).
[45]*See* Chapter 14 at note 140.
[46]Section 3(3) of the Act, 16 U.S.C. § 1532(3), provides as follows:

> The terms "conserve," "conserving," and "conservation" mean to use and the use of all methods and procedures which are necessary to bring any endangered species or threatened species to the point at which the measures provided pursuant to this Act are no longer necessary. Such methods and procedures include, but are not limited to, all activities associated with scientific resources management such as research, census, law enforcement, habitat acquisition and maintenance, propagation, live trapping, and transplantation, and, in the extraordinary case where population pressures within a given ecosystem cannot be otherwise relieved, may include regulated taking.

Procedures for Listing Species and Designating Their Critical Habitats

Section 4 of the Act sets forth criteria and procedures for making the key determination from which all other consequences of the Endangered Species Act flow, the decision to list a species as endangered or threatened. Because of the importance of this determination, Congress has enacted increasingly detailed prescriptions for the listing process. The 1966 Act required only that the Secretary of the Interior consult with affected states and seek the advice of other interested persons prior to determining that any species was threatened with extinction.[47] The 1969 Act added a comparable requirement to consult with affected foreign governments and specified that listings be done in accordance with the notice and comment requirements of informal rulemaking under the Administrative Procedure Act.[48]

The 1973 Act embellished these procedures still further, amplifying the requirements that the Secretary consult with affected states and foreign nations, and authorizing discretionary hearings. The procedures applicable to the listing of species under the 1973 Act, as originally enacted, remained essentially the ordinary procedures of informal rulemaking. With respect to critical habitat designations, newly introduced by the 1973 Act, no procedures were specified. Later amendments in 1978 and 1982 revised applicable procedures still further.

The remainder of this part of the chapter analyzes the current listing and critical habitat designation process. Earlier requirements no longer applicable are addressed only to the extent that they shed light on the meaning of the current requirements.

Authority to list species as endangered or threatened resides exclusively with the Secretary of the Interior, except as to those species over which the Secretary of Commerce was given authority by an executive reorganization in 1970.[49] For these, the Secretary of Commerce may determine

[47]1966 Act § 1(c) (repealed 1973).

[48]The court in Colorado River Water Conservation District v. Andrus, Civil Action 78-A-1191 (D. Colo. Dec. 28, 1981), held that informal rulemaking requirements also applied to listings under the 1966 Act but that republication under the 1969 Act of a species listed without compliance with those requirements pursuant to the 1966 Act cured the procedural defect of the original listing.

[49]*See* Reorg. Plan No. 4 of 1970, 35 Fed. Reg. 15627, 84 Stat. 2090 (1970). The Secretary of Commerce has jurisdiction over most marine species, including fish that migrate from freshwater to marine waters (anadromous fish). There are some notable exceptions to this general rule. For example, the Secretary of the Interior has jurisdiction over sea otters and marine birds. The listed species over which the Secretary of Commerce has jurisdiction are found in 50 C.F.R. §§ 222.23(a) and 227.4 (1996). Although the Secretary of Agriculture has no authority with respect to the listing of plants, Agriculture is charged with the enforce-

to list a species or to change its status from threatened to endangered, and the Secretary of the Interior has the ministerial duty to effectuate the listing or change its status.[50] If, however, the Secretary of Commerce recommends the removal of any species from a protected status or the downgrading of a species from endangered to threatened, the Secretary of the Interior must concur in that determination before it becomes effective.[51] The functions of the Secretary of the Interior have been delegated to the United States Fish and Wildlife Service and those of the Secretary of Commerce to the National Marine Fisheries Service.

Listings, delistings, and changes in status may be initiated by the appropriate Secretary or by petition from any interested person.[52] In the event of a petition, the Secretary must, to the maximum extent practicable, determine within ninety days whether the petition presents substantial information that the petitioned action may be warranted.[53] If the Secretary so determines, then a broader review of the status of the species concerned must be undertaken, and the Secretary must determine within twelve months of the receipt of the petition whether formally to propose the petitioned action.[54] The Secretary may decline to propose the requested action only by publishing a written finding either that the action is not warranted, or that it is warranted but that resources are insufficient to proceed immediately with the proposal because of other pending proposals on which the Secretary is making expeditious progress.[55] A decision to decline proposing a petitioned action on either of these bases is subject to judicial review.[56] In addition, a decision to decline proposing a petitioned action because of other pending proposals must be reviewed at least annually to determine whether a formal proposal should then be made.[57]

Once the Secretary decides to propose a species for listing, the Secretary must publish notice of the proposal in the Federal Register, give actual notice to appropriate state and local governments, publish a summary of the proposal in a local newspaper, notify appropriate scientific organiza-

---

ment of the Act and CITES insofar as they pertain to the importation and exportation of terrestrial plants. 16 U.S.C. § 1532(15).

[50] *Id.* § 1533(a)(2)(A).

[51] *Id.* § 1533(a)(2)(B).

[52] The 1973 Act's provision allowing a private petition to initiate listing was one of several innovations designed to encourage public participation in its implementation. The 1982 Amendments of this provision gave citizen participation even greater influence.

[53] 16 U.S.C. § 1533(b)(3)(A).

[54] *Id.* § 1533(b)(3)(B). Prior to the 1982 Amendments, there was no deadline for completion of the status review. The twelve-month deadline in the 1982 Amendments was intended to force the Secretary to conclude these reviews in an expeditious manner.

[55] *Id.*

[56] *Id.* § 1533(b)(3)(C)(ii).

[57] *Id.* § 1533(b)(3)(C)(i).

tions, and, if requested to do so, hold a public hearing.[58] If the species occurs in a foreign nation or is taken on the high seas by citizens of a foreign nation, the Secretary must try to notify the foreign nation and solicit its views.[59]

After proposing a listing action, the Secretary must make a final listing determination within one year. Although the deadline may be extended by six months if more time is needed to resolve substantial disagreements about the proposal, the Secretary ultimately must decide either that the species should be listed as threatened or endangered, or that it should not be listed.[60]

The Secretary must designate critical habitat at the time of listing, to the extent designation of critical habitat is "prudent" and "determinable" at that time.[61] Amendments in 1978 had required the Secretary to designate critical habitat at the time of listing unless the designation would be imprudent because it would expose a species to added risk, such as enabling unscrupulous collectors to locate individuals. Thus the Secretary could delay listing or withdraw the proposed listing at the expiration of an applicable deadline solely because critical habitat was not yet determinable. The 1982 amendments permitted listing without simultaneous designation of critical habitat if critical habitat was not determinable at that time.[62]

The easing of listing procedures in the 1982 amendments was intended to break the near stranglehold placed on listings by the 1978 amendments. The duty to designate critical habitat concurrently with listing had posed a major barrier to listing because of the related duty, also introduced by the 1978 amendments, to balance economic and biological considerations when designating critical habitat. The objective of this balancing was to exclude from the designation any area for which the Secretary "determines that the benefits of such exclusion outweigh the benefits of specifying the area as part of the critical habitat."[63] The difficulty of applying this requirement will be discussed later.

The Secretary's final listing decision must be made "solely on the basis of the best scientific and commercial data available to him."[64] If those

[58] *Id.* §§ 1533(b)(5)(A), (C), (D), and (E). State conservation agencies occupy a sort of "first among equals" position among those commenting on listing proposals in that, if the Secretary issues a final rulemaking inconsistent with a state agency's recommendations, the Secretary must furnish to the agency a written justification of the reason for doing so. *Id.* §1533(h). The Secretary is not, however, bound to follow state agencies' recommendations, and the written justification may simply be a copy of the final rulemaking.

[59] *Id.* § 1533(b)(5)(B).

[60] *Id.* § 1533(b)(6).

[61] *Id.* §§ 1533(a)(3) and (b)(6)(C).

[62] *Id.* § 1533(b)(6)(C)(ii).

[63] *Id.* § 1533(b)(2).

[64] *Id.* § 1533(b)(1)(A).

data support the determination that a species is endangered or threatened because of any "natural or manmade factors affecting its continued existence," the Secretary must list it.[65] The 1982 Amendments and their legislative history were emphatic that the listing process was to be an impartial and objective inquiry, free of economic or other extraneous considerations, particularly the "regulatory impact analyses" of the Reagan administration.[66] As discussed later, these other considerations come into play in other parts of the Act.

Nearly all of the procedural requirements applicable to listing can be bypassed temporarily in the case of an emergency posing a significant risk to the well-being of any species.[67] Although originally limited to fish and other wildlife, this emergency listing authority was expanded in 1979 to encompass plants as well. Emergency regulations continue in effect for 240 days and expire automatically unless within that time they are promulgated pursuant to nonemergency procedures. Only two procedural requirements must be met with respect to emergency regulations. First, detailed reasons why such regulations are necessary must be published with the regulations. Second, if the regulation pertains to a resident species, the conservation agency of each state in which the species is known to occur must be given actual notice of the regulation.[68]

Although the provisions of the Act pertaining to citizen petitions restrict

[65]*Id.* § 1533(a)(1)(E).

[66]It is evident from both the House and Conference Committee reports that the 1982 revision of listing procedures was intended as a direct slap at the performance of the Reagan Administration, and in particular its application to the listing process of "regulatory impact analyses" under Executive Order 12291, 46 Fed. Reg. 13193 (1981). The House report stated:

> The Committee strongly believes that economic considerations have no relevance to determinations regarding the status of species and intends that the economic analysis requirements of Executive Order 12291, and such statutes as the Regulatory Flexibility Act and the Paperwork Reduction Act not apply. The Committee notes, and specifically rejects, the characterization of this language by the Department of the Interior as maintaining the status quo and continuing to allow the Secretary to apply Executive Order 12291 and other statutes in evaluating alternatives to listing. The only alternatives involved in the listing of species are whether the species should be listed as endangered or threatened or not listed at all. Applying economic criteria to the analysis of these alternatives and to any phase of the species listing process is applying economics to the determinations made under Section 4 of the Act and is specifically rejected by the inclusion of the word "solely" in this legislation.

H.R. Rep. No. 97–567, Pt. 1, 97th Cong. 2d Sess. at 20 (1982), *reprinted in* 1982 U.S.C.C.A.N. 2807, 2820. The Conference Committee report endorsed the above conclusion. *See* H.R. Conf. Rep. No. 97–835, 97th Cong., 2d Sess. 20 (1982), *reprinted in* 1982 U.S.C.C.A.N. 2860.

[67]16 U.S.C. § 1533(b)(7).

[68]Although the Act does not define the term "resident species," regulations interpret this term to include any "species which exists in the wild in that State during any part of its life." *See* 40 Fed. Reg. 8566, 8567 (1975) and 41 Fed. Reg. 24354, 24355 (1976).

the Secretary's discretion somewhat, the Secretary still enjoys broad discretion in deciding when to consider the status of any unlisted species. The Act requires the Secretary to "give consideration" to species that have been identified as needing protection by state or foreign conservation agencies or international agreements,[69] but it establishes no priorities among potential candidates for addition to either list. Amendments adopted in 1979, however, required the Secretary to develop a system for assigning priorities to species for review.[70] Four years later, the Secretary of the Interior published a priority system that first considers the degree of threat a species faces, then the immediacy of that threat, and finally the species' taxonomic distinctiveness.[71] Once listed, the status of a species must be reviewed at least quinquennially to assure that listings continue to be based upon the best available information.[72]

Judicial Review of the Listing Process

Only a few decided cases have involved challenges to the decision to list or decline to list a particular species. For the most part, these cases allege a procedural irregularity in the listing process rather than a direct challenge to the substance of the listing decision. A prominent exception was a successful challenge to the Fish and Wildlife Service's failure to list the northern spotted owl.

In *Northern Spotted Owl v. Hodel*, the court held that because the Service had provided no explanation for its finding that the owl should not be listed, and because the Service had "disregarded all the expert opinion on population viability, including that of its own expert, that the owl is facing extinction," its decision not to list the owl was arbitrary and capricious.[73] Rather than order the Service to list the owl, however, the court remanded the matter to the agency for reconsideration. The Service's listing of the owl as a threatened species in 1990 added fuel to a rapidly growing conflict over logging in the Pacific Northwest that continues today.[74]

[69]16 U.S.C. § 1533(b)(1)(B). Although Appendix I of CITES is comprised of species "threatened with extinction," the Fish and Wildlife Service has taken the view that the appearance of a species there does not *ipso facto* qualify it for listing as an endangered species under the Endangered Species Act. *See* 41 Fed. Reg. 24062 (1976). Similarly, although Section 12 of the Act required the Smithsonian Institution to report to Congress on plants needing federal protection, 16 U.S.C. § 1541, the Service treated the Smithsonian's report as an ordinary petition for the listing of the species identified in it. *See* 41 Fed. Reg. 24524 (1976).

[70]16 U.S.C. § 1533(h).

[71]48 Fed. Reg. 43098 (1983).

[72]16 U.S.C. § 1533(c)(2).

[73]716 F. Supp. 479, 483 (W.D. Wash. 1988).

[74]For a discussion of litigation concerning spotted owls and Northwest forests under the

An unsuccessful challenge to the emergency listing of the Mojave population of desert tortoise produced a few noteworthy holdings in *City of Las Vegas v. Lujan.*[75] The court held that "what might constitute arbitrary and capricious action or an unacceptable explanation for a regulation in the normal course of events might well pass muster under the emergency provisions."[76] Thus, the court's scrutiny of the Secretary's decision was even less demanding than in the ordinary rulemaking context, where the Secretary already benefits from considerable judicial deference. The court also rejected the argument that an emergency listing must be invalidated if the listing will not affect the factor giving rise to the emergency. In this case, the emergency listing had been prompted by an outbreak of a virulent disease among the tortoise population, and the plaintiffs argued unsuccessfully that since listing would have no effect upon the spread of the disease, the emergency listing was unlawful.

In *United States v. Guthrie,* the defendant in a criminal prosecution for buying endangered turtles sought to introduce evidence that the listing of the species was in error because a recent DNA study had allegedly shown that the species was not a valid species at all.[77] The court rejected the proffered evidence, noting that the defendant had neither opposed the listing when it occurred nor petitioned to change it in the seven ensuing years. The court reasoned that to permit

> a challenge to an agency regulation on the grounds of new scientific evidence to be made collaterally in a criminal prosecution would deprive the courts of the expertise of the administrative agency, and would prevent the agency from fulfilling its function.[78]

Guthrie, the defendant in the case, apparently had a penchant for novel ideas. The court's opinion revealed that he planned to eradicate the species in the wild and then apply for a government grant to reintroduce it, using captive specimens that he possessed prior to the listing and other specimens illegally obtained after listing.

Even when courts find a procedural irregularity in a listing decision, they have allowed the listing to remain in effect while the government remedies its procedural failing. The courts in both *Idaho Farm Bureau Federation v. Babbitt*[79] and *Endangered Species Committee of the Building Industry*

National Forest Management Act, the National Environmental Policy Act, and other laws, see Chapter 9 at text accompanying notes 105–166 and Chapter 10 at text accompanying notes 106–122.

[75] 891 F.2d 927 (D.C. Cir. 1989).

[76] *Id.* at 932.

[77] 50 F.3d 936 (11th Cir. 1995).

[78] *Id.* at 944.

[79] 58 F.3d 1392 (9th Cir. 1995).

Association of Southern California v. Babbitt[80] found that the government had failed to make certain information available to the public during the listing process, but they refused to suspend the listings' effectiveness while the government corrected its errors. The court in the Idaho Farm Bureau case also held that the government's failure to make a final decision within the time period specified in the law did not invalidate the listing ultimately made. The court noted that the purpose of the statutory deadlines was to expedite the listing process, not to act as a bar to listings made after the deadline. The court observed that a less drastic remedy was available to anyone unhappy with the government's failure to meet the deadlines: simply file a citizen suit to compel a decision.[81]

This is exactly what environmental plaintiffs did to compel an overdue designation of critical habitat for the northern spotted owl. In *Northern Spotted Owl v. Lujan*, the Fish and Wildlife Service had listed the owl as a threatened species but had declined to designate its critical habitat on the basis that critical habitat was not then determinable.[82] The statute authorizes the Service to delay for up to twelve months designating critical habitat if it is not determinable at the time of listing. The court, however, held that the Service had to do more than simply declare that critical habitat was not determinable. Rather, the agency had to explain and justify that conclusion. The Service had neither justified its conclusion nor made any effort to secure the information needed to make a determination prior to the listing decision. Rather than remand to the agency for an explanation, the court ordered the agency to propose critical habitat by a certain date.

Whether the requirements of the National Environmental Policy Act (NEPA)[83] apply to listing and critical habitat designation has been the subject of several cases, with varying results. In *Pacific Legal Foundation v. Andrus*, the Sixth Circuit Court of Appeals held that listing decisions were exempt from NEPA's requirements.[84] The principal rationale for the court's decision was the fact that the Secretary has no discretion to withhold listing of an otherwise eligible species because of environmental considerations. Beyond that, however, the court believed that listing species served the environmental purposes of NEPA, and thus "preparing an impact statement is a waste of time."[85] A district court in Oklahoma reached a contrary result in an unreported decision,[86] but the Tenth Circuit Court

[80]852 F. Supp. 32 (D.D.C. 1994).
[81]58 F.3d at 1401.
[82]758 F. Supp. 621 (W.D. Wash. 1991).
[83]42 U.S.C. §§4321–47.
[84]657 F.2d 829 (6th Cir. 1981).
[85]*Id.* at 836.
[86]Glover River Org. v. Department of Interior, Civ. No. 78-202-C (E.D. Okla. Dec. 12, 1980).

of Appeals later reversed on the ground that the plaintiff lacked standing to challenge the listing.[87]

The Ninth and Tenth Circuits have reached opposite conclusions regarding NEPA's applicability to critical habitat designations. The Ninth Circuit, in *Douglas County v. Babbitt*, held that compliance with NEPA procedures was not required when designating critical habitat.[88] The Tenth Circuit, in *Catron County Board of Commissioners v. U.S. Fish and Wildlife Service*, ruled otherwise.[89] Even if NEPA applies, there is good reason to believe that it will rarely, if ever, require the preparation of a full-blown environmental impact statement, for reasons discussed elsewhere.[90] If, however, an environmental impact statement must accompany the designation of critical habitat, the Secretary will be hard pressed to comply with that requirement and at the same time comply with the requirement of *Northern Spotted Owl v. Lujan* that critical habitat be designated concurrently with listing if critical habitat is determinable at that time.

Recovery Plans

The purpose of the Endangered Species Act is to bring about recovery of the species it protects. To this end, it prohibits a variety of actions detrimental to listed species, charges federal agencies with special conservation duties, authorizes a program of cooperation with the states and the acquisition of land and water, and generally vests the Secretaries of Interior and Commerce with a broad range of powers. Because accomplishing the recovery goal usually requires many different actions over a long period of time, the U.S. Fish and Wildlife Service began the practice of preparing written "recovery plans" to guide recovery efforts. Thus, the plans' origins were administrative rather than legislative. As originally enacted, the Endangered Species Act made no mention of recovery plans.

In 1978, Congress added to the Act the first reference to recovery plans. The amendment simply directed the Secretary to "develop and implement plans for the conservation and survival" of listed species "unless he finds that such a plan will not promote the conservation of the species."[91] It said nothing about the plans' contents or the procedures by which they were to be developed. Later amendments added further detail, including requirements that each plan contain "site-specific management actions . . . necessary to achieve the plan's goals" and "objective, measurable cri-

[87] *Id.*, 675 F.2d 251 (10th Cir. 1982).
[88] 48 F.3d 1495 (9th Cir. 1995).
[89] 75 F.3d 1429 (10th Cir. 1996).
[90] *See* text accompanying notes 296–323, *infra*.
[91] 16 U.S.C. § 1533(f)(1).

teria'' for determining when a species has recovered and should be taken off the list of protected species.[92]

There has been very little litigation to date concerning recovery plans. In *Oregon Natural Resource Council v. Turner*,[93] plaintiffs filed suit nearly four years after the Bradshaw's desert parsley was listed as an endangered species. Efforts to prepare a recovery plan were already underway by the time the suit was filed, but a final plan was not completed until a year later. Completion of the plan led to dismissal of the case before the court made any dispositive ruling. The opinion ultimately entered in the case was in response to plaintiffs' motion to recover attorney fees on the theory that their action had sufficiently hastened the plan's preparation that they should be deemed to have prevailed in the case.

In considering plaintiffs' claim, the court held that because the statute fixes no deadline within which a recovery plan must be prepared,

> plaintiffs would have to show that the Secretary had determined affirmatively not to develop a recovery plan, or possibly that the time lapse after listing and before development and publication of a recovery plan was so great and so unreasonable as to amount to a complete failure to fulfill a duty.[94]

Since in this case the government was in the process of preparing a plan when the suit was filed, the only issue was whether its delay had been unreasonable. In deciding that issue, the court considered several factors, including the consequences of delay. These it found to be minor, since

> the development and publication of a recovery plan in and of itself would not have afforded the endangered species any additional protection. The recovery plan presents a guideline for future goals but does not mandate any actions, at any particular time, to obtain those goals.[95]

This reasoning leads to an anomalous result: The duty to prepare a recovery plan is mandatory, but recovery plans themselves offer only discretionary guidance. This characterization is consistent with the holding in an earlier case, *National Wildlife Federation v. National Park Service*, in which the court rejected the contention that once a recovery plan is developed, the Secretary and the National Park Service ''cannot later selectively decide which provisions to go forward with.''[96]

[92] *Id.* §§ 1533(f)(1)(B)(i) and (ii).
[93] 863 F. Supp. 1277 (D. Or. 1994).
[94] *Id.* at 1282.
[95] *Id.* at 1284.
[96] 669 F. Supp. 384, 388 (D. Wyo. 1987).

The only significant decision to the contrary is *Sierra Club v. Lujan.*[97] That action involved several species dependent upon flow from the Edwards underground aquifer at San Marcos and Comal Springs in west Texas. Groundwater pumping lowered the aquifer level, reducing and potentially eliminating flows at the two springs. The Fish and Wildlife Service had prepared a recovery plan for the endangered species at San Marcos Springs, but not for the species at Comal Springs. The court distinguished *National Wildlife Federation* as holding only that "the Secretary can temporarily delay implementation of a recovery plan."[98] Here, by contrast, "[f]or eight years, the Federal Defendants failed to implement the existing San Marcos Recovery Plan, and they failed to develop a plan for Comal Springs."[99] Sharply rebuking the government's inaction, the court held:

> The Court does not conclude the Federal Defendants must, without exception, immediately implement every step in every recovery plan. . . . [H]owever, the Federal Defendants may not arbitrarily, for no reason or for inadequate or improper reasons, choose to remain idle. Inaction eviscerates the recovery planning provisions of the ESA and amounts to an abdication of the Federal Defendants' statutory responsibility to plan for the survival and recovery, not the extinction, of endangered and threatened species.[100]

Whether the Edwards Aquifer case will prove to be a watershed or an aberration in the development of jurisprudence relating to recovery plans remains to be seen. For now, it stands alone as a forceful holding that recovery plans give rise to serious obligations on the part of the agency required to promulgate them.[101]

Section 9 Prohibitions

Listing species as threatened or endangered and designating critical habitat merely set the stage for the duties the Endangered Species Act imposes on federal agencies and other parties. These duties are both significant and controversial. They are found in section 9, which applies to all persons subject to the jurisdiction of the United States, and section 7,

[97]36 Envt'l Rep. Cas. (BNA) 1533 (W.D. Tex. Feb. 1, 1993), *appeal dismissed*, 995 F.2d 571 (5th Cir. 1993).

[98]*Id.* at 1541.

[99]*Id.* at 1541–42.

[100]*Id.* at 1551.

[101]For further discussion of recovery planning, see Jason M. Patliss, *Recovery, Conservation, and Survival Under the Endangered Species Act: Recovering Species, Conserving Resources, and Saving the Law*, 17 Pub. Land & Resources L. Rev. 55 (1996); Federico Cheever, *The Road to Recovery: A New Way of Thinking About the Endangered Species Act*, 23 Ecology L.Q. 1 (1996); Endangered Species Recovery: Finding the Lessons, Improving the Process (Tim W. Clark et al., eds., 1994).

which applies only to federal agencies. The following discussion focuses on section 9 prohibitions.

The Prohibition against Taking and the Controversy over the Definition of "Harm"

Section 9 of the Act sets forth two separate sets of prohibitions, one applicable to endangered fish or wildlife and the other to endangered plants. For fish and wildlife (but not for plants) section 9 makes it unlawful for any person subject to the jurisdiction of the United States to "take" any endangered species within the United States or its territorial sea, or upon the high seas.[102] One commentator has characterized this prohibition, with some hyperbole, as "perhaps the most powerful regulatory provision in all of environmental law."[103]

Section 3 of the Act defines the term "take" as follows: "The term 'take' means to harass, harm, pursue, hunt, shoot, wound, kill, trap, capture, or collect, or to attempt to engage in any such conduct."[104] For the most part, this definition is unexceptional, encompassing the sorts of hunting, shooting, trapping, and collecting activities traditionally subject to regulation in wildlife conservation legislation. In other respects, however, the definition introduced new terms, not ordinarily subsumed within prior notions of what it means to "take" wildlife. For example, the terms "wound" and "kill" implied an intent to prohibit activities based on their consequences rather than on the intent of the persons responsible for them. More significantly, the terms "harass" and "harm" suggested a much broader scope for the Act's taking prohibition than in earlier laws.

Congress did not define what it meant by the various verbs in the definition of "take." It left that task to the Secretary, who was to develop implementing regulations. In 1975, the Secretary of the Interior set the stage for a controversy that would find its way to the Supreme Court two decades later. In that year, he promulgated regulations defining the term "harm" to include "environmental modification or degradation." Specifically, the regulations defined harm as:

> [A]n act or omission which actually injures or kills wildlife, including acts which annoy it to such an extent as to significantly disrupt essential behavioral patterns, which include, but are not limited to, breeding, feeding or sheltering; significant environmental modification or degradation which has such effects is included within the meaning of "harm."[105]

[102] 16 U.S.C. §§ 1538(a)(1)((B) and (C)).

[103] J.B. Ruhl, Section 7(a)(1) of the "New" Endangered Species Act: Rediscovering and Redefining the Untapped Power of Federal Agencies' Duty to Conserve Species, 25 Envt'l L. 1107, 1115 (1995).

[104] 16 U.S.C. § 1532(18).

[105] 40 Fed. Reg. 44412, 44416 (1975), current version at 50 C.F.R. § 17.3 (1995).

The consequence of this definition was that many land use activities, including land clearing, logging, grazing, and draining or filling ponds and wetlands, potentially were subject to regulation and proscription as an unlawful "taking" of endangered wildlife.

It was not until 1979, however, that an opportunity arose for judicial scrutiny of the taking prohibition's scope. In *Palila v. Hawaii Department of Land and Natural Resources*, the Sierra Club and others charged that the state of Hawaii was taking the palila, an endangered bird dependent upon native forests for nesting sites, by maintaining a population of feral sheep and goats for sport hunting in native Hawaiian forest on state land.[106] The sheep and goats adversely affected the native forest through grazing. The trial court found that destruction of the native forest by the feral animals threatened the palila's survival and that, under the Fish and Wildlife Service's definition of "harm," this constituted a prohibited taking.[107] The court of appeals affirmed on the basis that maintaining the sheep and goats endangered the palila.

After the *Palila* decision, the Fish and Wildlife Service sought to narrow its definition of harm, eliminating from it any reference to environmental modification or disruption of essential behavior patterns.[108] This proposal, in effect an administrative attempt to overrule the decision in *Palila*, drew a storm of protest. Ultimately, the Service backed off its proposal and promulgated the following new definition of harm:

> "Harm" in the definition of "take" in the Act means an act which actually kills or injures wildlife. Such act may include significant habitat modification or degradation where it actually kills or injures wildlife by significantly impairing essential behavior patterns, including breeding, feeding or sheltering.[109]

Neither the *Palila* case nor others that followed it seriously questioned whether the Secretary's regulatory definition of "harm" was in fact authorized by the Endangered Species Act. A direct challenge to the regulation eventually found its way to the Supreme Court in 1995 in *Babbitt v. Sweet Home Chapter of Communities for a Great Oregon*.[110] By a vote of 6–3, the Court upheld the regulation as within the agency's authority under the Endangered Species Act. Although the Court wrote three separate opinions in the case, none provided a very convincing interpretation of the regulation itself. Thus, the Court left considerable room for further judicial elaboration of what activities fall within the scope of the harm prohibition.

[106]471 F. Supp. 985 (D. Haw. 1979), *aff'd*, 639 F.2d 495 (9th Cir. 1981).
[107]471 F. Supp. at 991, 995.
[108]46 Fed. Reg. 29490 (1981).
[109]46 Fed. Reg. 54748, 54750 (1981).
[110]115 S. Ct. 2407 (1995).

The principal issue that divided the Court was whether an activity, to constitute harm, must be *intended* to affect wildlife. The majority held that the actor's intent is irrelevant. Justice Scalia, in dissent, insisted that the statute requires a specific intent to do harm to an animal because, in his view, all the other verbs in the definition of "take" are "affirmative acts . . . directed immediately *and intentionally* against a particular animal."[111] Scalia's fundamental premise is clearly erroneous and his conclusion is undermined by one of the very examples he gives.

As to Justice Scalia's premise, all the other verbs in the definition of "take" do not in fact refer to conduct intentionally directed at wildlife. The terms "kill" and "wound," for example, refer to actions having specific consequences, irrespective of the actor's intent, as several earlier cases had held.[112] Scalia attempts to rebut the argument that his very narrow view of "harm" would add nothing to what is already prohibited by other verbs in the definition of "take" by asserting that it would be "harm" to destroy an animal's "habitat in order to take it (as by draining a pond to get at a turtle)."[113] If by "get at" the turtle, Scalia means to capture or kill it, then it is the capturing or killing that violates the take prohibition, not the draining of the turtle's pond. If, on the other hand, to "get at" the turtle means to get rid of it by leaving it without the ability to elude predators or secure food, then the "harm" that Scalia acknowledges comes not directly and immediately from draining the pond, but indirectly through a predator or starvation. Elsewhere, however, Scalia insists that the harm prohibition cannot apply to "indirect" injuries.[114]

It is precisely this issue of the directness of the linkage between the habitat destruction and the resulting injury or death of an endangered animal that confused the Court and that is likely to generate future litigation. The majority in *Sweet Home* rejected the argument that "harm" requires a direct and immediate injury or death. Instead, it held that the regulation "should be read to incorporate ordinary requirements of proximate causation and foreseeability."[115] "Proximate causation" is a concept that derives from the common law of torts. Justice O'Connor devoted

[111] *Id.* at 2423 (emphasis added).

[112] In National Wildlife Federation v. Burlington Northern Railroad, Inc., 23 F.3d 1508, 1509 (9th Cir. 1994), the Ninth Circuit Court of Appeals held that grizzly bear fatalities that occurred as a result of the bears being accidentally struck by trains "constituted a prohibited 'taking' within the meaning of the ESA." In United States v. Glenn-Colusa Irrigation Dist., 788 F. Supp. 1126, 1133 (E.D. Cal. 1992), the court held that "[t]here is no genuine question that a taking—harming or killing—of winter-run salmon is occurring" as a result of fish being sucked out of a river by irrigation pumps or entrained against a screen designed to keep fish from passing through the pumps. In neither case was there any intent to kill, or even to affect, an endangered species or any other wildlife.

[113] 115 S. Ct. at 2424.

[114] *Id.* at 2429.

[115] *Id.* at 2412, n. 9, and 2414, n. 13.

much of her concurring opinion to a discussion of this concept, which she characterizes as "not . . . susceptible of precise definition."[116] Nevertheless, at its core is the notion that persons should be responsible for consequences of their actions that are reasonably foreseeable.

In the context of tort law, which involves liability for damage to people or their property, the "reasonable person" test applies. The law imposes liability if a reasonable person should have foreseen the consequences of his or her actions. The difficulty of transferring this concept from tort law to endangered species conservation is that what is easily foreseeable to those with a modicum of training in natural history may not be foreseen by those who hold widely prevalent, but erroneous, views of ecology and animal behavior.

This difficulty is illustrated by one of Justice O'Connor's own examples. In explaining her view of proximate causation, she notes that "the landowner who drains a pond on his property, killing endangered fish in the process, would likely satisfy any formulation of the principle."[117] This is so, presumably, because every school child knows that fish cannot long survive out of water. Recall, however, that Justice Scalia offered the example of a landowner who drained a pond with an endangered turtle in it. Again, as every school child knows, turtles *can* survive out of water. Thus, draining the pond will not kill the turtle in it, at least not directly and immediately. Many (perhaps most) people believe that the suddenly pondless turtle will simply go find another pond. To them, the death of the turtle may not be a foreseeable consequence of draining the pond.

To a biologist, however, the death of the turtle as a consequence of draining the pond seems virtually assured. First, the turtle will not necessarily go off in search of a new pond; it may instead simply bury itself in the mud, hoping the sudden drought will prove short-lived. If it does wander elsewhere, the path to the nearest pond is likely fraught with many hazards: predators, road crossings, a desiccating sun, and an absence of suitable food. Even assuming the turtle survives these perils and finds another pond, the carrying capacity of that pond for turtles may already have been reached. Thus, a biologist may view the death of a turtle from draining its pond as a highly likely and predictable result, yet the turtle's death may be entirely unforeseen by the ordinary person with no biology training.

[116] *Id.* at 2420. In United States v. Glenn-Colusa Irrigation Dist., 788 F. Supp. 1126 (E.D. Cal. 1992), the Irrigation District argued that, under principles of proximate causation according to California law, it was not its irrigation pumps that proximately caused the death of salmon entrained against the screens through which the pumps drew water. Rather, it was the screens themselves, which had been installed by the state of California to mitigate fish losses through pumping. The court characterized this effort to shift responsibility from the Irrigation District to the state as "brash" and "absurd." *Id.* at 1133.

[117] 115 S. Ct. at 2420.

The above example illustrates yet another issue that split the *Sweet Home* court: the nature of the injury that must result from habitat destruction in order to trigger the harm prohibition. The challenged definition applies to significant habitat modification "where it actually kills or injures wildlife by significantly impairing essential behavioral patterns, including breeding, feeding or sheltering." The majority opinion asserts that under this definition, "the Government cannot enforce the §9 prohibition until an animal has actually been killed or injured."[118] The majority does not resolve the dispute between Justice O'Connor and Justice Scalia as to whether the required "injury" must be a physical injury, the proof of which, presumably, is the dead or bleeding body of a protected animal.

Justice O'Connor, in her concurring opinion, argues for a broad notion of injury. In her view, "to make it impossible for an animal to reproduce is to impair its most essential physical functions and to render that animal, and its genetic material, biologically obsolete. This, in my view, is actual injury."[119] Thus, under O'Connor's view, no dead or bleeding body need be produced, for the sort of "injury" the regulation addresses is something quite different. It is the significant impairment of the animal's essential behavioral patterns (its ability to secure food, find shelter, and reproduce) that in and of itself constitutes injury. This view can be readily reconciled with the language of the regulation (actual injury occurs "by significantly impairing essential behavioral patterns").[120] Further, viewing impairment of essential behavioral patterns as the injury, rather than some

[118]*Id.* at 2415. This statement, though dictum, is probably limited to enforcement by way of criminal or civil penalty actions. With respect to injunctive actions, whether by the government under 16 U.S.C. § 1540(e)(6) or by private citizens under 16 U.S.C. § 1540(g)(1)(A), it is a more doubtful proposition. In Forest Conservation Council v. Rosboro Lumber Co., 50 F.3d 781, 784 (9th Cir. 1995), the Ninth Circuit Court of Appeals held only three months prior to *Sweet Home* that "[s]o long as some injury to wildlife occurs, either in the past, present, or future, the injury requirement . . . would be satisfied. . . . [A] showing of an imminent injury to wildlife suffices."

[119]115 S. Ct. at 2419.

[120]The language of the regulation, however, is difficult to reconcile with the Interior Department's explanation of it. The explanation suggests that "harm" requires *both* significant impairment of essential behavioral patterns *and* actual injury to a protected animal. *See* 46 Fed. Reg. 54748, 54750 (1981). If that were the intent, however, the regulation should have been written to refer to habitat degradation that "actually kills or injures wildlife *and significantly impairs*" behavior rather than to habitat degradation that "actually kills or injures wildlife *by significantly impairing*" behavior. Two cases that appear to adopt the view that impairment of essential behavioral patterns, without more, constitutes injury are Forest Conservation Council v. Rosboro Lumber Co., 50 F.3d 781 (9th Cir. 1995), and Marbled Murrelet v. Pacific Lumber Co., 880 F. Supp. 1343 (N.D. Cal. 1995). In contrast, the court in Swan View Coalition, Inc. v. Turner, 824 F. Supp. 923, 939 (D. Mont. 1992), after concluding that "excessive road densities in the [Flathead National] Forest are currently impairing essential behavioral patterns of the grizzly bear," went on to question whether such impairment "is a sufficient basis upon which to infer that death or injury is necessarily occurring."

subsequent physical injury, prevents the harm prohibition from becoming entirely superfluous of the take definition's prohibition of killing or wounding.

Moreover, although Justice O'Connor curiously failed to recognize it, the above interpretation avoids the practical and conceptual difficulties of applying the "proximate cause" standard.[121] Habitat modification, by itself, almost never directly kills or injuries wildlife. Instead, by destroying food sources, eliminating shelter, or exposing the animal to increased predation or other risks, it increases the likelihood that some other factor will intervene to the animal's detriment. If the impairment of essential behavioral patterns is itself the injury, then it truly may be said that this injury often does directly and immediately result from habitat destruction (and is likely to be foreseen even by those without biological training). Moreover, Justice Scalia's insistence that " 'proximate' causation simply *means* 'direct' causation"[122] can be accommodated under this interpretation, for the injury (significant impairment of essential behavioral patterns) results directly from the habitat modification, without any intervening agent.

This interpretation also makes clear that the "significant impairment of essential behavior" test applies to individual living animals, not to the species as a whole. All members of the Court seem to embrace this view, although their opinions sometimes carelessly refer interchangeably to effects upon individual animals and effects upon whole populations or species. This confusion stems in part from the fact that section 9 literally prohibits taking a "species" rather than an individual animal of a protected species. Justice Scalia is right in his view (perhaps unduly generous to the legislative drafters) that here the term "species" was merely "shorthand" for any member of an endangered species.[123] A similar confusion

[121]The practical difficulties can be illustrated with a variation on Justice Scalia's own example of turtles in a drained pond. If the pond has two turtles, one of which is promptly crushed by a car on the road adjacent to the drained pond, establishing that draining the pond proximately and foreseeably caused the death of the turtle would not appear to be an insuperable challenge. If the other turtle took off in the opposite direction, however, only to be crushed weeks later on a roadway many miles distant, the practical problems of finding the turtle, establishing from whence it came, and proving the foreseeability of its death would be overwhelming. Yet, with respect to both turtles, the landowner's actions were identical. Lest this hypothetical seem far fetched, see National Wildlife Federation v. Burlington Northern Railroad, Inc., *supra* note 112, at 1511, n. 5 (evidence of a grizzly bear having been struck and killed by a car adjacent to the site of a massive railway spill of corn three years earlier deemed "too tenuous" to reopen record on the continuing effect of the spill on bears).

[122]115 S. Ct. at 2429 (emphasis in original).

[123]*Id.* at 2423, n. 2. The majority opinion's suggestion to the contrary, *id.* at 2413, n. 10, cannot be taken seriously. All the prohibitions of section 9, not only against taking, but also against importing, possessing, selling, and transporting, refer to "species" rather than indi-

was evident both in the Ninth Circuit Court of Appeals decision, *Palila v. Hawaii Department of Land and Natural Resources* ("*Palila II*"),[124] and in Justice O'Connor's characterization of that decision in *Sweet Home.*

Palila II found harm to the endangered palila, an Hawaiian bird, on the strength of expert testimony that the damage to the bird's habitat wrought by feral sheep was such that the sheep "must be removed to ensure the survival of the Palila."[125] The court of appeals sustained the trial court's order directing removal of the sheep in order to halt "habitat destruction that could result in extinction," but it expressly refrained from deciding "whether harm includes habitat degradation that merely retards recovery."[126] It is unclear from the opinion whether the court's finding of harm was premised upon significant impairment of essential behavior of birds then living, or whether the court expected the injury to manifest itself only in the future, with respect to birds not yet living. Justice O'Connor thought *Palila II* had been wrongly decided, because the habitat destruction "did not proximately cause actual death or injury to identifiable birds; it merely prevented the regeneration of forest land not currently inhabited by actual birds."[127] This characterization of the facts is not entirely accurate, however, for the case involved the effects of grazing on both unoccupied areas and on the ability of occupied habitat to remain suitable for the palila.

The Justices' opinions in *Sweet Home* raise as many questions as they answer with regard to the meaning of the harm regulation. Those questions might be answered through new legislation, clarifying regulations, or, as the Court itself suggests, "case-by-case resolution and adjudication."[128] The remainder of this discussion analyzes some of the cases that have probed other aspects of the taking prohibition.

The "harm" definition's requirement that there be "actual death or injury" to a protected animal was explored in a case not involving habitat modification, *American Bald Eagle v. Bhatti.*[129] In that case, plaintiffs sought to halt a planned deer hunt on public land because there was at least a "one in a million risk of harm" to bald eagles as a result of their ingesting a lead slug from an unretrieved deer carcass.[130] The court declined to stop the hunt, noting the absence of any evidence that past deer hunts, either on the land in question or anywhere else, had ever been shown to

---

vidual animals, yet none of these prohibitions would make sense if literally applied to "species" rather than to individual animals.

[124] 852 F.2d 1106 (9th Cir. 1988).

[125] *Id.* at 1109.

[126] *Id.* at 1110.

[127] 115 S. Ct. at 2421.

[128] *Id.* at 2418.

[129] 9 F.3d 163 (1st Cir. 1993).

[130] *Id.* at 165.

cause harm to an eagle. The court held, consistent with the regulation, that "there must be actual injury."[131] Unlike habitat destruction, which can directly cause injury by impairing essential behaviors, the only injury possible from the deer hunt was lead poisoning from ingestion of a spent slug. The court regarded the risk of that injury as too speculative.

A similar rationale underlay the decision in *National Wildlife Federation v. Burlington Northern Railroad, Inc.*[132] A derailment had caused a huge grain spill in an area frequented by grizzly bears. At least five bears were later struck and killed by Burlington Northern trains in the immediate vicinity of the spill and two other bears were struck elsewhere. The plaintiff sought to compel the railroad to reduce train speed in the spill area to prevent further bear fatalities. The court concluded that the prior fatalities had constituted "a prohibited taking"[133] but refused to grant prospective relief because the spill had been cleaned up, the rail bed had been stabilized, and no further bear fatalities had occurred in three years. According to the court, "[w]hile we do not require that future harm be shown with certainty before an injunction may issue, we do require that a future injury be sufficiently *likely.*"[134]

The court in *Burlington Northern* readily concluded that the seven bear fatalities constituted "a prohibited taking." Could it have concluded that the spill itself, even before any subsequent collisions with bears attracted to the corn, constituted a taking? Under the rationale suggested above in the discussion of *Sweet Home,* if the spill significantly modified habitat and significantly impaired essential behavioral patterns, including feeding, the spill itself (or at least the failure promptly to clean it up) might have been regarded as a prohibited taking. However, the court held that the first of these requirements (significant modification of habitat) had not been met. The trial court had concluded that the derailments "did alter, to a certain extent, grizzly bear habitat,"[135] but the appellate court took note of evidence that "the spills had not caused a significant impact on the

[131] *Id.* at 166. For a somewhat similar case in which the evidence of likely injury was established, *see* National Wildlife Federation v. Hodel, 23 Envt'l Rep. Cas. (BNA) 1089 (E.D. Cal., Aug. 26, 1985).

[132] *See* note 112, *supra.*

[133] 23F.2d at 1509.

[134] 23 F.3d at 1512 (emphasis added). *See also* National Wildlife Federation v. National Park Service, 669 F. Supp. 384 (D. Wyo. 1987), in which the court refused to hold the Park Service liable for the taking of grizzly bears as a result of its operation of a campground in Yellowstone National Park. The essence of the claim was that the operation of the campground attracted bears to it and that as a result of the conflicts that ensued, bears would be shot to protect campers. The court held that because, under the first year of an interim management plan, no bear mortalities had occurred, it was "hard pressed to understand how . . . the continued operation of [the campground] . . . would constitute a taking." *Id.* at 389.

[135] *Quoted in* 23 F.3d at 1510.

grizzly bear habitat in the Northern Continental Divide Grizzly Bear Ecosystem," and that the impacts were of a "localized nature."[136] The appellate court thus appeared to measure the "significance" of the habitat modification with respect to the entire habitat of the affected grizzly bear population, an approach that may be appropriate for a wide-ranging species without fixed territories but of dubious value for other species.

Another interesting question that arises from the *Burlington Northern* court's conclusion that the seven fatalities following the spill constituted "a prohibited taking" concerns who might be held liable for those fatalities. In that case, the train that spilled the grain and the trains that later struck the bears all belonged to the same railroad company. But who would have been responsible for the taking if a Burlington Northern train had spilled the grain and another train, owned by a different company, struck and killed a bear attracted to the spilled grain? In other words, can responsibility for the same taking be imposed on two independent, unrelated actors? A case suggesting such a possibility is *Defenders of Wildlife v. EPA*,[137] which involved a challenge to EPA's registration of strychnine for control of small mammals. The plaintiff offered evidence that use of strychnine for this purpose would result in some incidental taking of endangered species that fed upon the carcasses of poisoned animals. Thus, the court concluded, "EPA's registrations constituted takings of endangered species."[138] A similar, earlier case held that the Fish and Wildlife Service's migratory bird hunting regulations, which authorized use of lead shot to hunt waterfowl, constituted a taking of bald eagles because the eagles could be expected to suffer lead poisoning as a result of feeding on waterfowl with embedded lead shot.[139]

In both of the above cases, the mere enabling of others to engage in activities that result in taking were held themselves to be takings. This result is not as questionable as it may at first appear, for it is consistent with the *Sweet Home* court's conclusion that "harm" can occur indirectly, through a foreseeable chain of causation. In the two cases cited, however, the practical effect of the rulings can be minimized, since under section

Id. at 1511. Other cases that reject a claim of taking on the basis of insignificant effects include Pyramid Lake Paiute Tribe of Indians v. U.S. Dep't of Navy, 898 F.2d 1410, 1420 (9th Cir. 1990) (evidence did not show that defendant's diversion of water caused spawning problems of endangered fish, particularly where others, including plaintiff itself, also diverted water), and Fund for Animals, Inc. v. Florida Game and Fresh Water Fish Comm'n, 550 F. Supp. 1206, 1210 (S.D. Fla. 1982) (noise of airboats was no different than that caused by other noise sources and was only temporary).

[137]882 F.2d 1294 (8th Cir. 1989).

[138]*Id.* at 1301.

[139]National Wildlife Federation v. Hodel, *supra* note 131. The facts of this case were quite different from those of American Bald Eagle v. Bhatti, *supra* note 129. At least 96 bald eagle deaths had been caused by lead poisoning since 1976, including 23 in the preceding year. Thus, the risk of further eagle deaths was not remote and speculative, but virtually assured.

7(b)(4) of the Act the Secretary may authorize some taking of protected species incidental to federal actions that do not jeopardize the species' continued existence.

Not every habitat modification impairs essential behavioral patterns, as illustrated by *Morrill v. Lujan.*[140] The court in that case rejected a claim that construction of a commercial facility on apparently unoccupied habitat of an endangered species of beach mouse would lead to greater use by humans and cats of nearby occupied mouse habitat, thereby taking the mouse. The court viewed the causal connection between the planned construction and the anticipated harm to the mouse as simply too tenuous.

The Fish and Wildlife Service, however, when it earlier consulted on an application for a federal permit to construct bulkheads in connection with the development, apparently considered the effects of both the proposed development and other foreseeable development. On this basis, the Service concluded that the project would jeopardize the mouse's continued existence. In so doing, the agency acted in accord with *National Wildlife Federation v. Coleman*[141] and its own regulations.[142] In response to the jeopardy finding, the developer dropped from his plans construction of the bulkheads that required a federal permit, leaving section 9's taking prohibition as the only potential Endangered Species Act obstacle. As discussed above, the court refused to consider the impact of other foreseeable developments in determining the likelihood of a taking.

Can a taking ever occur as a result of inaction? A case challenging U.S. Forest Service management practices in Texas has sometimes been interpreted as so holding.[143] In the authors' view, however, that interpretation is dubious. Both the trial and appellate courts lumped together a variety of actions (clearcutting of foraging and nesting habitat of the red-cockaded woodpecker or "RCW") and inactions (failure to control midstory hardwood trees that encroach upon the pines in which the woodpeckers nest). The court of appeals reasoned that the government

had permitted clearcutting within two hundred feet of RCW cavity trees. The government also does not dispute that it did not remove midstory hardwood . . . thus leading to RCW abandonment of cavity trees. Such a course of conduct certainly

[140]802 F. Supp. 424, 431 (S.D. Ala. 1992).

[141]529 F.2d 359 (5th Cir.), *cert. denied sub nom.* Boteler v. National Wildlife Federation, 429 U.S. 979 (1976).

[142]*See* 50 C.F.R. § 402.02 (1995) (defining "effects of the action").

[143]Sierra Club v. Lyng, 694 F. Supp. 1260 (E.D. Tex. 1988), *aff'd in part sub nom.* Sierra Club v. Yeutter, 926 F.2d 429 (5th Cir. 1991). For an article in which the author asserts that the case holds that inaction can be a prohibited taking, see S. P. Quarles, *et al.*, *Sweet Home* and the Narrowing of Wildlife 'Take' Under Section 9 of the Endangered Species Act, 26 Envt'l L. Rep. 10003, 10013 (1996).

impairs the RCW's "essential behavioral patterns, including . . . sheltering," . . . and thus results in a violation of section 9.[144]

If the phrase "such a course of conduct" refers to the totality of Forest Service actions and inactions, as seems likely, the court's conclusion is not especially noteworthy. Whether inaction alone (failure to control hardwoods) would have constituted a taking was not at issue in the case and no court has yet found inaction alone to be a taking.

In summary, notwithstanding the Supreme Court's decision upholding the government's broad regulatory definition of harm, considerable uncertainty exists with respect to the scope of the taking prohibition as applied to habitat modification. Among the questions left unanswered are: when is habitat modification or degradation "significant"? how likely must the impairment of essential behavioral patterns be to constitute "actual injury"? what, besides breeding, feeding, or sheltering, is included in the term "essential behavioral patterns"? For regulated interests, these are not trivial questions, for their vulnerability to criminal prosecution or civil injunctive suit may hinge on the answers. The government need not await case-by-case adjudication to determine the answers. It can use its rulemaking authority to provide the guidance that landowners and other regulated interests desire.[145]

A noteworthy example of the government's power to go beyond merely defining "harm" is the set of regulations requiring shrimp nets to be equipped with special gear designed to keep sea turtles from being captured and drowned in those nets. The regulations are significant because the prohibited action of using a shrimp net without the required gear rarely results in the taking of a sea turtle for any individual fishing vessel. However, because nearly 20,000 shrimp boats fish throughout much of the year in the Gulf of Mexico and the South Atlantic, the taking of a substantial number of sea turtles is a statistical certainty. The regulations are designed to effectuate a taking prohibition that is otherwise unenforceable as a practical matter. Authority for these regulations derives both from section 4(d) of the Act (authorizing regulations "necessary and advisable . . . for the conservation" of threatened species),[146] and section 11(f) (authorizing "such regulations as may be appropriate to enforce" the Act).[147]

The sea turtle regulations were upheld in *Louisiana ex rel. Guste v. Ver-*

[144]926 F.2d at 938.

[145]It should be noted that the National Marine Fisheries Service has not adopted the Fish and Wildlife Service regulation at issue in *Sweet Home.*

[146]16 U.S.C. § 1533(d).

[147]*Id.* § 1540(f).

ity.[148] Significantly, the challenge attacked the sufficiency of the evidence of the need for regulations, not the agency's authority to adopt them. In much the same way, the government could, and probably should, remove much of the ambiguity surrounding the scope of the taking prohibition by promulgating regulations clearly defining what kinds of habitat modification are likely to result in taking.

The Prohibition against Taking through Harassment

One of the other novel elements of the Act's definition of "take" was its prohibition against harassing endangered species. The administrative definition of this term has many features in common with that of the harm definition. The term "harass" means

> an intentional or negligent act or omission which creates the likelihood of injury to wildlife by annoying it to such an extent as to significantly disrupt normal behavioral patterns which include, but are not limited to, breeding, feeding or sheltering.[149]

Like the definition of "harm," this definition focuses on impacts to behavioral patterns as the key concern and appears to treat these impacts, if sufficiently grave, as constituting "injury." In two respects, it appears to encompass a broader set of activities than does the harm definition. First, it requires only a "likelihood of injury" (i.e., a likelihood of disrupting behavioral patterns) rather than "actual" death or injury (i.e., actual impairment of essential behavioral patterns). Second, it encompasses not only overt acts, but also omissions. The legislative history suggests that the term "harass" was intended to address actions that might have only a transitory impact on the affected species.[150] To date, few courts have addressed the scope of the harassment prohibition, and those that have suggest substantial overlap with the harm prohibition.[151]

[148]853 F.2d 322 (5th Cir. 1988). In 1988 Congress mandated a study of the sea turtle regulations by the National Academy of Sciences. The Academy's resulting report strongly affirmed the need for the regulations to conserve sea turtles. See National Research Council, Decline of the Sea Turtles: Causes and Prevention (1990).

[149]50 C.F.R. § 17.3 (1995).

[150]The House Report accompanying the legislation noted that the prohibition against harassment "would allow . . . the Secretary to regulate or prohibit the activities of birdwatchers where the effect of those activities might disturb the birds and make it difficult for them to hatch or raise their young." H. R. Rep. No. 93–412, 93d Cong., 1st Sess. (1973), *reprinted in* Sen. Comm. on Envt'l and Public Works, 97th Cong., A Legislative History of the Endangered Species Act of 1973 as Amended in 1976, 1977, 1978, 1979, and 1980, at 140, 150 (1982) (hereinafter "Legislative History").

[151]*See* Marbled Murrelet v. Pacific Lumber, Co., *supra* note 120, at 1365–67.

Other Prohibitions Applicable to Endangered Wildlife

In addition to prohibiting taking, section 9 prohibits a variety of activities relating to trade or commerce in endangered wildlife. For example, it prohibits both importing and exporting such wildlife.[152] The Act's broad definition of "import,"[153] not limited by the term's meaning under the customs laws of the United States, encompasses even the unscheduled landing at a U.S. airport of a flight that originated in one foreign country and was destined for another.[154] Also prohibited are the sale or offer for sale of any endangered wildlife species in interstate or foreign commerce, and the delivery, receipt, carriage, transportation, or shipment of any such species in the course of a commercial activity.[155] The possession, sale, delivery, carriage, transportation, or shipment of an illegally taken wildlife specimen is prohibited whether or not it occurs in the course of a commercial activity.[156] Finally, section 9 prohibits violating any regulation promulgated for either an endangered or threatened species.[157]

Prohibitions Applicable to Endangered Plants

The same prohibitions as are applicable to endangered animals apply to endangered plants, with one very important exception: the broad prohibition against taking endangered animals does not apply to plants. Originally, the Act contained no taking prohibition at all with regard to plants. In 1982, however, Congress added a limited prohibition against removing endangered plants from federal lands.[158] Congress expanded this provision still further in 1988, by prohibiting malicious damage to or destruction of, as well as removal of, endangered plants on federal land.

The 1988 amendments also gave limited protection to plants on nonfederal land for the first time. They made it a violation of the Act to "remove, cut, dig up, or damage or destroy" any endangered plant on nonfederal land "in knowing violation of any law or regulation of any

[152]16 U.S.C. § 1538(a)(1)(A).

[153]*Id.* § 1532(10).

[154]United States v. 3,210 Crusted Sides of Caiman Crocodilus Yacare, 636 F. Supp. 1281, 1284 (S.D. Fla. 1986).

[155]16 U.S.C. §§ 1538(a)(1)(E) and (F).

[156]*Id.* § 1538(a)(1)(D).

[157]*Id.* § 1538(a)(1)(G).

[158]1982 Amendments, § 9(b)(1) (current version at 16 U.S.C. §1538(a)(2)(B)). The prohibition applies only if one "remove[s] and reduce[s] to possession" the protected plants. Accordingly, removing endangered plants incidental to development activities on federal land is not prohibited. The common law viewed animals in their wild state as incapable of private ownership, but never took this view of plants. Because of this difference, the constitutional hurdles facing a prohibition against taking plants may be greater than those facing a prohibition against taking wild animals.

state or in the course of any violation of a state criminal trespass law.''[159] These changes make federal protection of endangered plants on nonfederal land dependent upon underlying state law, in a manner similar to the Lacey Act.

Prohibitions Applicable to Threatened Species: Section 4(d)

Section 9 sets forth an array of prohibitions that apply automatically to endangered species. None of these prohibitions, however, applies automatically to threatened species. Rather, section 4(d) authorizes the Secretary to ''issue such regulations as he deems necessary and advisable to provide for the conservation'' of threatened species, including regulations that prohibit any or all of the activities prohibited for endangered species.[160] The Secretary has implemented this authority in two ways. Occasionally, ''special regulations'' applicable only to a particular threatened species have been promulgated. In addition, however, the Secretary has promulgated ''generic'' regulations applicable to all threatened species except those having their own special regulations.

The Secretary's authority to promulgate generic regulations, rather than species-by-species regulations, was challenged in the same case that challenged the Secretary's definition of ''harm'' as including habitat modification. Although the three circuit court of appeals judges who heard that case eventually issued six different opinions on the ''harm'' issue, they were unanimous in upholding the Secretary's authority to issue either generic or special regulations, or both, for threatened species.[161]

Challenges to regulations promulgated pursuant to section 4(d) have fared poorly. For example, in *Cayman Turtle Farm, Ltd. v. Andrus*, plaintiff contended that the Secretary of Commerce's regulations for the threatened green sea turtle were too restrictive in that they prohibited importation of turtles the plaintiff had raised in captivity.[162] The plaintiff argued that the evidentiary basis in the rulemaking record for the restriction did not support the Secretary's determination that it was necessary. The court deferred to the Secretary's expert judgment and rejected the plaintiff's claim.

In *Louisiana ex rel. Guste v. Verity*, the Fifth Circuit Court of Appeals rejected a similar challenge to the regulations requiring shrimp nets to be equipped with special gear to reduce the capture and drowning of sea turtles.[163] Louisiana challenged not only the adequacy of the evidentiary

[159]Pub. L. No. 100–478, § 1006, 102 Stat. 2306, 2308 (1988) (codified at 16 U.S.C. § 1538(a)(2)(B)).

[160]16 U.S.C. § 1533(d).

[161]Sweet Home Chapter of Communities for a Great Oregon v. Babbitt, 1 F.3d 1, 6 (D.C. Cir. 1993).

[162]478 F. Supp. 125 (D.D.C. 1979), *aff'd without opinion* (D.C. Cir. Dec. 12, 1980).

[163]853 F.2d 322 (5th Cir. 1988).

basis for the regulations, but also the Secretary's alleged failure to demonstrate that they would halt the decline of sea turtles. In effect, the state argued that unless the regulations could be shown to halt the decline, they were not "necessary" for conservation of the species. The court rejected both of these contentions, holding that "Congress simply presumes that prohibited takings will deplete the species," and that courts must honor its presumption.[164]

In *Sierra Club v. Clark,* the court held that the Secretary had no authority to promulgate special regulations that allowed sport trapping of threatened gray wolves in Minnesota.[165] The court focused its attention on the fact that the Secretary's authority under section 4(d) was to promulgate regulations that were necessary and advisable for the "conservation" of threatened species. The term "conservation," in turn, is defined broadly in the Act, except in one critical respect. That definition includes the use of a broad range of tools, but provides that "regulated taking" can be used only "in the extraordinary case where population pressures within a given ecosystem cannot be otherwise relieved."[166] The court invalidated the regulations because the Secretary had not made any finding that population pressures existed.

The *Sierra Club v. Clark* opinion itself suggests that the court's holding may not be applicable to species reintroduced into the wild as "experimental populations" under the authority of section 10(j) of the Act.[167] The court in a later case also construed the decision narrowly, finding it inapplicable to situations not involving activities specifically directed at the threatened species.[168]

Penalties

Penalties for violating the Act's prohibitions differ depending on whether the violation concerns a threatened or endangered species, the violator's state of knowledge, and, for some purposes, the violator's business. The stiffest penalties may be imposed against those who knowingly violate the Act's prohibitions with respect to endangered species; they can be imprisoned for a year and fined $50,000.[169] Knowing violations of the

[164]*Id.* at 333.

[165]755 F.2d 608 (8th Cir. 1985).

[166]16 U.S.C. § 1532(3).

[167]755 F.2d at 618. For a discussion of experimental populations, see text accompanying notes 199–203, *infra.*

[168]Pacific Northwest Generating Coop. v. Brown, 822 F. Supp. 1479, 1509 (D. Or. 1993), *aff'd* 38 F.3d 1058 (9th Cir. 1994).

[169]16 U.S.C. §1540(b)(1). The 1978 Amendments lowered the culpability standard for criminal violators from "willful" to "knowing." Pub. L. No. 95–632, § 6(3), 92 Stat. 3751, 3761 (1978).

Act's prohibitions pertaining to threatened species can be punished by imprisonment for six months and a fine of up to $25,000.[170] In addition, the Secretary must suspend for up to a year, or cancel, any federal fishing or hunting permit of any person convicted of a criminal violation of the Act.[171] A broad range of discretionary administrative sanctions also may be imposed against those convicted of criminal violations, including the immediate revocation or suspension of any federal lease, permit, or license authorizing the import or export of wildlife or plants, the operation of a quarantine station, or the use of federal lands.[172]

Persons who knowingly violate the Act's prohibitions with respect to endangered species are also subject to a civil penalty of up to $25,000 per violation.[173] Half that amount may be imposed for threatened species violations.[174] The foregoing civil penalties also may be assessed against persons engaged in business as importers or exporters of wildlife or plants whether their violations are knowing or unknowing.[175] For others, the maximum civil penalty that may be imposed for unknowing violations is $500.[176]

The Act authorizes a third type of penalty that potentially is of even greater magnitude than the criminal and civil penalties. Not only the wildlife or plant specimens involved in an unlawful act, but the guns, equipment, vessels, aircraft, and vehicles used to aid such an action are subject to forfeiture.[177] The Act's forfeiture provision was construed in an unusual factual context in *Carpenter v. Andrus.*[178] In that action the plaintiff had shot a leopard under authority of a Kenyan license. He instructed a shipper to ship the skull and skin to Haiti without shipping it through the United States. Contrary to his instructions, the skull and skin were shipped by common carrier into the United States where they were seized. Because section 11(e)(4)(A) subjects to forfeiture any wildlife imported contrary to the provisions of the Act, the court had to consider whether the products had been unlawfully imported. Despite the unqualified language of section 9(a)(1)(A) prohibiting importation of any endangered species into the United States,[179] the court reasoned that the provision was not

[170] *Id.*

[171] *Id.* § 1540(b)(2).

[172] *Id.* Section 9(d) of the Act requires importers and exporters of most wildlife and plants to be licensed by the Secretary. *Id.* § 1538(d).

[173] *Id.* § 1540(a)(1).

[174] *Id.*

[175] *Id.*

[176] *Id.*

[177] *Id.* § 1540(e)(4). For a case upholding the authority to seize products derived in part from endangered species, *see* Delbay Pharmaceuticals, Inc. v. Department of Commerce, 409 F. Supp. 637 (D.D.C. 1976).

[178] 485 F. Supp. 320 (D. Del. 1980).

[179] 16 U.S.C. § 1538(a)(1)(A)(1976).

intended to impose on common carriers a "strict duty to inspect their freight to assure that it did not contain any endangered species."[180] Accordingly, if the common carrier had no duty to inspect, then the products were not imported contrary to the provisions of the Act and thus not subject to forfeiture.

The extraordinary circumstances of the *Carpenter* case probably account for the court's strained result. The question whether a common carrier could be held to a standard of strict liability was not directly at issue because no action for recovery of civil penalties had been brought against the carrier. Moreover, the maximum penalty that could be imposed against a common carrier for an unknowing violation is a modest, and presumably insurable, $500. Curiously, despite the court's holding that the products had not been unlawfully imported and were not subject to forfeiture, the court declared that the plaintiff "is not entitled to keep the property in the United States, for that would contravene the policy of the Endangered Species Act."[181] That holding of the *Carpenter* case was legislatively ratified in the 1982 amendments, which except from the prohibition against importation the sort of inadvertent, noncommercial transshipments that occurred there.[182]

Exceptions to the Act's Prohibitions

The Act provides for a number of exceptions to its prohibitions. The list of exceptions has expanded steadily since 1973 from an original four to at least a dozen today. The following analysis discusses each of them in roughly the order in which they were enacted.

The Original Exceptions

Originally, the Act provided special exceptions for certain specimens already possessed at the time of its enactment, for specified cases of economic hardship, for certain Alaskan residents, and for scientific or propagation purposes. The first of these exempts certain activities with respect to "any fish or wildlife held in captivity or in a controlled environment" on the date of the Act's enactment or the species' listing, provided it was not then or subsequently held for commercial purposes.[183] However, the

[180]485 F. Supp. at 322.

[181]*Id.* at 324.

[182]1982 Amendments § 6(5) (codified at 16 U.S.C. § 1539(i)).

[183]Pub. L. No. 93–205, § 9(b), 87 Stat. 884, 894 (1973) (current version at 16 U.S.C. § 1538(b)(1)). This exemption originally applied to any action otherwise prohibited by section 9. The 1982 Amendments limited it, however, to import and export activities involving endangered species as well as any activity involving threatened species. This limitation makes sense in light of the clarification, also added by the 1982 Amendments, that the exemption is negated if the specimen was held for commercial purposes on the relevant date or anytime

captive progeny of exempt animals are not exempt from the Act's proscriptions.[184] To provide a flexible means of regulating captive progeny, as well as other captive endangered wildlife, the Fish and Wildlife Service has listed as threatened certain captive populations of otherwise endangered species. In so doing, the Service relies upon the Act's broad definition of "species" to include discrete populations of vertebrates.

The 1973 Act also gave the Secretary limited discretion to exempt otherwise prohibited activities to avoid "undue economic hardship" to persons who entered into contracts with respect to a species prior to the first published notice that it would be considered for listing.[185] This authority is of little practical significance, since the exemption cannot extend beyond one year from the date of the first notice, and, except in emergency situations, listings are seldom finalized within a one-year period.[186]

The third of the original exemptions applies to certain Alaskan natives and non-native permanent residents of Alaskan native villages, who may take listed species "primarily for subsistence purposes" and may sell nonedible byproducts of listed species in interstate commerce "when made into authentic native articles of handicrafts and clothing."[187] If the Secretary determines, however, that Alaskan native taking of a listed species "materially and negatively affects" it, the Secretary may prescribe regulations to restrict native taking for whatever period is necessary.[188]

The Supreme Court left undecided in *United States v. Dion* the issue of whether the Endangered Species Act applies to American Indians with treaty hunting rights.[189] The Court was able to sidestep the issue because

thereafter. 1982 Amendments, § 9(b). That clarification was needed in light of the holding in United States v. Molt, 10 Envt'l L. Rep. (Envt'l L. Inst.) 20777, 20778 (3rd Cir. July 17, 1980), that ambiguity as to the exemption rendered the Act's prohibition against importation unconstitutionally vague. The Molt decision was at odds with United States v. Kepler, 531 F.2d 796 (6th Cir. 1976), which interpreted the original commercial activity limitation on the exception to apply to a holding for commercial purposes after the Act's enactment.

[184]In Cayman Turtle Farm, Ltd. v. Andrus, *supra* note 162, the plaintiff alleged that the Act gave the Secretary no authority over animals raised in captivity. The court rejected that claim on the basis that the Act's limited exemption for captive animals implied that other captive animals were subject to regulation.

[185]16 U.S.C. § 1539(b).

[186]A further curiosity of this exemption is that, although the Act defines "undue economic hardship" to include the curtailment of subsistence taking under certain circumstances, *id.* § 1539(b)(2), such hardship does not qualify for the exemption unless the additional requirement of a contract is met. *Id.* § 1539(b)(1). The court in Delbay Pharmaceuticals, *supra* note 177, held that endangered species products lawfully imported pursuant to a hardship permit issued under the 1969 Act could not be sold in interstate commerce after passage of the 1973 Act if they were held for commercial purposes at the time of passage.

[187]16 U.S.C. § 1539(e)(1). Compare this exemption closely with the Alaskan native provisions of the Marine Mammal Protection Act and the subsistence priority provisions of the Alaska National Interest Lands Conservation Act, discussed in Chapters 5 and 8, respectively.

[188]*Id.* § 1539(e)(4).

[189]476 U.S. 734 (1986).

the case involved the prosecution of an Indian for taking bald eagles in violation of both the Bald and Golden Eagle Protection Act and the Endangered Species Act. The Court found that Congress intended the former statute to abrogate Indian treaty hunting rights, rendering it unnecessary to consider whether Congress also intended the Endangered Species Act to abrogate treaty rights.

In *United States v. Nuesca*, the Ninth Circuit Court of Appeals rejected a claim of native Hawaiians that they had aboriginal hunting rights to which the Endangered Species Act did not apply.[190] The court also rejected the constitutional claim that the statutory exemption of Alaskan natives deprived native Hawaiians of the equal protection of the laws.

Finally, the 1973 Act authorized the Secretary to permit any otherwise prohibited act "for scientific purposes or to enhance the propagation or survival of the affected species."[191] Applications for these permits, as well as applications for hardship exemptions, ordinarily must be noticed in the Federal Register with the opportunity for public comment on them. Although these provisions of the Act show an intent to promote scientific research and to avoid certain types of economic hardship, the Act is also clear that the conservation interests of listed species take priority over other concerns. The Secretary may grant permits and exemptions only if they will "not operate to the disadvantage" of the affected species.[192]

Exemptions Added in 1976 and 1978

The various exceptions set forth in the 1973 Act have all been retained, and Congress has added numerous others in later amendments. In 1976, Congress created a special exemption for certain sperm whale oil and scrimshaw products held within the United States on the effective date of the 1973 Act.[193] Originally, this exemption was to last only three years, since Congress anticipated that the stocks on hand would be exhausted within that period. That expectation was not realized, however, and Congress has repeatedly extended the exemption. In 1988, in the most recent amendments, Congress authorized the Secretary to renew certificates of exemption for up to five more years.

Also exempted from most of the Act's restrictions are raptors held in captivity (whether for commercial or noncommercial purposes) as of the

[190]945 F.2d 254 (9th Cir. 1991).

[191]Pub. L. No. 93–205, § 10(a), 87 Stat. 884, 896 (1973) (current version at 16 U.S.C. § 1539(a)).

[192]16 U.S.C. § 1539(d)(2). Legislative history indicates that this requirement was intended "to limit substantially the number of exemptions that may be granted." H. R. Rep. No. 93–412, 93d Cong. 1st Sess. at 17 (1973), *reprinted in* Legislative History at 156.

[193]16 U.S.C. § 1539(f). These products did not qualify for the "controlled environment" exemption discussed at text accompanying notes 183–84, *supra*, since at the time of the law's enactment, they were held for commercial purposes.

date of the 1978 amendments, and their captive progeny.[194] The Secretary of the Interior is, however, authorized to require the maintenance and production of records appropriate for administering the exemption.[195] Similarly, articles more than 100 years old and composed in whole or in part of listed species are exempted from certain of the Act's requirements, subject to such documentation standards as the Secretary of the Treasury may impose.[196]

A "self-defense" provision, an exemption of sorts, was added in 1978. This provision makes it a defense to any criminal prosecution under the Act that an action in violation of the Act was taken in defense of the defendant's or anyone else's physical well-being.[197] Civil penalties still may be assessed in cases of self-defense, however, and the exemption does not apply to defense of property. In *Christy v. Hodel*, a rancher who had suffered losses of sheep from grizzly bear predation killed a grizzly, an endangered species, and was assessed a civil penalty.[198] Christy challenged the civil penalty. The Ninth Circuit Court of Appeals rejected Christy's contentions that the statute's failure to exempt actions taken in defense of one's property deprived the plaintiff of property without due process of law and constituted a taking of property for which the government must provide compensation.

The 1982 Amendments: Experimental Populations

The 1982 amendments to the Act added three new types of exemptions. The most novel of these was the authority to list as "experimental populations" certain populations of otherwise endangered species and to apply fewer restrictions to those populations.[199] The impetus for this change was the awareness that while intentionally establishing a population of an endangered species in a new area may be an important conservation measure, the Act's stringent prohibitions may discourage the cooperation of private landowners or other government agencies in these situations. To

[194]16 U.S.C. § 1538(b)(2).

[195]*Id.* § 1538(b)(2)(B).

[196]*Id.* § 1539(h). The 1978 amendments added both the raptor and antique articles exemptions and provided exemptions from the requirements of both the Act and CITES. The 1982 Amendments modified the raptor exemption so that CITES requirements would remain applicable but made no comparable change in the antique articles exemption. The latter exemption is curiously structured. On its face, it provides an exemption for qualified antiques from both import and interstate commerce prohibitions. However, to be a qualified antique, an article must be entered into the United States through a specially designated customs port. Thus, the exemption does not apply generally to all antiques, but only to those lawfully imported since 1978.

[197]*Id.* § 1540(b)(3).

[198]857 F.2d 1324, 1331, 1335 (9th Cir. 1988), *cert. denied*, 490 U.S. 1114 (1989).

[199]16 U.S.C.. § 1539(j).

foster cooperation, Congress introduced the category of "experimental populations."

Although the rationale for the experimental population provision is easily understood, the provision itself is complicated. First, not every population purposely established through human effort qualifies as an "experimental population." Only populations established, with the Secretary's prior permission, outside the species' current range, are "experimental populations" within the meaning of the Act.[200] In addition, before allowing establishment of a new population, the Secretary must determine by regulation whether the population is "essential to the continued existence" of the species.[201] This determination dictates whether section 7 will apply to the new population. Experimental populations essential to the survival of a species are fully protected by section 7, while those not essential are unprotected by section 7, except when they occur on units of the National Park or National Wildlife Refuge Systems.[202]

Whether essential to the survival of the species or not, any experimental population is to be separately listed as a threatened species.[203] Thus the

[200] *Id.* § 1539(j)(1).

[201] *Id.* § 1539(j)(2)(B). Populations established prior to the 1982 Amendments with the authorization of the Secretary may be declared experimental populations if the Secretary undertakes a comparable rulemaking. *Id.* §1539(j)(3). Unless and until the Secretary does so, however, the population remains fully subject to the other provisions of the Act. *See* H. R. Conf. Rep. No. 97–835, 97th Cong., 2d Sess. at 34 (1982), *reprinted in* 1982 U.S.C.C.A.N. 2860, 2875. Although a new Section 11(g)(1)(C) of the Act, added by the 1982 Amendments, authorizes citizen suits to compel the Secretary to undertake nondiscretionary listing actions mandated by Section 4, no provision of the Act authorizes suits to compel the Secretary to determine whether previously established populations are "experimental populations."

[202] 16 U.S.C. § 1539(j)(2)(C). Nonessential populations outside those areas are treated as species proposed to be listed, making them subject to the conferral requirements of Section 7(a)(3). *See* text accompanying notes 251–252, *infra.* Even though an experimental population may not be essential to the continued existence of a species, it is not illogical to apply Section 7 to the population in some fashion. An experimental population, once separately listed, becomes a "species" for purposes of the Act. Section 7(a)(3) thus applies directly to it, independently of other populations of the same species.

[203] 16 U.S.C. § 1539(j)(2)(C). The language is unartful ("each member of an experimental population shall be treated as a threatened species"), but the legislative history is clear that it is the population itself, and not its individual members, that is to be treated as threatened. *See* H. R. Conf. Rep. No. 97–835, 97th Cong., 2d Sess. at 34 (1982), *reprinted in* 1982 U.S.C.C.A.N. 2860, 2875. Treating an experimental population as a separately listed species has an interesting consequence for the application of Section 7 to nonexperimental populations of the same species. When determining whether a federal action will jeopardize the continued existence of a listed species, the Secretary must evaluate the effects of the action solely on the nonexperimental populations, since the experimental populations are treated as separately listed "species." This seems to be the intended result, since nothing in the legislative history of the 1982 Amendments suggests that establishing an experimental population reduces Section 7 obligations toward naturally occurring populations of a listed spe-

Secretary can appropriately tailor the prohibitions applicable to that population. This flexibility, together with the partial waiver of section 7 for nonessential populations, was intended to give the Secretary a greater chance of gaining the cooperation of landowners and other government agencies in establishing new populations of endangered species. That is not to say, however, that the regulations applicable to experimental populations are to be developed through bargaining. Like any other threatened species regulation, they must be developed through rulemaking and must include the measures that are necessary and advisable for conservation of the species.

The 1982 Amendments: Incidental Takings

The remaining two exceptions added by the 1982 amendments pertain to the incidental taking of listed species. As the earlier discussion of the Act's taking prohibition noted, neither direct physical injury nor intent to cause injury to a listed species is required by the Act's broad definition of "take." Thus a wide range of common activities, including forest practices, land clearing, hydropower operation, commercial fishing, and others, could result in the unintentional, incidental taking of listed species. All takings are prohibited and punishable by significant civil and criminal penalties, but the Act's potential for influencing activities that result in incidental taking depends to a large extent on the public perception of the likelihood of prosecution. Not surprisingly, the government rarely prosecuted, and the Act's theoretically absolute prohibition against any form of taking was seriously undermined by nonenforcement.

In 1982, Congress authorized the Secretary to permit otherwise prohibited takings of endangered species if they are "incidental to, and not the purpose of, the carrying out of an otherwise lawful activity."[204] On its face, this exception (found in section 10 of the Act) seems to ease the Act's restrictions because it permits what was previously prohibited. In fact, however, this provision likely increased the Secretary's leverage over activities that incidentally take endangered species because it substituted a flexible regulatory authority for a threat of prosecution that few found credible.

To receive an incidental take permit under section 10, an applicant must submit to the Secretary a "conservation plan."[205] The Secretary must find that the plan includes measures to mitigate any incidental takings

---

cies. Of course, if the establishment of an experimental population brings about the recovery of a species, then it may be appropriate to delist all populations.

[204]1982 Amendments § 6(1) (codified at 16 U.S.C. § 1539(a)(1)(B)). The Secretary's authority with respect to threatened species is flexible enough to permit regulations that authorize incidental take. It is probably for this reason that the permit authority added in 1982 applies only to endangered, and not threatened, species.

[205]16 U.S.C. § 1539(a)(2)(A). These plans are often called "habitat conservation plans" or "HCPs," although the word "habitat" does not appear in the statute.

that occur and that the applicant will provide adequate funding for the plan.[206] In addition, the Secretary must find that the impacts of any incidental takings will be minimized and mitigated "to the maximum extent practicable,"[207] and that the taking will not "appreciably reduce the likelihood of the survival and recovery of the species in the wild."[208] The last requirement is the same as the administrative standard for "jeopardy" developed under section 7;[209] thus the Secretary may not permit an incidental taking that jeopardizes the continued existence of an endangered species.

The first permit authorizing incidental take of an endangered species under the 1982 amendments was challenged unsuccessfully in *Friends of Endangered Species, Inc. v. Jantzen.*[210] The permit authorized incidental take of the endangered mission blue butterfly in the course of developing land on San Bruno Mountain, south of San Francisco. The plaintiff mainly attacked the adequacy of the field studies underlying the government's decision to issue the permit and its failure to prepare an environmental impact statement under the National Environmental Policy Act. The court declined to disturb the Secretary's decision to issue the permit, notwithstanding differing expert opinions as to the impact of the proposed development on the butterfly. Nor did the court find that an environmental impact statement was required, in light of the mitigation measures required under the permit.

Issuance of an incidental take permit has become an increasingly common means of resolving the conflict between endangered species conservation needs and private landowners' economic objectives. Though there is a growing amount of commentary on this conservation tool, there have been no other reported cases.[211]

The 1982 Amendments: Incidental Taking by Federal Agency Activities

The final exception added by the 1982 amendments tries to harmonize the taking prohibition of section 9 with the federal agency duties of sec-

[206]*Id.*, §§ 1539(a)(2)(B)(ii), and (iii).

[207]*Id.* § 1539(a)(2)(B)(ii).

[208]*Id.* § 1539(a)(2)(B)(iv).

[209]See discussion of the prohibitory commands of section 7(a)(2), *infra*, at text accompanying notes 232–299.

[210]760 F.2d 976 (9th Cir. 1985).

[211]For additional reading on habitat conservation plans, see Comment, *Habitat Conservation Planning under the Endangered Species Act: No Surprises and the Quest for Certainty*, 67 Colo. L. Rev. 371 (1996); T. Beatley, Habitat Conservation Planning: Endangered Species and Urban Growth (1994); J.B. Ruhl, *Regional Habitat Conservation Planning under the Endangered Species Act: Pushing the Legal and Practical Limits*, 44 Sw. L.J. 1393 (1991); and M. Bean, S. Fitzgerald, and M. O'Connell, Reconciling Conflicts Under the Endangered Species Act (World Wildlife Fund 1991).

tion 7. Some agency actions that satisfy the section 7 requirement to refrain from jeopardizing the continued existence of a listed species may nonetheless cause the incidental taking of one or more individuals of that species. To ensure that section 9 does not prohibit federal actions that satisfy the section 7 jeopardy standard, in 1982 Congress amended section 7 to waive effectively the taking prohibition for these actions, provided the agency comply with specified measures to minimize incidental takings.[212] A proper understanding of this provision requires a fuller appreciation of the requirements of section 7, addressed in the next part.

The Affirmative Duties of Section 7(a)(1)

Section 7(a)(1) contains two closely related provisions. The first is directed to the Secretaries of Interior and Commerce, the agency heads charged with principal responsibility for implementing the federal endangered species program. They are to review other programs administered by them and to "utilize such programs in furtherance of the purposes of this [Act]."[213] The second provision is directed to all other agencies and requires them to "utilize their authorities in furtherance of the purposes of this [Act] by carrying out programs for the conservation of endangered species and threatened species."[214] After more than two decades, these duties remain relatively unexplored. According to one commentator, "section 7(a)(1) has been the monumental underachiever of the ESA family."[215] The same commentator, however, characterizes section 7(a)(1) as a "sleeping giant" that "has the potential to eclipse all other ESA programs."[216] The case law to date bears out the former characterization but gives little support for the latter assertion.

The first few cases to address the affirmative duties of section 7(a)(1) involved the Secretary of the Interior's administration of the migratory waterfowl hunting program. In *Defenders of Wildlife v. Andrus*, plaintiff alleged that the challenged regulations, insofar as they permitted migratory bird hunting during twilight hours, failed to provide adequate assurance that endangered birds would not be mistakenly killed as a result of hunter misidentification.[217] The court found that the regulations' administrative record was "virtually barren of any information regarding the impact of the contested shooting hours on birds that should not be taken"[218] and that there was a "substantial argument," presented through the plaintiff's

[212] 16 U.S.C. §§ 1536(b)(4) and (o)(2).
[213] *Id.* § 1536(a)(1).
[214] *Id.*
[215] Ruhl, *supra* note 103, at 1128.
[216] *Id.* at 1109, 1110.
[217] 428 F. Supp. 167, 169 (D.D.C. 1977).
[218] *Id.* at 169.

affidavits, "that the destruction of protected species may be considerable."[219]

The *Defenders* court concluded that under the Endangered Species Act the Secretary must "do far more than merely avoid the elimination of protected species"; rather, the Secretary has "an affirmative duty to increase the population of protected species."[220] Since the rulemaking process had not adequately focused on this duty, the court found the resulting regulations to be arbitrary and unlawful.

In *Connor v. Andrus*, the Secretary had restricted duck hunting in parts of three states to insure that there would be no mistaken shooting of the endangered Mexican duck.[221] The restriction was challenged as inadequately supported by the administrative record. Although the *Connor* court professed to agree with *Defenders* that the Secretary had an affirmative duty to restore endangered species, in its view the administrative record failed to show that the hunting ban would serve that duty.[222] Thus the *Connor* court, like that in *Defenders*, searched the administrative record for support of the challenged regulations, but found none. The superficial difference in the approach of the two courts was that one required affirmative record evidence to support the absence of restrictions while the other required evidence to support restrictions. In fact, both courts seem to have been moved by extrarecord evidence produced by the plaintiffs regarding whether the challenged regulations were necessary to prevent harm to endangered species.[223]

A third case to consider how section 7(a)(1) affected the Secretary's duty to manage waterfowl hunting was *National Wildlife Federation v. Hodel*.[224] The Secretary had postponed for a year the effective date of a requirement that waterfowl hunters use nontoxic shot rather than lead shot, despite evidence of significant risk to bald eagles from lead poisoning as a result of the eagles ingesting lead shot while feeding on ducks. The court did not dispute the government's contention that it could "pick and choose between a number of different" means of carrying out the mandatory duty imposed by section 7(a)(1).[225] The court held, however, that the government had neither clearly identified the factors it considered in making its choice nor articulated a rational connection between those factors and the choice it made.

None of these cases imparts any great significance to section 7(a)(1). Instead, in each case, the result reached is unsurprising under ordinary

[219] *Id.* at 170.
[220] *Id.*
[221] 453 F. Supp. 1037, 1041 (W.D. Tex. 1978).
[222] *Id.* at 1041.
[223] *See* 428 F. Supp. at 170 and 453 F. Supp. at 1041 n. 2.
[224] 23 Envt'l Rep. Cas. (BNA) 1089 (E.D. Cal. Aug. 26, 1985).
[225] *Id.* at 1092.

principles of judicial review of administrative agency action. The courts in *Defenders of Wildlife* and *National Wildlife Federation* may have been somewhat more demanding than usual in insisting upon a clear basis for agency action, but if so, the difference was one of degree, not kind.

Two cases involving California and Nevada's Truckee River have more deeply probed the nature of the section 7(a)(1) duty. The Truckee River empties into Pyramid Lake on the reservation of the Pyramid Lake Tribe of Paiute Indians. Pyramid Lake is the sole home of an endangered fish, the cui-ui, and one of a few places where a threatened fish, the Lahontan cutthroat trout, occurs. The cui-ui's endangered status resulted from diversions of Truckee River water for irrigation and for municipal and industrial use. Water diversion has caused the lake level to drop considerably, so the cui-ui are seldom able to swim out of the lake, into the Truckee River, in order to spawn.

Upstream from Pyramid Lake, the Interior Department manages Stampede Dam and Reservoir. When the Department refused to sell water from the reservoir to the cities of Reno and Sparks, Nevada, the cities sued. The cities argued that there was sufficient water available to supply the cities without jeopardizing the continued existence of the fish, which would have violated section 7(a)(2) of the Act. The Department of the Interior, however, argued that merely avoiding jeopardy to the fish would not fully discharge its duties. In addition, the agency argued, it had a duty under section 7(a)(1) "to replenish the species so that they are no longer endangered or threatened with extinction."[226] The court agreed. Even if section 7(a)(1) thus provides a source of authority for actions beneficial to protected species, however, later Truckee River litigation dispelled the notion that it might *compel* any particular beneficial action.

In *Pyramid Lake Paiute Tribe of Indians v. U.S. Department of the Navy*, the Tribe challenged agricultural leasing practices at nearby Fallon Naval Station.[227] To reduce dust that interfered with flight training, the Navy annually leased out areas near the runway for planting irrigated crops. The water needed to irrigate the crops was diverted from the Truckee River, and as a result, flows into Pyramid Lake were reduced. The Tribe argued that the Navy's legitimate concern with reducing dust could be addressed through other means that used less water, including planting different crops. Its theory of the duty imposed by section 7(a)(1) was that

> if an alternative to the challenged action would be equally as effective at serving the government's interest, and at the same time would enhance conservation to

[226]Carson-Truckee Water Conservancy Dist. v. Watt, 549 F. Supp. 704, 708–709 (D. Nev. 1982), *aff'd sub nom.* Carson-Truckee Water Conservancy Dist. v. Clark, 741 F.2d 257 (9th Cir. 1984), *cert. denied,* 470 U.S. 1083 (1985).

[227]898 F.2d 1410 (9th Cir. 1990).

an equal or greater degree than does the challenged action, then the agency must adopt the alternative.[228]

This standard (except for the portion pertaining to alternatives that enhance conservation "to an equal" degree) seems like a sensible and workable approach. The Ninth Circuit, however, declined to embrace it.

The court's opinion noted that the standard the Tribe offered "would work to divest an agency of virtually all discretion in deciding how to fulfill its duty to conserve."[229] Although the Navy had not followed the agricultural leasing approach urged by the Tribe, it had foregone a plan to lease a larger area and was in the process of studying water conservation alternatives. In short, it had at least done something, unlike the Federal Emergency Management Authority in *Florida Key Deer v. Stickney*, which was held to have violated its duty under section 7(a)(1) because it had "failed to consider or undertake *any* action to fulfill its mandatory obligations under section 7(a)(1)."[230]

Perhaps recognizing that an obligation merely to do something, rather than nothing, would rob section 7(a)(1) of any practical value, the Ninth Circuit went on to say in the Paiute Tribe case that even if the court were to adopt the "stringent standard" urged by the Tribe, it would find for the Navy. The district court had concluded that the Tribe's recommendations would have an "insignificant effect upon the availability of water in the lower Truckee River for the preservation of the cui-ui."[231] Thus, the court appeared to leave open the possibility of reading section 7(a)(1) to prevent a federal agency from choosing an alternative that is effective at serving the agency's purpose but significantly less effective at promoting species conservation than an alternative that it did not pursue.

The Prohibitions of Section 7(a)(2)

Though the precise nature of the affirmative obligations imposed on federal agencies by section 7(a)(1) remains unclear and relatively unexplored, a substantial body of case law examines the prohibitory commands of section 7(a)(2). The latter section obliges every federal agency to

insure that any action authorized, funded, or carried out by such agency . . . is not likely to jeopardize the continued existence of any endangered species or threat-

[228] *Id.* at 1417.
[229] *Id.* at 1418.
[230] 864 F. Supp. 1222, 1238 (S.D. Fla. 1994) (emphasis in original).
[231] *Quoted in* 898 F.2d at 1418.

ened species or result in the destruction or adverse modification of [critical] habitat of such species.[232]

These duties are to be carried out "in consultation with and with the assistance of the Secretary." As originally enacted, the Endangered Species Act neither defined the key terms in this provision nor specified procedures for its implementation. Later amendments have added considerable detail to the provision's procedural aspects but have left largely intact the substantive commands of avoiding jeopardy to the continued existence of listed species and avoiding the adverse modification of critical habitat. The discussion that follows examines these closely interrelated commands.

The Duty to Avoid Jeopardizing the Continued Existence of Listed Species: The Tellico Dam Decision

The section 7(a)(2) directive that federal agencies ensure that their actions not jeopardize the continued existence of listed species has been the single most significant provision of the Endangered Species Act. The Supreme Court established the power and scope of this provision in forceful terms less than five years after Congress passed the Act.

In *TVA v. Hill*, plaintiffs challenged completion of Tellico Dam on the Little Tennessee River.[233] The dam threatened to destroy virtually the entire known habitat of a recently discovered fish, the endangered snail darter. Though the district court agreed that the dam would surely jeopardize the snail darter's continued existence, it refused to apply section 7 retroactively to halt a project that was substantially completed at the time the species was discovered and that would entail the loss of several million dollars if not completed.[234] Although the court acknowledged that balancing the interests between wildlife conservation and development was

[232]16 U.S.C. § 1536(a)(2). Originally, the Act required federal agencies to insure that each of their actions "does not jeopardize" the continued existence of listed species. The change to the current "is not likely to jeopardize" standard was made as a result of the Endangered Species Act Amendments of 1979, Pub. L. No. 96–159, § 4(1)(C), 93 Stat. 1225, 1226 (1979). The legislative history reveals that this change was not intended to alter the obligations of federal agencies. The most useful discussions of that legislative history are found in E. Erdheim, *The Wake of the Snail Darter: Insuring the Effectiveness of Section 7 of the Endangered Species Act*, 9 Ecol. L. Q. 629, 655 (1981), and O. A. Houck, *The "Institutionalization of Caution" Under § 7 of the Endangered Species Act: What Do You Do When You Don't Know?* 12 Envt'l L. Rep. (Envt'l L. Inst.) 15001, 15006–07 (1982). Although the court in Roosevelt Campobello International Park Comm'n v. EPA, 684 F.2d 1041, 1048 (1st Cir. 1982), stated that this change "softened the obligation" of federal agencies, it went on to impose duties more far reaching than any previously imposed by any court in an Endangered Species Act case. *See* text accompanying notes 263–267 *infra*.

[233]437 U.S. 153 (1978).

[234]Hill v. TVA, 419 F. Supp. 753 (E.D. Tenn. 1976).

a legislative rather than a judicial function, it asserted that Congress had already done the balancing by appropriating funds for the dam's construction after the fish's existence had become known.[235] The court of appeals reversed the district court's decision, holding that the lower court had no choice but to enjoin completion of the dam in light of its finding that its completion would jeopardize the snail darter's continued existence.[236]

The Supreme Court agreed to review the case. Attorney General Griffin Bell argued the case personally. The Attorney General's presence in a Supreme Court case usually means that the United States attaches great importance to the issue involved, but in this remarkable case, Attorney General Bell's presence instead reflected bitter divisions within the federal government. The government's brief set forth the TVA position that the appellate court's decision was wrong and should be reversed. Appended to it, however, were the separate and diametrically opposite views of the Secretary of the Interior.

The Supreme Court, by a 6–3 majority, held that section 7 prohibited TVA from closing the gates on Tellico Dam, even though that action was the culmination of a process that had begun years prior to passage of the Endangered Species Act.[237] Chief Justice Burger, who wrote for the majority, reasoned that "[o]ne would be hard pressed to find a statutory provision whose terms were any plainer than those in § 7,"[238] and concluded that the final step of gate closure was a "federal action" subject to the prohibition against jeopardy. The provision's language was reinforced, in the Chief Justice's view, by the overall structure of the Act, which showed a "plain intent," reflected "in literally every section of the statute," to "halt and reverse the trend toward species extinction, whatever the cost."[239] Implicit in the Chief Justice's argument was the notion that, whenever a federal agency has a choice that affects the well-being of a listed species, the exercise of that authority constitutes an "action" subject to the requirements of section 7.

Justice Powell argued in a strong dissent that section 7 applies only to "actions with respect to which the agency has reasonable decisionmaking alternatives still available" and that abandonment of the Tellico project was not a reasonable alternative.[240] The factual predicates of both Burger's and Powell's arguments apparently were erroneous.[241] Nevertheless, Jus-

[235] *Id.* at 762.

[236] Hill v. TVA, 549 F.2d 1064 (6th Cir. 1977).

[237] TVA v. Hill, 437 U.S. 153 (1978).

[238] *Id.* at 173.

[239] *Id.* at 184.

[240] *Id.* at 205.

[241] Justice Powell's assumption that abandonment of the Tellico project was not a reasonable alternative (at least in economic terms) was subsequently rejected when the project and its alternatives were carefully reconsidered. By a unanimous vote, the newly created Endan-

tice Burger's opinion was the majority opinion and became the law of the case.

Congressional Response to the Tellico Decision

Congress responded swiftly to the Supreme Court's decision, but with surprising restraint, by enacting the Endangered Species Act Amendments of 1978.[242] Despite extraordinary pressure to exempt Tellico Dam from the requirements of the Act, Congress chose instead to adopt a formal process for considering the exemption of any project presenting an "irresolvable conflict" with section 7. Tellico Dam was to be considered immediately for an exemption pursuant to an expedited and truncated procedure.[243] The determining factors were to be whether there were "no reasonable and prudent alternatives to the agency action" and whether the benefits of the action clearly outweighed the benefits of alternatives that did not jeopardize endangered species.[244]

The main reason Congress revised section 7 in 1978 was to add the exemption process, but this amendment was not the only change Congress made with regard to federal actions. The 1978 amendments also formalized the process of consultation between the Secretary and the federal agencies whose actions are subject to section 7,[245] prescribed certain re-

---

gered Species Committee disapproved an exemption for Tellico Dam under the standards Congress had just set. *See* text accompanying note 331 *infra.* The faulty premise of Chief Justice Burger's reasoning was that "by definition, any *prior* action of a federal agency which *would* have come under the scope of the Act must have already *resulted* in the destruction of an endangered species or its critical habitat" 437 U.S. at 186 n. 32 (emphasis in original). In fact, the adverse effects of many actions do not fully occur right away. Where the adverse impacts have not yet been fully realized, the undoing of the action may avoid the jeopardy that is otherwise certain. In light of the delayed impact on endangered species of many activities, what are federal agencies' duties with respect to projects already completed? This question was presented in Colorado River Water Conservation District v. Andrus, Civil Action 78–A-1191 (D. Colo. Dec. 28, 1981). There the plaintiffs somewhat disingenuously sought the dismantling of every dam on the Colorado River because of the dams' impacts on endangered fish. The court denied the plaintiffs' motion for summary judgment "because there is still a genuine issue of material fact as to which dams adversely affect the endangered fishes, and whether shutting down the dams will enhance or endanger the fishes population." (slip opinion at 23–24). Because of the erroneous factual premise of Chief Justice Burger's reasoning, TVA v. Hill offers little guidance as to how courts should handle such claims. In the authors' view, Justice Powell's formulation is more instructive. Thus an agency with reasonable alternatives as to how to operate a completed project should be required to choose a mode of operation that avoids jeopardy to a listed species, but the agency should not be required to terminate the project or choose a mode of operation that negates its purpose.

[242]Pub. L. No. 95–632, 92 Stat. 3751 (1978).

[243]Grayrocks Dam and Reservoir, halted in Nebraska v. Rural Electrification Administration, 12 Envt'l Rep. Cas. (BNA) 1156 (D. Neb. Oct. 3, 1978), was also required to be considered immediately for an exemption.

[244]16 U.S.C. §§ 1536(h)(1)(A)(i) and (ii).

[245]*Id.* § 1536(a).

quirements for the "biological opinions" that result from that consultation process,[246] created a duty to conduct "biological assessments" prior to beginning new projects,[247] and restricted federal agencies' ability to commit resources to projects after consultation had begun.[248]

As a result of the Tellico experience, Congress sought through the 1978 amendments to ensure that the consultation process would be as effective as possible in identifying ways to avoid conflicts between proposed federal actions and listed species or their critical habitats. Key to that goal was insuring that the process was begun at the earliest possible time. Section 7(c) requires, with respect to any federal action for which no actual construction had begun and no contract for construction had been entered into as of the time of the 1978 amendments, that a "biological assessment" be conducted if the Secretary advises that any listed or proposed-to-be-listed species occurs in the area of the proposed action.[249] The purpose of a biological assessment is to determine whether a listed or proposed species is "likely to be affected by such action."[250]

If, through the biological assessment process or otherwise, a species that has been *proposed* to be listed as threatened or endangered is found likely to be jeopardized by a federal action, the federal agency proposing the action has a duty to "confer" with the Secretary.[251] The law does not expressly describe what product or action is expected to result from this conferral process. It is apparent, however, that if the Secretary is not persuaded that the responsible agency will modify the action to avoid jeopardy, the Secretary should invoke the power to list the species in an emergency rulemaking.[252]

If a *listed* species is likely to be affected by a proposed federal action, the proposing agency must "consult" with the Secretary pursuant to sec-

[246] *Id.* § 1536(b).

[248] *Id.* § 1536(d).

[247] *Id.* § 1536(c).

[249] *Id.* § 1536(c)(1). The biological assessment may be done either by the federal agency involved or by any person eligible to apply for an exemption for the action under 16 U.S.C. § 1536(g)(1). If done by the latter, it must be carried out in cooperation with the Secretary and under the supervision of the relevant federal agency. *Id.* § 1536(c)(2).

[250] *Id.* § 1536(c)(1). In Thomas v. Peterson, 753 F.2d 754 (9th Cir. 1985), the court enjoined construction of a Forest Service road because the agency had failed to prepare a biological assessment. The court held that "failure to prepare a biological assessment for a project in an area in which it has been determined that an endangered species may be present cannot be considered a *de minimis* violation of the ESA." *Id.* at 763. In contrast, in State of Idaho by and through Idaho Public Utilities Comm'n v. ICC, 35 F.3d 585 (D.C. Cir. 1994), the court held that delay beyond the statutorily prescribed time period for preparing a biological assessment was inconsequential. In Bays Legal Fund v. Browner, 828 F. Supp. 102 (D. Mass. 1993), the court held that the biological assessment requirement was met by a draft environmental impact statement, even though that had not been the purpose of the statement.

[251] 16 U.S.C. § 1536(a)(4).

[252] *See* text accompanying notes 67–68, *supra*.

tion 7(a)(2).[253] The consultation process normally is to be completed within 90 days[254] and is to result in a "biological opinion" detailing how the proposed action will affect the species or its critical habitat.[255] If the action is likely to jeopardize the continued existence of the species or result in the destruction or adverse modification of critical habitat, the biological opinion also must suggest reasonable and prudent alternatives, if any, to avoid these effects.[256] These detailed requirements with respect to the consultation process and the necessary elements of a biological opinion largely codified administrative practice prior to the 1978 amendments.

Litigation under Section 7(a)(2)

In *TVA v. Hill*, the Supreme Court resolved the central issue concerning section 7(a)(2) and held that its requirements are mandatory and admit of no exceptions. Apart from creating a rigorous and rarely used exemption process, nothing Congress did in 1978 or later changed that central fact. Since *TVA v. Hill*, lower courts have been called upon to resolve a host of other issues not addressed by the Supreme Court. The remainder of this section examines these other issues and cases.

In *TVA v. Hill*, TVA did not contest the Fish and Wildlife Service's biological opinion that completion of Tellico Dam would jeopardize the snail darter's continued existence and destroy its critical habitat. Rather, the disputed issue was whether section 7 applied to a project that was substantially completed before the species it threatened was listed. In most cases, however, the effect of a proposed federal action on a listed species will be much less certain. A key question that arises is whose judgment controls—that of the agency proposing the project, or that of the Secretary, expressed through the biological opinion?

[253] 16 U.S.C. § 1536(a)(2).

[254] The Secretary and the federal agency may agree to a longer period. However, if the proposed action is the issuance of a permit or license, the extension cannot exceed 60 days without the applicant's consent. *Id.* §§ 1536(b)(1)(A) and (B).

[255] *Id.* § 1536(b)(3)(A). The 1982 Amendments authorize "advance consultation" for certain actions that require a federal license or permit, prior to the actual filing of the permit application. The purpose of this provision is to allow prospective applicants to learn at the earliest possible time whether their proposed actions will encounter section 7 difficulties. If the advance consultation results in an opinion from the Secretary that no section 7 problems are likely to arise, the agency need not engage in formal consultation when the permit application is filed, provided that both the action and the information about its impacts have not changed in the interim. If, on the other hand, advance consultation results in an opinion that the action is likely to jeopardize a listed species or adversely modify its critical habitat, the prospective applicant must apply for the permit and secure final agency action on it before seeking an exemption from the Act. *Id.* § 1536(b)(3)(B).

[256] *Id.* § 1536(b)(3)(A).

This question was addressed in one of the earliest section 7 cases, *National Wildlife Federation v. Coleman.*[257] The case concerned construction of a segment of Interstate Highway 10 in Mississippi and its effects upon the endangered Mississippi sandhill crane. The Secretary had steadfastly insisted that, without the relocation of a proposed interchange and elimination of certain "borrow pits," construction could not be done in compliance with section 7.[258] The testimony of a Department of the Interior biologist at trial suggesting that construction of the highway was not the major threat to the cranes cast sufficient doubt on the credibility of that position to persuade the trial court to find no violation of section 7. The court of appeals, however, while recognizing that section 7 "does not give the Department of Interior a veto over the actions of other federal agencies,"[259] and stating that its scope of review was limited to considering whether the agency had made a "clear error of judgment,"[260] held that the plaintiffs had sustained their burden of showing that the Department of Transportation had not taken the actions necessary to ensure the cranes' survival and the preservation of their critical habitat.

The court enjoined further work on part of the highway "until the Secretary of the Department of the Interior determines that the necessary modifications are made in the highway project to insure that it will no longer jeopardize the continued existence . . . or destroy or modify critical habitat of the Mississippi sandhill crane."[261] The defendants filed a petition for rehearing or clarification, arguing that the relief was inconsistent with the court's statement that the Secretary of the Interior could not exercise a veto power over the actions of other agencies.[262] The court denied the petition.

In *Roosevelt Campobello International Park Commission v. EPA*, both the Fish and Wildlife Service and the National Marine Fisheries Service had issued biological opinions concluding that the Environmental Protection Agency could not issue a permit authorizing construction of an oil refinery in Maine without violating its duty to ensure no jeopardy to endangered bald eagles and whales.[263] Despite these opinions, EPA issued the permit following an adjudicatory hearing in which an EPA administrative law judge

[257]529 F.2d 359 (5th Cir.), *cert. denied sub nom.* Boteler v. National Wildlife Federation, 429 U.S. 979 (1976).

[258]On the eve of trial, the Secretary published an emergency determination of the cranes' critical habitat, which the proposed highway segment crossed. 40 Fed. Reg. 27501 (1975).

[259]529 F.2d at 371.

[260]*Id.* at 372.

[261]*Id.* at 375.

[262]At least one commentator took the same view. *See* Comment, *Implementing § 7 of the Endangered Species Act of 1973: First Notices from the Courts,* 6 Envt'l L. Rep. (Envt'l L. Inst.) 10120 (1976).

[263]684 F.2d 1041 (1st Cir. 1982).

concluded that the refinery would not jeopardize those species. Plaintiffs challenged issuance of the permit.

A critical factual issue in the case was the likelihood of an oil spill as a result of tanker traffic servicing the refinery. The administrative law judge concluded that the risk of such a spill was extremely small. The court of appeals did not directly dispute the reasonableness of this conclusion in light of the evidence presented at the administrative hearing. Indeed, the court acknowledged that "[w]ere the issue whether, by a preponderance of the evidence, it had been established that [tankers] could make the transit through Head Harbor Passage to Eastport with reasonable safety, the ALJ's decision might be accepted."[264] The court, however, focused upon the section 7(a)(2) requirement that "each agency shall use the best scientific and commercial data available" in fulfilling its duty to insure no jeopardy to listed species.[265] That requirement, scarcely noticed when it was added by the 1979 amendments, obliged EPA to develop other information that was not in the hearing record but that the record revealed "would contribute a more precise appreciation of risks of collision and grounding"; this information, in the court's view, "obviously represent[s] as yet untapped sources of 'best scientific and commercial data.' "[266]

Roosevelt Campobello, while affirming the principle that the action agency, and not the Secretary, has the ultimate authority to determine whether to go forward with a proposed action, erects a major hurdle for an agency that chooses to do so in the face of an adverse biological opinion. Not only must the agency rely upon convincing evidence to rebut the Secretary's opinion, but it must also affirmatively seek out the best evidence relevant to the jeopardy issue that is capable of being acquired.[267] This requirement gives agencies a strong incentive to pursue the reasonable and prudent alternatives specified in biological opinions.

In determining whether a proposed federal action is likely to violate section 7, should the inquiry focus narrowly on the direct and immediate effects of the action or on its reasonably foreseeable consequences? This question arose in the early case of *National Wildlife Federation v. Coleman,* discussed above, and was resolved in favor of a broad inquiry.[268] In that case, the highway alone was not likely to jeopardize the sandhill cranes'

[264] *Id.* at 1054–55.

[265] *Id.* at 1052–53.

[266] *Id.* at 1055.

[267] Unlike Roosevelt Campobello, courts have often rejected challenges to biological opinions finding that proposed federal actions would not result in jeopardy to the continued existence of a listed species, even when those opinions have admittedly been based on weak or disputed information. *See, e.g.,* Stop H-3 Ass'n v. Dole, 740 F.2d 1442 (9th Cir. 1984), and Greenpeace Action v. Franklin, 14 F.3d 1324 (9th Cir. 1993).

[268] National Wildlife Federation v. Coleman, *supra* note 141.

continued existence, but private developments anticipated in the vicinity of the proposed interchange likely would. The Fish and Wildlife Service did not limit its inquiry to the highway alone; the agency considered the probable development that the federal action would likely promote. The Fifth Circuit Court of Appeals upheld the Service's broad inquiry.

A somewhat similar case, *Riverside Irrigation District v. Andrews*,[269] arose a decade later in the Tenth Circuit Court of Appeals. To build a dam and reservoir on a tributary of the South Platte River, the Riverside Irrigation District needed a permit from the U.S. Army Corps of Engineers to deposit dredge and fill material in a navigable waterway. The Corps had previously issued a general "nationwide" permit authorizing activities of the sort proposed, provided that no individual activity adversely affected a listed species or its critical habitat. If adverse effects were expected, a prospective permittee would need to seek an individual permit from the Corps rather than proceed under authority of the nationwide permit.

The Corps determined that because the dam would reduce downstream flows to the detriment of the whooping crane and its critical habitat, Riverside could not proceed under the nationwide permit. Riverside brought suit, challenging the Corps' authority to consider anything more than "the direct on-site effects of the discharge."[270] The court, relying upon *Coleman*, disagreed and held that "the Corps was required to consider all effects, direct and indirect, of the discharge for which authorization was sought."[271]

When a proposed federal action is a discrete, physical activity, such as building a highway or dam, applying section 7 is relatively easy and straightforward. Section 7 is not limited to such activities, however. Rather, it applies broadly to all federal actions, without limitation. How to apply it to federal actions that are of a programmatic or planning nature, or that are part of a sequence of related actions, is more challenging. Several courts have grappled with this problem.

Among the first cases of this nature were two concerning the application of section 7 to oil and gas leasing on the outer continental shelf, *North Slope Borough v. Andrus*,[272] and *Conservation Law Foundation v. Andrus*.[273] Outer continental shelf leasing is, by statute, segmented into three distinct phases: (1) lease sale and preexploration, (2) exploration, and (3) production. Federal approval is required prior to each phase. The district court in *North Slope Borough* rejected the government's argument that each discrete phase should be treated as a separate federal "action" for pur-

[269]758 F.2d 508 (10th Cir. 1985).
[270]*Id.* at 512.
[271]*Id.* at 513.
[272]642 F.2d 589 (D.C. Cir. 1980).
[273]623 F.2d 712 (1st Cir. 1979).

poses of section 7 consultation.[274] Rather, it concluded that the "action" was the lease sale and all resulting activities.[275] "Otherwise," in the court's view, "any statute providing an agency with maximum flexibility and planning ability . . . would also tacitly relieve that agency from much of the scrutiny required by the [Endangered Species Act]."[276] The court of appeals was "in qualified agreement" with this aspect of the district court's ruling.[277]

In *North Slope Borough*, the Secretary of Commerce had insufficient information to say whether the total action, inclusive of all three stages, was likely to jeopardize the continued existence of the endangered bowhead whale. The Secretary could conclude, however, that the initial stage of lease sale and preexploration activities would not have that effect, provided that certain conditions were met. Accordingly, the district court reasoned, despite its holding on the scope of the action to be considered in a biological opinion, "[w]hen no definitive biological opinion can be issued because of inadequate information, the [Endangered Species Act] does not require that the government halt all activities, unless the intermediate activities violate §7(a)(2)."[278] The court of appeals apparently concurred.[279]

This reasoning allows a segmented action to proceed incrementally through its various phases toward completion. The plaintiffs in *North Slope Borough* and *Conservation Law Foundation* argued that such an approach would violate the section 7(d) prohibition against irreversible commitments of resources. Both courts disagreed, emphasizing the Secretary of the Interior's authority to modify or cancel leases at any stage of the process if evidence of jeopardy to endangered species were developed.[280] Without extended discussion, the court in yet a third oil-leasing case reached the same result.[281]

The holdings in *North Slope Borough* and *Conservation Law Foundation* were followed in a Ninth Circuit Court of Appeals outer continental shelf leasing case, *Village of False Pass v. Clark*.[282] Four years later, however, in *Conner v. Burford*,[283] the Ninth Circuit declined to extend its holding in *False Pass* to on-shore oil and gas leasing under the Mineral Leasing Act of 1920.[284]

[274] 486 F. Supp. 332, 350 (D.D.C. 1980).
[275] *Id.* at 351.
[276] *Id.* at 350.
[277] 642 F.2d at 608.
[278] 486 F. Supp. at 357.
[279] *Cf.* 642 F.2d at 611.
[280] *Id.* at 610–11 and 623 F.2d at 714–15.
[281] California v. Watt, 520 F. Supp. 1359, 1387 (C.D. Cal. 1981).
[282] 733 F.2d 605 (9th Cir. 1984).
[283] 848 F.2d 1441 (9th Cir. 1988), *cert. denied sub nom.* Sun Exploration & Prod. Co. v. Lujan, 489 U.S. 1012 (1989).
[284] 30 U.S.C. §§ 181–287.

At issue in *Conner* were oil and gas leases on national forests in Montana. The Fish and Wildlife Service issued biological opinions concluding that "leasing itself was not likely to jeopardize the protected species," but the opinions "did not assess the potential impact that post-leasing oil and gas activities might have on protected species."[285] As in the outer continental shelf cases, the Service proposed successive consultations, each focused on one stage of activity. The court held that the Service's failure to prepare a comprehensive biological opinion that examined the impacts of both leasing and post-leasing activities violated its duty to base its opinion on the best data available. Even though information about post-leasing activities was limited and somewhat speculative, the court reasoned that enough was known about the location and vulnerability of the species concerned that the Service could have determined, at the leasing stage, whether "post-leasing activities in particular areas were fundamentally incompatible with the continued existence of the species."[286]

The *Conner* case seemed to stand for the proposition that unless a statute specifically prescribes a staged approach to agency decisionmaking, as the Outer Continental Shelf Leasing Act (OCSLA) does, the duty of the Secretary under section 7 is to prepare a comprehensive biological opinion that examines not only the effects of the immediate decision being proposed, but also those reasonably expected to follow. Nevertheless, in *Swan View Coalition, Inc. v. Turner*, the court upheld the Fish and Wildlife Service's issuance of a biological opinion concerning adoption of a forest plan for the Flathead National Forest, holding that the National Forest Management Act was "analogous to OCSLA" because of its "system of checks and balances in place which adequately safeguards endangered species."[287]

The Flathead plan did not authorize any site-specific development activities. It did, however, set resource production objectives for the forest, determine which areas of the forest would be subject to what types of development, and determine the suitability of lands for timber production. The Fish and Wildlife Service examined none of these in its biological opinion but found that adoption of the plan would not jeopardize the continued existence of any listed species. It did, however, focus on certain standards and guidelines that prescribed how endangered species would be addressed in implementing the plan. The court held that the law required no more, "because the standards and guidelines operate as parameters within which all future development must take place. If a development project cannot be maintained within those parameters, the safeguard mechanisms in the Plan will prevent such development from going forward."[288]

[285]848 F.2d at 1452.
[286]*Id.* at 1454.
[287]824 F. Supp. 923, 934 (D. Mont. 1992).
[288]*Id.* at 935.

The endangered species standards and guidelines included in the forest plan at issue in *Swan View Coalition* pertained only to those species that were listed as threatened or endangered at the time the plan was approved. What are the Forest Service's obligations if, after plan approval, another species that occurs in the forest is listed as threatened or endangered? This question was presented in *Pacific Rivers Council v. Thomas.*[289] The Forest Service argued that approval of the plan was a completed "federal action" and that it therefore had no duty to reopen consultation on the plan itself. The agency asserted that each ongoing and future timber, road, or range project was a separate federal action that may or may not require its own consultation, depending on the effect it had on listed species.

The court disagreed. It concluded that because the plan has "an ongoing and long-lasting effect even after adoption," it "represent[s] ongoing agency action" for which consultation must be reinitiated.[290] Until the Forest Service resumed consultation on the plan, it was obligated to halt all ongoing and future projects that may affect a listed species, even those the Forest Service itself had concluded were not likely to adversely affect the species. Once consultation resumed, however, the court indicated that it would then resolve whether ongoing and announced projects could proceed consistent with section 7(d) obligation that an agency make no irretrievable commitment of resources while engaged in consultation. Relying upon *Lane County Audubon Society v. Jamison,*[291] the court concluded, however, that "timber sales constitute *per se* irreversible and irretrievable commitments of resources."[292]

Similar issues were presented in *Sierra Club v. Marsh.*[293] The Army Corps of Engineers had consulted with the Fish and Wildlife Service concerning a combined highway and flood control project that would destroy more than forty acres of marsh habitat used by two endangered bird species. The Service concluded that the project would result in jeopardy to these species unless the Corps acquired and protected nearly 200 acres of nearby privately owned marsh habitat. The Corps agreed to do so but was unable to acquire the mitigation land quickly without conditions that reduced its conservation value. The Corps declined the Service's request to reinitiate consultation in light of the changed circumstances.

The Sierra Club sued, contending that the Corps had a duty to reinitiate consultation and to stop work on the project until it had done so. The court agreed, explaining its ruling as follows:

[289] 30 F.3d 1050 (9th Cir. 1994).
[290] *Id.* at 1053.
[291] 958 F.2d 290 (9th Cir. 1992).
[292] 30 F.3d at 1057.
[293] 816 F.2d 1376 (9th Cir. 1987).

We do not hold that every modification of or uncertainty in a complex and lengthy project requires the action agency to stop and reinitiate consultation. . . . [However], [t]he creation and management of a refuge for the birds is the most important of many modifications the Service considered absolutely necessary to insure that the project was not likely to jeopardize their continued existence. . . . The institutionalized caution mandated by section 7 of the ESA requires the [Corps] to halt all construction that may adversely affect the habitat until it insures the acquisition of the mitigation lands or modifies the project accordingly.[294]

The Duty to Avoid Adverse Modification of Critical Habitat

When the Endangered Species Act was enacted in 1973, it introduced the novel concept of "critical habitat," widely heralded as one of the significant innovations of the new law. Ironically, however, more than two decades later, there is considerable truth in one commentator's observation that "critical habitat has turned out to be an agony."[295] It remains one of the Act's most contentious, ambiguous, and confusing concepts. There is no clear, consistent, and shared understanding of what it means or what role it is to play in the Act's administration.

In *Douglas County v. Babbitt*, for example, the Ninth Circuit Court of Appeals reasoned that designation of critical habitat "prevents human interference with the environment" and is therefore exempt from the National Environmental Policy Act.[296] However, when designating the critical habitat at issue in that case, the Fish and Wildlife Service asserted that the designation does not prevent human interference, but rather allows at least some timber harvest within the designated area.

The court in *Catron County Board of Commissioners v. U.S. Fish and Wildlife Service*,[297] a Tenth Circuit case that reached exactly the opposite NEPA holding from that of *Douglas County*, shared the same underlying assumption about the effect of a critical habitat designation. The court reasoned, without ever referring to the Secretary's implementing regulations, and contrary to the government's denial, that the impact of the critical habitat designation "will be immediate and the consequences could be disastrous."[298]

It is no small wonder that "[t]here is no doubt that critical habitat designations are a red flag to the development community and that community's representatives in Congress."[299] Yet, the agencies chiefly respon-

[294]*Id.* at 1388–89.

[295]O. A. Houck, *The Endangered Species Act and Its Implementation by the U.S. Departments of Interior and Commerce*, 64 U. Colo. L. Rev. 277, 297 (1993). The Houck article provides an extensive analysis of Endangered Species Act implementation over two decades.

[296]48 F.3d 1495, 1506 (9th Cir. 1995), *cert.denied*, 116 S. Ct. 698 (1996).

[297]75 F. 3d 1429 (10th Cir. 1996).

[298]*Id.* at 1436.

[299]Houck, *supra* note 295, at 308.

sible for implementing the Act have never interpreted critical habitat as the *Catron County* and *Douglas County* courts suggest. Moreover, Congress has obscured, rather than clarified, the concept, and the courts (including both *Catron County* and *Douglas County*) have never given more than superficial attention to the duties that arise from the designation of critical habitat.

First Efforts to Clarify the Duty with Regard to Critical Habitat

As originally enacted, the Endangered Species Act referred to critical habitat only once. Section 7 directed all federal agencies

> to insure that actions authorized, funded, or carried out by them do not jeopardize the continued existence of . . . endangered species and threatened species or result in the destruction or modification of habitat of such species which is determined by the Secretary, after consultation as appropriate with the affected States, to be critical.[300]

Congress neither explained the relationship between the two duties imposed on federal agencies nor defined the key terms that comprised these duties: critical habitat, "jeopardize the continued existence," and "destruction or modification." It left these tasks to the Secretary.

Although stated as two separate duties, the duties to avoid jeopardizing the survival of species and destroying or modifying their critical habitat clearly overlapped to a considerable degree. Actions that destroy habitat deemed "critical" for a listed species would seem, by ordinary usage, certain to jeopardize that species' continued existence. Actions that merely "modify" critical habitat might not jeopardize the continued existence of the same species, but what sort of modifications did Congress have in mind? A literal interpretation would be that any modification, no matter how minor, was prohibited. It is doubtful that Congress could have intended such a result, however, for if it had, it would not have needed to prohibit the destruction of critical habitat. Further, a literal reading of the prohibition against modifying critical habitat would bar actions that resulted in modifications beneficial to the species.

The Secretary first sought to clarify the twin duties of section 7 with guidelines issued in April 1976.[301] The guidelines defined critical habitat as "any air, land, or water area including any elements thereof which the Secretary . . . has determined is essential to the survival of wild populations of a listed species or to its recovery." The guidelines then defined the two federal agency duties under section 7 in a way that made them largely

[300]Pub. L. No. 93–205, § 7, 87 Stat. 884, 892 (1973) (current version at 16 U.S.C. § 1536(a)(2)).

[301]Guidelines to Assist Federal Agencies in Complying with Section 7 of the Endangered Species Act, issued by the Director of the U.S. Fish and Wildlife Service on April 22, 1976.

overlapping. An action would be deemed to jeopardize the continued existence of a listed species if it

> reasonably would be expected to result in the reduction of the reproductive ability, numbers, or distribution of a listed species to such an extent that the loss would pose a threat to the continued survival or recovery of these [*sic*] species in the wild.

An action would be deemed to destroy or adversely modify critical habitat (the guidelines put the adjective "adverse" in parentheses before the word "modification") if it

> would have a deleterious effect upon any of the constituent elements of critical habitat which are necessary to the survival or recovery of such species, *and such effect is likely to result in a decline in the numbers of the species* (emphasis added).

Thus, from the very outset, the Secretary took the view that critical habitat modification *per se* was not prohibited, but only modification that was likely to cause a decline in species numbers. Arguably, the decline need not be so serious as to pose a threat to the survival or recovery of the species (the jeopardy standard), and to that extent, at least, the 1976 guidelines *may* have intended that the duty toward critical habitat be more exacting than the duty of avoiding jeopardy to the species.

Merging Section 7's Duties: The 1978 Regulations

Whether and precisely how these two duties may have varied was of little consequence at the time, since only one critical habitat had yet been designated in 1976.[302] Moreover, the guidelines were intended to serve only as interim guidance pending the promulgation of regulations. Regulations were published in January 1978, and they drove the twin duties of section 7 still closer.[303] The jeopardy definition appeared to be more demanding than under the earlier guidelines. It now referred to actions, the effect of which was to "reduce the reproduction, numbers, or distribution of a listed species to such an extent as to appreciably reduce the likelihood of the survival and recovery of that species in the wild." It was no longer necessary that the action "pose a threat to continued survival or recovery;" it was now sufficient that the impact of the action on survival and recovery prospects simply be "appreciable."

By strengthening the jeopardy standard, however, the 1978 regulations further blurred the distinction between that duty and the duty to avoid destruction or modification of critical habitat. The latter duty was rede-

[302]The first critical habitat designation was done for the snail darter on an emergency basis. *See* 41 Fed. Reg. 13926 (1976).

[303]43 Fed. Reg. 870 (1978).

fined to prohibit any alteration of critical habitat that "appreciably diminishes the value of that habitat for survival and recovery of a listed species." Further, the 1978 regulations redefined "critical habitat" in a way that appeared to broaden it. Whereas the earlier guidelines had focused on areas and constituent elements "essential to the survival" of a species, the later regulations abandoned this concept in favor of areas and constituent elements "the loss of which would appreciably decrease the likelihood of the survival and recovery of a listed species or a distinct segment of its population."

The effect of these changes was to create a circular loop by which the twin duties of section 7 effectively became merged into one. An action would be deemed to modify or destroy critical habitat only if it appreciably diminished the value for survival and recovery of a listed species of any habitat that had been designated critical. An area could only be designated critical habitat, however, if its loss would appreciably decrease the likelihood of survival and recovery of that species or a distinct segment of its population. Any action that had the above-described effect upon critical habitat would, inescapably, also appreciably reduce the likelihood of survival and recovery of the species. In this manner, the twin duties of section 7, already substantially overlapping under the 1976 guidelines, essentially lost their separate identities in 1978.

The 1978 Amendments: Sowing the Seeds of Further Confusion

The ink on the 1978 regulations had hardly dried before Congress amended the Act. One amendment added the first statutory definition of critical habitat and imposed a requirement to consider economic factors when designating it. Unfortunately, much of what Congress had to say only compounded the preexisting confusion. A close examination of the legislative history is necessary to an understanding of what Congress intended in the 1978 amendments.

Legislative action in 1978 began in the Senate. Amendments brought to the Senate floor by Senators Culver (D-IA) and Baker (R-TN) made only one very minor and unexplained change relating to critical habitat. This change was to insert the adjective "adverse" before the word "modification" in section 7's prohibition against federal actions that cause the destruction or modification of critical habitat. This change was of little consequence, since the Secretary's 1976 guidelines and 1978 regulations already imputed the modifier "adverse" to the statutory prohibition against modifying critical habitat. The report accompanying the Senate amendments, however, highlighted another critical habitat concern.

According to the report,

> [i]t has come to the committee's attention that under present regulations the Fish and Wildlife Service is now using the same criteria for designating and protecting

areas to extend the range of an endangered species as are being used in designation and protection of those areas which are truly critical to the continued existence of a species. . . . There seems to be little or no reason to give exactly the same status to lands needed for population expansion as is given to those lands which are critical to a species' continued survival.[304]

The committee report singled out a proposed critical habitat for the grizzly bear that encompassed several million acres. The committee observed that "[t]he goal of expanding existing populations of endangered species in order that they might be delisted is understandable."[305] (It would have been more accurate to say that this is the goal of the Act itself.) Nevertheless, the committee expressed concern that "increas[ing] the amount of area involved in critical habitat designation . . . increases proportionately the area that is subject to the regulations and prohibitions which apply to critical habitats,"[306] surely a tautology.

The committee's concern was heightened by the fact that "[i]n many cases the Fish and Wildlife Service has been unable to explain fully or predict what the impacts of a critical habitat designation are going to be."[307] Rather than attempt to resolve these concerns, however, the committee punted. It simply directed the Service to "examine this ambiguity in its regulatory process for critical habitat designation."[308]

When the committee's amendments came to the Senate floor, however, some were unwilling to wait for an administrative resolution. Senator James McClure (R-ID) offered an amendment that defined critical habitat very similarly to the way in which the Act now defines it. The McClure amendment sought to distinguish between areas occupied by a species at the time of its listing and areas not then occupied. As to the occupied areas, McClure's amendment defined critical habitat as those areas on which are found physical or biological features (1) essential to the conservation of the species, and (2) which require special management considerations or protection. As to areas not occupied by the species at the time of its listing, and "into which the species can be expected to expand naturally," McClure's amendment would have allowed these also to be included as critical habitat, but only if the Secretary determines, at the time of listing, "that such areas are essential for the conservation of the species." A final element of McClure's amendment specified that critical habitat "will not include the entire geographical area which can be oc-

[304]Sen. Rep. No. 95–874, 95th Cong., 2d Sess., *reprinted in* Legislative History at 947–948.
[305]*Id.*at 948.
[306]*Id.*
[307]*Id.*
[308]*Id.*

cupied'' by the species, ''except in those circumstances determined by the Secretary.''[309]

McClure's amendment was puzzling because of its many seeming inconsistencies. Oddly, it seemed to impose a lower standard for designating critical habitat outside of currently occupied range than within it (*i.e.*, for areas outside current range, no finding that such areas ''require special management consideration or protection'' was necessary). Further, the effort to ensure that critical habitat not include all habitat that could be occupied was subject to a standardless exception that gave the Secretary the very power that the rest of the provision was intended to take away.

The discussion of the McClure amendment on the Senate floor was very brief and unilluminating because the amendment had been worked out in advance with Senator Culver's staff and was agreed to without debate. What little Senator McClure said about his amendment clearly shows that he was concerned about the grizzly bear and its potential for expansion into areas not occupied at the time of its listing. What McClure apparently sought to do was to force the government to make a critical habitat determination at the time of listing, to discourage in a rather unspecific way the inclusion within that habitat of areas beyond those occupied by the species at that time, and to forbid the inclusion of any areas into which the species could not be expected to expand naturally.

Moreover, McClure may have intended his amendment to give the government only one bite at the apple, at least with respect to those areas not occupied by the species at the time of its listing. Literally, with respect to these areas, the amendment required that the Secretary's determination that they were essential to the conservation of the species be made at the time of listing. In describing his amendment, McClure said that it dealt ''primarily, and most importantly, [with] the extension of the area of the critical habitat once established.''[310] Although it is far from clear, McClure seems to have intended that any extension of critical habitat once it has been established must be limited to those areas that were occupied by the species at the time of its listing.

Later in the floor debate, Senator Garn (R-UT), apparently unaware that the McClure amendment had been approved, offered his own, nearly

[309]Legislative History at 1065. The reference in McClure's amendment to unoccupied areas ''into which the species can be expected to expand naturally'' is not reflected in the law today. It apparently sought to distinguish between areas into which ''natural expansion'' was possible, and other areas that the species was unlikely to occupy except as a result of active reintroduction efforts. The latter, under this part of McClure's amendment, would not have been eligible for critical habitat designation. Elsewhere in the same amendment, however, McClure gave the Secretary the authority to designate the entire area that can be occupied by the species, apparently including reintroduction areas.

[310]*Id.*

identical amendment.[311] The Garn amendment had two additional features, however. One would have required that critical habitat be designated at the time a species is listed, except when "an emergency exists because no critical habitat information is available or there are other contingencies." The other would have required an analysis of the effects of critical habitat designation on environmental quality and economic development. Upon realizing that much of the amendment had already been adopted as part of the McClure amendment, Garn modified his amendment to limit it to the above requirement that critical habitat be designated contemporaneously with listing.[312] Inexplicably, the economic analysis requirement was dropped.

Senate action in 1978 was completed while the House was still deeply divided over the ESA's future. Chairman Leggett (D-CA) and Representative Dingell (D-MI) each had offered bills that split Leggett's subcommittee of the Merchant Marine and Fisheries Committee almost evenly. Unable to resolve its differences in a very contentious public markup session, the subcommittee retreated to a closed meeting room and made a deal. That deal was embodied in H.R. 14104, subsequently introduced by Leggett, Dingell, and others. The bill included a definition of critical habitat that subtly changed the Fish and Wildlife Service's then-existing regulatory definition. The bill also directed the Secretary to "consider the economic impact, and any other relevant impacts, of specifying any particular area as critical habitat" for any listed *invertebrate* species; it also authorized the Secretary to "exclude any such area from the critical habitat if he determines that the benefits of such exclusion outweigh the benefits of specifying the area as part of the critical habitat."[313] Except for its limitation to invertebrates, and a subsequent qualifier intended to prevent exclusion of areas needed to prevent extinction, this latter provision remains in the Act today.

The committee's proposed definition of critical habitat sought to narrow the Fish and Wildlife Service definition. In 1978, the Service's definition included areas, "the loss of which would appreciably decrease the likelihood of conserving a listed species." The committee believed this definition potentially encompassed "all areas the loss of which would cause any decrease in the likelihood of conserving the species so long as that decrease would be capable of being perceived or measured," a fact that "could conceivably lead to the designation of virtually all of the habitat of a listed species."[314] To avoid this result, the committee's amend-

[311] *Id.* at 1107–1108.

[312] *Id.*

[313] Legislative History at 665.

[314] H. R. Rep. No. 95–1625, 95th Cong., 2d Sess. (1978), *reprinted in* Legislative History at 749.

ment replaced the word "appreciably" with the word "significantly." The committee was clear that it thought this change narrowed the scope of the definition.

The committee's other change, authorizing the exclusion of areas from critical habitat because of economic considerations, was necessarily based on a belief that federal agency duties under section 7 with respect to critical habitat differ in some manner from the duty to avoid actions that jeopardize the continued existence of a listed species. If actions that impermissibly affected critical habitat always also jeopardized species survival, there never could be any economic impact associated solely with critical habitat designation.

The committee's accompanying report made explicit what was only implicit in its proposed amendment. The report noted that even if economic considerations resulted in the exclusion of all areas from critical habitat, "the act would still be in force to prevent any taking or other prohibited act described in [section 9]. In addition, agencies would still be prohibited from taking an action which would jeopardize the existence of the . . . species."[315] Thus, the committee clearly believed that the duties associated with critical habitat went beyond the duty of avoiding jeopardy to the survival of a species. The committee did not address, however, in what ways the critical habitat duty differed from the duty to avoid jeopardy.

Congress accepted the committee's proposed changes with only one modification: the authority to exclude areas from critical habitat because of economic impacts was broadened to encompass all species, not just invertebrates. The conference committee accepted this provision from the House bill and embraced a modified version of the Senate bill's definition of critical habitat. Specifically, the conference committee dropped the requirement that areas not occupied at the time a species was listed could be included within critical habitat only if they were designated at the time of listing. The resulting language is the language of the Act today.

The 1986 Regulations

The Secretary did not promulgate until 1986 new regulations to implement the major changes Congress made to section 7 in 1978.[316] The 1986 regulations did not attempt to elaborate upon the statutory definition of critical habitat. A charitable interpretation would be that the Secretary concluded there was no need to elaborate upon what Congress had said. A more cynical view would be that the Secretary concluded it was futile to try to reconcile the inconsistencies in the statutory definition.

The 1986 regulations made virtually no substantive change in the preexisting definitions of "destruction or adverse modification" and "jeop-

[315] *Id.* at 741.
[316] 51 Fed. Reg. 19926 (1986).

ardize the continued existence of." Apart from minor grammatical improvements, the new definitions merely added the word "both" immediately before the phrase "the survival and recovery" in each definition. The addition of this single word apparently was intended to emphasize the fact that jeopardy to the continued existence of a species, or adverse modification of its critical habitat, required an appreciable impact on survival *and* recovery, not survival *or* recovery. Though this issue has aroused the passions of many, in the authors' view, the distinction is inconsequential.[317] The Fish and Wildlife Service itself has belatedly come to recognize this.[318] Of more importance is the fact that the 1986 regulations left unchanged the merging of the two separate duties of section 7 toward species and their critical habitats. Because these appear in the statute as two separate duties, one presumes that Congress intended them to have different meanings. The Secretary's 1986 definitions, however, effectively reject any duty toward critical habitat that is more far-reaching than the duty to avoid jeopardy.

The Fish and Wildlife Service Reconsiders

The Secretary is beginning to reconsider the view that the two section 7 duties are essentially redundant, as a result of litigation to force the Fish and Wildlife Service to designate critical habitat for the northern spotted owl and other species. Some of the Service's challengers apparently believe that the Service's interpretation of critical habitat is much too restrictive and that they can force a change in the interpretation through litigation. In response, the Service is beginning to articulate a new view of critical habitat.

The first clear evidence of this change came in the designation of critical habitat for the northern spotted owl, an action that was ordered by the court in *Northern Spotted Owl v. Lujan.*[319] The Service's critical habitat designation was accompanied by a lengthy explanation of what the concept meant. The Service stated that "[t]he Act's definition of critical habitat indicates that the purpose of critical habitat is to contribute to a species' conservation, which by definition equates to recovery."[320] At

[317]The Secretary acknowledged in the preamble to the 1980 regulations that "it is difficult to draw clear-cut distinctions" between impacts to survival and impacts to recovery, and that "[i]f survival is jeopardized, recovery is also jeopardized." *Id.* at 19934. The authors submit that the converse is also true: If prospects for the recovery of a species are diminished, so too are prospects for its survival (since survival prospects are at least partly a function of the likelihood of recovery). For that reason, the semantic distinction between "survival or recovery" and "survival and recovery" is meaningless; both can be expressed with the single word "conservation."

[318]*See* text accompanying notes 320–321 *infra.*

[319]758 F. Supp. 621 (W.D. Wash. 1991).

[320]57 Fed. Reg. 1822 (1992).

any given time, the Service explained, a species occupies a point somewhere on a hypothetical linear continuum between extinction and recovery. The Service believes that an action affecting the critical habitat of a species close to the extinction end of the continuum will likely violate both the jeopardy prohibition and the critical habitat duty. On the other hand, the same or a similar action affecting a species closer to the "recovery" end of the continuum may violate the critical habitat duty, even though it does not violate the jeopardy prohibition. In the Service's words, "the adverse modification standard may be reached closer to the recovery end of the survival continuum, whereas the jeopardy standard traditionally has been applied nearer to the extinction end of the continuum."[321] According to this view, section 7's duties with respect to critical habitat are more protective than its "no jeopardy" duty.

The Service's explanation may comport with actual practice. Unfortunately, however, it does not comport with the language of the Service's regulatory definitions or bear up particularly well under close scrutiny. First, with respect to the regulatory definitions, the definition of "adverse modification or destruction" requires that the modification diminish the value of the habitat "for *both* the survival and recovery" of the species. The Service, in explaining its critical habitat designation for the northern spotted owl, describes how the loss of a single "critical habitat unit" could "preclude recovery *or* reduce the likelihood of survival of the species," and suggests that this would violate section 7 critical habitat duties.[322] The problem, however, is that the Secretary, in the 1986 regulations, specifically rejected the "survival *or* recovery" formulation. If the loss of a critical habitat unit affects only one or the other, but not both, its loss does not meet the regulatory definition.

A further problem with the Service's new explanation is that it suggests that whether or not a particular action will "jeopardize the continued existence" of a species depends in some way on whether that species is close to the extinction end or the recovery end of the hypothetical continuum. Nothing in the Service's definition of "jeopardize the continued existence" supports this view. That definition focuses entirely on whether the effect of the action on the likelihood of survival and recovery is "appreciable" (i.e., capable of being perceived or measured). If the effect of any action is to reduce the likelihood of survival and recovery, then it really does not matter at all whether the species is close to extinction or close to recovery; in either case it constitutes jeopardy. If the Service simply means that, as a practical matter, actions that negatively affect a species on the brink of extinction are more likely to be recognized as "apprecia-

[321] *Id.*
[322] *Id.* at 1823 (emphasis added).

bly diminishing" its prospects for survival and recovery than are actions negatively affecting a species close to recovery, that observation is probably correct. What is unclear, however, is why the same observation should not also be true of actions negatively affecting critical habitat.

The Service's recent effort to articulate a new understanding of critical habitat thus is not very convincing. Part of the problem stems from the assumed dichotomy between "likelihood of survival" and "likelihood of recovery." The Service presumes that actions can negatively affect the first without negatively affecting the second, and vice versa. On its face, this seems illogical. Any action that reduces the likelihood of survival logically must also reduce the likelihood of recovery, for any action that makes it more likely that a species will go extinct necessarily makes it less likely that it will eventually recover. It also seems tautological that any action that makes it less likely that a species will recover necessarily means that the species is more likely to remain in danger of extinction and hence less likely to survive. Building an understanding of critical habitat around this false dichotomy thus seems doomed to failure. A more defensible approach might be constructed on the admittedly shaky foundation of the 1978 amendments to the Act, which focus on occupied and unoccupied habitat.

Making Sense of Critical Habitat

Is there an understanding of critical habitat that can make sense of the 1978 legislative history? Probably not, since Congress seemed only dimly aware of the details of the Act's implementation and failed to make clear its own conception of how critical habitat was to fit in with the rest of the statutory structure. There is one possible approach that reconciles what Congress seemed to be striving for in 1978 with what the Service now appears to desire. This approach would distinguish between occupied and unoccupied habitat. Occupied critical habitat can be considered as that which is minimally necessary to preserve the species' status quo. For this occupied habitat, the twin duties of section 7 are essentially the same: Any habitat alteration that appreciably diminishes the likelihood of survival is, by virtue of that fact, an alteration that adversely modifies critical habitat. Unoccupied critical habitat is habitat, the protection of which is needed to improve the species' status quo; that is, it is primarily needed for the recovery of the species. For this habitat, the impact of any proposed alteration can be evaluated without primary reference to its effect on the likelihood of survival. Instead, proposed alterations of this habitat would be evaluated with primary reference to the impact they would have on the ability of the habitat to contribute to the recovery of the species as envisioned by the designation of that habitat. Anything that forecloses or significantly reduces that ability would be deemed to destroy or adversely modify it, whether or not it could be said to appreciably diminish the species' likelihood of survival.

Unfortunately, the above formulation is not entirely satisfactory because the distinction between "occupied" and "unoccupied" habitat is not so neat, as the case of the northern spotted owl illustrates. The owl apparently utilizes some habitat for nesting and roosting, other habitat for foraging, and still other habitat for "dispersing." It "occupies," at least transiently, all of these habitats, and the habitats themselves may be transitory in that they will change in character over time. In fact, the Service's designation of critical habitat for the owl includes some relatively "degraded" habitats that are expected to "recover" and become of more value to the owl. The Service "envisions that, as habitat within critical habitat begins to recover . . . increasing levels of [timber] harvest will be allowed within critical habitat."[323] These "degraded" habitats, because of the role they are intended to play in the species' recovery, are functionally analogous to the "unoccupied" habitats described in the above formulation. Because their protection is aimed primarily at improving, rather than maintaining, the owl's status quo, actions that modify them should be evaluated with respect to the actions' impact on the ability of the habitats to contribute to the recovery of the species as envisioned.

This view of critical habitat may encounter conceptual obstacles similar to those that have confronted other views. It cannot be entirely squared with existing regulatory definitions, though it is at least faithful to the fact that the definition of "jeopardize the continued existence of" requires some reduction of species numbers, distribution, or reproduction, while the definition of "destruction or adverse modification" contains no such requirement. Further, it is probably closer to what Congress had in mind in 1978 and it would make comprehensible the duty to consider the economic impact of critical habitat designation. The focus of economic analysis would properly be on the habitats selected as critical because of their potential contribution to recovery. It is these habitats for which meaningful choice may be most likely. Already occupied habitats, in which the jeopardy prohibition would be essentially coterminous with the critical habitat duties, need not be part of the economic analysis, since for those habitats the jeopardy prohibition would independently proscribe any activity that might adversely modify critical habitat.

Embracing an understanding of critical habitat like that suggested above would also make compliance with NEPA more intelligible if, as the *Catron County* case holds, the designation of critical habitat is not exempt from NEPA. The focus of inquiry in any NEPA analysis would be on the areas of unoccupied habitat designated within the critical habitat.

[323] *Id.* at 1825.

The Exemption Process

Congress quickly responded to the Supreme Court's 1978 decision in *TVA v. Hill* that section 7 barred the completion of federal projects that jeopardize endangered species, "whatever the cost." Later that year, Congress amended the Act to allow some balancing of economic, social, and ecological values. The amendments added an elaborate and stringent process for exempting federal actions from the section 7 prohibitions.

The exemption process was intended to be a last-resort option, available only after all other avenues for avoiding conflicts have been exhausted. It may be initiated only after the consultation process has been completed and a final biological opinion issued.[324] If the action involves the issuance of a federal license or permit, however, the exemption process may not be initiated until after final agency action is taken on the permit or license application, and then only if the permit is denied "primarily because of the application of section [7(a)] to such agency action."[325]

Those entitled to seek an exemption include the federal agency proposing the action, the governor of the state in which the action will occur, if any, and, in the case of actions involving the proposed issuance of federal licenses or permits, the permit or license applicant.[326] The exemption application is submitted to the Secretary, who must within twenty days make certain threshold determinations as to whether the Act's requirements with respect to biological assessments, good faith consultation, and the irreversible commitment of resources have been met.[327] If the Secretary determines that they have not been met, the application must be denied.[328]

If the threshold requirements are met, the Secretary must hold a formal hearing and, on the basis of that hearing, prepare a report summarizing the evidence with respect to each of the factors that must be addressed in any final exemption decision.[329] The Secretary's report is submitted to a seven-member "Endangered Species Committee."[330] This Committee

[324]16 U.S.C. §§ 1536(g)(2)(A) and (B).

[325]*Id.* §§ 1536(g)(2)(A) and 1532(12).

[326]*Id.* § 1536(g)(1). The qualifying language "if any" in reference to the state in which the action occurs appears to reflect the congressional belief that section 7 applies extraterritorially. For a useful discussion of this question, see Erdheim, *supra* note 232, at 668–73. The Supreme Court left this issue unresolved in Lujan v. Defenders of Wildlife, 504 U.S. 555 (1992).

[327]16 U.S.C. § 1536(g)(3)(A). Prior to the 1982 Amendments, the functions of the Secretary described here were to be undertaken by a three-member "Endangered Species Review Board."

[328]*Id.* § 1536(g)(3)(B). A decision to deny is considered a final agency action subject to judicial review.

[329]*Id.* § 1536(g)(4).

[330]The Committee is composed of the Secretaries of the Interior, Agriculture, and Army;

has the ultimate responsibility for granting or denying an exemption. It may grant an exemption only if at least five members determine that:

1. there are no reasonable and prudent alternatives to the agency action;
2. the benefits of such action clearly outweigh the benefits of alternative courses of action consistent with conserving the species or its critical habitat, and such action is in the public interest;
3. the action is of regional or national significance; and
4. neither the Federal agency concerned nor the exemption applicant made any irreversible or irretrievable commitment of resources prohibited by subsection (d).[331]

The Committee also must establish "reasonable mitigation and enhancement measures" as conditions to any exemption it grants.[332] A decision granting or denying an exemption is subject to review at the behest of any person in any appropriate United States Court of Appeals.[333]

The normal exemption process may be bypassed in three instances. First, the Secretary of State may review applications for exemption and, if he certifies in writing that carrying out the action would violate an international treaty or other international obligation of the United States, the Endangered Species Committee may not grant the exemption.[334] Conversely, if the Secretary of Defense determines that any action must be exempted "for reasons of national security," then the Committee must exempt it.[335] Finally, in emergency situations involving a presidentially declared disaster area, the President may make the threshold and final determinations otherwise required of the Secretary and the Committee.[336] None of these special provisions has been used.

The exemption process has rarely been used. Apart from Tellico Dam itself, few federal actions have been considered for an exemption and only two exemptions have been granted.[337] The main reasons for the low num-

---

the Administrators of the Environmental Protection Agency and the National Oceanic and Atmospheric Administration; the Chairman of the Council of Economic Advisors; and a presidentially appointed representative from any affected state. *Id.* § 1536(e)(3). The Committee is sometimes referred to as the "God Committee" because of its power to authorize projects that are likely to jeopardize the continued existence of a species.

[331] *Id.* § 1536(h)(1)(A).

[332] *Id.* § 1536(h)(1)(B).

[333] *Id.* § 1536(n).

[334] *Id.* § 1536(i).

[335] *Id.* § 1536(j).

[336] *Id.* § 1536(p).

[337] Exemptions were granted to the Grayrocks Dam in Wyoming in 1979 and to 13 Bureau of Land Management timber sales in Oregon in 1992. In the Grayrocks Dam situation, parties to litigation regarding the dam (see note 243, *supra*) had reached a settlement that allowed the dam to go forward, and the Endangered Species Committee endorsed the settlement.

ber of applications probably include the small number of jeopardy opinions issued,[338] the stringent substantive standards for the grant of an exemption, and the complexity of the process. A likely additional factor is that most institutions, public or private, recognize that merely by seeking an exemption they risk being perceived as hostile to endangered species conservation. As long as public support for conservation is believed to be high, there is an incentive to compromise and avoid the need for an exemption.

Enforcement

The Act authorizes two types of enforcement, public and private. Public enforcement, through which criminal or civil penalties are imposed against violators, is the responsibility of the Secretaries and the Department of Justice. The Act also authorizes the Attorney General to seek injunctive relief for violations of the Act.[339]

The Act's liberal citizen suit provision allows private citizens to enjoin violations of the Act. This provision authorizes any person to file a suit to enjoin any other person, including any agency of the United States, from violating any provision of the Act.[340] It also authorizes suits in the nature of mandamus to compel the Secretary to carry out any nondiscretionary duty with respect to the listing of species under section 4 of the Act.[341] The citizen suit provision has served as the basis for all of the litigation to compel federal agencies to carry out their obligations under section 7.

Although the citizen suit provision authorizes "any person" to commence an action to enjoin a violation of the Act, not all courts have construed this provision to confer automatic standing on a plaintiff alleging

---

With respect to the BLM timber sales, conservation groups challenged the exemption decision. They claimed that President George Bush and his staff had engaged in *ex parte* communication with some Committee members in violation of the Administrative Procedure Act. Portland Audubon Society v. Endangered Species Committee, 984 F.2d 1534 (9th Cir. 1993) (holding that the Committee and the President are subject to the APA *ex parte* restrictions). During the appeal of the exemption decision, the sales remained enjoined in a case brought under the National Environmental Policy Act. See discussion of this litigation in Chapter 10 at text accompanying notes 117–119. The BLM withdrew the exemption application and the sales after Bill Clinton became President in 1993.

An exemption application was filed on behalf of the same oil refinery at issue in *Roosevelt Campobello International Park Commission v. EPA, supra* note 263. However, that application was dismissed on the grounds that it was prematurely filed in light of the pendency of EPA review proceedings. *See* Pittston Co. v. Endangered Species Committee, 14 Envt'l Rep. Cas. (BNA) 1257 (D.D.C. March 21, 1980).

[338] *See* O. A. Houck, *The Endangered Species Act and Its Implementation by the U.S. Departments of Interior and Commerce,* 64 U. Colo. L. Rev. 277, 318 (1993).

[339] 16 U.S.C. § 1540(e)(6).

[340] *Id.* § 1540(g).

[341] *Id.* § 1540(g)(1)(C).

a violation. In fact, three courts have denied standing to plaintiffs who had failed to demonstrate any actual injury as a result of the actions they challenged.[342] Each of those cases involved a challenge to a listing, and together they may suggest judicial reluctance to entertain suits brought to negate the Act's purposes. In the ordinary case, however, where plaintiffs can demonstrate a likelihood of injury to their interests, the Act's citizen suit provision should confer automatic standing on plaintiffs who invoke it.

In *Bennett v. Plenert*, a case that the Supreme Court has agreed to review, the Ninth Circuit denied standing to ranch operators and irrigation districts that sought to challenge a Fish and Wildlife Service biological opinion proposing to use reservoir water to protect listed fish species.[343] The court found that because the plaintiffs had no interest in preserving the listed fish, they were not within the "zone of interests" protected by the Act. This dubious holding goes well beyond the earlier cases noted above. Whatever the ultimate outcome in *Bennett*, other statutes may confer standing to challenge actions taken pursuant to the ESA. In *Douglas County v. Babbitt*, the court found that counties that objected to a critical habitat designation for the spotted owl because of its economic impact had procedural standing under the National Environmental Policy Act and the Administrative Procedure Act to challenge the Secretary's failure to prepare an environmental impact statement for the designation.[344]

Before bringing a citizen suit under the Act, prospective plaintiffs must give written notice of the alleged violation to the Secretary and to any alleged violator of the Act.[345] The purpose of requiring notice to the Secretary is to enable the government to take appropriate action against the alleged violator or, if the Secretary is the alleged violator, to correct the alleged violation.

Although the statute specifies that, for most violations, notice must be given at least 60 days in advance of filing suit,[346] many early decisions did not insist on strict compliance with this requirement.[347] In 1989, however,

[342]Glover River Organization v. Department of the Interior, 675 F.2d 251 (10th Cir. 1982); Pacific Legal Foundation v. Andrus, 13 Envt'l Rep. Cas. (BNA) 1266 (M.D. Tenn., May 23, 1979), *aff'd on other grounds*, 657 F.2d 829 (6th Cir. 1981); Colorado River Water Conservation Dist. v. Andrus, Civ. No. 78–A-1191 (D. Colo. Dec. 28, 1981).

[343]63 F.3d 915 (9th Cir. 1995).

[344]48 F.3d 1495, 1500–01 (9th Cir. 1995), *cert. denied*, 116 S. Ct. 698 (1996).

[345]16 U.S.C. § 1540(g)(2).

[346]An exception to the 60–day-notice requirement applies to actions "respecting an emergency posing a significant risk to the well-being of any species." *Id.* § 1540(g)(2)(C).

[347]*See, e.g.*, Sierra Club v. Froehlke, 534 F.2d 1289 (8th Cir. 1976); Village of Kaktovik v. Corps of Engineers, 12 Envt'l Rep. Cas. (BNA) 1740 (D. Alas. Dec. 29, 1978); Fund for Animals v. Andrus, 11 Envt'l Rep. Cas. (BNA) 1189 (D. Minn. July 14 and Aug. 30, 1978); Libby Rod & Gun Club v. Poteat, 457 F. Supp. 1177 (D. Mont. 1978), *aff'd in part, rev'd in*

the Supreme Court construed the prior notice requirements of a nearly identical citizen suit provision in another law as "mandatory conditions precedent to commencing suit."[348] Two years later, in *Sierra Club v. Yeutter*, the Fifth Circuit Court of Appeals ruled that although the notice requirement of the Endangered Species Act was likewise mandatory, it was not jurisdictional and therefore could be waived by the government's failure to object in a timely manner to the lack of prior notice.[349]

The citizen suit provision authorizes the award of attorneys' fees and other costs of litigation to any party whenever a court determines such an award to be appropriate.[350] Accordingly, successful plaintiffs are not automatically entitled to fees and costs, but neither are unsuccessful plaintiffs barred from them. In the *North Slope Borough* case, unsuccessful plaintiffs initially were awarded their costs and fees on the rationale that their suit, even though unsuccessful, was a "prudent and desirable effort to achieve an unfulfilled objective of the Act."[351] The court of appeals, however, denied the award on the grounds that the suit "was not so 'exceptional' and such a 'substantial contributions to the statutory goals' of the underlying acts that an award is appropriate."[352]

Federalism under the Endangered Species Act

When Congress passed the Marine Mammal Protection Act of 1972 and the Endangered Species Act of 1973, thus prohibiting the taking of all marine mammals and endangered species, it gave to the federal government a power that for the most part had previously been exercised only by the states. In each Act, however, Congress avoided ousting the states altogether from their established jurisdiction by creating a cooperative program in which both the states and the federal government would play important roles. Though only a little more than a year separated the two Acts, there was a marked difference in both the content and the manner of implementing the cooperative programs they established. Chapter 5 described the federal/state relationship under the Marine Mammal Protection Act.[353] The next part of this chapter will compare that relationship with its counterpart under the Endangered Species Act.

The first major difference between the two Acts was that the Endangered Species Act's preemption of state authority over the taking of spe-

---

part on other grounds, 594 F.2d 742 (9th Cir. 1979); National Wildlife Federation v. Hodel, *supra* note 131, at 1092, n. 6.

[348]Hallstrom v. Tillamook County, 493 U.S. 20, 31 (1989).

[349]926 F.2d 429, 435–436 (5th Cir. 1991).

[350]16 U.S.C. § 1540(g)(4).

[351]North Slope Borough v. Andrus, 515 F. Supp. 961, 965 (D.D.C. 1981).

[352]Village of Kaktovik v. Watt, 689 F.2d 222, 224 (D.C. Cir. 1982).

[353]*See* Chapter 5 at text accompanying notes 146 *et seq.*

cies was neither immediate nor total. Rather, the 1973 Act provided for a transition period of up to fifteen months during which federal restrictions on the taking of resident endangered and threatened species would be inapplicable.[354] In addition, it expressly permits the states to enforce laws or regulations pertaining to taking that are more restrictive than the exemptions or permits provided for in the Act.[355]

At the expiration of the transition period, the Act's restriction on taking became fully applicable to all listed species. However, like the Marine Mammal Protection Act, the Endangered Species Act offers the states an opportunity to continue participating in management and conservation of resident endangered and threatened species and to receive federal financial assistance to do so. The 1973 Act provides a mechanism for state participation, the signing of a "cooperative agreement."[356]

Cooperative agreements under the 1973 Act are unlike "cooperative arrangements" under the Marine Mammal Protection Act. The latter simply provide for delegation to the states of administration and enforcement of the federal Act without entitling the states to receive federal matching grants.[357] Yet cooperative agreements are also unlike the more complex mechanism provided by the Marine Mammal Protection Act for approval of state laws relating to the taking of marine mammals. Under that scheme, state laws can be approved only after a lengthy and detailed administrative process involving a formal public hearing.[358] Cooperative agreements under the Endangered Species Act, on the other hand, are *required* to be signed whenever the Secretary determines, without any public hearing, informal rulemaking procedure, or even public notice, that a state's proposed program for conservation of endangered species meets certain criteria specified in section 6(c). The Act contemplates separate cooperative agreements with respect to plants and animals. The criteria

[354] 16 U.S.C. § 1535(g). The inapplicability of federal restrictions on taking was subject to three exceptions. The first, so large as nearly to swallow the rule, provided that federal taking restrictions were applicable, even during the transition period, to those species listed on Appendix I of CITES or otherwise protected by international agreement. Although CITES does not restrict the taking of any species on its appendices, this exception apparently was designed to aide enforcement of CITES' export restrictions. A second exception allowed the federal taking restrictions to become operative upon request by a state. *Id.* § 1535(g)(2)(B)(i). The final exception permitted the Secretary to impose taking restrictions unilaterally for a limited period in emergency situations. *Id.* § 1535(g)(2)(B)(ii).

[355] *Id.* § 1535(f). This provision was intended to allow the state of Alaska to restrict the taking of endangered and threatened species by Alaskan natives who would otherwise be exempt from federal regulation. *See* H.R. Conf. Rep. No. 93–740, 93d Cong., 1st Sess. at 27–28 (1973), *reprinted in* 1973 U.S.C.C.A.N. 2989, 3006, and text accompanying notes 187–188, *supra.*

[356] 16 U.S.C. § 1535(c).

[357] *See* Chapter 5 at text accompanying note 148.

[358] *Id.* at text accompanying notes 162–170.

for each are identical except that the authority to acquire habitat is a necessary element of a state program only with respect to animals.[359]

What the states receive in return for meeting the criteria specified in the Act is eligibility for federal financial assistance of up to three-fourths of the costs of approved programs.[360] Financial assistance is allocated among eligible states at the Secretary's discretion, based upon criteria specified in the Act.[361] The states also receive a very limited exemption from the requirement that they obtain a federal permit before taking any protected species pursuant to their conservation programs.[362] Thus, while the procedures for states to play a role under the Endangered Species Act are substantially simpler than under the Marine Mammal Protection Act, the substance of that role is much narrower.[363]

The extent to which the Endangered Species Act preempts state authority over the species it protects is addressed, somewhat confusingly, in section 6(f).[364] This section distinguishes between state authority over the taking of endangered species and state authority over trade involving them. With respect to taking, section 6(f) authorizes state laws or regulations that are "more restrictive than the exemptions or permits" provided for in the federal Act, but it preempts those that are less restrictive. If this were the only provision of the Act on this topic, there would be little reason for confusion. However, section 4(d), authorizing the Secretary to promulgate regulations necessary for the conservation of threatened species, concludes with the statement that "such regulations shall apply in any State which has entered into a cooperative agreement pursuant to [section 6(c)] only to the extent that such regulations have also been adopted by such State."[365]

Moreover, section 6(g)(2) provides that the Act's taking prohibitions "shall not apply with respect to the taking of any resident endangered species or threatened species . . . within any state . . . which is then a party to a cooperative agreement with the Secretary pursuant to [section 6(c)]."[366] This apparent waiver of the taking prohibition is subject to two

[359] *Compare* 16 U.S.C. §§ 1535(c)(1) and (2). Prior to the 1978 Amendments, Section 6(c) appeared to envision cooperative agreements only with respect to animals.

[360] *Id.* § 1535(d)(2), as amended by 1982 Amendments § 3(1). Where two or more states have a common interest in one or more listed species, they may jointly enter into an agreement with the Secretary, for which the federal share of funding may be as high as 90 percent. *Id.* § 1535(d)(2)(ii).

[361] *Id.* § 1535(d)(1).

[362] *See* 50 C.F.R. §§ 17.21(c)(5) and 17.31(b)(1995).

[363] In addition to cooperative agreements, the Act provides that the Secretary may enter into agreements with the states for the administration and management of areas established for the conservation of endangered and threatened species. 16 U.S.C. § 1535(b).

[364] *Id.* § 1535(f).

[365] *Id.* § 1533(d).

[366] *Id.* § 1535(g)(2).

important exceptions. First, to the extent that the taking is contrary to state law, it remains subject to the federal taking prohibition. Second, the waiver does not apply to any species included on Appendix I of CITES or "otherwise specifically covered by any other treaty or Federal law."

The apparent conflict between these provisions was addressed in *Swan View Coalition, Inc. v. Turner.*[367] In that case, the environmental plaintiffs alleged that Forest Service road building practices violated the Act's taking prohibition by significantly degrading the habitat of grizzly bears and wolves in Montana. An intervenor-defendant argued that because Montana was a party to a cooperative agreement under section 6(c), and because Montana state law did not recognize habitat modification as a form of prohibited taking, the federal prohibition was inapplicable under sections 4(d) and 6(g)(2) of the Act. The court acknowledged that the intervenor had "raised compelling arguments on this issue," but in light of the "clear language of §6(f) . . . combined with the overwhelming priority Congress has given to the preservation of threatened and endangered species, the court must conclude that the less restrictive takings provisions under Montana law are preempted."[368] The court in *United States v. Glenn-Colusa Irrigation District* reached the same result without analysis.[369]

The result reached in both *Swan View Coalition* and *Glenn-Colusa Irrigation District* probably is correct, although the language of the Act's various provisions seems irreconcilable. The references in sections 4(d) and 6(g)(2) to cooperative agreements under section 6(c) have remained unchanged since the Act's beginning. However, the requirements to enter into a section 6(c) cooperative agreement have been relaxed significantly during that time. Today, a cooperative agreement may encompass only plants, only wildlife, only some species of plants, or only some species of wildlife that are federally listed in the state. It seems unlikely that if Congress intended the language of sections 4(d) and 6(g)(2) to take precedence over that of section 6(f), it would have made no changes in the former when it eased the requirements to enter into cooperative agreements. Further, in entering into these agreements, the Department of the Interior routinely has secured from the states an acknowledgement that, to the extent their laws are less restrictive than the federal Act, they are preempted.

With respect to state authority over trade in endangered or threatened species, section 6(f) provides that any state law or regulation concerning import, export, or interstate or foreign commerce in the species is void to the extent it permits what is prohibited by the Act or prohibits what is

[367] 824 F. Supp. 923 (D. Mont. 1992).
[368] *Id.* at 938.
[369] 788 F. Supp. 1126 (E.D. Cal. 1992).

"authorized pursuant to an exemption or permit" under the Act.[370] In *Man Hing Ivory and Imports, Inc. v. Deukmejian*, the Ninth Circuit Court of Appeals held that a California law banning importation and sale of African elephant ivory was unlawful to the extent that it purported to prohibit activities authorized by a permit issued pursuant to regulations under the Endangered Species Act.[371] Although the federal permit conditioned its validity upon compliance with applicable state law, the court held that this was a boilerplate provision common to all Fish and Wildlife Service permits and intended to refer only to "state or federal health, quarantine, customs, and agricultural laws."[372]

The court held that the Endangered Species Act did not preempt a New York law prohibiting the sale of wild birds, even by persons licensed under the federal Act to engage in the business of importing wildlife, in *Cresenzi Bird Importers, Inc. v. State of New York*.[373] These business licenses, required by section 9(d) of the Act,[374] were neither "permits" nor "exemptions" under section 6(f), which voids state laws that prohibit acts "authorized pursuant to an exemption or permit."

The International Component

The Endangered Species Act directed the President to begin implementing the Convention on Nature Protection and Wildlife Preservation in the Western Hemisphere, some three decades after the Convention's signing. In addition, it directs the Secretary to encourage foreign nations to establish and carry out endangered species programs of their own and authorizes both financial assistance and the loan of federal wildlife personnel for these purposes.[375] The 1973 Act also authorizes the Secretary to conduct law enforcement investigations and research abroad.[376] Finally, and most significantly, the Act implements the Convention on International Trade in Endangered Species of Wild Fauna and Flora (CITES).[377]

The Bobcat Debate

An early source of controversy over the 1973 Act's CITES provisions was the 1977 decision by CITES parties to add all unprotected species of

[370] 16 U.S.C. § 1535(f).

[371] 702 F.2d 760 (9th Cir. 1983). This case was decided prior to the African elephant's being placed on CITES Appendix I in 1989, thus banning all trade in elephant ivory.

[372] *Id.* at 765.

[373] 658 F. Supp. 1441 (S.D.N.Y. 1987), *aff'd*, 831 F.2d 410 (2d Cir. 1987).

[374] 16 U.S.C. § 1538(d).

[375] *Id.* § 1537(a)-(c).

[376] *Id.* § 1537(d).

[377] For a detailed discussion of CITES' framework and provisions, see Chapter 14 at text accompanying notes 135 *et seq.*

cats to Appendix II. While an Appendix II listing does not preclude international commercial trade, it does require that the Scientific Authority of the country of export determine that the export of an Appendix II specimen "will not be detrimental to the survival of that species."[378] The application of that standard to the export of bobcats from the United States was the focus of a 1981 decision of the United States Court of Appeals for the District of Columbia, *Defenders of Wildlife v. Endangered Species Scientific Authority*.[379]

The Endangered Species Scientific Authority was, until 1979, the designated "Scientific Authority" of the United States under CITES.[380] In 1978 it published certain guidelines governing how it would make the "no detriment" determination.[381] The guidelines identified population trend, total harvest, harvest distribution, and habitat evaluation information as the principal information upon which these determinations would be made. Applying those guidelines, the Scientific Authority approved the export in 1979 of bobcats taken in thirty-four states and the Navajo Nation.[382]

Defenders of Wildlife challenged both the guidelines and the decision to authorize bobcat export. The district court rejected most of the plaintiff's claims, enjoining export only from part or all of seven states. The court of appeals, in a decision reached after export for the season in question was over, struck down the guidelines but remanded to the district court for a determination whether the administrative record, considered in light of the appellate court's decision, supported no detriment findings for any of the states in question.[383]

The court of appeals' opinion addressed a number of important questions. The court rejected the plaintiff's claim that listing a species on Appendix II obligated the Scientific Authority to make no detriment findings with respect to each subspecies of the listed species prior to authorizing export.[384] The court also rejected the claim that state-by-state determinations were unlawful.[385] With respect to the challenged guidelines, however, the court determined[386] that they were unlawful because

[378]CITES, art. IV, § 2.

[379]659 F.2d 168 (D.C. Cir.), *cert. denied sub nom.* Int'l Ass'n of Fish and Wildlife Agencies v. Defenders of Wildlife, 454 U.S. 963 (1981).

[380]*See* Exec. Order No. 11,911, 3 C.F.R. § 112 (1976 Compilation), *reprinted in* 16 U.S.C. § 1537, revoked by Exec. Order No. 12, 608, 52 Fed. Reg. 34617 (1987).

[381]43 Fed. Reg. 15098 (1978).

[382]44 Fed. Reg. 55540 (1979).

[383]659 F.2d at 183.

[384]*Id.* at 179–80.

[385]*Id.* at 180.

[386]The court first held that CITES provides a source of rights enforceable by private parties in court. *Id.* at 175. Interestingly, the court did not rely upon the Act's citizen suit provision to reach this conclusion.

they did not require "(1) a reliable estimate of the number of bobcats and (2) information concerning the number of animals to be killed in the particular season."[387]

The bobcat decision quickly created a major controversy. Critics of the decision charged that it placed impossibly onerous duties on state and federal decisionmakers and that these duties were likely to be imposed pursuant to other wildlife statutes as well. Neither fear appeared justified. First, the court of appeals expressly disavowed the need to have "some kind of head count," insisting instead only on "a reasonably accurate estimate."[388] Second, the requirement of reasonably accurate estimates has not imposed insuperable burdens in other statutory contexts. For example, the Marine Mammal Protection Act mandates population estimates of overlapping and often very similar porpoise stocks distributed over several million square miles of the Pacific Ocean, and courts have not set aside the administratively developed estimates.[389]

Nonetheless, in 1982, Congress nullified the court's decision by providing that in carrying out the specified duties with respect to Appendix II species, the Secretary is not required to make population estimates and may not require the states to make these estimates.[390] In 1984, a court of appeals confirmed that the 1982 amendments did indeed eliminate the requirements for population estimates and kill levels.[391]

The Exemption for Captive-Bred Specimens

Another early decision interpreting CITES was *Cayman Turtle Farm, Ltd. v. Andrus.*[392] In that case the court upheld threatened species regulations against a challenge that the regulations should have provided an exception for the plaintiff's captive-reared specimens. The court considered the narrow exception in CITES for specimens of Appendix I species "bred in captivity"[393] and the restrictive definition of that term agreed upon by CITES parties[394] in concluding that the challenged regulations were consistent with the limited exception and therefore lawful.[395]

[387] *Id.* at 178.

[388] *Id.*

[389] See the discussion of the tuna-porpoise controversy in Chapter 5.

[390] 16 U.S.C. § 1537a(c)(2).

[391] Defenders of Wildlife v. Endangered Species Scientific Authority, 725 F.2d 726 (D.C. Cir. 1984).

[392] 478 F. Supp. 125 (D.D.C. 1979), *aff'd without opinion* (D.C. Cir. Dec. 12, 1980).

[393] CITES, art. VII, § 4.

[394] Proceedings of the Second Meeting of the Conference of the Parties, Conf. Doc. 2.12 (1979).

[395] 478 F. Supp. at 131 n. 2. Even had the challenged regulations been more restrictive than required by CITES, "stricter domestic measures" are expressly authorized by CITES Article XIV. 478 F. Supp. at 131–32. *See also* United States v. Crutchfield, 26 F.3d 1098 (11th Cir. 1994) (noting that the key issue in the case below was whether CITES-listed species

Penalties for CITES Violations

Enforcement of CITES has generally focused on forfeiture rather than on civil or criminal penalties.[396] The legislative history of CITES supports this focus:

> Under the amendment provided in H.R. 14104 a tourist who unknowingly imports a listed species into the United States could not be fined more that $500. The committee assumes that in most cases, law enforcement officials will seek forfeiture of the item rather than impose a civil penalty.[397]

In *United States v. 3,210 Crusted Sides of Caiman Crocodilus Yacare,*[398] the court held that, where a permit violates CITES requirements, the entire shipment is subject to forfeiture, not just the number of specimens exceeding the amount declared on the CITES permit.[399] Although courts have imposed significant criminal and civil penalties for violations of the Endangered Species Act's CITES provisions,[400] many administrative law judges are, in practice, quite lenient on persons violating CITES.[401]

The Panda Case

Perhaps the best-known CITES case is *World Wildlife Fund v. Hodel.*[402] In this case, the Toledo Zoo had received a permit from the Fish and Wildlife

---

at issue were illegally imported without the required CITES permits or whether the specimens were the captive-bred progeny of legally imported "pre-act" specimens).

[396] *See, e.g.,* United States v. One Handbag of Crocodilus Species, 856 F. Supp. 128 (E.D.N.Y. 1994) (forfeiture sought for 57 items under three captioned actions). *See generally* J. Z. Brooks, *A Survey of the Court Enforcement of International Wildlife Trade Regulations Under United States Law,* 17 Wm. & Mary J. Envt'l L. 145 (1993); M. Alagappan, Comment, *The United States' Enforcement of the Convention on International Trade in Endangered Species of Wild Fauna and Flora,* 10 Nw. J. Int'l L. & Bus. 541 (1990).

[397] H.R. Rep. No. 95–1625, 95th Cong., 2d Sess. at 26 (1978), *reprinted in* 1978 U.S.C.C.A.N. 9453, 9476. *See also* H.R. Conf. Rep. No. 95–1894, 95th Cong. 2d Sess. at 26, *reprinted in* 1978 U.S.C.C.A.N. 9453, 9493; Rittenberry v. United States Fish and Wildlife Service, 2 O.R.W. 208G (FWS App. 1980).

[398] 636 F. Supp. 1281 (S.D. Fla. 1986).

[399] *Id.* at 1287.

[400] *See, e.g.,* United States v. Asper, 753 F. Supp. 1260 (M.D. Pa. 1990). This case also presents an interesting application of the United States Sentencing Guidelines in the context of a CITES violation. For another case applying the Sentencing Guidelines to a CITES violation, *see* United States v. Stubbs, 11 F.3d 632 (6th Cir. 1993). Of particular interest is U.S.S.G. § 2Q2.1(b)(3)(B) (1992), which states: "If the offense involved . . . fish, wildlife or plants that are listed in Appendix I to [CITES], increase by four levels." Other potentially applicable provisions of the Sentencing Guidelines concern increases in sentencing for crimes involved commercial purposes (U.S.S.G. § 2Q2.1(b)(1)) and for leaders or organizers of criminal activity involving more than five persons (U.S.S.G. § 3B1.1).

[401] *See* Alagappan, *supra* note 396, at 550.

[402] 1988 WL 66193 (D.D.C. 1988). For a more detailed discussion of this case, see J. Z. Brooks, *supra* note 396.

Service to import two giant pandas from China for an exhibit in which the pandas were to be placed on public exhibition for 200 days.[403] On the day before the pandas were scheduled to leave China, World Wildlife Fund (WWF) sought a preliminary injunction to prevent their shipment. The injunction was denied, and the pandas were shipped and placed on public exhibition. WWF quickly filed an amended complaint and a motion for an order enjoining further public exhibition of the pandas, directing the Service to confiscate the pandas, directing return of the pandas to China, and declaring that the import permit was invalid.

The court focused on three issues arising out of article III(3) of CITES and the Endangered Species Act: (1) whether the importation would enhance the survival of giant pandas; (2) whether the importation would be detrimental to the survival of giant pandas; and (3) whether the importation was "primarily for commercial purposes." Because the standards of review for agency action and for granting a preliminary injunction are relatively difficult to meet, the court found for the Service on the first two issues. However, with regard to commercial purpose, the court found no evidence in the record that the Service had ever considered the commercial nature of the zoo's exhibit. The court found significant the fact that the Toledo Zoo was charging the public an additional fee to view the pandas. The court held that charging this additional fee violated CITES' prohibition on the importing of an Appendix I species for primarily commercial purposes and enjoined the Toledo Zoo from collecting the additional fee to view the pandas.[404]

The impact of this decision was considerably greater than might first be apparent from the limited result in the plaintiff's favor. As a result of this controversy, the Service revised its panda import permit policy to address WWF's concerns. China subsequently announced that it would no longer export to the United States either giant pandas or golden monkeys, another endangered species.[405] In addition, the case heightened public awareness of the plight of pandas and other endangered animals subjected to international trade, as well as awareness of CITES itself.

CONCLUSION

The Endangered Species Act begins with a congressional finding that various plant and animal extinctions have occurred in the United States "as a consequence of economic growth and development untempered by

[403]The giant panda is classified as an Appendix I species under CITES, and consequently, international trade in pandas is strictly regulated.

[404] 1988 WL 66193 at 4–5.

[405]K. D. Hill, *The Convention on International Trade in Endangered Species: Fifteen Years Later*, 13 Loy. L. A. Int'l & Comp. L.J. 231, 255–56, n. 158 (1990).

adequate concern and conservation.''[406] Implicit in this finding is the recognition that if further extinctions are to be averted, it will be necessary to ''temper'' economic growth and development. The experience of the past quarter decade has demonstrated repeatedly how difficult that task is and how politically volatile are the efforts to accomplish it.

In late 1996, the number of species on the endangered and threatened list in the United States surpassed 1,000. The total has been steadily growing. The factors that resulted in the endangerment of these species have often been at work for many decades; reversing their effects will likely take as long or longer. As a result, it is much too soon to characterize the Endangered Species Act as either a success or a failure. It is, however, not too soon to conclude that the law is unlikely to accomplish its goals for most of the species that it is intended to conserve without new conservation tools and a congressional commitment of resources to match the rhetoric that Congress used when it enacted the law.

The new tools most urgently needed are positive incentives for private landowners to carry out practices that will benefit, not merely avoid harming, imperiled species. Such incentives would not only help further the Act's goal of recovering species, but they would make it possible for species conservation to be an objective of private landowners, rather than a restraint upon their objectives.

A similar clear need is for earlier action, before species reach the point at which ultimate recovery is virtually impossible. The federal endangered species program was clearly intended to serve as a sort of emergency room; in practice, it has more often served as an intensive care unit or even a hospice. The category of ''threatened species'' was intended to accomplish the goal of prompting earlier attention to declining species, but it has been sparingly used and often for species whose true situation is little different from those that are designated as ''endangered.''

Finally, unless Congress provides funding commensurate with the task of recovering the large number of imperiled species already on the endangered and threatened list, little progress is likely to be made for many of them. There is probably no other federal conservation program for which the disparity between the goals that Congress has proclaimed and the resources that Congress has made available to pursue those goals is so great. The tragedy of that fact is that with the loss of each species, our nation's natural heritage is impoverished.

[406] 16 U.S.C. § 1532(1).

Part III

WILDLIFE, LAND, AND WATER

In Part II, we examined the evolution of species conservation laws. In this part we explore the most important laws that affect wildlife habitat on federal lands. We also examine the federal laws governing water resource development and federal conservation programs for private lands. The division of topics between species conservation in Part II and land and water conservation in Part III is not meant to suggest a dichotomy between these two realms. Rather, it is simply a tool to organize an almost overwhelming body of information into manageable units.

By now there is widespread agreement that a conservation strategy must transcend individual species and focus on ecosystems, the habitat in which they live. Wildlife is dependent upon the health of its habitat: habitat loss or alteration is a significant factor in the decline of over 85 percent of species listed under the Endangered Species Act. In addition, the sheer number of species at serious risk of extinction forces us to deal with communities of species dependent on the same habitat, rather than each one individually.

We cannot, however, relegate individual species concerns to the periphery of resource planning in favor of looking at the "whole system." We know far too little about ecosystems to replace decades of experience in species management with vague notions of ecosystem management. This emerging concept, also called "landscape planning," "watershed analysis," and the like, is an elusive one. Reduced to its simplest terms, ecosystem management acknowledges the interconnectedness of seemingly isolated actions and attempts to coordinate those actions in a way that will minimize their impacts on ecosystem function. But it is a process, not a goal. Recognizing connections in nature does not, by itself, assure healthy ecosystems or viable wildlife populations; these will result only if we choose to make them overriding goals of land and water management.

Scientific complexity alone makes managing ecosystems to conserve biodiversity a daunting enterprise. In the last twenty years, we have moved from viewing ecosystems as stable, closed, and internally regulated, to a new picture of more open systems in constant flux, usually without long-term stability, and affected by a series of human and other stochastic factors. Resource conservation and ecosystem change consequently are characterized by uncertainty. Moreover, ecosystems cover large areas, their boundaries are vague and fluid, and they are infinitely complex as well as dynamic.

A further complication arises from the fact that the goals of ecosystem management must be defined from the social and economic perspective, as well as the scientific. Human social and economic systems favor predictability. Governmental units have conspicuous artificial boundaries, and there often is little communication among governing entities. Our political system and the regulatory agencies have trouble dealing with complexity and change. Nature is full of both.

Were society to overcome these obstacles and agree on conservation of biodiversity as a high priority for land and water management, the public lands would seem to be a logical starting point for planning efforts. These lands are relatively undeveloped, and they are owned by all the people of the nation. In contrast to private lands, public lands are physically and politically easier to manage for biodiversity. In addition, the laws under which they are managed emphasize land use planning.

The federal public lands cover over a third of the country, concentrated in the western states; there, they comprise half the land base. They include some of the most important lands in the nation for wildlife. Almost 50 percent of threatened and endangered species are found on federal lands. State lands make an important contribution to wildlife habitat as well, and their exclusion from this book does not reflect a judgment that they are unimportant for sustaining wildlife populations. Rather, it is simply because the focus of this book is federal law.

The study of wildlife law on federal lands reveals that despite some progress, wildlife conservation is not a commanding force in federal land management. It is one of several management objectives on all the major federal land classifications: national wildlife refuges, national parks, wilderness areas, national forests, and national resource lands. It is not the exclusive, or even the dominant, goal on any lands but the national wildlife refuges. Critics would argue that even on many wildlife refuges, managers permit secondary uses like livestock grazing, military exercises, and oil development to subvert the conservation goal.

Chapter 8 covers what we have called the "conservation lands and waters": national wildlife refuges, national parks, wilderness areas, and national marine sanctuaries. These reserve systems are governed by different agencies and different statutes, but the laws all emphasize conservation of the area in a natural state.

The National Wildlife Refuge System is the only system that has wildlife conservation as its primary objective. It is a collection of units that were designated as refuges for reasons ranging from waterfowl production for hunters to endangered species conservation. In national parks, the primary values are scenery, recreation, and wildlife. The objective of wilderness designation is the preservation of wildness and solitude; a wilderness area is one "where the earth and its community of life are untrammeled by man." The ban on roads and structures in wilderness means that these areas contain some of the least fragmented wildlife habitats in the nation. Finally, although land conservation historically has received much greater emphasis than has marine conservation, one national program of relatively recent vintage has as its primary objective the conservation of marine resources—the national marine sanctuary program.

Even if the conservation lands were managed primarily for wildlife conservation, they are not complete ecosystems within themselves, capable of sustaining native wildlife populations. The effects on resources within a reserve of activities outside the boundary—such as logging and development—are a recurring theme in the cases dealing with conservation lands, as is the effect of reserve management on lands outside their borders. Moreover, if biological evolution is to continue, there must be large habitat areas, or at least hospitable connections between smaller natural areas, that allow animals to move freely to breed and for young to disperse to new territories. Ecosystem processes, such as hydrologic cycles, similarly may require conservation of rivers or other features that extend outside a park or refuge boundary. The conservation lands represent only parts of larger ecosystems that include adjacent multiple-use public lands, as well as nearby private lands.

Chapters 9 and 10 trace the development of wildlife conservation on multiple-use lands, the national forests and national resource lands. Wildlife conservation on these lands historically has taken a back seat to commodity uses, primarily logging on national forests, and livestock grazing on national resource lands. Change has begun, however, and the laws and cases discussed in these chapters show a growing demand for ecologically sound land management on multiple-use lands as well as conservation lands.

Chapter 11 examines federal laws governing water resource development. Rivers are the lifeblood of fish and wildlife and the ecological communities of which they are a part. Rivers, too, respect no artificial boundaries. And they, like the land, are coveted by humans for commercial purposes—energy, irrigation, navigation, and others. The laws governing water projects have shown great promise for protecting the integrity of rivers for fish and wildlife in the face of human demands. The discussion of these laws in Chapter 11 shows that as yet, the laws have not lived up to their promise.

Chapter 12 examines federal programs for private land conservation that can result in significant habitat conservation. Incorporating private land as well as public land and rivers in a comprehensive conservation strategy is particularly important in the eastern United States, where public lands are rare and private inholdings within the public land boundary can be substantial (for example, about 50% in the South). Regulation is not the only tool available to improve the odds for wildlife on private lands, and the incentive programs discussed in Chapter 12 may be critical in creating an integrated wildlife conservation strategy.

The analyses in these chapters conclude that current laws governing most federal lands permit managers to set conserving biodiversity as an overarching goal—but with few exceptions they do not *require* managers to do so. Federal resource managers have long enjoyed a great deal of discretion in decisionmaking. But this nearly unfettered discretion is starting to be limited in significant ways. Tracing the evolution of the law affecting wildlife on federal public lands reveals a concerted effort by citizen groups to narrow agency discretion and to require better planning, more informed decisionmaking, and consideration of a broader range of goals and concerns than resource extraction and exploitation.

The record also shows that the efforts, with a few notable exceptions, have met with some success in Congress but little in the courts. In the vast majority of cases, courts are unwilling to overturn agency decisions, whether the challenge is by a conservation group or a resource user group.

To be sure, there have been some striking successes that have caused major shifts in federal wildlife management. The best known is the series of cases in the Pacific Northwest in the late 1980s and early 1990s in which a court enjoined most logging in old growth forests on the country's most productive national forests because the Forest Service had failed to manage for viable populations of the northern spotted owl. The court required the agency to revise its owl management plan three times before lifting the injunction. The court's third order was based on the agency's failure to manage for a diversity of species, in addition to the spotted owl.

These cases resulted in perhaps the most high-profile and comprehensive federal land management plan in history: a document prepared under order from the President of the United States, drafted by six federal agencies working together, covering hundreds of old-growth associated species, in forests ranging from the Canadian border to northern California. Jack Ward Thomas, a wildlife biologist and head of the team that prepared the plan, became the first biologist appointed Chief of the Forest Service. It is noteworthy that this landmark ecosystem management plan was a result not of the Endangered Species Act, but primarily of the National Forest Management Act and the National Environmental Policy Act.

The success conservationists attained in these and some other cases in the 1980s and early 1990s has met fierce political opposition. In the 1995 and 1996 sessions of Congress, bills were introduced to eliminate the National Forest Management Act's species diversity requirement (on which the spotted owl lawsuits were based), elevate grazing to even greater prominence on the public lands, divest major national park lands, change wildlife refuge law to permit more extractive uses, open the Arctic National Wildlife Refuge to oil and gas development, and restore historic high logging levels to the Tongass National Forest in Alaska, among others. Congress passed a rider to the 1995 Rescissions Bill overturning much of the management plan for old growth forests, leading to civil disobedience and hundreds of arrests in Oregon and Washington when the forests once again began to fall.

The past decade has been a dynamic one for wildlife conservation on federal lands. We have moved into a time of transition, one in which our knowledge of ecosystem dynamics and our concerns about species decline are growing rapidly. Wildlife concerns are receiving ever-increasing attention from land managers, begetting intense political reaction from those who have been content with past management policies. It is hard to predict when equilibrium will be restored, or where wildlife conservation on federal lands will be when it is.

Chapter 8

CONSERVATION LANDS AND WATERS

This chapter discusses designation and management of federal lands devoted primarily to conservation. These include the National Wildlife Refuge System, the only federal lands that were acquired or reserved expressly for wildlife conservation, the National Park System, and the Wilderness Preservation System. The latter two systems are managed for conservation of several resources, of which wildlife is one. The chapter also discusses federal designation and management of marine waters for conservation through the National Marine Sanctuaries Program.

THE NATIONAL WILDLIFE REFUGE SYSTEM

The National Wildlife Refuge System is the only extensive system of federal lands managed chiefly for wildlife conservation. Its purpose is to develop "a national program of wildlife and ecological conservation and rehabilitation" through the "restoration, preservation, development and management of wildlife and wildlands habitat."[1] The Refuge System comprises nearly 91 million acres, almost 76 million of which are in Alaska.

The origins of the Refuge System can be traced to the turn of the century, when presidential proclamation established the first federal wildlife refuges. Soon Congress also became involved in creating refuges, first by authorizing the President in 1905[2] and 1906[3] to designate certain areas

[1]50 C.F.R. § 25.11(b) (1995).

[2]Act of Jan. 24, 1905, ch. 137, 33 Stat. 614 (current version at 16 U.S.C. §§ 684–686).

[3]Act of June 29, 1906, ch. 3593, 34 Stat. 607 (current version at 16 U.S.C. §§ 684–686).

as wildlife ranges. In 1908, Congress established a National Bison Range in Montana by its own action.[4]

The Migratory Bird Treaty Act, passed in 1918, provided the stimulus for a systematic program of refuge acquisition. The Act's failure to authorize acquisition of migratory bird habitat came to be recognized as a serious shortcoming. To provide the needed authority, Congress passed the Migratory Bird Conservation Act (hereinafter Conservation Act) in 1929.[5]

The Conservation Act established a Migratory Bird Conservation Commission to review and approve proposals of the Secretary of the Interior for the purchase or rental of areas under the Act.[6] Although several subsequent statutes—including the Fish and Wildlife Coordination Act,[7] the Fish and Wildlife Act of 1956,[8] the Land and Water Conservation Fund Act,[9] and the Endangered Species Acts of 1966, 1969, and 1973[10]—contained acquisition authority, the Conservation Act continues to be a major source of authority for wildlife refuge acquisition.[11]

The Migratory Bird Hunting Stamp Act, enacted in 1934, assured a steady source of funding for refuge acquisition under the Conservation Act.[12] The Hunting Stamp Act, however, almost assured a refuge system keyed principally to the production of migratory waterfowl. A 1958 amendment provided that funds derived from the Act could be used to

[4]Act of May 23, 1908, ch. 192, 35 Stat. 267 (current version at 16 U.S.C. § 671).

[5]16 U.S.C. §§ 715–715d, 715e, 715f–715k, 715n–715r.

[6]The Secretaries of both Interior and Agriculture serve on the Migratory Bird Conservation Commission, as does the Secretary of Transportation and two members each from the Senate and the House of Representatives. For purposes of considering the acquisition of areas within a particular state, the ranking officer of the game agency of that state is an *ex officio* member of the commission. 16 U.S.C. § 715a. Unlike most other statutes authorizing federal land acquisition, the Conservation Act requires that the Secretary obtain the legislative consent of the state in which the lands to be acquired are located prior to acquisition. *Id.* § 715f.

[7]The Fish and Wildlife Coordination Act provides that lands acquired or administered by a federal agency pursuant to its authority may be made available to the Secretary of the Interior for administration by him "where the particular properties have value in carrying out the national migratory bird management program." 16 U.S.C. § 663(b). The Act is discussed in detail in Chapter 11.

[8]16 U.S.C. § 742f(a).

[9]*Id.* § 460*l*-4 through 460*l*-11.

[10]The acquisition authority of the various Endangered Species Acts is discussed in Chapter 7.

[11]Despite the many sources of acquisition authority, the Advisory Committee on Wildlife Management reported to Secretary of the Interior Udall in 1968 that "there is not support nor clear authority for the Bureau of Sport Fisheries and Wildlife to extend the refuge system in relation to wildlife needs other than for migratory birds." The Committee's report, commonly known as the Leopold Report, is reprinted in Department of the Interior, Fish and Wildlife Service, Final Environmental Statement, Operation of the National Wildlife Refuge System, app. W (1976).

[12]16 U.S.C. §§ 718–718b and 718c–718h.

acquire "waterfowl production areas," which are small wetland or pothole areas.[13]

Scope of Federal Authority to Acquire Refuge Lands

Only a few cases have addressed the scope of the federal government's authority to acquire lands for wildlife refuge purposes. *Swan Lake Hunting Club v. United States* was one of the most significant of these.[14] In *Swan Lake,* the court rejected a contention that the Conservation Act did not authorize the Secretary to acquire lands by condemnation. The court also rejected efforts to construe narrowly the terms of legislative and gubernatorial consents to federal acquisition under the Conservation Act.

The United States Supreme Court first considered the scope of federal refuge acquisition authority in *United States v. Little Lake Misere Land Co.*[15] In *Little Lake,* the United States had acquired under the Conservation Act parcels of land in Cameron Parish, Louisiana, for Lacassine National Wildlife Refuge. The sellers had retained mineral rights for ten years. Shortly after the acquisition, Louisiana passed a law providing that in any sale of land to the United States, retained mineral rights were imprescriptible. Some thirty years after the purchase, the United States brought suit to quiet title to the land. The Supreme Court, while acknowledging that state law ordinarily governs land transactions to which the United States is a party, held that it did not govern in this case because the acquisitions were ones "arising from and bearing heavily upon a federal regulatory program."[16] Further, in fashioning a federal rule of law to govern the acquisition, the Court rejected the customary route of "borrowing" state law because the state law in question was "plainly hostile to the interests of the United States."[17]

Two decisions of the Eighth Circuit Court of Appeals substantially extended the *Little Lake* rule. In *United States v. Albrecht,* the government sought to enforce an easement against the drainage of a pothole area in North Dakota.[18] The defendant argued that the purported easement was not a recognized property interest under North Dakota law. Following

[13]Acquisition of waterfowl production areas is expressly exempted from the legislative consent requirement of the Conservation Act. However, the Wetlands Loan Act of 1961 required consent of the governor or the head of the appropriate state agency prior to acquisition of such an area. 16 U.S.C. § 715k-5. This requirement apparently only formalized prior administrative practice under the Conservation Act. *See* 107 Cong. Rec. 17172 (1961) (remarks of Senator Hruska). For the Supreme Court's interpretation of this requirement, see text accompanying notes 22–26 *infra.*

[14]381 F.2d 238 (5th Cir. 1967).

[15]412 U.S. 580 (1973).

[16]*Id.* at 592.

[17]*Id.* at 597.

[18]496 F.2d 906 (8th Cir. 1974).

Little Lake, the Court held that the question should be decided by federal law because it bore heavily upon a federal regulatory program, acquisition of pothole areas under the Hunting Stamp Act. The Court likewise refused to "borrow" North Dakota law on the following grounds:

> We fully recognize that laws of real property are usually governed by the particular states; yet the reasonable property right conveyed to the United States in this case effectuates an important national concern . . . and should not be defeated by any possible North Dakota law barring the conveyance of this property right. . . . We, therefore, specifically hold that the property right conveyed to the United States in this case, whether or not deemed a valid easement or other property right under North Dakota law, was a valid conveyance under federal law and vested in the United States the rights as stated therein.[19]

The *Albrecht* decision demonstrated an extraordinary solicitousness toward federal land acquisition in furtherance of wildlife conservation programs. In effect, the *Albrecht* court was prepared to recognize any "reasonable" federal property right, wholly without regard to state law.[20]

Another opportunity to examine the clash between state and federal interests came only a few years later. In 1977, North Dakota passed laws placing strict limitations on the acquisition of easements by the federal government. In addition, piqued by controversy over the Bureau of Reclamation's Garrison Diversion project, the governor purported to revoke the authority predecessors had granted to the federal government to acquire easements under the Wetlands Loan Act of 1961.[21] The United States sued to declare invalid the North Dakota law and the governor's revocation.

The Supreme Court upheld the federal government's claims in *North Dakota v. United States.*[22] The Court held that once given, gubernatorial consent to easement acquisition under the Wetlands Loan Act cannot be revoked, at least not so long as the United States did not unreasonably delay in carrying out its acquisitions.[23] The two dissenting Justices would have remanded to determine whether the federal government had in fact delayed unreasonably.[24]

As to the legislative restrictions enacted by North Dakota, the Court held that they could not be applied to land acquisitions carried out pur-

[19] *Id.* at 911.

[20] *Compare* the Supreme Court's decision in Cappaert v. United States, 426 U.S. 128 (1976), holding that the United States need not comply with state law to reserve unappropriated groundwater when setting aside lands under the Act for the Preservation of American Antiquities for the purpose of preserving the endangered Devil's Hole pupfish.

[21] *See* note 13, *supra.*

[22] 460 U.S. 300 (1983).

[23] *Id.* at 314–15.

[24] *Id.* at 321 (O'Connor, J., dissenting).

suant to previously given gubernatorial consents.[25] The Court expressly refrained from deciding whether any of the state's restrictions—requiring concurrence by county commissions, limiting the duration of easements, and prohibiting certain types of easements—might be valid if applied to easements acquired pursuant to later gubernatorial consents.[26]

The North Dakota wetland easements continued to create controversy. In *United States v. Vesterso,* three members of a North Dakota county water board were convicted of damaging federal property by draining wetland areas to alleviate flooding on adjacent properties.[27] Two of the board members owned land that benefitted from the drainage project. The defendants argued that they were not "persons" subject to the Refuge Administration Act because they were acting in their official capacity. The court of appeals rejected this argument and affirmed the convictions, finding that "[t]he Act clearly uses the term 'person' in such a way as to include public officials within its meaning."[28] Rather than fine or jail the defendants, however, the court placed them on probation, which would terminate if they restored the wetlands to their former condition.

Administration of the National Wildlife Refuge System: Legislative Authority

Three statutes constitute the basic authority within which the National Wildlife Refuge System operates. These are the National Wildlife Refuge Administration Act of 1966,[29] the Refuge Recreation Act of 1962,[30] and the Refuge Revenue Sharing Act of 1964.[31] Over the years Congress has considered several bills whose purpose is to give the Refuge System its own "organic act," but none has been enacted.

Until 1966, there was no single law governing administration of the many federal wildlife refuges. The numerous administrative units, known variously as "game ranges," "wildlife ranges," "wildlife management areas," "waterfowl production areas," and "wildlife refuges," were under the jurisdiction of the Fish and Wildlife Service, or, in a few cases, the joint jurisdiction of the Fish and Wildlife Service (FWS) and the Bureau of Land Management (BLM). Each unit was, however, governed by the

[25] *Id.* at 317.

[26] *Id.* at 316.

[27] 828 F.2d 1234 (8th Cir. 1987). By acquiring an easement, the government purchases the landowner's promise not to drain, fill, or otherwise destroy wetland areas on land subject to the easement.

[28] *Id.* at 1242. The Act defines "person" as "any individual, partnership, corporation, or association." 16 U.S.C. § 668ee(a).

[29] 16 U.S.C. §§ 668dd and 668ee.

[30] *Id.* §§ 460k–460k-4.

[31] *Id.* § 715s.

often vague standards of the law or executive order by which it was created.[32]

The National Wildlife Refuge System Administration Act of 1966[33] introduced a measure of rationality into this system by consolidating the varied units into a single National Wildlife Refuge System. The Act, however, did little to spell out standards to guide System administration. It (1) placed restrictions on the transfer, exchange, or other disposal of lands within the system; (2) clarified the Secretary's authority to accept donations of money to be used for land acquisition; and (3) most importantly, authorized the Secretary to "permit the use of any area within the System for any purpose, including but not limited to hunting, fishing, public recreation and accommodations, and access whenever he determines that such uses are compatible with the major purposes for which such areas were established."[34] This authorization of "compatible" uses thus made clear that the national wildlife refuges are not to be managed as "single-use" lands, but more properly as "dominant use" lands.[35]

Similarly, although the Conservation Act provided that all refuges acquired pursuant to its authority be operated as "inviolate sanctuar[ies],"[36] later enactments rendered this language largely meaningless. Congress authorized the Secretary to permit public hunting on an ever-increasing proportion of the total area of any refuge, if "compatible with the major purposes for which such areas were established."[37]

The other major law governing refuge administration is the Refuge Recreation Act of 1962, which authorizes the Secretary to permit public recreation

> when in his judgment public recreation can be an appropriate incidental or secondary use: *Provided,* That such public recreation use shall be permitted only to the extent that it is practicable and not inconsistent with . . . the primary objectives for which each particular area is established.[38]

The Secretary also may permit recreational uses "not directly related to the primary purposes and functions of the individual areas" of the System,

[32]The Refuge Recreation Act and the Refuge Revenue Sharing Act also governed the refuges after their passage in 1962 and 1964, respectively. See discussion of these laws *infra.*

[33]16 U.S.C. §§ 668dd and 668ee. The National Wildlife Refuge System Administration Act was the name given in 1969 to sections 4 and 5 of the Endangered Species Preservation Act of 1966, Pub. L. No. 89–669, 80 Stat. 926.

[34]16 U.S.C. § 668dd(d)(1).

[35]See discussion of litigation over secondary uses of refuges, *infra* at text accompanying notes 59–97.

[36]16 U.S.C. 715d(2).

[37]*Id.* § 668dd(d)(1)(A). See discussion of litigation regarding hunting, *infra* at text accompanying notes 79–89.

[38]*Id.* § 460k.

but only upon an express determination that they "will not interfere with the primary purposes" of the refuges and that funds are available for their development, operation, and maintenance.[39] The Refuge Recreation Act also authorizes the Secretary to acquire lands for various specified purposes, including conserving threatened and endangered species.[40]

The Refuge Revenue Sharing Act of 1964 governs the disposition of revenue from commercial activities on refuges. It provides that the net receipts from the "sale or other disposition of animals, timber, salmonid carcassas [*sic*], hay, grass, or other products of the soil, minerals, shells, sand, or gravel, from other privileges, or from leases for public accommodations or facilities . . . in connection with the operation and management" of areas of the National Wildlife Refuge System shall be paid into a special fund. The fund is used for public schools and roads in the counties in which refuges having revenue-producing activities are located.[41]

Administration of the National Wildlife Refuge System: Litigation

Considerable controversy has arisen over refuge management.[42] Conflicts have concerned FWS's jurisdiction over certain refuges, the Secretary's authority to transfer lands from the System, and, most frequently, his or her authority to permit or prohibit particular uses under the compatibility standard.

Agency Jurisdiction

Consolidation of so many disparate units into a single National Wildlife Refuge System did not change the fact that many of the units had been established for widely varying purposes. In particular, the System included units expressly established both for wildlife protection *and* for livestock grazing. Reflecting this duality of purpose, certain "game ranges" were jointly administered until 1975 by FWS and BLM. In that year, the Secretary of the Interior directed that BLM become sole manager of three game ranges.[43] The prospect that FWS, whose paramount mission is wildlife conservation, would be divested of its authority in favor of the multiple-use-oriented BLM so alarmed wildlife proponents that they both filed

[39] *Id.*

[40] *Id.* § 460k–1.

[41] *Id.* § 715s. The only reported litigation concerning the Refuge Revenue Sharing Act is Cameron Parish Police Jury v. Hickel, 302 F. Supp. 689 (W.D. La. 1969), and Watt v. Alaska, 451 U.S. 259 (1981), discussed at text accompanying notes 103–105 *infra.*

[42] *See generally* Richard J. Fink, *The National Wildife Refuges: Theory, Practice, and Prospect,* 18 Harv. Envt'l L. Rev. 1, 38–77 (1994).

[43] Memorandum from the Under Secretary of the Interior to the Directors of the Bureau of Land Management and the Fish and Wildlife Service, dated Feb. 5, 1975.

a lawsuit and simultaneously introduced legislation to reverse the Secretary's decision.

Both the lawsuit and the legislative efforts were successful in 1976. In *Wilderness Society v. Hathaway,* the District Court for the District of Columbia enjoined the transfer of authority to the BLM on the ground that "the Secretary is required to exercise his discretion and authority with respect to the administration of game ranges and wildlife refuges through the Fish and Wildlife Service."[44] One month later, Congress confirmed this result by amending the Refuge Administration Act to direct that all units of the System be administered through FWS and that all units then within the System remain so except under certain limited conditions.[45]

With the exception of one district court whose decision was later reversed, courts have resolved jurisdictional conflicts under the Refuge Administration Act in favor of FWS. In *Trustees for Alaska v. Watt,* the Secretary of the Interior had delegated responsibilities pertaining to oil and gas exploration and development in the Arctic National Wildlife Refuge to the United States Geological Survey (USGS).[46] The USGS was to develop initial guidelines for exploration, review and approve exploration plans, and, after a period of five years, recommend whether additional exploration or development was warranted.[47] The USGS had the exclusive responsibility for the first and last of these. Its approval of exploration plans was subject to concurrent approval by FWS.

The court concluded that this arrangement violated the Refuge Administration Act's requirement that FWS administer all units of the Refuge System. To "administer" a refuge is to "manage" it, the court reasoned, and the Service's duty to manage required it "to control and direct the Refuge by regulating human access in order to conserve the entire spectrum of wildlife found in the Refuge."[48] Even preparing a report to Congress that "supplies information essential to determining whether development activity will be permitted within the Refuge" was, in the court's view, "a Refuge administration function."[49] As to the requirement that FWS concur in the approval of any exploration plan, the court con-

[44]Civil No. 75–1004 (D.D.C. Jan. 26, 1976).

[45]Act of Feb. 27, 1976, Pub. L. No. 94–223, 90 Stat. 199 (codified at 16 U.S.C. § 668dd). The game range controversy is described at greater length in Comment, *National Game Ranges: The Orphans of the National Wildlife Refuge System,* 6 Envt'l L. 515 (1976).

[46]524 F. Supp. 1303 (D. Alaska 1981), *aff'd,* 690 F.2d 1279 (9th Cir. 1982). For further discussion of oil and gas leasing efforts on ANWR, see discussion *infra* at text accompanying notes 134–145.

[47]The Alaska National Interest Lands Conservation Act of 1980 (ANILCA), 16 U.S.C. § 3142(d), (e), (h), imposed these responsibilities. ANILCA is discussed further at text accompanying notes 106–145 *infra.*

[48]524 F. Supp. at 1309.

[49]*Id.* at 1310.

cluded that this was the very sort of "joint administration" of a refuge that the 1976 amendments prohibited.[50] The Ninth Circuit subsequently affirmed the district court's decision and embraced its opinion.

In *Schwenke v. Secretary of the Interior*,[51] the Montana district court regarded the 1976 amendments as a mere "shuffling of administration from one agency to another"[52] that did nothing to alter the fact that public grazing has "a co-equal priority with wildlife conservation"[53] on the Charles M. Russell National Wildlife Refuge and was to be administered under the Taylor Grazing Act.[54] The Ninth Circuit disagreed and reversed the district court's decision.[55] The appellate court found that the executive order creating the refuge had established a limited priority for wildlife over livestock in access to forage, and, more importantly, that "Congress clearly wanted the Russell Range administered by the Fish and Wildlife Service because of its underlying mission to protect wildlife."[56] The court held that the 1976 amendments changed the statute governing grazing on the Refuge from the Taylor Grazing Act to the Refuge Administration Act.

Authority to Exchange Refuge Lands

Conservation groups also have challenged the Secretary's authority under the Refuge Administration Act to exchange refuge lands for other lands. In *Sierra Club v. Hickel*, plaintiffs sought to void a land exchange between the Secretary and two public utility companies, the effect of which was to allow a nuclear power plant to be built immediately adjacent to a wildlife refuge.[57] In the court of appeals, the three judges who decided the case were unable to agree on a single rationale. Two judges believed the challenge was barred by the doctrine of sovereign immunity, emphasizing that the land exchange had already been completed. One of them also believed exchange of land was so committed to the Secretary's discretion that there could be no judicial review of an exchange decision. The dissenting judge argued that the exchange was reviewable and that the proper standard of review was the compatibility of the result with the purposes of the refuge. Had the challenge been initiated prior to the exchange's completion, Judge McCree, who rested his opinion solely on

[50] *Id.* at 1309–10.

[51] Civil Act. 79–133–BLG (D. Mont. Jan. 14, 1982), *rev'd*, 720 F.2d 571 (9th Cir. 1983).

[52] *Id.* slip op. at 9.

[53] *Id.* at 1.

[54] 43 U.S.C §§ 315 *et seq.* See Chapter 10 at text accompanying notes 11–15 for a discussion of the Taylor Grazing Act.

[55] Schwenke v. Secretary of the Interior, 720 F.2d 571 (9th Cir. 1983).

[56] *Id.* at 577.

[57] 467 F.2d 1048 (6th Cir. 1972).

sovereign immunity grounds, would have had to face the issue of reviewability. In the authors' opinion, the dissenting judge was correct, both as to the decision's reviewability and as to the standard of review.[58]

Secondary Uses: In General

The Secretary's authority to permit secondary uses that are "compatible with the major purposes" of the refuges has been the greatest source of controversy over refuge management.[59] Several lawsuits and government reports[60] have highlighted the conflicts between wildlife conservation and secondary uses, culminating in a broad-based 1993 settlement between Secretary Bruce Babbitt and conservation groups that has led to a significant decrease in incompatible refuge uses.[61]

In the 1993 settlement, the Secretary agreed to (1) put in place a process for determining the compatibility of secondary uses on refuges, including written determinations and NEPA compliance for all secondary uses of the Refuge System over which it has control; (2) terminate within one year all but a small group of activities on specific refuges which it has determined to be incompatible; and (3) not to permit any new secondary use unless it has made a written compatibility determination. While the Secretary did not agree to any substantive action that was not already required by law, the settlement evidences a serious commitment to protect refuge resources and establishes a process that enables greater public oversight of refuge uses. A 1995 FWS report describing the agency's compli-

[58]For discussion of a challenge to a land exchange decision under the Alaska National Interest Lands Conservation Act (ANILCA), see *infra* at text accompanying notes 124–133.

[59]Since the primary purposes of most refuges are wildlife conservation and production, 50 C.F.R. § 25.11(b) (1995), human uses of refuges not related to wildlife conservation are referred to as "secondary uses." Not all secondary uses are subject to the requirement that FWS make a compatibility determination before permitting the use. *See* Fink, *supra* note 42, at notes 530–39 and accompanying text. The Fink article at 63–76 includes an extensive discussion of secondary uses of refuges.

[60]*See* U.S. Fish and Wildlife, Dep't of the Interior, Compatibility Task Group, Secondary Uses Occurring on National Wildlife Refuges (1990) (hereinafter "Compatibility Task Group Report"); U.S. General Accounting Office, National Wildlife Refuges: Continuing Problems with Incompatible Uses Call for Bold Action (1989); U.S. Dep't of the Interior, Final Recommendations on the Management of the National Wildlife Refuge System (1979). The 1979 report led to a lawsuit when Secretary James Watt disregarded a recommendation restricting the use of chemical pesticides on refuges. The suit was settled when the Secretary agreed to abide by the recommendation. *See* Environmental Defense Fund v. Watt, 18 Envt'l Rep. Cas. (BNA) 1336 (E.D.N.Y. Oct. 22, 1982) (decision on application for attorneys' fees).

[61]*See* National Audubon Society v. Babbitt, Civil No. C92–1641 (W.D. Wa. Oct. 20, 1993). A companion case to enjoin the Navy from using the Copalis National Wildlife Refuge off the coast of Washington state for aerial target practice was settled when the Navy voluntarily stopped the practice and Secretary Babbitt rescinded the Navy's letter of permission to use the refuge for this purpose. Defenders of Wildlife v. Lujan, Civil No. C92–1643 (W.D. Wa. April 11, 1994).

ance with the agreement showed progress in curtailing incompatible uses.[62]

With one exception in which landowners challenged FWS restrictions on public use, plaintiffs in cases prior to the 1993 settlement have been conservation and humane groups seeking to halt on particular refuges recreational and commercial activities that the groups deemed incompatible with wildlife conservation. Most courts have deferred without extensive analysis to a FWS determination that a use is compatible.[63] They have, however, required FWS to make such a determination before permitting a secondary use.[64] In two instances courts even have rejected a compatibility determination as arbitrary and capricious.[65] The following section examines the major cases.

Secondary Uses: Public Access

One of the first challenges regarding refuge use concerned the Secretary's authority to restrict public access to a refuge. In *Coupland v. Morton*, a group of North Carolina property owners sued to invalidate regulations severely restricting their right to drive over the beaches of Virginia's Back Bay National Wildlife Refuge in order to reach their properties.[66] The landowners claimed the regulations were unauthorized and the accompanying environmental impact statement was inadequate because it failed to consider the environmental effects of requiring an alternate and substantially longer means of access outside the refuge.

As to the NEPA claim, the court not only found the impact statement adequate, but stated in *dictum* its belief that "the United States, as the owner of the Back Bay National Wildlife Refuge, should be able to exclude a trespasser without filing an environmental impact statement."[67] As to

[62]In 1995, FWS issued a required report on its compliance with the settlement. Fish and Wildlife Service, U.S. Dep't of the Interior, Audubon et al. v. Babbitt—Final Report (December 1994). For example, the report indicates that of 83 activities deemed incompatible in the earlier Task Force report that were within FWS' authority to control, the agency had terminated 20 and modified 48.

[63]*See, e.g.*, Animal Lovers Volunteer Ass'n v. Cheney, 795 F. Supp. 994 (C.D. Cal. 1992), in which plaintiffs challenged oil and gas production and a red fox control program on Seal Beach National Wildlife Refuge in California. *See also* Humane Society of the United States v. Lujan, 768 F. Supp. 360 (D.D.C. 1991) and Friends of Animals v. Hodel, 1988 WL 236545 (D.D.C. 1988), discussed *infra* at 89.

[64]*See* The Wilderness Society v. Babbitt, 5 F.3d 383 (9th Cir. 1993), discussed *infra* at 93. Also note the settlement agreement in National Audubon Society v. Babbitt, discussed *supra* at text accompanying note 61.

[65]National Audubon Society v. Hodel, 606 F. Supp. 825 (D. Alaska 1984), and Defenders of Wildlife v. Andrus, 11 Envt'l Rep. Cas. (BNA) 2098 (D.D.C. July 14, 1978) and 455 F. Supp. 446 (D.D.C. 1978), discussed *infra* at text accompanying notes 70–78.

[66]5 Envt'l L. Rep. (Envt'l L. Inst.) 20504 (E.D. Va.), *aff'd*, 5 Envt'l L. Rep. (Envt'l L. Inst.) 20507 (4th Cir. 1975).

[67]5 Envt'l L. Rep. (Envt'l L. Inst.) at 20506. For a similar case reaching the same conclusion

the validity of the regulations, the court found "that the continued and rapidly escalating use of the Refuge beach as a traffic corridor for land developers and land owners . . . is inimical to the use of the property as a wildlife refuge and is a depredation of the purpose of the property as a wildlife refuge."[68] Thus the court affirmed a potentially far-reaching power of the Secretary to protect the national wildlife refuges, but for the most part FWS has declined to exercise this power.[69]

Secondary Uses: Boating

While the *Coupland* case affirmed the Secretary's authority under the Refuge Administration Act to prohibit incompatible refuge uses, most cases have addressed the question of whether refuge laws impose judicially enforceable limitations upon the Secretary's authority to permit secondary uses. One of the first cases to explore this issue was *Defenders of Wildlife v. Andrus,* in which the plaintiff challenged Fish and Wildlife Service regulations governing motor boating and water skiing within Ruby Lake National Wildlife Refuge.[70] The Ruby Lake refuge provides "one of the most important habitats and nesting areas for over-water nesting waterfowl in the United States."[71] The regulations permitted boats of any horsepower to use refuge waters, subject to speed limits.

The court focused on three requirements of the Refuge Recreation Act: (1) that any authorized recreational use be an "appropriate incidental or secondary use"; (2) that use be compatible with the primary purposes of the refuge; and (3) that before permitting any recreational use that is "not directly related to primary purposes of individual areas" of a refuge, the Secretary must determine "that such recreational use will not interfere with the primary purposes for which the areas were established."[72] With

---

with respect to National Park Service limitations on private vehicle use of Fire Island National Seashore, see Biderman v. Secretary of Interior, 7 Envt'l Rep. Cas. (BNA) 1279 (E.D.N.Y. July 19, 1974), *aff'd sub nom.* Biderman v. Morton, 5 Envt'l L. Rep. (Envt'l L. Inst.) 20027 (2d Cir. 1974).

[68] 5 Envt'l L. Rep. (Envt'l L. Inst.) at 20506. Although the court's reasoning would seem to be based on the "compatibility" requirement of the Administration Act, in fact the court never mentioned that Act but upheld the regulations as a proper exercise of authority under the Migratory Bird Conservation Act.

[69] The Service's reluctance to restrict refuge uses may stem in part from its experience following the decisions in Coupland. After having its initial regulations declared valid, FWS revised them to permit the North Carolina property owners greater use. 41 Fed. Reg. 22361 (1976) and 41 Fed. Reg. 31537 (1976). Apparently the political pressures for relaxing the original rules were so great that Interior was forced to yield much of the authority it had just won. *See* Comment, *The Back Bay Wildlife Refuge "Sand Freeway" Case: A Legal Victory in Danger of Political Emasculation,* 5 Envt'l L. Rep. (Envt'l L. Inst.) 10148 (1975). In 1980 Congress passed a law permitting certain full-time residents of the area south of the refuge to commute across it daily. *See* Act of July 25, 1980, Pub. L. No. 96–315, § 3, 94 Stat. 958.

[70] 11 Envt'l Rep. Cas. (BNA) 2098 (D.D.C. July 14, 1978).

[71] *Id.* at 2100.

[72] 16 U.S.C. § 460k.

respect to each of these three requirements, the court held that the Secretary bore the burden of proof to demonstrate compliance.[73] In its initial decision, the court held that the Secretary had violated the last of these requirements because (1) he had failed to make the required determination, and (2) since the challenged regulations represented a balancing of economic, political, and recreational interests against the primary purpose of the refuge, they were beyond the Secretary's authority.[74]

Five days after this decision, the Secretary promulgated substantially identical regulations accompanied, however, by an express determination that the permitted uses would not interfere with the primary purposes of the refuge. The original plaintiff sued again, and the court found the Secretary's determination to be arbitrary and capricious.[75] The court also found that "the degree and manner of boating use" permitted by the regulations violated each of the three requirements of the Refuge Recreation Act identified in the earlier decision, that is, that such use is "not incidental or secondary use, is inconsistent, and would interfere with the Refuge's primary purpose."[76] The court reached these conclusions after finding that proposed speed limits were "obviously unenforceable" and that the Secretary's failure to rely instead upon horsepower limits was "completely contrary to all reason and the facts of the record."[77]

The Ruby Lake Refuge case was hailed as a "landmark."[78] Though the two decisions are limited by their facts, they appeared to be significant because they placed judicially enforceable limits upon the Secretary's discretion to permit recreational uses of wildlife refuges. Courts in later cases however, have not chosen to apply the same rigorous scrutiny to other refuge management actions.

Secondary Uses: Hunting

Hunting was first permitted on a wildlife refuge in 1924.[79] From then until 1949, hunting was permitted primarily on units where it was a tra-

[73]11 Envt'l Rep. Cas. (BNA) at 2101.

[74]*Id.* at 2101–02.

[75]Defenders of Wildlife v. Andrus, 455 F. Supp. 446, 449 (D.D.C. 1978).

[76]*Id.* By limiting this finding to the "degree and manner of boating" permitted by the regulations, the court backed away from the suggestion in its initial opinion that motor boating *per se* might be inconsistent with refuge purposes under the Refuge Recreation Act. 11 Envt'l Rep. Cas. (BNA) at 2102.

[77]455 F. Supp. at 449. The boating controversy did not end with this decision. In 1984, FWS issued a rule opening the boating season two weeks early in alternate years for a five-year period, ostensibly to permit research on the effects of motorized boating on duck brood survival. Conservation groups filed a lawsuit, which was dismissed when FWS agreed to withdraw the new rule and reinstate the former restrictions. Dennis Drabelle, *The National Wildlife Refuge System,* in Audubon Wildlife Report 151, 167 (National Audubon Society 1985).

[78]George C. Coggins and Michael E. Ward, *The Law of Wildlife Management on the Federal Public Lands,* 60 Ore. L. Rev. 59, 112 (1981).

[79]Jennifer A. Heck, Congressional Research Service, *National Wildlife Refuges: Places to Hunt?* at CRS-3 (1992).

ditional activity on the land in question before it became part of the refuge system. In 1949, Congress authorized the Secretary to permit public hunting on up to 25 percent of any unit acquired under the Conservation Act.[80] The proportion was increased to 40 percent in 1958.[81] Finally, the National Wildlife Refuge System Administration Act of 1966 authorized the Secretary to permit "compatible" hunting on any unit of the National Wildlife Refuge System, retaining the 40 percent limitation only with respect to migratory waterfowl.[82] Regulations provide that an area will be opened to hunting "dependent upon the provisions of law applicable to the area" and upon a determination by the Secretary that opening the area to hunting "will be compatible with the principles of sound wildlife management and will otherwise be in the public interest."[83]

In 1984, the Secretary issued regulations opening to hunting numerous refuges that previously had been closed to hunting and foregoing annual review of hunting regulations. This action prompted a lawsuit by the Humane Society, based on claimed violations of the Refuge Administration Act, the Refuge Recreation Act, the National Environmental Policy Act (NEPA), and the Endangered Species Act (ESA).[84] The district court dismissed the Society's claims for lack of standing. The circuit court reversed the trial court's standing determination and remanded the case for further proceedings.[85] The court did not reach the merits of the plaintiffs' claims.[86]

The Humane Society later challenged the FWS' 1989 decision to permit

[80]Act of Aug. 12, 1949, ch. 421, § 2, 63 Stat. 600. "Waterfowl production areas" acquired with duck stamp revenues are, by definition, open to hunting.

[81]Act of Aug. 1, 1958, Pub. L. No. 85–585, § 2, 72 Stat. 486.

[82]The Fish and Wildlife Improvement Act of 1978, Pub. L. No. 95–616, § 6, 92 Stat. 3114 (codified at 16 U.S.C. § 668dd(d)(1)(A) (1985)), authorized the Secretary to permit migratory waterfowl hunting on more than 40 percent of any given refuge if the Secretary finds that to do so would be beneficial to the species. The principal purpose of this amendment was to reduce waterfowl mortality from communicable diseases by controlling bird concentrations in refuges. *See* Sen. Rep. No. 1175, 95th Cong., 2d Sess. 5, reprinted at 1978 U.S. Code Cong. & Ad. News 7641, 7644.

[83]50 C.F.R. § 32.1 (1995).

[84]Humane Society of the United States v. Clark, Civil Action No. 84–3630 (D.D.C. July 25, 1986).

[85]Humane Society of the United States v. Hodel, 840 F.2d 45 (D.C. Cir. 1988).

[86]In holding that the Humane Society had standing to challenge the Secretary's action, the court did comment on the purpose of the Refuge Acts:

> Protecting the human aesthetic interest in viewing live animals and birds plainly bears a plausible relationship to the policies embodied in the two Refuge Acts. The two Refuge Acts have as their principal aim the protection of wildlife. *See* Sen. Rep. No. 1858, 87th Cong., 2d Sess. 1 (1962); H.R. Rep. No. 1473, 87th Cong., 2d Sess. 3 (1962) (on Refuge Recreation Act); H.R. Rep. No. 1168, 89th Cong., 1st Sess. 2 (1965) (on Refuge Administration Act).

Id. at 61.

public deer hunting on the Mason Neck National Wildlife Refuge in Virginia.[87] The Mason Neck refuge was established in 1969 as a bald eagle sanctuary and was closed to hunting until 1989. The agency's stated purpose of opening the refuge to limited hunting was to control the white-tailed deer population. The district court found that the agency "took account of the relevant factors"[88] and that the decision was not arbitrary and capricious.

Plaintiffs challenged a decision to permit hunting in only one other case, in which the district court simply followed the court's decision in the Mason Neck Refuge case and rejected the plaintiffs' claims.[89] To date, in no case has a challenge to hunting on a wildlife refuge succeeded.

Secondary Uses: Grazing

Livestock grazing and farming are the principal agricultural uses of refuge lands.[90] A 1990 report revealed that refuge managers consider about 40 percent of livestock grazing that occurs on refuges to be harmful to wildlife.[91]

In 1991, conservation groups filed suit to halt cattle grazing on Oregon's Hart Mountain National Antelope Refuge until FWS prepared a compatibility determination and an environmental impact statement (EIS) on the impacts of grazing. The suit was dismissed when the Service agreed to prepare both documents and to refrain from issuing any grazing permits until thirty days after adoption of a management plan. The district court later denied plaintiffs' application for costs and attorneys' fees as prevailing party under the Equal Access to Justice Act.[92]

The Ninth Circuit reversed.[93] In determining that the plaintiffs were entitled to fees, the court found that (1) the plaintiffs' claim had a basis in law, citing the Refuge Administration Act's requirement that FWS make a compatibility determination before permitting secondary uses; and (2) the government's action in permitting grazing was not substantially justified.

The court based the latter conclusion on its finding that FWS had "renewed annual grazing permits without regard to the incompatibility of grazing to the Refuge's purposes."[94] As early as 1989, the Refuge manager had reported that grazing was damaging fish and wildlife habitat. Under these circumstances, the court said, "the Service had a duty to investigate

[87]Humane Society of the United States v. Lujan, 768 F. Supp. 360 (D.D.C. 1991).
[88]*Id.* at 364.
[89]Friends of Animals v. Hodel, 1988 WL 236545 (D.D.C. 1988).
[90]Compatibility Task Group Report, *supra* note 60, at 22 (Table 4).
[91]*Id.* at 25.
[92]28 U.S.C. § 2412.
[93]The Wilderness Society v. Babbitt, 5 F.3d 383 (9th Cir. 1993).
[94]*Id.* at 388.

the compatibility of grazing with the Refuge's purposes prior to permitting grazing on the Refuge.''[95] The dissenting judge argued that, based on the ''ongoing activity'' exception to NEPA's EIS requirement, FWS was not required to halt the practice before making a compatibility determination. The majority declined to apply NEPA's ongoing activity exemption to the Refuge Administration Act in the absence of any support for the proposition that ''a particular exception to NEPA is generally applicable to all environmental statutes.''[96] The court, however, limited its finding to the facts of the case before it and did not address ''the separate issue whether a compatibility determination must be completed for all ongoing activities.''[97]

Applicability of NEPA to Refuge Budget Proposals

One U.S. Supreme Court decision concerned the question of whether the National Environment Policy Act required the preparation of an environmental impact statement in connection with the annual budget proposals for financing the National Wildlife Refuge System. In *Sierra Club v. Andrus*, the Supreme Court ruled that NEPA did not require an EIS.[98] The Court reasoned that requests for appropriations were neither ''proposals for legislation'' nor ''proposals for . . . other major federal actions'' within the meaning of NEPA.

The Court relied upon the traditional distinction that Congress has drawn between ''legislation'' and ''appropriations'' and concluded that preparation of an environmental impact statement was ill suited to the budget process.[99] In reaching this conclusion, the Supreme Court rejected the court of appeals' rationale that requests for appropriations might require EISs when such requests would bring about a significant change in the status quo.[100] Importantly, however, the Court did conclude that if, in response to budget cuts, the Fish and Wildlife Service were to revise its ongoing programs in ways that significantly affected the quality of the environment, impact statements would be required for such revisions.[101]

[95] *Id.* at 389.

[96] *Id* at 389, n.2.

[97] *Id.* The FWS issued its EIS in 1994. U.S. Dep't of the Interior, Fish and Wildife Service, Final Environmental Impact Statement for the Hart Mountain National Antelope Refuge Comprehensive Management Plan (May 1994). The preferred alternative entirely eliminated livestock grazing.

[98] 442 U.S. 347 (1979).

[99] *Id.* at 359–64.

[100] Sierra Club v. Andrus, 581 F.2d 895, 903 (D.C. Cir. 1978).

[101] 442 U.S. at 363. For a case involving the application of NEPA to a transfer of federal land owned by the General Services Administration into the National Wildlife Refuge System, see New England Power Co. v. Goulding, 486 F. Supp. 18 (D.D.C. 1979).

Distribution of Revenue from Oil and Gas Leases

The only other Supreme Court opinion concerning the National Wildlife Refuge System addressed the relationship between the Refuge Revenue Sharing Act and the Mineral Leasing Act of 1920[102] for refuges reserved from the public domain. *Watt v. Alaska* concerned the distribution of revenues from oil and gas leases on the Kenai National Moose Range (now known as the Kenai National Wildlife Refuge).[103] Since 1954 these receipts had been distributed on the basis of section 5 of the Mineral Leasing Act to provide 90 percent of lease revenues to the state of Alaska and 10 percent to the United States Treasury. However, the Refuge Revenue Sharing Act provides for a payment of 25 percent of the net receipts from the sale of various resources to the counties in which refuge lands are located and the remainder to the U.S. Treasury.[104] In 1964, the Refuge Revenue Sharing Act was amended so as to include "minerals" in the list of resources subject to such payments.

The issue in *Watt v. Alaska* was thus whether the 1964 amendment superseded the revenue distribution formula of the 1920 Act. The Supreme Court held that it did not. Instead, the Court reasoned that the addition of the term "minerals" in 1964 constituted a mere "perfecting" amendment, which brought the Refuge Revenue Sharing Act into conformity with the 1947 Mineral Leasing Act for Acquired Lands.[105] The effect of the Court's holding is to require that net receipts from mineral leases on refuges comprised of reserved lands be distributed according to the Mineral Leasing Act, whereas the same receipts from leases on refuges comprised of acquired lands must be distributed according to the Refuge Revenue Sharing Act.

THE REFUGE SYSTEM AND THE ALASKA NATIONAL INTEREST LANDS CONSERVATION ACT (ANILCA)

National wildlife refuges in Alaska are subject to a law that does not apply to the rest of the System, the Alaska National Interest Lands Conservation Act of 1980 (ANILCA).[106] ANILCA added 53.7 million acres to

[102]30 U.S.C. §§ 181–287 (Mineral Leasing Act); 16 U.S.C. § 715s (Refuge Revenue Sharing Act).

[103]451 U.S. 259 (1981).

[104]A different, optional formula, revised by amendments in 1978, governs payments to counties in which refuges comprised of acquired lands are located. The 1978 amendments also added salmonid carcasses to the list of refuge resources subject to shared payments. *See* 16 U.S.C. § 715s(a) and (c).

[105]30 U.S.C. §§ 351–359.

[106]Pub. L. No. 96–487; 94 Stat. 2371 (1980) (partially codified in 16 U.S.C. §§ 3101 *et seq.*, and in scattered sections of 16, 43 U.S.C.).

the Refuge System, nearly tripling its size.[107] The law also introduced major new management standards for refuges within Alaska.

Statutory Framework

ANILCA's wildlife refuge provisions were unlike earlier refuge system legislation in several key respects. First, the Act required comprehensive land use planning for each Alaskan unit and established a timetable for completing plans.[108] This was the first time Congress had required planning for wildlife refuges.[109] Recognizing the large size and unspoiled condition of the Alaska refuges, Congress wished to take advantage of the "opportunity to manage these areas on a planned ecosystem-wide basis with all of their pristine ecological processes intact."[110] Unlike other laws mandating land use planning on public lands, such as the National Forest Management Act (NFMA) and the Federal Land Policy and Management Act (FLPMA),[111] however, the Act does not give much guidance to planners and sets no substantive requirements for plans.

Second, the Act set forth a statement of the purposes of each unit and, in part, established priorities among these purposes. The general purposes are to conserve fish and wildlife populations and habitats "in their natural diversity," to fulfill the nation's international treaty obligations with respect to fish and wildlife and their habitats, and to ensure water quality and quantity within refuges in a manner consistent with wildlife conservation.[112] The Act designates for each unit key wildlife species to be conserved, but Congress also expressed the desire to "conserve the entire spectrum of plant and animal life" found in each refuge.[113]

A further purpose of each refuge unit other than Kenai National Wildlife Refuge is "to provide . . . the opportunity for continued subsistence uses by local residents."[114] This purpose, however, is to be fulfilled "in a manner consistent with" fish and wildlife conservation and fulfillment of international treaty obligations.[115] Thus, in the event of conflict between subsistence use and either of the other two purposes, the first is subordinate to the latter two.

[107] *See* George C. Coggins *et al.*, Federal Public Land and Resources Law 144 (3d ed. 1993).

[108] Pub. L. No. 96–487, § 304(g); 94 Stat. 2394–95 (1990) (uncodified).

[109] For a discussion of planning efforts applicable to the entire Refuge System, see Fink, *supra* note 42, at 50–52.

[110] Sen. Rep. No. 96–413, 96th Cong., 2d Sess. 174, *reprinted in* 1980 U.S.C.C.A.N. 5070, 5118.

[111] For a discussion of land use planning under NFMA, see Chapter 9. For a discussion of land use planning under FLPMA, see Chapter 10.

[112] Pub. L. No. 96–487, §§ 302, 303; 94 Stat. 2385–93.

[113] Sen. Rep. No. 96–413, *supra* note 110.

[114] Pub. L. No. 96–487, §§ 302, 303; 94 Stat. 2385–93.

[115] *Id.*

Overlaying this express statement of purposes for each refuge is title VIII of the Act, governing subsistence uses on all public lands in Alaska. That title declares a qualified policy that utilization of the public lands is to cause the least adverse impact possible on rural residents who depend upon subsistence uses of the resources of these lands, and a further policy that nonwasteful subsistence uses of fish and wildlife and other renewable resources shall be the priority consumptive uses of all these resources.[116] The substantive provisions that effectuate these policies, however, give rise to a number of ambiguities and potential inconsistencies.

Section 815(4) sets forth a general disclaimer against anything in Title VIII "modifying or repealing the provisions of any Federal law governing the conservation or protection of fish and wildlife," including the Refuge Administration Act.[117] That presumably leaves unaffected the Secretary's duty to determine that particular uses, including hunting, are compatible with the purposes of a refuge before he or she may allow them. Despite that disclaimer, section 816(b) provides that "[n]otwithstanding any other provision of this Act or other law, the Secretary . . . may temporarily close any public lands . . . to subsistence uses of a particular fish or wildlife population only if necessary for reasons of public safety, administration, or to assure the continued viability of such population."[118] Read alone, this provision would seem to override the Refuge Administration Act. Read in conjunction with section 815(4), it leaves the Secretary's closure authority in considerable confusion.

The Secretary's authority to restrict subsistence taking on refuges short of total closure is similarly unclear. Section 804 provides that "[w]henever it is necessary to restrict the taking of populations of fish and wildlife on [federal] lands for subsistence uses in order to protect the continued viability of such populations, or to continue such uses," the Secretary shall do so through the application of specified criteria.[119] Arguably, the Secretary's authority to restrict subsistence taking is limited by this section to those instances in which such restrictions are necessary to protect the "continued viability" of fish or wildlife populations, the same standard that governs closure authority. However, sections 302 and 303 clearly make subsistence uses of refuges secondary to conserving fish and wildlife populations in their natural diversity and fulfilling the nation's international treaty obligations. Moreover, to the extent that sections 302 and 303 make subsistence uses even a secondary purpose of Alaskan refuges, this purpose is limited to use by "local residents," whereas Title VIII defines "subsistence uses" more broadly to refer to uses by "rural Alaska

[116]16 U.S.C. § 3112.
[117]*Id.* § 3125(4).
[118]*Id.* § 3126(b).
[119]*Id.* § 3114.

residents."[120] Thus the Secretary's authority to restrict subsistence uses on Alaskan refuges may in fact be greater than what section 804 implies.[121]

The detailed provisions of the subsistence title represent a major novel aspect of legislatively mandated management of the National Wildlife Refuges. Other provisions of ANILCA are similarly unique. For example, section 304(d) obliges the Secretary to continue to permit the exercise of preexisting commercial fishing rights or privileges and various uses of refuge lands incidental to those rights and privileges.[122] The Secretary is not obliged, however, to permit a "significant expansion" of commercial fishing activities within refuges. Finally, section 304(f) authorizes the Secretary to enter into cooperative management agreements with nearby landowners for management of lands outside the refuges in ways compatible with the major purposes of such refuges.[123]

Authority to Exchange Alaska Refuge Lands

In *National Audubon Society v. Hodel*, conservation groups successfully challenged a land exchange approved by Secretary James Watt pursuant to ANILCA.[124] The Secretary had proposed to convey to Alaska native corporations St. Matthew Island, a designated wilderness within the Alaska Maritime National Wildlife Refuge in the Bering Sea. The corporations planned to develop the island as a support base for oil and gas development.

Section 1302 of ANILCA provides that an exchange must be for equal value, except that it may be made for other than equal value if "the Secretary determines it is in the public interest."[125] The proposed exchange was not for equal value and thus was valid only if it was in the public

[120] *Id.* § 3113.

[121] Alaska Fish and Wildlife Federation v. Dunkle, 829 F.2d 933 (9th Cir. 1987), *cert. denied,* 485 U.S. 988 (1988), discussed in Chapter 4, sheds no light on this question. In that case, the court found that the Migratory Bird Treaty Act does not permit closed season subsistence hunting of migratory birds on wildlife refuges in Alaska. Thus, the court found it unnecessary to determine whether ANILCA "provides an independent basis for restricting subsistence hunting. Having determined that the MBTA requires the government to comply with its international treaty obligations, we do not need to consider whether ANILCA also places this obligation on the government in its operation of the Yukon Delta National Wildlife Refuge." 829 F.2d at 935 n.1.

[122] Pub. L. No. 96–487, § 304(d); 94 Stat. 2393.

[123] *Id.* § 304(f); 94 Stat. 2393–3294.

[124] 606 F. Supp. 825 (D. Alaska 1984).

[125] 16 U.S.C. § 3192(h)(1). Note that in 1988, Congress amended this section to provide that the Secretary was not authorized to convey, "by exchange or otherwise, lands or interest in lands within the coastal plain of the Arctic National Wildlife Refuge . . . without prior approval by Act of Congress." Pub. L. 100–395, § 201, 102 Stat. 981 (1988), codified at 16 U.S.C. § 3192(h)(2). See discussion of ANILCA's provisions regarding oil and gas leasing on the coastal plain, *infra* at text accompanying notes 134–145.

interest. The Secretary argued that his Public Interest Determination (PID) was committed to agency discretion by law and thus was unreviewable. The court disagreed, holding that a PID is subject to review under the arbitrary and capricious standard because the Secretary's authority is subject to the constraints of ANILCA and there is an administrative record for the PID that the court can review.[126]

In reviewing the PID, the court first concluded that the benefits to the United States from the exchange were slight. The government acquired nondevelopment easements in lands that already were subject to the Refuge Administration Act's compatibility requirement. In an important ruling that affects two million acres of inholdings,[127] the court held that lands that had been part of the refuge system in 1971 and conveyed to Alaska native corporations under the Alaska Native Claims Settlement Act or ANILCA remain subject to laws and regulations governing the refuge system.[128] Thus, the court concluded, the nondevelopment easements added little protection.[129] The court found that other rights acquired in the exchange also were of questionable value to the purposes of ANILCA.[130]

The court then found the Secretary's conclusion that development of St. Matthew Island posed little risk to its natural resources to be unsupported. The FWS assessment had stated that the proposed development would cause serious harm to wildlife, adversely affecting half a million nesting seabirds and many whale species.[131] The court concluded that "[t]he record thus demonstrates that, directly contrary to the Secretary's findings, overall national wildlife conservation and management objectives will not be advanced either in the short or long terms under the exchange."[132] Accordingly, the Secretary's decision was arbitrary and capricious and a "clear error of judgment."[133] The court invalidated the exchange and issued a preliminary injunction against construction on St. Matthew Island.

[126]606 F. Supp. at 834–35.

[127]Dennis Drabelle, *The National Wildlife Refuge System*, in Audubon Wildlife Report 151, 176 (National Audubon Society 1985).

[128]606 F. Supp. at 838.

[129]The court's conclusion appears to be correct on the facts of this case. *See* 606 F. Supp. at 838–39. In light of the long history of conflicts over FWS' compatibility determinations and the courts' deference to the agency on this issue, however, in some situations nondevelopment easements could add substantial protection.

[130]The court accepted the Secretary's determination of the factors that are relevant to a PID. Thus, the court accepted that the economic benefits of oil and gas development are in the public interest under ANILCA. 606 F. Supp. at 835.

[131]*Id.* at 843–44.

[132]*Id.* at 845.

[133]*Id.* at 846.

ANILCA and the Arctic National Wildlife Refuge (ANWR)

The 19 million-acre Arctic National Wildlife Refuge (ANWR) is subject to additional provisions of ANILCA. The largest deposit of oil and gas discovered in North America, the Prudhoe Bay oil field, is sixty miles west of ANWR. The oil and gas industry has been particularly interested in exploring the Refuge's 1.55 million–acre coastal plain, its most biologically productive area.[134]

Section 1003 of ANILCA prohibits all "leasing or other development leading to production of oil and gas" on ANWR unless authorized by Congress.[135] Section 1002 of ANILCA mandated a continuing resource assessment of the coastal plain.[136] That section also directed the Secretary to prepare and submit a report to Congress containing, among other information, (1) specific information about potential oil and gas production and fish and wildlife within ANWR's coastal plain; and (2) recommendations concerning possible exploration, development, and production of oil and gas within the coastal plain.[137] The report was due by September 2, 1986.

On October 2, 1986, five environmental groups filed an action for declaratory and injunctive relief, seeking a declaration requiring the Secretary to submit an EIS pursuant to section 102(2)(C) of NEPA[138] before he submitted the 1002 report to Congress.[139] The Trustees also sought a mandatory injunction requiring the Secretary to circulate the EIS for public comment prior to submitting the report and EIS to Congress. The district court granted the requested relief, and the Ninth Circuit affirmed.[140]

Six months after the court of appeals' decision in December 1986, the EIS process was completed, and the Secretary submitted the EIS and the 1002 report to Congress. Among the report's significant findings was that the marginal probability of economic success in tapping oil reserves in

[134] *See* Natural Resource Defense Council v. Lujan, 768 F. Supp. 870, 873 (D.D.C. 1991). The plain is used by many species of fish and wildlife. It is the main calving grounds for the migratory Porcupine caribou herd. *Id.*

[135] Pub. L. No. 96–487, § 1003, 94 Stat. 2371, 2452 (1980) (codified at 16 U.S.C. § 3143).

[136] *Id.* § 1002(c), 16 U.S.C. § 3142.

[137] *Id.* § 1002(h).

[138] 42 U.S.C. § 4332(2)(C).

[139] Trustees for Alaska v. Hodel, 806 F.2d 1378 (9th Cir. 1986).

[140] The court rejected the defendants' argument that the report fell within NEPA's exemption for legislative proposals, and thus they did not have to circulate it for public comment prior to submitting it to Congress. *Id.* at 1383–84. The court found that the report was a "study process" required by statute, which is not subject to the exemption. The dissent argued that it was not a "study process." *Id.* at 1384.

ANWR was 19 percent.[141] This assessment was raised to 46 percent in a 1991 update of the report.[142] The Secretary did not prepare a supplemental EIS or circulate the 1991 Overview for public comment.

Conservation groups challenged the adequacy of the 1987 report and its 1991 update, as well as the failure to prepare a supplemental EIS for the 1991 Overview. In *Natural Resources Defense Council v. Lujan*, the district court held that the adequacy of the report itself was not subject to judicial review but that the defendants had violated NEPA by failing to supplement the EIS when the significant new information in the 1991 Overview became known.[143] The court ordered the defendants to prepare a supplemental EIS and circulate it for public comment.

In the late 1980s, while litigation over the 1002 report was proceeding, debate raged in Congress over competing bills to open portions of ANWR to oil and gas leasing or to designate the entire coastal plain as wilderness.[144] In 1992, the conflict became dormant after Congress deleted proposals to open ANWR to leasing from national energy legislation and the newly elected Clinton administration declared its opposition to opening ANWR to oil and gas exploration.[145] With the change in control of Congress in 1994, however, efforts to open ANWR to oil exploration have been renewed.

THE NATIONAL PARK SYSTEM

Purposes and Origin

The National Park System has substantial importance for wildlife because it preserves large blocks of relatively undisturbed habitat, administered to preserve its natural values and characteristics. The system encompasses over 83 million acres. It is comprised of many diverse units known variously as national parks, national monuments, national seashores, national lakeshores, national wild rivers, national preserves, and still others.[146] The National Park Service in the Department of the Interior manages all these units.

Congress created the first national park, Yellowstone, in 1872 as a

[141]Bureau of Land Management, Overview of the 1991 Arctic National Wildlife Refuge Recoverable Petroleum Resource Update (1991) (hereinafter "1991 Overview").

[142]*Id.*

[143]768 F. Supp. 870 (D.D.C. 1991).

[144]Fink, *supra* note 42, at 33.

[145]Natural Resources Defense Council v. Lujan was dismissed as moot because Congress was no longer considering legislation to open the plain to development. Fink, *supra* note 42, at 33, n. 218.

[146]Despite the diversity of names, the units comprising the system are generally subject to a uniform set of rules.

"pleasuring ground for the benefit and enjoyment of the people."[147] By 1916, there were thirteen national parks and nineteen national monuments, and the responsibility for their administration was scattered among several government agencies. To provide more cohesive management for these areas, Congress passed the National Park Service Act (hereafter Organic Act) in 1916.

The Organic Act sets forth the system's purposes and the general standards governing its administration.[148] The act explicitly recognizes wildlife conservation as among the parks' chief purposes:

> [T]he fundamental purpose of the said parks, monuments, and reservations . . . is to conserve the scenery and the natural and historic objects and the wild life therein and to provide for the enjoyment of the same in such manner and by such means as will leave them unimpaired for the enjoyment of future generations.[149]

There are, inevitably, conflicts among these purposes. Human use and enjoyment of the parks has long been recognized as a fundamental purpose of their creation, and the number of visitors to the parks has soared. To accommodate these visitors, an extensive road system and numerous facilities have been built that impinge upon wildlife needs.[150] It is unclear to what extent the Secretary may be compelled to conserve park wildlife when its conservation conflicts with other uses. It is, however, well-established that the Secretary has discretion to do so except when Congress explicitly provides otherwise.[151]

The Duty to Protect Park Resources

A series of cases in the 1970s involving California's Redwood National Park addressed the issue of whether the Organic Act's statement of purpose imposes any judicially enforceable duties on the Secretary of the Interior.[152] While not shedding a great deal of light on this question, the cases resulted in some important amendments to the act.

[147] 16 U.S.C. § 21.

[148] 16 U.S.C. §§1 and 2–3. Many of the system's units are administered under the terms of the special statutes that established them.

[149] *Id.* §1.

[150] However, some parks (or portions of them), are also designated wilderness. These areas are managed as wilderness; roads and structures are not permitted. See discussion of wilderness management *infra* at text accompanying notes 261–291.

[151] *See, e.g.*, Michigan United Conservation Clubs v. Lujan, 949 F.2d 202 (6th Cir. 1991), discussed *infra*.

[152] *See* Sierra Club v. Andrus, 487 F. Supp. 443 (D.D.C. 1980), *aff'd sub nom.* Sierra Club v. Watt, 659 F.2d 203 (D.C. Cir. 1981); Sierra Club v. Dep't of the Interior, 424 F. Supp. 172 (N.D. Cal. 1976); Sierra Club v. Dep't of the Interior, 398 F. Supp. 284 (N.D. Cal. 1975); Sierra Club v. Dep't of the Interior, 376 F. Supp. 90 (N.D. Cal. 1974).

In 1974, the Sierra Club challenged the adequacy of the Secretary's efforts to address threats to the park that arose from logging on private lands on its periphery.[153] The legislation that established the park authorized the Secretary to take certain actions to deal with this kind of problem. Those actions included modifying park boundaries, acquiring interests in land beyond the boundaries, and negotiating agreements with adjoining landowners to minimize harm to park resources.

The Sierra Club contended that the Secretary's exercise of his discretion with respect to actions protecting the park was judicially reviewable. The district court agreed[154] and, in a later ruling, held that the Secretary's actions had been arbitrary and capricious.[155] Both decisions rested in part upon the court's finding that, in addition to the express statutory authorization to the Secretary, he had a further "trust obligation" to protect park resources.[156]

Prompted by the court's decision, the Department of the Interior submitted to the Office of Management and Budget (OMB) proposed legislation that would give the Department express regulatory power over peripheral timber operations, proposed timber-harvesting guidelines to timber companies managing adjoining lands, and requested that the Justice Department initiate litigation against certain timber companies to restrain timber practices endangering the park.

OMB disapproved the Department of the Interior's request for new legislative authority, the timber companies declined to accept the harvesting guidelines, and the Justice Department had not yet decided whether to initiate litigation, when the court entered a third decision.[157] The court held that, in light of the Department of the Interior's efforts, "primary responsibility for the protection of the Park rests, no longer upon the Department of the Interior, but squarely upon Congress."[158] Thus the Department of the Interior was in effect deemed to lack power to compel actions necessary to protect the park from outside threats.

Congress responded in 1978 by amending the Organic Act. The Redwood National Park Expansion Act substantially expanded Redwood National Park and created a "Park Protection Zone" in which the Secretary could acquire additional lands under certain conditions.[159] Most important, the law also provided a new, rigorous standard governing the exercise of the Secretary's discretion in managing *all* park areas. The Act directed that

[153]Sierra Club v. Dep't of the Interior, 376 F. Supp. 90 (N.D. Cal. 1974).
[154]*Id.*
[155]Sierra Club v. Dep't of the Interior, 398 F. Supp. 284 (N.D. Cal. 1975).
[156]376 F. Supp. at 93, 95–96.
[157]Sierra Club v. Dep't of the Interior, 424 F. Supp. 172 (N.D. Cal. 1976).
[158]*Id.* at 75.
[159]Pub. L. No. 95–250, § 101, 92 Stat. 163.

> [t]he authorization of activities shall be construed and the protection, management, and administration of [national park] areas shall be conducted in light of the high public value and integrity of the National Park System and shall not be exercised in derogation of the values and purposes for which these various areas have been established, except as may have been or shall be directly and specifically provided by Congress.[160]

The House report accompanying this legislation described the intent of this provision to be "to afford the highest standard of protection and care to the land within Redwood National Park."[161] In fact, because the provision quoted above applies to all units of the National Park System, it reflects a directive that the "highest standard of protection and care" be given to the entire System.[162]

Notwithstanding the apparent intent of the 1978 amendments, the court in *Sierra Club v. Andrus* upheld the Secretary's failure to take discretionary action in defense of Redwood Park resources.[163] There, the plaintiff challenged, among other things, the Secretary's failure to join pending adjudications of water rights in the state of Utah so as to protect claimed federal reserved water rights in various Western parks. The plaintiff claimed that the 1978 amendments obligated the Secretary to participate in pending adjudications. The court disagreed, largely because of its conclusion that the pending adjudication did not present "a real and immediate water supply threat" to the resources of the various parks.

The court also expressed the view that the 1978 amendments "eliminated 'trust' notions in National Park System management," in response to the earlier Redwoods litigation.[164] The court concluded that the Secretary had "broad discretion in determining what actions are best calculated to protect Park resources," although such discretion is "not unlimited."[165] This decision left considerable uncertainty as to the precise nature of the Secretary's authority and duties with respect to the administration of the National Park System and the appropriate role for the courts to play in reviewing Secretarial action.[166]

[160]16 U.S.C. § 1a-1.

[161]H.R. Rep. No. 95–581, pt. I, 95th Cong., 1st Sess. 21, *reprinted in* 1978 U.S.C.C.A.N. 463, 468.

[162]For further discussion of these amendments, see Robert B. Keiter, *Taking Account of the Ecosystem on the Public Domain: Law and Ecology in the Greater Yellowstone Ecosystem*, in Environmental Policy and Biodiversity 111, 118–19 (R. Edward Grumbine ed., 1994).

[163]487 F. Supp. 443 (D.D.C. 1980), *aff'd sub nom.* Sierra Club v. Watt, 659 F.2d 203 (D.C. Cir. 1981).

[164]487 F. Supp. at 448.

[165]*Id.*

[166]Although the Sierra Club did not appeal that portion of the district court's decision refusing to order the Secretary to join the pending state water adjudications, he did in fact later join these proceedings.

Courts in subsequent decisions have deferred to the Secretary's discretion when the Park Service has permitted activities that potentially conflict with wildlife conservation. In *National Wildlife Federation v. National Park Service*, conservationists challenged the Service's decision to continue operating the Fishing Bridge campground in Yellowstone National Park despite possible impacts on the threatened grizzly bear.[167] The court found that the decision to keep the campground open at a reduced level, pending preparation of an EIS regarding the effects of the campground on the grizzly, violated neither the Organic Act nor the Endangered Species Act.[168] Citing several cases, the court found that "the Park Service is empowered with the authority to determine what uses of park resources are proper and what proportion of the park's resources are available for such use."[169]

Another court upheld a management plan for the Cape Cod National Seashore that permitted off-road vehicle (ORV) use despite alleged threats to the Seashore's intertidal areas.[170] The court declined to issue an injunction against implementation of the plan because experts disagreed as to the likely ecological effects of ORV use on the Seashore; the court found that the Secretary's decision was supported by evidence and was not arbitrary and capricious.[171]

Discretionary Authority to Protect Park Resources

In the cases discussed above, environmental plaintiffs unsuccessfully challenged the Secretary's failure to take particular actions to protect park resources. However, the Secretary's duty to leave the parks "unimpaired for the enjoyment of future generations" has often been held to be an affirmative source of authority for actions to protect park resources. For example, in *Cappaert v. United States*, the Supreme Court held that in reserving a tract of land in Nevada as part of Death Valley National Monument in order to protect a rare species of fish, the federal government also implicitly reserved the groundwater appurtenant to the land. Had they not reserved the groundwater, the Court reasoned, the purpose of preserving the monument unimpaired for future generations would be frustrated.[172]

[167]669 F. Supp. 384 (D. Wyo. 1987).

[168]See Chapter 7 for a discussion of the Endangered Species Act.

[169]669 F. Supp. at 391.

[170]Conservation Law Foundation of New England, Inc. v. Clark, 590 F. Supp. 1467 (D. Mass. 1984), *aff'd*, 864 F.2d 954 (1st Cir. 1989).

[171]590 F. Supp. at 1484.

[172]426 U.S. 128 (1976). The tract of land involved in *Cappaert* was reserved pursuant to the Act for the Preservation of Antiquities, 16 U.S.C. §§431–433. That Act authorizes the reservation of land containing historic landmarks, historic and prehistoric structures, and other

In *New Mexico State Game Commission v. Udall*,[173] the court of appeals held that the authority to preserve park resources, coupled with the express authority elsewhere in the 1916 Act to destroy "such animals . . . as may be detrimental to the use of any of said parks, monuments or reservations,"[174] permitted the Park Service to take deer in Carlsbad Caverns National Park solely for research purposes and without compliance with state game laws.[175] Almost twenty-five years later, the court in *Wilkins v. Lujan* cited the *Udall* decision when it upheld a Park Service plan to remove wild horses from a Missouri park.[176] The Park Service had determined that among other problems, the horses were destroying native plants and competing with native wildlife for forage. The district court had issued an injunction against the horse removal plan, but the appellate court reversed and vacated the injunction. Noting that the horses were considered an exotic species, and that their continued presence in the park was in conflict with the park's purpose "to maintain, rehabilitate, and perpetuate the park's natural resources," the court held that the action was not arbitrary and capricious.[177]

It is clear that the Secretary has authority to regulate hunting, trapping, and fishing within parks. Varying approaches to regulation of these activities have been taken over the years. An early directive from the Secretary to the Park Service's first director provided that "[h]unting will not be permitted in any national park."[178] Beginning in the late 1930s, however, Congress authorized hunting and trapping in the legislation creating certain "nontraditional" park units such as national seashores and lakeshores. In other units, deemed "recreational" by the Park Service, the Service on its own initiative authorized hunting and trapping.[179]

The 1978 amendments to the Organic Act led the Park Service to conclude that, as the court in *National Rifle Ass'n v. Potter* characterized it,

---

objects of historic or scientific interest." The *Cappaert* court rejected the contention that the Antiquities Act was intended to protect only archaeological sites and held that the "rare inhabitants" (referring to the endangered Devil's Hole pupfish) of the land in question were " 'objects of historic or scientific interest.' " 426 U.S. at 142.

[173] 410 F.2d 1197 (10th Cir.), *cert. denied*, 396 U.S. 961 (1969).

[174] 16 U.S.C. § 3.

[175] *See also* United States v. Vogler, 859 F.2d 638 (9th Cir. 1988), *cert. denied*, 488 U.S. 1006 (1989) (Park Service had authority to regulate access to and mining of claims in national parks); United States v. Moore, 640 F. Supp. 164 (S.D. W. Va. 1986) (Park Service had authority to prohibit state from spraying pesticide to eliminate black flies inside boundaries of a national river; black flies are "wildlife," and Park Service regulation prohibits taking of wildlife).

[176] 995 F.2d 850 (8th Cir. 1993), *cert. denied*, 114 S. Ct. 921 (1994).

[177] *Id.* at 853. Note also that a Park Service regulation prohibits introducing species into a park ecosystem. 36 C.F.R. § 2.1(a)(2) (1994).

[178] Quoted in National Rifle Association v. Potter, 628 F. Supp. 903, 905 (D.D.C. 1986).

[179] In 1964 the Park Service had divided the system into three management categories—national, historical, and recreational. 628 F. Supp. at 905.

"Congress conceived of the park system as an integrated whole, wherein the Park Service was to permit hunting and trapping only where it had been specifically authorized, or discretion given it to do so, by Congress in the applicable enabling act."[180] In 1983, the Park Service issued a rule prohibiting hunting in all units of the System, except where Congress has specifically mandated hunting or authorized it as a discretionary activity,[181] and later the Service issued a rule prohibiting trapping except where Congress has specifically mandated it.[182] The National Rifle Association challenged these prohibitions as an abuse of the Secretary's discretion. The court found that the Service's reading of the law was reasonable and upheld the regulation.[183]

The regulation's distinction between hunting and trapping was at issue in *Michigan United Conservation Clubs v. Lujan.*[184] Congress had specifically authorized hunting, but made no mention of trapping, in the legislation creating two national lakeshores. Applying its regulation, the Park Service prohibited trapping in the lakeshores while permitting hunting. Trappers challenged this action as an abuse of discretion, arguing that by authorizing hunting, Congress also intended to permit trapping. The district court rejected this argument, and the Sixth Circuit affirmed. In a significant ruling, the appellate court agreed with and adopted the findings in *Potter* that "unlike national forests, Congress did not regard the National Park System to be compatible with consumptive uses."[185] The court continued:

> Notwithstanding that the goals of user enjoyment and natural preservation may sometimes conflict, the NPS may rationally conclude, in light of the Organic Act and its amendments, that its primary management function with respect to wildlife is preservation unless Congress has declared otherwise.[186]

In contrast to hunting and trapping, sport fishing is generally permitted throughout the system, in accordance with state law.[187] Commercial fish-

[180]628 F. Supp. at 906.

[181]48 Fed. Reg. 30252 (1983); 36 C.F.R. § 2.2(b) (1996).

[182]51 Fed. Reg. 33263 (1986); 36 C.F.R. § 2.2(b)(3) (1996). Regulations also prohibit the "feeding, touching, teasing, frightening or intentional disturbing of wildlife nesting, breeding or other activities." *Id.* § 2.2(a).

[183]The court's opinion includes a thorough review of the Organic Act, including its 1970 and 1978 amendments.

[184]949 F.2d 202 (6th Cir. 1991).

[185]949 F.2d at 207.

[186]*Id.* The court then considered whether the lakeshores' enabling legislation showed Congress' intent to distinguish between hunting and trapping and concluded that it did, stating that "the court cannot but find that Congress considers the two activities to be distinct." *Id.* at 208, quoting *Potter, supra* note 178, at 911.

[187]36 C.F.R. § 2.3(a) (1996).

ing is prohibited unless it is specifically authorized by federal statutory law.[188] Further, the Secretary has discretion to regulate fishing to protect diminishing fish populations. In *Organized Fishermen of Florida v. Hodel,* the court upheld regulations imposing bag limits on fish caught in Everglades National Park, establishing sanctuaries in the park for endangered species, and prohibiting all commercial fishing in the park after 1985.[189]

Courts also have held that the Secretary has authority to regulate activities on nonfederal waters that are within park boundaries. In *United States v. Brown,* the court upheld the Secretary's authority to prohibit hunting on waters within the boundaries of Voyageurs National Park, when such waters arguably had not been ceded to the federal government but had instead remained the property of the state.[190] Although the court found—as an alternative holding—that the state had ceded jurisdiction over both lands and waters within the park, it held that even if the state had not ceded jurisdiction over the waters, the federal government could regulate activities on those waters, including hunting, as a proper exercise of its powers under the Property Clause. The court deemed the statutory provision implementing that authority to be the general authority of the Secretary of the Interior to promulgate "such rules and regulations as he may deem necessary or proper for the use and management of the parks."[191]

In so ruling, the court purported to decide the question left open in *Kleppe v. New Mexico*[192] "whether the Property Clause empowers the United States to enact regulatory legislation protecting federal lands from interference occurring on non-federal public lands, or, in this instance, waters."[193] Notwithstanding this decision, the significance of the case is limited both by the fact that the court's finding was an alternative holding and by the further fact that the waters at issue were within the boundaries of the national park, albeit arguably not a part of the park.

The controversy concerning hunting in Voyageurs National Park was not resolved in the *Brown* case. Duck hunters remained disgruntled with the hunting prohibition, and in an attempt to settle the dispute, Congress enacted the Boundary Revision Act.[194] The Act ceded certain park lands back to the state to satisfy the duck hunters, but it precluded all taking of wildlife on the lands except waterfowl. The Secretary approved a state wildife plan for the area that permitted trapping, and conservationists

[188] *Id.* § 2.3(d)(4).

[189] 775 F.2d 1544 (11th Cir. 1985), *cert. denied,* 476 U.S. 1169 (1986).

[190] 552 F.2d 817 (8th Cir.), *cert. denied,* 431 U.S. 949 (1977).

[191] 16 U.S.C § 3.

[192] 426 U.S. 529 (1976).

[193] 552 F.2d at 822.

[194] Act of Jan. 3, 1983, Pub. L. No. 97–405, 96 Stat. 2028, 16 U.S.C. § 160a-1.

challenged his action.[195] The court held in *Voyageurs National Park Ass'n v. Arnett* that the Secretary had violated the Boundary Revision Act.[196]

Activities Outside Park Boundaries

The Redwood National Park cases in the 1970s illustrated the conflicts that may arise from impacts on park resources of activities outside park boundaries. Just as activities outside a park affect park resources, wildlife does not always stay in the park. The interaction of park wildlife and livestock outside park boundaries can lead to major controversy. For example, ranchers hotly contested a Fish and Wildlife Service plan to reintroduce wolves to Yellowstone National Park because of concerns about predation on livestock outside park boundaries.[197]

Predators are not the only animals that cause concern among ranchers. Animals such as bison need large areas of habitat, and they move from higher elevations into lowlands during the winter in search of forage. Ranchers are concerned that bison may contaminate their cattle with the microorganism brucella (which causes a disease called brucellosis) when they leave the park.[198] Bison-cattle interaction was the subject of several cases in which conservationists charged federal officials with failing to comply with the National Environmental Policy Act (NEPA) for various actions relating to shooting or capturing Yellowstone National Park's bison outside the park boundaries.[199]

In *Fund for Animals v. Lujan,* plaintiffs alleged that Park Service officials' failure to prevent bison from leaving the park, where state game officers shot them, required an EIS.[200] The Montana district court rejected the plaintiffs' claim. However, in *Fund for Animals v. Espy,* conservationists succeeded in obtaining a preliminary injunction under NEPA against a plan to capture pregnant bison outside the boundaries of Yellowstone National Park and transport the bison to Texas for a research project regarding

[195]The Boundary Revision Act required the Secretary to approve the state's wildlife management plan prior to transferring the land to the state.

[196]609 F. Supp. 532 (D. Minn. 1985). The court's ruling was based on the Boundary Revision Act's express prohibition on taking wildlife except waterfowl; the case is not contrary to the cases discussed above in which courts refused to compel the Secretary to protect park resources under the Organic Act.

[197]For discussion of reintroduction of endangered species, see Chapter 7.

[198]Despite these concerns, there are no confirmed instances of free-ranging bison contaminating cattle with brucella, according to the court in Fund for Animals, Inc. v. Espy, 814 F. Supp. 142, 144 (D.D.C. 1993).

[199]The last remaining wild herd of 2,400 bison is found in Yellowstone. *Id.*

[200]The court granted summary judgment for the defendants in 1985 in an unpublished opinion. The decision is discussed in a related case, Fund for Animals, Inc. v. Lujan, 962 F.2d 1391 (9th Cir. 1992).

the transmittal of brucellosis from bison to cattle.[201] The proposal contemplated

> the capture of up to 60 free-roaming, wild pregnant bison in Montana and their transportation by truck approximately 2000 miles to Texas. There, half of them would be artificially infected with the brucella organism and corralled in close proximity to cattle for a period of time. . . . After the study, the infected bison would be slaughtered."[202]

The court found that the Secretary of Agriculture had "authorized a study which is arguably detrimental to the environment of Yellowstone National Park and its vicinity" without complying with NEPA and issued the injunction.[203]

Conflicts between activities outside the parks and conservation of park resources have been increasing, with rapid expansion of development adjacent to parks.[204] Along with development, the realization has grown that it is extremely difficult to protect the resources within a park from the effects of activities outside park boundaries. Conflicts of this nature are likely to increase and give rise to still further litigation in the future.[205]

THE WILDERNESS PRESERVATION SYSTEM

The Wilderness Act of 1964

It has been said that "without wildlife wilderness is mere scenery."[206] Conversely, although the Wilderness Act of 1964 hardly mentions the word "wildlife," wilderness is vital to the long-term survival of many spe-

[201] 814 F. Supp. 142 (D.D.C. 1993).

[202] *Id.* at 145. The suit was against the Department of Agriculture, which had agreed to fund the project. The Park Superintendent had refused to allow bison to be removed from the park for the project. However, bison were being attracted out of the park by baited traps and other means.

[203] *Id.* at 150.

[204] *See, e.g.*, United States v. So. Fla. Water Management Div., 28 F.3d 1563 (11th Cir. 1994) (Federal government brought suit against local water district for alleged contamination of national wildlife refuge and national park; resulted in agreement that designated procedures for restoration and preservation of the refuge).

[205] For a discussion of external threats to park wildlife, see George Cameron Coggins, *Protecting the Wildlife Resources of the National Parks from External Threats*, 22 Land & Water L. Rev. 1 (1987). For a general discussion of external threats to park resources, see Our Common Lands: Defending the National Parks (Daniel J. Simon ed., 1988). *See also* Keiter, *supra* note 162, at 116–19.

[206] C. Schoenfeld & J. Hendee, Wildlife Management in Wilderness 23 (1978) (quoting L. Crisler, Arctic Wild (1958)).

cies.[207] Wilderness is the only legislative land designation that protects habitat from most forms of development, including road building. By recognizing the value of remaining areas of federal land "where the earth and its community of life are untrammeled by man, where man himself is a visitor who does not remain,"[208] and by directing that they be managed for "the preservation of their wilderness character,"[209] the Act assures that substantial areas of public land will be spared from forms of use and development most damaging to wildlife.

The Wilderness Preservation System currently includes 94 million acres of public lands, of which almost 57 million are in Alaska. Unlike other federal land systems, there is no single agency responsible for the administration of the Wilderness System. Rather, the Forest Service, the Bureau of Land Management (BLM), the Fish and Wildlife Service, and the National Park Service retain jurisdiction over those areas of land that were under their jurisdiction at the time the Act was passed and that Congress has added to the Wilderness System.[210] Wilderness designation simply adds management directives to lands that are in the public domain for other purposes.

Statutory Framework

The Wilderness Act states that its purposes are "supplemental" to those for which the forests, parks, and refuges were originally established.[211] The Act does, however, impose significant management directives. It prohibits commercial enterprises and permanent roads within wilderness areas and sharply restricts temporary roads, motorized and mechanical transport, and structures.[212] The Act does not prohibit all commercial uses of wilderness areas. For example, grazing may continue in those wilderness areas of the national forests where it was established prior to the Act's passage, and the President may authorize specific water and power projects that are "in the public interest."[213] More significantly, the Wilderness Act authorized the continued application of mining and mineral-leasing

[207] *See* Forest Service, U.S. Dep't of Agriculture, Wilderness Management 216–19 (Misc. Pub. No. 1365) (1978).

[208] 16 U.S.C. § 1131(c).

[209] *Id.* § 1131(a).

[210] *Id.* § 1131(b).

[211] *Id.* § 1133(a). At the time of the Act's passage in 1964, it applied only to units of the National Forest, National Park, and National Wildlife Refuge Systems. The Federal Land and Policy Management Act of 1976 directed that the National Resource Lands managed by BLM be reviewed for wilderness suitability, 43 U.S.C. § 1782, and lands under BLM management have been added to the System.

[212] *Id.* § 1133(c).

[213] *Id.* § 1133(d)(4). For a discussion of grazing in wilderness areas, see Mitchell P. McClaran, *Livestock in Wilderness: A Review and Forecast*, 20 Envt'l L. 857 (1990).

laws to national forest wilderness areas for twenty years, until the end of 1983.[214]

The Wilderness Act's only express mention of wildlife is one sentence stating that nothing in the Act "shall be construed as affecting the jurisdiction or responsibilities of the several States with respect to wildlife and fish in the national forests."[215] Congress elaborated on wildlife management in wilderness in the House Committee Report on the California Wilderness Act of 1984. Congress recognized that "certain wildife management activities were compatible, and sometimes essential, elements in the management of certain wildlife populations in many wilderness areas."[216] Water supplies may be maintained and developed with mechanical equipment when essential to wildlife survival,[217] and researchers may use motorized transport for wildlife studies.[218] Roads are not permitted, however, for wildlife management purposes.[219]

An important feature of the Wilderness Act is that only Congress can designate a wilderness area[220] or modify wilderness boundaries.[221] In other words, public land managers cannot alter the status and extent of designated wilderness areas. The agencies' role in wilderness designation is to inventory, evaluate, and ultimately recommend to the President that particular areas be designated or not. The President, in turn, makes a recommendation to Congress, which it is free to accept or reject. Congress also may modify application of the Wilderness Act's management constraints with respect to a specific area.

Litigation

Wildlife management issues have not figured prominently in litigation under the Wilderness Act. Nevertheless, because commercial and recreational activities in wilderness can have major impacts on scarce undis-

[214] *Id.* § 1133(d)(3). The Act permits development of valid mining claims existing prior to that date, subject to some restrictions. This exemption has been attacked as fundamentally inconsistent with the paramount wilderness values the Act is intended to serve. *See* Izaak Walton League of America v. St. Clair, 353 F. Supp. 698 (D. Minn. 1973), *rev'd on other grounds*, 497 F.2d 849 (8th Cir. 1974). *See also* Comment, *Closing the Mining Loophole in the 1964 Wilderness Act*, 6 Envt'l L. 469 (1976); Comment, *Geothermal Leasing in Wilderness Areas*, 6 Envt'l L. 489 (1976); and Comment, *Geothermal Energy Exploitation in Wilderness Areas: The Courts Face a Hot Issue*, 4 Envt'l L. Rep. (Envt'l L. Inst.) 10119 (1974). See discussion of mineral leasing litigation *infra* at 228–257.

[215] 16 U.S.C. § 1133(d)(7).

[216] H.R. Rep. No. 40, 98th Cong., 1st Sess. 42 (1983).

[217] *Id.* at 45.

[218] *Id.* at 46.

[219] *Id.*

[220] 16 U.S.C. §§ 1132(b), (c).

[221] *Id.* § 1132(e).

turbed wildlife habitat, litigation involving these activities has important consequences for wildlife.

Most of the early cases concerned the designation process and the restrictions that apply to potential wilderness areas prior to final designation as wilderness or nonwilderness. Several cases addressed the question of whether an EIS was required for certain activities in potential wilderness areas or those designated nonwilderness. Mineral development in potential wilderness areas has been an especially vexing question. More recently, controversy has centered on wilderness management.

The Designation Process and Management of Potential Wilderness Areas

The 1971 decision in *Parker v. United States* established that if the President and Congress are to have a meaningful opportunity to consider adding an area to the Wilderness System, the Forest Service must not authorize activities that destroy the area's wilderness value prior to the final decision.[222] Section 603(c) of the Federal Land Policy and Management Act (FLPMA) codified the *Parker* result for BLM lands in 1976.[223]

The Forest Service in 1977 initiated a "Roadless Area Review and Evaluation," commonly known by its acronym RARE II, to speed the consideration of eligible areas and thus to remove use restrictions on areas found ineligible. As a result of RARE II, all remaining roadless areas in the National Forest System were assigned to one of three categories: (1) those appropriate to be recommended for wilderness designation, (2) those inappropriate for recommendation, and (3) those requiring further study, called "further planning" areas. The RARE II programmatic EIS declared that in further planning areas, development that might reduce the land's wilderness potential was prohibited.[224]

In *California v. Bergland,* the state of California successfully challenged the RARE II programmatic EIS.[225] The court found inadequate the EIS analysis of the impact on wilderness quality of forty-seven California roadless areas deemed inappropriate for wilderness designation. The court

[222]448 F.2d 793 (10th Cir. 1971), *cert. denied,* 405 U.S. 989 (1972).

[223]Section 603(c), 43 U.S.C. § 1782(c), provides that the BLM must manage areas potentially eligible for wilderness designation "so as not to impair the suitability of such areas for preservation as wilderness" and "to prevent unnecessary or undue degradation" of the areas. See text accompanying notes 228–254 *infra* for a discussion of litigation under this section regarding mineral leasing and road building in wilderness study areas.

[224]Note, however, that courts have held that the Forest Service may allow development in areas contiguous to further planning areas that were not administratively designated "primitive areas" prior to the Wilderness Act's passage. *See* Wilson v. Block, 708 F.2d 735 (D.C. Cir. 1983) (ski area); Big Hole Ranchers Association, Inc. v. Forest Service, 686 F. Supp. 256 (D. Mont. 1988) (road construction and logging).

[225]483 F. Supp. 465 (E.D. Cal. 1980), *aff'd in part, rev'd in part sub nom.* California v. Block, 690 F.2d 753 (9th Cir. 1982).

required the Forest Service to prepare site-specific EISs for the individual California areas.

In 1984, Congress resolved much of the controversy over roadless areas not recommended for wilderness designation by enacting wilderness legislation for twenty states. Areas not designated wilderness were legislatively "released" from further wilderness review during preparation of forest plans under the National Forest Management Act.[226] Courts have held, however, that release from wilderness consideration did not preclude judicial review of the Forest Service's obligation to prepare an EIS evaluating the impact of proposed activities on the roadless character of a nonwilderness area.[227]

Mineral leasing has engendered the lion's share of litigation involving activities in potential wilderness areas. In 1980, the court in *Mountain States Legal Foundation v. Andrus*[228] considered the relationship between RARE II, the Wilderness Act, FLPMA, and the Mineral Leasing Act of 1920.[229] Under the Mineral Leasing Act, the Secretary of the Interior has authority to lease for development public lands containing oil and gas deposits. In *Mountain States*, Secretary Andrus had declined to take action for several years on pending oil and gas applications in further planning areas on Forest Service lands.[230] The plaintiff sued to force action before the Wilderness Act's twenty-year grace period for oil and gas leasing expired. Represented by soon-to-be Secretary of the Interior James Watt, the plaintiff charged that Secretary Andrus's nonaction constituted a "withdrawal" within the meaning of section 103(j) of FLPMA[231] and as such required prior congressional approval.[232] The Wyoming district court agreed and ordered the Secretary either to notify Congress of the withdrawal or to cease withholding the land from oil and gas development for the purpose of preserving its wilderness character.[233]

[226]Charles C. Wilkinson and H. Michael Anderson, Land and Resource Planning in the National Forests 351 (1987) (hereinafter Wilkinson and Anderson). See *id.* at 345–52 for a discussion of the RARE process.

[227]*See* National Audubon Society v. Forest Service, 4 F.3d 832 (9th Cir. 1993), *modified*, 46 F.3d 1437 (1994) (logging in roadless area released by the Oregon Wilderness Act); City of Tenakee Springs v. Block, 778 F.2d 1402 (9th Cir. 1985) (logging in roadless area released by the Alaska National Interest Lands Conservation Act).

[228]499 F. Supp. 383 (D. Wyo. 1980).

[229]30 U.S.C. §§ 181–287.

[230]The opinion is clear that at least one of the areas involved was assigned to this category, 499 F. Supp. at 387, but is less clear as to the others.

[231]43 U.S.C. § 1702(j).

[232]*Id.* § 1714(c).

[233]499 F. Supp. at 397. The withdrawal issue was not addressed in Southern Appalachian Multiple Use Council, Inc. v. Bergland, 15 Envt'l Rep. Cas. (BNA) 2049 (W.D.N.C. April 16, 1981), which upheld the Secretary of Agriculture's authority to manage wilderness study areas as designated wilderness. The court stated that the Secretary had stopped timber cutting, mining, and other activities in these areas. *Id.* at 2053.

The Ninth Circuit several years later rejected the district court's reasoning in *Mountain States.* In *Bob Marshall Alliance v. Hodel,* the court found that declining to grant oil and gas lease applications is a "legitimate exercise of the discretion granted to the Interior Secretary" under the Mineral Leasing Act and does not constitute an unauthorized administrative withdrawal of the lands.[234]

The court in *Bob Marshall Alliance* also considered how NEPA applies to the granting of mineral leases in further planning areas. The leases at issue were in the Deep Creek Further Planning Area of Montana's Lewis and Clark National Forest, bounded by wilderness areas and wilderness study areas and "home to a large and unique wildlife population."[235] The BLM had issued nineteen oil and gas leases, covering Deep Creek's entire 42,000 acres. Some of the leases contained a "no surface occupancy" (NSO) stipulation, which prohibited any surface-disturbing activity. Others did not contain this stipulation, but rather permitted the Secretary to impose reasonable conditions on surface-disturbing activities. The BLM had prepared environmental assessments (EAs) for the leases, but no EISs.

Following its prior decision in *Conner v. Burford,*[236] the court ruled that NEPA required the BLM to prepare EISs for the leases issued without an NSO stipulation.[237] In addition, NEPA required BLM to consider a "no action" alternative for *all* of the leases, including those issued with the NSO stipulation.[238] The court enjoined all activities in Deep Creek under existing leases and the issuance of any new leases, pending compliance with NEPA.[239]

[234]852 F.2d 1223, 1229–30 (9th Cir. 1988).

[235]852 F.2d at 1225. The court stated that the area was important for four threatened or endangered species (grizzly bear, gray wolf, peregrine falcon, and bald eagle) and abounded in bighorn sheep, elk, mule and white-tailed deer, black bear, moose, mountain goat, and mountain lion.

[236]836 F.2d 1521 (9th Cir. 1988). Conner concerned national forest lands. See Chapter 7 for a discussion of Conner.

[237]*Accord,* Sierra Club v. Peterson, 717 F.2d 1409 (D.C. Cir. 1983). In that case, the court explained that the government had given up the power to prevent surface disturbances by issuing leases without strict conditions; thus the finding of no significant impact was arbitrary and capricious. In Cabinet Mountains Wilderness v. Peterson, 685 F.2d 678 (D.C. Cir. 1982), the same court had concluded that the Forest Service's approval of exploratory mineral drilling in the Cabinet Mountains Wilderness Area did not require an EIS. The area supported populations of the threatened grizzly bear, but the court found that the mitigation measures imposed on the project on the recommendation of the Fish and Wildlife Service would avoid a significant environmental impact. For a discussion of the Cabinet Mountains case, see Comment, *Claiming the Cabinets: The Right to Mine in Wilderness Areas,* 7 Pub. Land L. Rev. 45 (1986).

[238]The court found that NEPA's requirement to consider alternatives applies even where an EIS is not required. 852 F.2d at 1229. Thus, the EAs for the NSO leases should have considered the "no action" alternative.

[239]In Wilderness Society v. Robertson, 824 F. Supp. 947 (D. Mont. 1993), the court held

A question not addressed in *Mountain States* or *Bob Marshall Alliance* was whether and to what extent the government can regulate the lessee's activities after a lease is issued. With respect to potential wilderness areas on BLM lands, FLPMA provides statutory standards. Section 603(c) requires the Secretary of the Interior to manage potential wilderness areas "so as not to impair the suitability of such areas for preservation as wilderness."[240] This nonimpairment duty is "subject, however, to the continuation of existing mining and grazing uses and mineral leasing" which was being conducted when FLPMA was approved.[241] Section 603(c) also requires the Secretary to "take any action required to prevent unnecessary or undue degradation of the lands and their resources."[242]

The court in *Utah v. Andrus*[243] considered the Secretary's authority to limit access to mining claims on state and federal lands under section 603. The court concluded that this section imposes two separate management standards that depend upon the nature and date of the use in question.[244] Uses not occurring at the time of FLPMA's enactment or not of types referred to in section 603(c) are subject to the nonimpairment standard;[245] uses of the types referred to there and occurring at the time of FLPMA's enactment are exempted from the nonimpairment standard but subject to the less restrictive undue degradation standard.[246] *Utah* further limited the nonimpairment exemption to "*actual* uses, not merely a statutory right to use," such as yet unexercised rights of access to mineral leases.[247] Similarly, the court of appeals in *Rocky Mountain Oil & Gas Ass'n v. Watt* held that only lessees who had acquired their leases prior to FLPMA were exempt from the nonimpairment standard and then only to the extent of their on-the-ground activities at the time of FLPMA's enactment.[248]

The *Rocky Mountain* court later addressed the conflict between FLPMA's

---

that the Forest Service did not have to prepare an EIS when determining the validity of mining claims under § 4 of the Wilderness Act, 16 U.S.C. § 1133(d). The court found that unlike issuing a lease, determining the validity of a claim was a nondiscretionary act.

[240]43 U.S.C. § 1782(c). Secretary Watt attempted to minimize the impact of this requirement by excluding lands from classification as Wilderness Study Areas on specious grounds. For example, in Sierra Club v. Watt, 608 F. Supp. 305 (E.D. Cal. 1985), the court held that the Secretary improperly excluded from wilderness review lands in which the United States held only a surface interest and others held mineral rights.

[241]*Id.*

[242]*Id.*

[243]486 F. Supp. 995 (D. Utah 1979).

[244]*Accord,* Sierra Club v. Hodel, 848 F.2d 1068, 1084 (10th Cir. 1988).

[245]*But see* Sierra Club v. Hodel, discussed at text accompanying notes 249–252 *infra.*

[246]486 F. Supp. at 1003.

[247]486 F. Supp. at 1006 (emphasis in original). However, even where the use is not an existing use, if such use flows from an existing right, the nonimpairment standard cannot be so restrictively applied as to "take" that right. *Id.* at 1011.

[248]696 F.2d 734, 750 (10th Cir. 1982).

"savings" provisions, which protect nonfederal rightholders' ability to enjoy property interests on or adjacent to Wilderness Study Areas (WSAs), and the section 603(c) nonimpairment and nondegradation provision. In *Sierra Club v. Hodel*, plaintiffs challenged the BLM's decision not to regulate proposed improvements to a county road passing between two WSAs.[249] The court upheld BLM guidelines that exempted "valid existing rights," such as rights of way, from the nonimpairment standard, even though the statute expressly exempted only existing mining, grazing, and mineral leasing uses. The court found that FLPMA is ambiguous regarding the relationship between section 701(h), which provides that "[a]ll actions by the Secretary shall be subject to valid existing rights,"[250] and the nonimpairment provision of section 603(c). The court held that the BLM was reasonable in resolving this ambiguity by treating rights of way in the same manner as the existing uses expressly exempted by Congress.[251]

The court, however, made it clear that an activity exempted from the nonimpairment duty remains subject to the nondegradation duty. The court upheld the district court's order that the county apply to the BLM for a permit to move a section of the road to an area on BLM land where the road would cause less degradation of the adjacent WSA's riparian habitat. The effect of the order was to "require BLM to specify where . . . the road should be located in order that it make the least degrading impact on the WSA."[252]

The early 1980s witnessed heated conflicts over mineral leasing because of the Wilderness Act's December 31, 1983 deadline for issuing leases on Forest Service lands.[253] To prevent Secretary Watt from issuing a host of eleventh-hour leases, Congress prohibited him from using appropriated funds for that purpose.[254]

For BLM lands, FLPMA provides that mineral leasing may continue until the wilderness or nonwilderness status of particular areas is resolved. Political controversy has prevented final decisions regarding millions of acres of BLM wilderness study areas. However, FLPMA's nonimpairment

[249]848 F.2d 1068 (10th Cir. 1988). The road was part of the right-of-way for the infamous Burr Trail.

[250]Pub. L. No. 94–579, § 701(h), 90 Stat. 2744, set out as a historical note following 43 U.S.C.A. § 1701(h).

[251]848 F.2d at 1087–88. The court noted that its opinion in Rocky Mountain did not imply that activities other than the three enumerated in section 603(c) could not be exempted; the court simply had not been called upon to consider "whether by implication other types of rights might fall within section 603(c)'s exemption." *Id.* at 1087.

[252]*Id.* at 1088.

[253]For a detailed discussion of mineral leasing on wilderness lands during this period, see Lawrence J. Cwik, *Oil and Gas Leasing on Wilderness Lands: The Federal Land Policy and Management Act, the Wilderness Act, and the United States Department of the Interior, 1981–1983*, 14 Envt'l L. 585 (1984).

[254]*See* Wilkinson and Anderson, *supra* note 226, at 369–70.

and nondegradation standards, as interpreted in *Utah*, *Rocky Mountain*, and *Sierra Club*, should provide a measure of interim protection for the wildlife and other wilderness values of these areas.

The only other activity in potential wilderness areas that has been the subject of litigation is snowmobile use. In *Voyageurs Region National Park Ass'n v. Lujan*, the court of appeals upheld the district court's refusal to enjoin snowmobile use on a peninsula in Voyageurs National Park in Minnesota, pending study of the area for wilderness designation.[255] In concluding that snowmobile use in the study area was not unlawful, the court relied heavily on the park's enabling legislation, which expressly provided for "winter sports, including the use of snowmobiles."[256]

The court also considered whether the Secretary's regulation governing areas under study for wilderness designation would permit snowmobiles. This regulation incorporates the nonimpairment standard of *Parker v. United States*[257] and provides that potential wilderness areas "shall be developed with a view to protecting such areas and preserving their wilderness character . . . in such manner as will leave them unimpaired for future use and enjoyment as wilderness, with inconsistent uses held to a minimum."[258] The Director of the National Park Service had concluded that snowmobile trails "would not constitute such a diminishment of the area as to preclude future designation as wilderness if snowmobile use were discontinued," since the trail surfaces would not be artificially hardened and would require only minimal clearing and signing.[259] The court found that the Service's decision, although "troublesome," was not arbitrary and capricious.[260]

Management of Designated Wilderness Areas

The 1984 designation of wilderness in twenty states, together with subsequent designations in other states, has changed the focus of wilderness litigation from the designation process to wilderness management. Cases have been brought by both sides: commercial and recreational users of wilderness and adjacent lands, on the one hand, and conservation inter-

[255] 966 F.2d 424 (8th Cir. 1992).

[256] 16 U.S.C. § 160h, cited in 966 F.2d at 426.

[257] See discussion of Parker *supra* at text accompanying notes 222–223.

[258] 43 C.F.R § 19.6 (1995).

[259] 966 F.2d at 426. The court distinguished this case from Parker on this ground—that unlike the timber harvest proposed in Parker, snowmobile use would not permanently change the area. *Id.* at 427.

[260] *Id.* at 428. Plaintiffs also had argued that snowmobile use would adversely affect bald eagle reproduction, wolf populations, and vegetation. The court observed that the regulations governing snowmobile use provided for closure of specified areas for wildlife management. *Id.*

ests on the other. Motorized access to mining claims, motorized boating and portaging, and timber cutting for pest control have been the subject of lawsuits. The most protracted litigation concerned the Forest Service's controversial program to control the pine beetle by timber harvesting in designated wilderness in Texas and the Southeast.[261]

In most cases courts have upheld the managing agency's decisions. For example, in *Clouser v. Espy*, the court found that (1) the Forest Service has statutory authority to regulate means of access to mining claims located within wilderness areas, and (2) the Service's decision to prohibit motorized access to a mining claim was not arbitrary and capricious.[262] The Wilderness Act authorizes the Secretary to regulate access to mining claims located in national forest wilderness areas "consistent with the preservation of the area as wilderness."[263] Regulations provide that mechanical transport or motorized equipment will be permitted only if "essential" to mining activities[264] or "customarily used with respect to other such claims."[265] The *Clouser* court upheld both the regulations and the Forest Service's decision that the regulations precluded motorized access to the claims at issue in the case.[266]

One case in which the appellate court did not defer to the agency's judgment was *Friends of the Boundary Waters Wilderness v. Robertson*.[267] In that case, in a 2–1 decision the appellate court reversed the district court's refusal to enjoin motorized portages in the Boundary Waters Canoe Area Wilderness. The appellate court relied on both the Wilderness Act and the Boundary Waters Canoe Area Wilderness Act to find that Congress intended to prohibit motorized portaging in most areas of the Boundary Waters Wilderness. The court's opinion demonstrates a high degree of concern for protection of wilderness character and is instructive in that regard.

The Boundary Waters Canoe Area Wilderness in northeastern Minnesota consists of over one million acres of streams, lakes, and forests. Prior to the area's designation as wilderness in 1978, boaters used trucks to transport motor boats between some of the lakes. Section 4(g) of the law designating the area as wilderness required the Secretary of Agriculture

[261]*See* Sierra Club v. Lyng, 663 F. Supp. 556 (D.D.C. 1987); Sierra Club v. Lyng, 662 F. Supp. 40, 41 (D.D.C. 1987); Sierra Club v. Block, 614 F. Supp. 488 (D.D.C. 1985); Sierra Club v. Block, 614 F. Supp. 134 (E.D. Tex. 1985).

[262]42 F.3d 1522 (9th Cir. 1994).

[263]16 U.S.C. § 1134(b).

[264]36 C.F.R. § 228.15(b) (1995).

[265]*Id.* § 228.15(c).

[266]The Forest Service had concluded that the scope of the proposed operation was small and that equipment could be transported using pack horses. 42 F.3d at 1536.

[267]978 F.2d 1484 (8th Cir. 1992), *cert. denied*, 113 S. Ct. 2962 (1993).

to terminate most motorized portages as of January 1, 1984, unless he determined that "there is no feasible nonmotorized means of transporting boats across the [named] portages."[268]

The dispute centered around interpretation of the word "feasible." The Forest Service had conducted tests to determine if teams of two or three people using portage wheels could transport motor boats across the portages. Twenty-six of the thirty-four teams successfully completed the portages. Nevertheless, the Chief of the Forest Service concluded that although nonmotorized portaging could be done, it was not feasible because of safety concerns.[269]

The district court accepted the Chief's determination, but the appellate court disagreed. The court of appeals found that "feasible" meant "physically possible" or "capable of being done," and that the tests had shown that nonmotorized portaging met this definition. In interpreting the word "feasible," the court pointed to the purposes behind both the Wilderness Act of 1964 and the Boundary Waters Canoe Area Wilderness Act: preventing further development and maintaining and restoring natural conditions. The court concluded that

> [p]rohibiting motorized portages, unless it is not "physically possible" to do so, is entirely consistent with the purposes announced in these Acts, and we are persuaded that the Chief's definition of "feasible" was overly restrictive and contrary to clear congressional intent and the plain meaning of the word "feasible."[270]

The court ruled that as a matter of law, the Forest Service erred in ordering that the portages remain open.[271] The dissent argued that congressional intent regarding the word "feasible" was unclear, and that the court should defer to the agency's reasonable interpretation of the word in this case.

Drawing lines on a map around wilderness does not, of course, isolate the wilderness from adjacent lands. Certain management actions in wilderness inevitably affect adjacent lands, and *vice versa.* Not surprisingly, wilderness management that affects adjacent lands has spawned litigation by property owners.

In one recent case, *Stupak-Thrall v. United States,* the court held that a prohibition on the use of houseboats and sailboats on Crooked Lake in the Sylvania Wilderness in Michigan, and a policy "discouraging" the use of electric fish finders, boom boxes, and other mechanical devices on the lake, did not unreasonably infringe on the riparian rights of people who

[268]Pub. L. No. 95–495, § 4(g), 92 Stat. 1649, 1651. The Boundary Water wilderness legislation permitted continued use of motor boats on certain lakes in the wilderness area.

[269]978 F.2d at 1485–86.

[270]*Id.* at 1487.

[271]*Id.* at 1488.

owned lakefront property outside the wilderness boundary.[272] The court acknowledged that the regulations affected the private owners' valid rights to use and enjoy the lake. However, in an important ruling that cited and extended the U.S. Supreme Court's reasoning in *Kleppe v. New Mexico*,[273] the court held:

> It is within the power of Congress under the Property Clause to set aside federal land as wilderness and to protect, preserve and, if necessary, restore the wilderness quality of that land by regulating private as well as federal property on lakes within a wilderness area.[274]

The court held that the Forest Service's restrictions were reasonable, in that they were "directed toward protecting and preserving the wilderness character of the area" and had "minimal impact" on the landowners.[275] This reasoning could be applied to other management actions in wilderness that affect adjacent landowners. Query, however, as to what constitutes "minimal impact" on adjacent landowners, and whether restrictions that have more than a minimal impact are similarly authorized.[276]

Perhaps the most significant litigation regarding actions in wilderness areas with impacts on adjacent lands was the series of cases regarding the Forest Service's pine beetle control program in Texas and the Southeast in the 1980s, which involved cutting timber in wilderness areas. In these cases, entitled *Sierra Club v. Block* and *Sierra Club v. Lyng*, one of the Forest Service's stated justifications for the cutting was to protect adjacent private timberlands from beetle infestations. Here the challenges came not from the private property owners, but from conservation groups objecting to timber cutting in wilderness to protect private lands.

Initially the Forest Service's justification for its pine beetle program was to protect the wilderness itself. Conservation groups argued that this rea-

[272] 843 F. Supp. 327 (W.D. Mich. 1994), *aff'd*, 70 F.3d 881 (6th Cir. 1995).

[273] 426 U.S. 529 (1976) (Property Clause authorizes the federal government to establish regulations to protect wild animals on federal land). See Chapter 2 for a discussion of Kleppe.

[274] 843 F. Supp. at 332. The Court of Appeals in *Stupak-Thrall* rejected the plaintiffs' contention that even if Congress had the authority to regulate private property on the lake, the Forest Service did not. The court held that

> Congress has authorized the Forest Service to regulate the Sylvania Wilderness, and that when the Forest Service acts to preserve "wilderness character" under the Wilderness Act, the scope of authority—except to the extent that Congress may expressly limit it—is coextensive with Congress's own authority under the Property Clause.

70 F.3d at 888.

[275] *Id.* at 334.

[276] In fact, the Court of Appeals in *Stupak-Thrall* noted that plaintiffs were really after "a bigger fish"—a regulation on motorboat use, which had not been issued when the case originally was brought. The court declined to rule on the motorboat issue, since plaintiffs had not exhausted their administrative remedies. *Id.* at 890.

soning was spurious, in that pine beetles are a natural part of the southeastern forest ecosystem and as such are not a threat to wilderness character.[277] Although the plaintiffs in those cases obtained preliminary injunctions on NEPA grounds rather than on the basis of Wilderness Act violations, in an EIS subsequent to the injunction proceedings, the Forest Service deleted wilderness protection as a justification for the beetle control program. In fact, the agency noted that the Forest Service manual requires that wilderness management not interfere with natural ecological processes and recommended a substantial reduction in tree cutting to control pine beetles.[278]

The Forest Service did not, however, halt the program completely. Instead, the agency proposed to continue cutting in order to protect the endangered red-cockaded woodpecker, whose nesting and foraging areas were damaged by the beetles, and to prevent beetle infestations from spreading to adjacent private timberland. Conservation groups again brought suit.

The Forest Service argued that it had authority to carry out its program under section 4(d)(1) of the Wilderness Act, which permits the Secretary to take such actions "as may be necessary in the control of fire, insects, and diseases."[279] Plaintiffs argued that the beetle control activities were not "necessary" within the meaning of section 4(d)(1) because the Forest Service had not demonstrated that the program was effective.[280] The court concluded that this section authorized the Forest Service to take action to control fire, insects, or diseases within wilderness to benefit commercial interests on adjacent land. The court stated, however, that the Secretary's discretion was limited when the purpose of control efforts was to protect commercial interests on adjacent land and the actions taken were inconsistent with the Wilderness Act's directive to preserve wilderness character.[281] Consequently, the court required the Forest Service to demonstrate that its beetle control efforts were "necessary to effectively control the threatened outside harm that prompts the action being taken."[282] The

[277]Sierra Club v. Block, 614 F. Supp. 488 (D.D.C. 1985) (Wilderness areas in Arkansas, Louisiana, and Mississippi); Sierra Club v. Block, 614 F. Supp. 134 (E.D. Tex. 1985) (Wilderness areas in Texas). As two commentators later noted, "The integral role played by insect infestations in the ecology of southeastern forests supports the argument that the Forest Service's attempt to control beetle infestations was the activity that impaired the wilderness character of the areas, not the infestations themselves." Daniel Rohlf and Douglas L. Honnold, *Managing the Balances of Nature: The Legal Framework of Wilderness Management*, 15 Ecology L.Q. 249, 267 (1988).

[278]*Id.*

[279]16 U.S.C. § 1133(d)(1).

[280]Sierra Club v. Lyng, 662 F. Supp. 40, 41 (D.D.C. 1987) (hereinafter "Lyng I").

[281]*Id.* at 42–43.

[282]*Id.* The court stated that "[o]nly a clear necessity for upsetting the equilibrium of the

court declined to resolve the case until the Forest Service completed its final EIS.

The final EIS provided additional information as to the efficacy of the beetle control program. The plaintiffs again argued that the Forest Service had not adequately demonstrated that the program was effective, but the court construed section 4(d)(1) to permit measures that "fall short of full effectiveness so long as they are reasonably designed to restrain or limit the threatened spread of beetle infestations from wilderness land onto the neighboring property."[283] Finding that the Forest Service's determination that the program was reasonably effective was not arbitrary and capricious, the court granted summary judgment to the agency.[284]

The *Lyng* cases also raised the issue of how to resolve an apparent conflict between preserving wilderness character and protecting endangered species. As noted above, the Forest Service gave as one reason for its beetle control program protecting the mature pine forest habitat of the red-cockaded woodpecker from destruction by beetles. The Forest Service perceived a dilemma between cutting the trees and protecting woodpeckers, or not cutting the trees and probably jeopardizing woodpeckers, but preserving wilderness integrity. The agency resolved the issue by limited tree cutting to preserve woodpecker habitat from beetle infestations.[285]

Two commentators later questioned why the Forest Service initially hesitated to adopt this strategy, when it had not hesitated to cut trees in wilderness to protect private commercial interests. They observed that

> [i]f section 4(d)(1) permits action within wilderness areas to protect commercial resources on nearby land, it must also allow wilderness managers to control fire, insects, and diseases when necessary to protect endangered species such as red cockaded woodpeckers, even if such efforts harm wilderness character.[286]

---

ecology could justify this highly injurious, semi-experimental venture of limited effectiveness." *Id.*

[283]Sierra Club v. Lyng, 663 F. Supp. 556, 560 (D.D.C. 1987) (hereinafter "Lyng II").

[284]663 F. Supp. at 560–61. *Accord*, Sierra Club v. Lyng, 694 F. Supp. 1260 (E.D. Tex 1988). *See* Rohlf and Honnold, *supra* note 277, at 265–70, for further analysis of the Lyng cases and their implications for wilderness management.

[285]See Rohlf and Honnold, *supra* note 277, at 267.

[286]*Id.* at 277. Rohlf and Honnold also observed:

> It may be tempting to argue that red-cockaded woodpeckers are an element of wilderness character and that attempting to save their habitat from destruction by beetle infestations, therefore, is consistent with preserving wilderness character. Pine beetle infestations, however, naturally cause localized changes in an area's plant and animal communities. If sufficiently large areas in the southeastern United States were designated as wilderness and properly managed, these localized changes would not significantly diminish suitable red-cockaded woodpecker habitat on an areawide basis. Actions that destroy wilderness character now may be necessary because of modifica-

In the later case of *B.H. McDaniel v. United States,*[287] owners of private land adjacent to a Forest Service wilderness area in Texas unsuccessfully sued the Forest Service under the Federal Tort Claims Act[288] for negligently failing to prevent the spread of a southern pine beetle infestation onto their land. The agency's final EIS for suppression of pine beetles, discussed above, had adopted a preferred alternative that permitted, but did not require, control of pine beetle infestations. The policy provided that the "no control" option would be used most frequently "to allow natural forces to play their role in the wilderness ecosystem."[289] The policy continued, "It is only when these natural forces are predicted to threaten an essential [red-cockaded woodpecker] colony or cause unacceptable damage to specific resources adjacent to the wilderness that control in wilderness *may* be taken."[290] Forest Service personnel were required to perform a biological evaluation and determine what action to take on a case-by-case basis. By the time the agency took control actions in this case, the infestation had spread to the McDaniels' property.

The court found that the Forest Service's actions fell within the "discretionary exception" to the Federal Tort Claims Act and thus plaintiffs could not maintain their suit.[291] It appears that if they follow proper procedures, wilderness managers may choose to let natural processes occur in wilderness areas to preserve wilderness character without fear of liability for damage to neighboring property.

Congressional designation of new wilderness areas is far from resolved as this book goes to press. It is likely, however, that wilderness management, rather than designation, will be the focus of much of the future Wilderness Act litigation. As managers struggle to preserve wilderness

---

> tion, fragmentation, and destruction of red-cockaded woodpecker habitat throughout the bird's range.

Id. at 277 n.161. For further discussion of wilderness management and endangered species protection, see *id.* at 275–77.

[287]899 F. Supp. 305 (E.D. Texas 1995).

[288]28 U.S.C. §§ 1346(b) *et seq.* The Act is a limited waiver of the government's sovereign immunity, which permits recovery of damages against the government for tortious actions in certain circumstances.

[289]899 F. Supp. at 310, quoting the April, 1987, Record of Decision for Suppression of the Southern Pine Beetle.

[290]*Id.* (Emphasis in original.)

[291]*Id.* at 312. The discretionary function exception, 28 U.S.C. § 2680(a), provides that there shall be no liability under the Federal Tort Claims Act on "[a]ny claim . . . based upon the exercise or performance or the failure to exercise or perform a discretionary function or duty on the part of a federal agency or an employee of the Government, whether or not the discretion involved be abused." The court held that "[t]he decisions of the [Forest Service] are precisely the type of administrative policy making and implementation which Congress intended to protect from judicial second-guessing when enacting the discretionary exception embodied in 28 U.S.C. § 2680(a)." 899 F. Supp. at 312.

character, perhaps their greatest challenge will be to maintain natural ecosystems on lands surrounded by an increasing array of commercial activities.

THE NATIONAL MARINE SANCTUARIES PROGRAM

Conservation of terrestrial and freshwater biological diversity has received far more attention than has conservation of marine biological diversity.[292] In recent years, however, scientists and policymakers have begun to recognize the threats human activities pose to marine ecosystems and to address these threats.[293] The Marine Protection Research and Sanctuaries Act of 1972[294] was an important early step toward conservation of marine ecosystems in the United States.[295] Amendments to the Act, particularly those enacted in 1992,[296] have emphasized the conservation role of marine sanctuaries and strengthened federal authority to protect sanctuary resources. The few court decisions interpreting the Act have consistently upheld administrative determinations and actions that protect resources. At the same time, however, amendments have increased the complexity of the designation process. Administrative reluctance to propose designations, coupled with meager funding,[297] has resulted in the designation of only fourteen sanctuaries in the Act's twenty-five years.[298]

[292] *See, e.g.*, Dennis D. Murphy & David A. Duffus, *Conservation Biology and Marine Biodiversity*, 10 Conservation Biology 311 (1996); Kerry E. Irish & Elliott A. Norse, *Scant Emphasis on Marine Biodiversity*, 10 Conservation Biology 680 (1996).

[293] *See, e.g.*, Nat'l Research Council, Understanding Marine Biological Diversity: A Research Agenda for the Nation 5–8 (1995); Global Marine Biological Diversity (Elliott A. Norse ed., 1993).

[294] Pub. L. No. 92–532, title III, 86 Stat. 1061. The current law is codified at 16 U.S.C. §§ 1431–1445a. Amendments in 1992 changed the Act's short title to the "National Marine Sanctuaries Act." Hereafter the Act is referred to as "the Act" or "the National Marine Sanctuaries Act."

[295] Congress has also addressed conservation of marine life in legislation directed at particular species. See Chapter 5 for a discussion of the Marine Mammal Protection Act and Chapter 6 for a discussion of the Magnuson Fishery Conservation and Management Act. For discussion of the international regime for conservation of the Antarctic marine ecosystem, see text accompanying notes 161–211 in Chapter 14.

[296] Oceans Act of 1992, Pub. L. No. 102–587, title II, 106 Stat. 5039.

[297] An independent review panel convened by NOAA to examine the marine sanctuary program in 1990 found that it should be funded at $30 million annually. H.R. Rep. No. 102–565, 102d Cong., 2d Sess. (1992), *reprinted in* 1992 U.S.C.C.A.N. 4264, 4274. At that time its annual budget was $4 million. While the budget is increasing, the greatest amount authorized by the 1996 amendments is $18 million, for fiscal year 1999. Act of Oct. 11, 1996, Pub. L. No. 104–283, § 313.

[298] *See* 15 C.F.R. Part 922 (1996) for a list of all marine sanctuaries and the regulations governing them.

The National Marine Sanctuaries Act

The National Marine Sanctuaries Act authorizes the Secretary of Commerce to "designate any discrete area of the marine environment as a national marine sanctuary and promulgate regulations implementing the designation."[299] The Secretary must first determine that the designation will "fulfill the purposes and policies" of the Act.[300] Among the Act's purposes is "to maintain, restore, and enhance living resources by providing places for species that depend upon these marine areas to survive and propagate."[301]

The Act specifies factors the Secretary must consider in determining whether an area meets the standards for designation as a marine sanctuary. The first factor is

[299]16 U.S.C. § 1433(a). "[M]arine environment" is defined as "those areas of coastal and ocean waters, the Great Lakes and their connecting waters, and submerged lands over which the United States exercises jurisdiction, including the exclusive economic zone, consistent with international law." *Id.* § 1432(3). The Secretary has delegated designation and management authority to the National Oceanic and Atmospheric Administration (NOAA).

[300]*Id.* § 1433(a)(1). The Secretary also must find that "the area is of special national significance due to its resource or human-use values" and that existing authorities "are inadequate or should be supplemented to ensure coordinated and comprehensive conservation and management of the area." *Id.* § 1433(a)(2).

[301]*Id.* § 1431(b)(9).

The other purposes and policies are as follows:

1. to identify and designate as national marine sanctuaries areas of the marine environment which are of special national significance;
2. to provide authority for comprehensive and coordinated conservation and management of these marine areas, and activities affecting them, in a manner which complements existing regulatory authorities;
3. to support, promote, and coordinate scientific research on, and monitoring of, the resources of these marine areas, especially long-term monitoring and research of these areas;
4. to enhance public awareness, understanding, appreciation, and wise use of the marine environment;
5. to facilitate to the extent compatible with the primary objective of resource protection, all public and private uses of the resources of these marine areas not prohibited pursuant to other authorities;
6. to develop and implement coordinated plans for the protection and management of these areas with appropriate Federal agencies, State and local governments, Native American tribes and organizations, international organizations, and other public and private interests concerned with the continuing health and resilience of these marine areas;
7. to create models of, and incentives for, ways to conserve and manage these areas;
8. to cooperate with global programs encouraging conservation of marine resources.

Id. § 1431(b).

the area's natural resource and ecological qualities, including its contribution to biological productivity, maintenance of ecosystem structure, maintenance of ecologically or commercially important or threatened species or species assemblages, maintenance of critical habitat of endangered species, and the biogeographic representation of the site.[302]

Further evidence that Congress intended marine sanctuaries to play a role in conserving natural ecosystems appears in the Act's "Findings," which state that "protection of these special areas can contribute to maintaining a natural assemblage of living resources for future generations."[303]

The Designation Process

The sanctuary designation process is spelled out in great detail in the Act and involves extensive consultation and notice requirements. First, the Secretary must consult with specified congressional committees, the heads of all interested federal agencies, officials of state and local government entities "that will or are likely to be affected by" establishment of the sanctuary, officials of any Regional Fishery Management Council established under the Magnuson Fishery Management and Conservation Act[304] that may be affected by the proposed designation, and other interested persons.[305] In addition, the Secretary must prepare a draft EIS that includes a "resource assessment report" and maps depicting the area's proposed boundaries.[306]

[302] *Id.* § 1433(b)(1)(A). The 1992 amendments inserted the phrase "maintenance of critical habitat for endangered species." Pub. L. No. 102–587, § 2103(b)(1).

[303] 16 U.S.C. § 1431(a)(6). The 1992 amendments added this paragraph. Pub. L. No. 102–587, § 2101(a)(4). Other "Findings" are as follows:

1. this Nation historically has recognized the importance of protecting special areas of its public domain, but these efforts have been directed almost exclusively to land areas above the high-water mark;
2. certain areas of the marine environment possess conservation, recreational, ecological, historical, research, educational, or esthetic qualities which give them special national, and in some cases international, significance;
3. while the need to control the effects of particular activities has led to enactment of resource-specific legislation, these laws cannot in all cases provide a coordinated and comprehensive approach to the conservation and management of special areas of the marine environment;
4. a Federal program which identifies special areas of the marine environment will contribute positively to marine resources conservation, research, and management;
5. such a Federal program will also serve to enhance public awareness, understanding, appreciation, and wise use of the marine environment.

Id. § 1431(a).

[304] 16 U.S.C. §§ 1801–1882. See Chapter 6 for a discussion of this statute.

[305] 16 U.S.C. § 1433(b)(2).

[306] *Id.* § 1434(a)(2). The resource assessment report must document "present and potential uses of the area, including commercial and recreational fishing, research and education,

The Secretary must publish in the Federal Register notice of the sanctuary proposal that includes the draft EIS, proposed regulations to implement the proposal, and a summary of a draft management plan for the area. At the same time notice is published in the Federal Register, the Secretary must submit to the appropriate congressional committees the notice and additional specified information, including a statement of the types of activities that will be subject to regulation to protect the area's values, a draft management plan, and an estimate of the annual cost of the proposed designation. The congressional committees may hold hearings and/or issue reports on the proposal, which the Secretary must consider. The Secretary also must publish notice of the proposal in newspapers in communities that may be affected by the proposal and hold at least one public hearing in the affected area.[307]

Potential restrictions on fishing within sanctuaries are of special concern to many. The Act directs the Secretary to work with fishery managers in developing regulations. The Act allows the appropriate Regional Fishery Management Council to prepare draft regulations for fishing within the Exclusive Economic Zone. The Council's draft regulations, or a Council determination that regulations are not necessary, "shall be accepted and issued as proposed regulations by the Secretary unless the Secretary finds that the Council's action fails to fulfill the purposes and policies of this chapter and the goals and objectives of the proposed designation."[308] The Secretary may prepare fishing regulations if the Council fails to do so or if the Secretary rejects the Council's regulations. This section also requires the Secretary to "cooperate with other appropriate fishery management authorities with rights or responsibilities within a proposed sanctuary at the earliest practicable stage in drafting any sanctuary fishing regulations."[309]

Within thirty months of the notice of proposed designation being published in the Federal Register, the Secretary must publish in the Federal Register and submit to Congress either a notice of designation or findings explaining why a designation has not been published.[310] The notice must include final regulations to implement the designation. The designation and regulations take effect "after the close of a review period of forty-five days of continuous session of Congress."[311] An exception to this occurs if

minerals and energy development, subsistence uses, and other commercial, governmental, or recreational uses." *Id.*, § 1433(b)(3).

[307] *Id.* § 1434(a).

[308] *Id.* § 1434(a)(5).

[309] *Id.* Other fishery management authorities may include international, state, or tribal.

[310] *Id.* § 1434(b)(1). The final notice may not be published sooner than 45 days after the proposal has been submitted to Congress, to enable the appropriate committees to hold hearings and/or issue reports on the proposal. *Id.* § 1434(a)(6) and (b)(1).

[311] *Id.* § 1434(b)(1). There is no procedure specified for Congress to disapprove a proposal

the sanctuary is within the seaward boundary of any state and the governor of the state certifies that the designation or any of its terms is unacceptable, "in which case the designation or the unacceptable term shall not take effect" in the area within the state's boundaries.[312] If the governor certifies that terms of the designation are unacceptable, and the Secretary believes that this action will affect the designation so that the "goals and objectives of the sanctuary cannot be fulfilled," the Secretary may withdraw the designation.[313] In other words, the governor of an affected state may exercise veto power over the designation or any of its terms.

Effects of Sanctuary Designation

Some of the effects of sanctuary designation vary, depending upon the regulations and management plan that apply to an individual sanctuary. The Act gives the Secretary broad discretion to regulate activities that are not compatible with resource protection.[314] Activities that often are regulated or prohibited within sanctuaries are harvesting of living marine resources, oil and gas development, waste discharge, vessel traffic, altering the seabed, and recreation.[315]

In addition to regulations that vary by sanctuary, there are some statutorily prescribed effects of designation. These include required federal interagency consultation, penalties for conducting prohibited activities, and liability for damage to sanctuary resources.

First, the 1992 amendments added an important provision requiring federal agency consultation with the Secretary for all agency actions that are likely to "destroy, cause the loss of, or injure any sanctuary resource."[316] These actions include those that take place inside or outside sanctuary boundaries, and private activities authorized by licenses, leases or permits.[317] The action agency must provide a statement describing the action and its potential effects on sanctuary resources no later than forty-five days before final approval of an action. If the Secretary finds that the action is likely to destroy or injure a sanctuary resource, the Secretary is to recommend "reasonable and prudent alternatives" that will protect sanctuary resources, such as by conducting the action elsewhere. If the

---

or any of its terms. "[D]isapproval of, or amendments to, sanctuary designations can be addressed through traditional legislative procedures." H.R. Rep. No. 102–565, 102d Cong., 2d Sess. (1992), *reprinted in* 1992 U.S.C.C.A.N. 4264, 4269.

[312] *Id.*

[313] *Id.* § 1434(b)(2).

[314] The Secretary is to facilitate public and private uses of the resources within sanctuaries "to the extent compatible with the primary objective of resource protection." 16 U.S.C. § 1431(b)(5).

[315] See 15 C.F.R. Parts 922 E-O (1996) for regulations applicable to marine sanctuaries.

[316] Pub. L. No. 102–587, § 2104(d), codified at 16 U.S.C. § 1434(d)(1)(A).

[317] *Id.*

head of the agency proposing the action decides not to follow the Secretary's recommendations, the agency head must provide a written statement explaining the reasons for this decision.[318]

Next, the Act prohibits certain activities in all marine sanctuaries. The 1992 amendments added a new section providing that it is unlawful to

> (1) destroy, cause the loss of, or injure any sanctuary resource managed under law or regulations for that sanctuary;
> (2) possess, sell, deliver, carry, transport, or ship by any means any sanctuary resource taken in violation of this section;
> (3) interfere with the enforcement of [the Act]; or
> (4) violate any provision of [the Act] or any regulation or permit issued pursuant to [the Act].[319]

Each violation is punishable by a civil penalty up to $100,000.[320] In addition, a vessel (including its equipment and cargo) used in connection with a violation of the Act is subject to forfeiture.[321] Moreover, the Act provides for a "rebuttable presumption that all sanctuary resources found on board a vessel that is used or seized in connection with a violation" of the Act were taken in violation of the Act.[322] Finally, the Act provides for strict liability for destruction or injury to sanctuary resources, including "the amount of response costs and damages" resulting from the destruction or injury.[323]

Litigation under the Act

There have been few reported cases under the National Marine Sanctuaries Act, and none before 1994. It is likely that litigation will increase as a greater share of administrative attention turns to managing sanctuaries. The cases to date have consistently upheld both the Secretary's regulations and other actions taken under the Act to protect sanctuary resources.

Two appellate courts have rejected challenges to regulations governing activities in marine sanctuaries. In *Craft v. National Park Service*, the Ninth Circuit Court of Appeals held that regulations prohibiting alteration of the seabed and removal of or damage to cultural and historical resources

[318] *Id.* § 1434(d). This process is somewhat similar to the consultation process under Section 7 of the Endangered Species Act. See Chapter 7 at text accompanying notes 253–256 for a discussion of that process.
[319] Pub. L. No. 102–587, § 2106, codified at 16 U.S.C. § 1436.
[320] 16 U.S.C. § 1437(c)(1).
[321] *Id.* § 1437(d)(1).
[322] *Id.* § 1437(d)(4).
[323] *Id.* § 1443(a)(1). The statute also authorizes the Secretary to "undertake or authorize all necessary actions to prevent or minimize" injury to sanctuary resources. *Id.* § 1443(b)(1).

were neither overbroad nor vague.[324] The government had assessed civil penalties against divers in the Channel Islands sanctuary in California who had removed artifacts from shipwrecks and excavated the seabed with hammers and chisels, and the divers had challenged the regulations. The district court upheld the regulations, and the Ninth Circuit affirmed.

In *Personal Watercraft Industry Ass'n v. Dep't of Commerce*, the District of Columbia Circuit Court of Appeals reversed a district court decision striking down as arbitrary and capricious a regulation limiting operation of personal watercraft in the Monterey Bay National Marine Sanctuary in California.[325] The district court had found the regulation to be arbitrary and capricious because the Secretary had singled out personal watercraft for restrictions, pending a decision on regulation of other vessels. The court of appeals found ample support in the record for the Secretary's decision, noting that "[r]egulations . . . are not arbitrary just because they fail to regulate everything that could be thought to pose any sort of problem."[326]

The regulation applied to "motorized personal water craft," defined as motorized vessels less than fifteen feet in length, capable of exceeding a speed of 15 knots, and capable of carrying not more than one operator and one other person.[327] The court found that "[t]he record is full of evidence that machines of this sort threatened the Monterey Bay National Marine Sanctuary."[328] It noted that the Secretary had singled out these craft for regulation because

> [t]he small size, maneuverability and high speed of these craft is what causes these craft to pose a threat to resources. Resources such as sea otters and seabirds are either unable to avoid these craft or are frequently alarmed enough to significantly modify their behavior such as cessation of feeding or abandonment of young.[329]

Many people who had commented on the regulation when it was proposed had described "instances of personal watercraft operators harassing sea otters and other marine mammals, disturbing harbor seals, [and] damaging the Sanctuary's kelp forests."[330] All commenters had recommended either prohibiting personal watercraft outright or restricting their use to specific areas. The regulation was intended, the court found, to prohibit

[324] 34 F.3d 918 (9th Cir. 1994).

[325] 48 F.3d 540 (D.C. Cir. 1995).

[326] *Id.* at 544. The court observed, "[N]othing in . . . the Marine Protection, Research, and Sanctuaries Act, or in the Administrative Procedure Act, or in any judicial decision, forces an agency to refrain from solving one problem while it ponders what to do about others." *Id.* at 546.

[327] *Id.* at 542.

[328] *Id.* at 545.

[329] *Id*, quoting 57 Fed. Reg. 43314 (1992).

[330] *Id.*

operation of personal watercraft in areas with high concentrations of wildlife and in other areas vulnerable to disturbance. Thus, the court found persuasive the evidence that the Secretary had a reasoned basis for regulating personal watercraft while gathering more information on whether and how to regulate other vessels.

The other two reported cases involve administrative actions to prevent or recover damages for injury to sanctuary resources. In *United States v. Fisher*, the Eleventh Circuit Court of Appeals upheld a district court's preliminary injunction prohibiting marine salvors from using prop wash deflectors in the Florida Keys Marine Sanctuary.[331] Prop wash deflectors are used to direct propeller wash to remove seabed sediments and expose underlying materials. The defendants had created at least 100 depressions in the seabed by using prop wash deflectors. The government filed suit to recover damages for injury to the seabed, to prevent further damage to sanctuary resources, and for forfeiture of the three vessels used in the salvage operations.

The government's motion for a preliminary injunction was referred to a magistrate. At the hearing on the motion, evidence showed that the defendants had created at least twenty-seven craters in one area of the Sanctuary that were more than thirty feet across and more than eighty feet deep. The government's experts explained that seagrasses were necessary to stabilize seabeds and preserve water quality.[332] They further testified that the defendants' depressions

> have injured and disrupted vital seagrass environments in the Sanctuary. [Citation omitted] This seagrass, anchored on the ocean's bottom, is an integral part of the coral reef ecosystem and serves as shelter, food, and habitat for myriad life forms present in the Sanctuary.[333]

One expert testified that the regrowth of the grass probably would not occur "within our lifetimes."[334]

The magistrate recommended that a preliminary injunction be granted, but it recommended that the scope of the injunction be narrowed to prevent the defendants' use of prop wash deflectors, rather than all salvage activities, in the Sanctuary.[335] The district court followed this recommendation and the appellate court affirmed.[336]

[331] 22 F.3d 262 (11th Cir. 1994).

[332] *Id.* at 265.

[333] *Id.* at 266.

[334] *Id.*

[335] *Id.*

[336] The court rejected the defendants' arguments that the sanctuary designation had not yet taken effect because the government had not prepared a management plan or an environmental impact statement, as the Secretary must do when proposing to designate a sanc-

The final case interpreting the Act is a district court case. In *United States v. M/V Miss Beholden*, the court granted the government's motion for summary judgment and held the defendants liable for damage to a coral reef that occurred when they intentionally grounded their vessel in the Florida Keys National Marine Sanctuary to avoid sinking during a storm.[337] In a case of first impression, the court held that the National Marine Sanctuaries Act imposes strict liability for damage to sanctuary resources.[338] The court noted that the liability provision in the Act was modeled on those of the Clean Water Act and the Comprehensive Environmental Response, Compensation, and Recovery Act and that it is well settled that those statutes impose strict liability.

The court rejected the defendants' proffered defenses under section 1443(a)(3) of the National Marine Sanctuaries Act, one of which is an exception to the imposition of strict liability. The first defense was that the grounding was caused by an act of God (the storm). The Act provides that a person is not liable if the damage was "caused solely by an act of God . . . and the person acted with due care."[339] The court noted that weather forecasts warned of an approaching storm as early as two days before the grounding, so defendants would have a hard time establishing that they had exercised due care. It is likely, however, that if defendants had been able to show due care, they would not have been liable under the Act.

The other defense raised was that damage to the reef was "negligible," a defense under section 1443(a)(3)(C). The court found that damage to over 1,000 meters of the reef was not negligible.

Conclusion

If the marine sanctuaries program were to live up to the ambitions embodied in the Act, it would offer substantial protection to marine wildlife habitats. The program's primary objective is resource protection and all activities in marine sanctuaries are to be compatible with that objective.[340]

---

tuary. The court found that the Florida Keys sanctuary designation had taken effect immediately because it had been designated directly by Congress through legislation, the Florida Keys National Marine Sanctuary and Protection Act, rather than by administrative action. *Id.* at 267. The court further found that the National Marine Sanctuaries Act does not require the government to prepare a management plan before bringing an enforcement action under 16 U.S.C. § 1443.

[337] 856 F. Supp. 668 (S.D. Fla. 1994).

[338] *Id.* at 670.

[339] 16 U.S.C. § 1443((a)(3)(A).

[340] *Id.* § 1431(b)(5). Thus, the marine sanctuary program is similar to the national wildlife refuge system. See text accompanying notes 1–145, *supra*, for a discussion of that system.

The experience of the last twenty-five years, however, has not been encouraging. None of the Act's protections come into play until a sanctuary is designated, and sanctuary designation appears to be a particularly notable victim of political wrangling. Designation has been hampered by both administrative and, more recently, congressional reluctance to assert federal authority over marine resources, despite the Act's plethora of measures to assure the participation of all interested parties in designation and management of sanctuaries.

In the second edition of this book, written ten years after the Act's passage, the program was characterized as "still very much a fledgling program with only six sanctuaries designated thus far." In the fifteen years since then, only eight more sanctuaries have been designated, only one of which—in American Samoa—was established during the eight years of the Reagan administration. Of the seven others, Congress directed NOAA to designate four and established two others by special legislation. In 1992, Congressman Dennis M. Hertel wrote:

> Although the National Marine Sanctuary Program has been in existence since 1972, efforts at site selection and designation have been slow and deliberate, often requiring Congressional intervention to effectuate final designations. . . . [I]nitially the Program survived in meager obscurity, suffering from indifference both in program and budget support. At the end of the last decade, a resuscitation occurred, and under the careful eye of Congress, incremental progress was achieved. Now, there is hope that the program will attain a higher level of visibility and respect, as preservation and conservation efforts get underway.[341]

Four years later, Congress expressly barred NOAA from designating a sanctuary in the Northwest Straits in Washington State without specific authorizing legislation.[342] The agency had designated one sanctuary in the intervening years.

Even assuming sanctuaries are designated expeditiously and managed with care, protection for sanctuary resources is not assured. Perhaps even more than the conservation lands discussed in this chapter, marine sanctuaries are threatened by activities outside their boundaries over which sanctuary managers have little or no control. Land development, logging,

[341]H.R. Rep. No. 102–565, 102d Cong., 2d Sess. (1992), *reprinted in* 1992 U.S.C.C.A.N. 4264, 4279 (additional views submitted by Dennis M. Hertel). Mr. Hertel was Chair of the Subcommittee on Oceanography, Great Lakes, and the Outer Continental Shelf of the House Merchant Marine and Fisheries Committee and sponsor of the bill that was the basis for the 1992 amendments to the Act.

[342]Act of Oct. 11, 1996, Pub. L. No. 104–283, Sec. 10, 110 Stat. 3363.

farming, and manufacturing are a few of the land-based activities that result in sediment and pollution in coastal waters, to the detriment of marine species.[343] It will be a challenge to effect the Act's lofty vision of marine conservation in an age of booming coastal development.

[343] *See* Norse, *supra* note 293, at 114–30. Laws such as the Clean Water Act, 33 U.S.C. §§ 1251–1387, the Coastal Zone Management Act of 1972, 16 U.S.C. §§ 1451–1464, and the Coastal Barrier Resources Act, 16 U.S.C. §§ 3501–3510, address some of the threats to marine systems from land-based activities.

Chapter **9**

MULTIPLE USE LANDS I: NATIONAL FORESTS

The national forests and national resource lands are federal lands managed under the principle of "multiple use." The Forest Service manages the national forests, while the Bureau of Land Management is responsible for the national resource lands. Under the multiple use principle, wildlife conservation, logging, water development, livestock grazing, mining, recreation, and other uses are all objectives of land management. In practice, however, commodity-oriented uses historically have predominated on the multiple use lands, often to the detriment of fish and wildlife.[1] This chapter examines the law applicable to wildlife and national forests, and Chapter 10 discusses wildlife and the national resource lands.

Passage of the National Forest Management Act of 1976 promised a new era in wildlife conservation on national forest lands. Real change did not begin, however, until the end of the 1980s. Fish and wildlife conservation made major strides in the early 1990s, followed by a political reaction that has left considerable doubt as to the future direction of wildlife management on national forests.

EARLY HISTORY

During the nineteenth century, the United States pursued an affirmative policy of disposing of lands it had obtained from other sovereigns as a result of its westward expansion. A variety of disposal laws encouraged homesteaders, miners, and others to convert useful areas of the public

[1] *See, e.g.*, Charles F. Wilkinson, Crossing the Next Meridian 20–21 (1992).

domain into private ownership.[2] Toward the end of that century, however, largely in reaction to widespread abuse of lands transferred to private owners, the policy of wholesale disposal began to be curbed.[3] The mechanism to accomplish this was the reservation or withdrawal of certain specially valuable lands from the operation of some or all of the disposal laws.

The Forest Reserve Act of 1891 provided authority for the first systematic withdrawal of federal lands. The Act authorized the President to "set apart and reserve . . . public lands wholly or in part covered with timber or undergrowth, whether of commercial value or not, as public reservations."[4] These public reservations, initially managed by the Department of the Interior's General Land Office,[5] were the country's first national forests.

The political pendulum soon began to swing the other way, however, as Western interests became angered by the magnitude of presidential withdrawals and chafed at the restrictions imposed by the Forest Reserve Act against timber cutting, trespassing, and mining.[6] The result was the Forest Service Organic Administration Act of 1897, which restricted the President's authority under the earlier act by providing that "[n]o public forest reservation shall be established, except to improve and protect the forest within the reservation, or for the purpose of securing favorable conditions of water flows, and to furnish a continuous supply of timber for the use and necessities of citizens of the United States."[7] This Act, which became known simply as the Organic Act, prescribed the standards that were to govern the management of the national forests. It said not a word about wildlife.

The Organic Act's language appeared to limit the purposes for which national forests could be established to three: (1) protection of the forest; (2) securing favorable waterflows; and (3) furnishing a supply of timber. It did not necessarily limit the purposes for which such forests, once established, could be managed. Nor did it amplify the meaning of the directive to "protect the forest," which could arguably have referred to more than just the trees.[8] Whatever the justification, it is clear that the

[2]The term "public domain," in its most technical meaning, refers solely to those federal lands originally acquired by the United States from another sovereign and continuously held since they were acquired. All other federal lands are referred to technically as "acquired lands." Although substantive consequences may attach to these distinctions, they rarely do for any purpose relating to wildlife. Accordingly, except where expressly noted, the terms "public lands" and "federal lands" refer to both public domain and acquired lands.

[3]M. Clawson, America's Land & Its Uses 25 (1972).

[4]Act of March 3, 1891, ch. 561, § 24, 26 Stat. 1103 (repealed 1976).

[5]Authority to manage the national forests was transferred to the Department of Agriculture by the Forest Reserve Transfer Act of 1905, 16 U.S.C. § 472.

[6]*See* Note, *The Multiple Use-Sustained Yield Act of 1960,* 41 Ore. L. Rev. 49, 57 (1961).

[7]Act of June 4, 1897, ch. 2, 30 Stat. 35 (current version at 16 U.S.C. § 475).

[8]In 1901, the Attorney General apparently took the view that protection of forest did not

Forest Service, almost from the outset, managed the forests for a number of uses in addition to those specified in the Organic Act. The Agriculture Appropriation Act of 1907, for example, included money for the care of fish stocks in waters of the national forests.[9] Other uses, such as grazing and general recreation, became well established in the early decades of Forest Service management.[10] Thus, without any express statutory authority, the Forest Service began what was in effect a policy of "multiple use" of the national forests, permitting as many varied uses of the forests as were compatible with the three statutorily expressed purposes.

Within this framework of discretionary multiple use, the management of wildlife soon became a controversial issue, not so much because of its conflict with the statutory purposes of timber, waterflow, and forest protection, but because of jurisdictional disputes between the federal government and the states over their respective authorities within national forests. That dispute, described at greater length in Chapter 2, resulted in a 1928 decision of the Supreme Court, *Hunt v. United States,* upholding the power of the Secretary of Agriculture to order the removal of deer threatening harm to Kaibab National Forest through overbrowsing.[11]

Six years later, buoyed by the *Hunt* decision, the Forest Service issued a regulation by which it took upon itself the authority to establish "hunting and fishing seasons . . . fix bag and creel limits, specify sex of animals to be killed, [and] fix the fees to be paid" for hunting and fishing in designated national forests.[12] In the same month, Congress authorized the President to establish within national forests "fish and game sanctuaries or refuges . . . devoted to the increase of game birds, game animals, and fish of all kinds naturally adapted thereto."[13] Within such sanctuaries, all "hunting, pursuing, poisoning, angling for, killing, or capturing by trapping, netting, or any other means" was prohibited,[14] except for "predatory animals . . . destructive to livestock or wild life or agriculture."[15] Most importantly, although such sanctuaries were to remain as parts of the

---

include protection of forest wildlife, for he advised the Secretary of the Interior that the Organic Act "declared the objects of forest reserves to be the protection of forests and . . . conferred specific powers upon the Secretary with reference to their control and management, but not including the power" to regulate the killing of big game. 23 Op. Att'y Gen. 589, 593 (1901). *But see* text accompanying notes 11–19 *infra.*

[9] 34 Stat. 1270.

[10] *See* Note, *supra* note 6, at 66–71.

[11] 278 U.S. 96 (1928).

[12] Regulation G-20-A, 1 Fed. Reg. 1259, 1266 (Aug. 15, 1936) (originally promulgated March 29, 1934).

[13] 16 U.S.C. § 694. In 1916, Congress had given the President similar authority with respect only to those national forests acquired rather than reserved from the public domain. *See* Act of Aug. 11, 1916, ch. 313, 39 Stat. 476 (current version at 16 U.S.C. § 683).

[14] 16 U.S.C. § 694a.

[15] *Id.* § 694b.

national forests, other uses could be made of them, but only "so far as such uses may be consistent with the purposes for which such fish and game sanctuaries or refuges are authorized to be established."[16]

Thus bold administrative and legislative strokes taken in a single month established federal authority to preempt states from wildlife regulation in the national forests and to designate within those forests areas where wildlife conservation was to be the dominant use. Little came of either authority, however; despite periodic calls for action, the authority to create sanctuaries within national forests has been essentially unutilized.[17] The administrative assertion of authority to regulate the taking of wildlife in national forests evoked such strong opposition from the states that it was never implemented and in 1941 was replaced with new regulations providing for cooperative agreements between the Forest Service and the states under which state law governs the taking of all game.[18] Those regulations have continued in effect unchanged since their promulgation more than five decades ago.[19]

THE MULTIPLE-USE SUSTAINED-YIELD ACT

So long as both public and private forests were abundant, there were few occasions for serious conflict among the various uses to which the national forests could be put. In the period following World War II, however, vastly increased demand for forest products and a similar dramatic increase in the demand for outdoor recreation created a great potential

[16] *Id.* § 694.

[17] *See* Guilbert, *Wildlife Preservation Under Federal Law,* in Federal Environmental Law at 594, n. 314 (E. Dolgin & T. Guilbert eds., 1974). The failure may be attributable in part to the complex and, since 1939, confused division of responsibilities. Although the President retains ultimate authority for designating sanctuaries, under the Act as originally passed this authority could be exercised only "upon the recommendation of the Secretary of Agriculture and Secretary of Commerce and with the approval of the State legislatures of the respective States in which said national forests are situated." In 1939, however, Reorganization Plan No. II, § 4(f), 53 Stat. 1433–1434, transferred to the Secretary of the Interior the functions of the Secretary of Agriculture relating to wildlife conservation. Accordingly, when the 1934 Act was codified, the Secretary of the Interior was substituted for the Secretary of Agriculture each time reference to the latter was made in the Act. Under the language of the statute as codified the Secretary of the Interior is responsible (with the Secretary of Commerce) for recommending to the President the designation of wildlife sanctuaries within national forests. Because these sanctuaries remain a part of the national forests, however, they continue to be under the Department of Agriculture's jurisdiction. For consistency, the codified statute gives the Secretary of the Interior authority to permit other compatible uses in all designated sanctuaries. It is hard to imagine a situation better calculated to provoke interagency conflicts.

[18] *See* Gottschalk, *The State-Federal Partnership in Wildlife Conservation,* in Council on Environmental Quality, Wildlife and America 291 (1978).

[19] 36 C.F.R. § 241.1–.3 (1996).

for conflict.[20] Necessarily, the question arose whether the Forest Service had discretion to permit forest uses other than those prescribed in the Organic Act. Congress answered that question in 1960 by passing the Multiple-Use Sustained-Yield Act.[21]

The Multiple-Use Sustained-Yield Act declared a "policy of the Congress that the national forests are established and shall be administered for outdoor recreation, range, timber, watershed, and wildlife and fish purposes."[22] Three of those purposes had not been explicitly included in the Organic Act, and one of the Organic Act's stated purposes (forest protection) was omitted in the new law.

The 1960 Act, however, declared its enumeration of purposes to be "supplemental to, but not in derogation of, the purposes for which the national forests were established" under the Organic Act."[23] Further, both the House and Senate committees that considered the legislation emphasized that "no resource would be given a statutory priority over the others."[24]

The 1960 law further directed the Secretary of Agriculture "to develop and administer the renewable surface resources of the national forests for multiple use and sustained yield of the several products and services obtained therefrom," giving "due consideration . . . to the relative values of the various resources in the particular areas."[25] While the Act includes a definition of "multiple use,"[26] it does not define "due consideration." Thus, while it is clear that "wildlife and fish purposes" are among the purposes for which the national forests are to be administered, it is far from clear how the Forest Service in any particular instance is to balance those purposes against others spelled out in the statute. Thus, the lan-

[20] G. Robinson, The Forest Service: A Study in Public Land Management 14 (1975).

[21] 16 U.S.C. §§ 528–531.

[22] *Id.* § 528.

[23] *Id.*

[24] *See* H.R. Rep. No. 86-1551, 86th Cong., 2d Sess. 2 (1960) at 3, *reprinted in* 1960 U.S.C.C.A.N. 2377, 2379.

[25] 16 U.S.C. § 529.

[26]

> The management of all the various renewable surface resources of the [lands] so that they are utilized in the combination that will best meet the needs of the American people; making the most judicious use of the land for some or all of these resources or related services over areas large enough to provide sufficient latitude for periodic adjustments in use to conform to changing needs and conditions; that some land will be used for less than all of the resources, and harmonious and coordinated management of the various resources, each with the other, without impairment of the productivity of the land, with consideration being given to the relative values of the various resources, and not necessarily the combination of uses that will give the greatest dollar return or the greatest unit output.

Id. § 531(a).

guage of the 1960 Act was not only vague because of its generality, but also compounded preexisting interpretational difficulties.

Before turning to an examination of how the multiple use standard for forest management has been interpreted and implemented, mention should be made of the Supreme Court's one opportunity to explore the purposes of the Organic Act and the subsequent expansion of those purposes in the Multiple-Use Sustained-Yield Act. In *United States v. New Mexico* the Supreme Court, by a 5–4 vote, concluded that the 1897 Act authorized the reservation of national forests for only two purposes, timber and waterflow, and not for "aesthetic, environmental, recreational, or wildlife-preservation purposes."[27]

At issue in the *New Mexico* case was whether the 1899 reservation of Gila National Forest in New Mexico implicitly reserved federal water rights sufficient to preserve a minimum instream flow for fish and wildlife conservation purposes. Had the Court been willing to conclude that the 1897 Act's first stated purpose, to "improve and protect the forest," was independent of its purposes to conserve water flows and furnish a supply of timber, it would have been obliged to consider whether "the forest" referred solely to the trees therein or some broader assemblage of life. Justice Powell's dissent took strong exception to the majority's refusal so to hold and concluded that "[i]t is inconceivable that Congress envisioned the forests it sought to preserve as including only inanimate components such as the timber and flora. Insofar as the Court holds otherwise, the 55th Congress is maligned and the Nation is the poorer, and I dissent."[28]

Less clear was the Court's resolution of the question whether passage of the Multiple-Use Sustained-Yield Act had the effect of reserving additional water rights for a broader array of uses, including fish and wildlife, for previously reserved national forests. The majority appeared to say that it did not have this effect,[29] although the dissent treats this as dictum.[30] The majority expressly declined to say whether the 1960 Act authorized the subsequent reservation of National Forests for which a broader array of reserved water rights would be implied.[31]

The multiple use standard embodied in the 1960 Act offered a standard against which one might hope to measure Forest Service decisions in resolving the many user conflicts that were then beginning to arise. That

[27]438 U.S. 696, 707–8 (1978).

[28]438 U.S. at 723–24 (Powell, J., dissenting). In the majority opinion, Justice Rehnquist argued in a footnote that although the 1897 Act authorized the establishment of national forests to improve and protect the forest *or* for timber and water supply, the word "or" really meant "or, in other words." *Id.* at 707 n.14.

[29]438 U.S. at 713–15.

[30]*Id.* at 718–19 n.1.

[31]*Id.* at 715 n.22.

hope, however, has been largely unfulfilled with regard to judicial review of Forest Service decisions.

Parker v. United States, a case in which the plaintiffs sought to enjoin a proposed timber sale in White River National Forest, was a promising beginning.[32] The court held that the Secretary's duty to give "due consideration to the relative values of the various resources" under the Multiple-Use Act was mandatory and reviewable. The *Parker* case was ultimately successful on other grounds,[33] however, and subsequent decisions under the Multiple-Use Act have failed to amplify the nature of the duty it imposes.

Dorothy Thomas Foundation v. Hardin, for example, involved a similar challenge to a proposed timber sale in Nantahala National Forest.[34] Because the Forest Service could show that it had given at least some consideration to each of the purposes specified by the 1960 Act, however, plaintiffs were unable to demonstrate that the resulting decision to sell the timber was arbitrary, capricious, and an abuse of discretion. Much the same result was reached in *Sierra Club v. Hardin,* which challenged a proposed timber sale of over one million acres, then the largest such sale in Forest Service history.[35] Although the district court was persuaded by plaintiff's evidence of the

> overwhelming commitment of the Tongass National Forest to timber harvest objectives in preference to other multiple use values, Congress has given no indication as to the weight to be assigned each value and it must be assumed that the decision as to the proper mix of uses within any particular area is left to the sound discretion and expertise of the Forest Service.[36]

Further emphasizing the degree of discretion to be exercised by the Forest Service, the court rejected the plaintiffs' contention that "due" consideration meant "equal" consideration, because the Act's definition of multiple use "clearly contemplates that some areas may be unsuited to utilization of all resources."[37] In fact, the court stated that the requirement of due consideration was "impossible to define" and not an "objective standard." Accordingly, the court "considered that evidence in the record of 'some' consideration was sufficient to satisfy the Act absent a showing that no *actual* consideration was given to other uses."[38]

More than a year after entry of the district court's judgment in *Sierra*

[32] 307 F. Supp. 685 (D. Colo. 1969).
[33] 448 F.2d 793 (10th Cir. 1971), *aff'g* 309 F. Supp. 593 (D. Colo. 1970).
[34] 317 F. Supp. 1072 (W.D.N.C. 1970).
[35] 325 F. Supp. 99 (D. Alas. 1971).
[36] *Id.* at 123.
[37] *Id.* at n. 48.
[38] *Id.* (emphasis in original).

Club, the plaintiffs discovered a highly critical report prepared by a team of independent environmental experts for the company to which the timber sale was to be made. The plaintiffs persuaded the court of appeals to order the trial court to consider their motion for a new trial on the basis of that evidence, because it was relevant to the question whether the Forest Service had in fact been aware of the ecological consequences of its proposed sale. In its "memorandum" decision,[39] the court of appeals stated that it would, for purposes of its order, accept the district court's interpretation that "due consideration" required only "some consideration," but

> with the caution that "due consideration" to us requires that the values in question be informedly and rationally taken into balance. The requirement can hardly be satisfied by a showing of knowledge of the consequences and a decision to ignore them.[40]

Despite the qualified promise that such language offered, subsequent litigation challenging Forest Service decisions has not elucidated the "due consideration" standard of the Multiple-Use Act.[41] Most litigation has been based on other statutes, especially the National Forest Management Act.[42]

THE WILDERNESS ACT AND THE SIKES ACT EXTENSION

The many attempts to assert judicial control over Forest Service discretion under the multiple use standard were complemented by legislative efforts to circumscribe that authority. The Wilderness Act of 1964 was the first successful result of those efforts.[43] Although the Multiple-Use Act expressly recognized the establishment of wilderness areas as consistent with

[39]Sierra Club v. Butz, 3 Envt'l L. Rep. (Envt'l L. Inst.) 20292 (9th Cir. March 16, 1973). The Ninth Circuit distinguishes between "memoranda" and "opinions." The former "will not be regarded as precedent in this Court and shall not be cited in this Court in briefs or oral argument." 9th Cir. R. 21(c). Thus the *Sierra Club* decision is of uncertain precedential value.

[40]*Id.* at 20293.

[41]*See, e.g.*, West Virginia Div. of Izaak Walton League of America, Inc. v. Butz, 522 F.2d 945 (4th Cir. 1975), Zieske v. Butz, 406 F. Supp. 258 (D. Alas. 1975), and Texas Committee on Natural Resources v. Butz, 433 F. Supp. 1235 (E.D. Tex.1977), *rev'd sub nom.* Texas Committee on Natural Resources v. Bergland, 573 F.2d 201 (5th Cir. 1978).

[42]See discussion of the National Forest Management Act *infra* at text accompanying note 63 *et seq.*

[43]16 U.S.C. §§ 1131–1136. For a discussion of the Wilderness Act and its origins, *see* McCloskey, *The Wilderness Act of 1964: Its Background and Meaning*, 45 Ore. L. Rev. 288 (1966). For a discussion of the Wilderness Act and national forest planning, see Charles F. Wilkinson & H. Michael Anderson, Land and Resource Planning in the National Forests 334–70 (1987). For further discussion of the Wilderness Act and wildlife, see Chapter 8.

its multiple use mandate, nothing compelled the Secretary of Agriculture to establish such areas or, once having established them, to keep them as wilderness areas. One of the major purposes of the Wilderness Act was to eliminate this discretion on the part of the Secretary and to place the ultimate responsibility for wilderness classifications on Congress.[44] While the importance of the Wilderness Act for wildlife interests is discussed more fully in Chapter 8, it is noteworthy here because it represents a direct and unequivocal limitation on Forest Service discretion.

In the same year Congress enacted the Wilderness Act, it also established the Public Land Law Review Commission and directed it to undertake a comprehensive study and evaluation of the nation's laws affecting the public lands. The Commission completed its report six years later.[45] One of the Commission's most significant and controversial recommendations was that Congress expressly authorize the Forest Service (and the Bureau of Land Management) to adopt a "dominant use" zoning system in which the "highest and best use of particular areas of land" would be recognized as "dominant over other authorized uses."[46] The Commission argued that such a system of dominant use was already employed; by giving it express legislative sanction, the endless squabbles over the authority to prefer a particular use in a particular area might be eliminated. Although those concerned with the interests of wildlife might be apprehensive that various commercial uses of the public lands would, in most circumstances, be preferred to wildlife uses,[47] the Commission apparently arrived at its ultimate dominant use recommendation after first concluding that in some circumstances the conservation of wildlife should take precedence over all compatible uses and that the protection of rare and endangered species should always take such precedence, absent "compelling circumstances."[48]

[44]The Wilderness Act declares that nothing therein "shall be deemed to be in interference with the purpose for which national forests are established as set forth in the [Organic Act] and the Multiple-Use Sustained-Yield Act." 16 U.S.C. § 1133(a)(1).

[45]Public Land Law Review Commission, One Third of the Nation's Land (Government Printing Office, 1970) (hereinafter "PLLRC Report").

[46]*Id.* at 48–50.

[47]Although the Commission argued that a dominant use system was already in existence, cases like Kisner v. Butz, 350 F. Supp. 310 (N.D.W. Va. 1972), suggest that wildlife was seldom accorded a dominant status. In *Kisner,* plaintiffs challenged the construction of a 4.3-mile connecting link between two segments of a forest road in Monongahela National Forest. The only apparent adverse effect of the construction was to endanger one of the few remaining black bear breeding habitats in the state. The Forest Supervisor decided to go ahead with construction because "to decide the road construction question solely upon consideration of the black bear habitat would in effect create a dominant or exclusive management plan" for the area in question. 350 F. Supp. at 314.

[48]P. Hagenstein, *One Third of the Nation's Land—Evolution of a Policy Recommendation,* 12 Nat. Res. J. 56, 66 (1972).

The Commission's dominant use recommendation was never enacted into law. Certain of its more specific recommendations pertaining to wildlife were, however. Among these was its recommendation that "[f]ormal statewide cooperative agreements should be used to coordinate public land fish and wildlife programs with the states."[49] The law embodying this recommendation, known as the Sikes Act Extension, was enacted in 1974.[50] It directs the Secretaries of Agriculture and the Interior, in consultation with the appropriate state fish and game agencies, to develop "comprehensive plan[s] for conservation and rehabilitation programs" to be implemented on public lands under their respective jurisdictions in each state.[51] In accordance with these comprehensive plans, the respective Secretaries are directed, in cooperation with the states, to "develop, maintain, and coordinate programs for the conservation and rehabilitation of wildlife, fish, and game."[52] Such programs must include "specific habitat improvement projects and related activities and adequate protection for species . . . considered threatened or endangered."[53] Finally, the Act provides that the state fish and game agency may enter into "cooperative agreements" with the Secretaries in connection with the programs to be implemented in that state.[54]

The Sikes Act Extension has been praised as an ambitious attempt to encourage wildlife habitat management on the public lands.[55] On the other hand, the Department of Agriculture characterized the legislation as "duplicative and totally unnecessary" because it simply restated authority already conferred by the Multiple-Use Act.[56] The latter authority, however, is purely discretionary, whereas the Sikes Act authority is mandatory. This difference notwithstanding, the impact of the Sikes Act Extension on wildlife conservation is unclear.[57]

[49]PLLRC Report, *supra* note 45, at 159.

[50]16 U.S.C. §§ 670g–670o.

[51]*Id.* § 670h(a)(1). The Secretary of the Interior has a similar duty with respect to lands under the jurisdiction of the National Aeronautics and Space Administration and the Energy Research and Development Administration. Comprehensive plans with respect to those lands require the prior written approval of the respective agency heads. *Id.* § 670h(a)(2).

[52]*Id.* § 670g(a).

[53]*Id.*

[54]*Id.* § 670h(c).

[55]Sen. Rep. No. 93–934, 93d Cong., 2d Sess., *reprinted in* 1974 U.S.C.C.A.N. 5790.

[56]*Id.* at 5802.

[57]*See* Wilkinson & Anderson, *supra* note 43, at 310:

> The impact of the Sikes Extension Act [sic] on Forest Service authority and policy is difficult to assess. . . . The involvement of state wildlife managers in the federal planning process may serve to enhance the status of wildlife in multiple-use decision-making. However, the communication may also serve to reinforce an historical emphasis on hunting and fishing, rather than management of non-game species, on both the state and federal levels. This will be determined by the policy direction of

The Act requires the development of "comprehensive plans," but it does not define them or otherwise prescribe their required contents in any detail. Nor does it require public participation in the development of plans. The Act requires only that the plans consist of "programs," which in turn must include habitat improvement projects and "adequate protection" for threatened and endangered species.[58] Cooperative agreements must contain certain other features, such as range rehabilitation, control of off-road vehicles, and other "necessary and appropriate" terms and conditions.[59] With or without a cooperative agreement, any state may agree with the Secretaries to require "[P]ublic land management area stamps" of anyone hunting, trapping, or fishing on public lands subject to a Sikes Act Extension program.[60] Revenues from the sale of such stamps go to the states and must be used for carrying out programs under the act.[61]

Overshadowing the Sikes Act Extension's failure to be very specific about what it requires is the fact that it expressly states that it is not to "be construed as limiting the authority of the Secretary of the Interior or the Secretary of Agriculture . . . to manage the national forests or other public lands for wildlife and fish and other purposes in accordance with the Multiple-Use Sustained-Yield Act of 1960 . . . or other applicable authority."[62] Thus, the Act does nothing to change the virtually unlimited discretion the land management agencies exercise in fulfilling their multiple use mandate.

THE NATIONAL FOREST MANAGEMENT ACT (NFMA)

The most significant legislative inroad into agency discretion to manage the national forests since the Wilderness Act is the National Forest Management Act of 1976.[63] Although criticism and dissatisfaction accompa-

---

each state agency and on the results of the [National Forest Management Act] planning process.

[58]Compare the Public Land Law Review Commission's recommendation that rare and endangered species "be given a preference over other uses of public lands." PLLRC Report, *supra* note 45, at 160.

[59]16 U.S.C. § 670h(c)(3).

[60]*Id.* § 670i.

[61]*Id.* § 670i(b). In 1988, Congress amended this subsection to permit the funds to be used to provide public access to program lands as well as for wildlife conservation. Compare the PLLRC Report recommendation that a "land use fee . . . be charged for hunting and fishing on all public lands open for such purposes." PLLRC Report at 169. Under the Sikes Act Extension, land management area stamps cannot be required in any state in which all federal lands comprise 60 percent or more of the total area of such state (Alaska, Idaho, Nevada, and Utah), though a substantially similar fee may be required. 16 U.S.C. § 670*l*.

[62]16 U.S.C. § 670h(c).

[63]Pub. L. No. 94–588, 90 Stat. 2949, codified at 16 U.S.C. §§ 1601–1614 (hereinafter "the Act" or "NFMA").

nied the multiple use management concept since its inception, the impetus for the 1976 Act was a series of decisions[64] declaring clear-cutting in national forests to violate the 1897 Organic Act's requirement that only such "dead, matured, or large growth" trees as were "marked and designated" could be cut.[65] In the process of amending these features of the Organic Act, Congress imposed a number of detailed standards governing Forest Service management.

The National Forest Management Act (NFMA) "called for a fundamental reshaping of national forest wildlife policy."[66] Judicial decisions under NFMA have compelled the Forest Service to alter both its planning processes and its outcomes. NFMA has led to dramatic changes in management of over 24 million acres of some of the most productive federal forest land in the nation and elevated the role of science in national forest management to unprecedented heights.[67]

The NFMA and Its Regulations

NFMA

The most significant provisions of NFMA are in the form of amendments to the Forest and Rangelands Renewable Resources Planning Act of 1974.[68] That earlier Act was designed to aid long-range planning for use of the renewable resources on national forests by directing the Secretary of Agriculture to prepare and update every five years a Renewable Resources Program providing for the protection, management, and development of the National Forest System.[69] As part of the Renewable Resources Program, the Secretary was to develop "land and resource management plans for units of the National Forest System"[70] using "a systematic interdisciplinary approach to achieve integrated consideration of physical, biological, economic, and other sciences."[71] Among the glaring inadequacies of the 1974 Act, however, was its failure to specify how land management plans were to be prepared and what they were to contain.

[64]West Virginia Division of Izaak Walton League of America, Inc. v. Butz, 522 F.2d 945 (4th Cir. 1975), Zieske v. Butz, 406 F. Supp. 258 (D. Alas. 1975), and Texas Comm. on Natural Resources v. Butz, 433 F. Supp. 1235 (E.D. Tex. 1977), *rev'd sub nom.* Texas Comm. on Natural Resources v. Bergland, 573 F.2d 201 (5th Cir. 1978).

[65]16 U.S.C. § 476 (1970) (repealed 1976).

[66]Wilkinson & Anderson, *supra* note 43, at 296.

[67]This land includes both national forests and national resource lands (managed by the Bureau of Land Management). See discussion of litigation regarding the old-growth forests of the Pacific Northwest, *infra* at text accompanying notes 105–166.

[68]Pub. L. No. 93–378, 88 Stat. 476 (1974).

[69]16 U.S.C. § 1602.

[70]*Id.* § 1604(a).

[71]*Id.* § 1604(b).

NFMA fills those gaps in copious detail. It directed the Secretary, within two years, to promulgate regulations describing the process of development and revision of land management plans.[72] To assist in the preparation of these regulations, the Secretary was required to appoint an advisory committee of non–Forest Service scientists to assure an effective interdisciplinary approach.[73] The Secretary's regulations form the basis for the development of land management plans applicable to individual units of the National Forest System.[74]

NFMA stated as a goal the incorporation of the standards and guidelines set forth in the regulations into all land management plans of the System by September 30, 1985.[75] As a part of that process of incorporation, the Secretary must provide for public participation through making plans available to the public, holding public meetings, or otherwise soliciting public comment.[76]

Section 6(g) of NFMA[77] includes several significant provisions for wildlife. First, the regulations must specify guidelines that (1) "provide for diversity of plant and animal communities based on the suitability and capability of the specific land area";[78] (2) "insure consideration of the economic and environmental aspects of various systems of renewable resource management, including the related systems of silviculture and protection of forest resources, to provide for outdoor recreation (including wilderness), range, timber, watershed, wildlife and fish";[79] and (3) "insure that timber will be harvested . . . only where . . . protection is provided for streams, streambanks, shorelines, lakes, wetlands, and other bodies of water from detrimental changes . . . where harvests are likely to seriously and adversely affect water conditions or fish habitat."[80] Further, clear-cutting or other even-aged management may be permitted only where "such cuts are carried out in a manner consistent with the protection of soil, watershed, fish, wildlife, recreation and esthetic resources."[81] Finally, NFMA requires that the regulations specify procedures that insure compliance

[72]*Id.* § 1604(g).

[73]*Id.* § 1604(h)(1). The regulations, codified at 36 C.F.R. Part 219 (1996), are discussed *infra* at text accompanying notes 87–102.

[74]For a summary of the Act's planning processes, see George Cameron Coggins, *The Developing Law of Land Use Planning on the Federal Lands,* 61 U. Colo. L. Rev. 307, 333–48 (1990). For a detailed discussion of forest planning, see Wilkinson & Anderson, *supra* note 43.

[75]16 U.S.C. § 1604(c). This goal was not met. *See* Wilkinson & Anderson, *supra* note 43, at 43–44 for a discussion of the reasons for delays in completion of forest plans.

[76]*Id.*§ 1604(d).

[77]*Id.* § 1604(g).

[78]*Id.* § 1604(g)(3)(B).

[79]*Id.* § 1604(g)(3)(A).

[80]*Id.* § 1604(g)(3)(E).

[81]*Id.* § 1604(g)(3)(F)(v).

with the National Environmental Policy Act in the preparation of land management plans.[82]

NFMA offers those interested in wildlife conservation important opportunities. First, although it recites at numerous places that multiple use management is to remain the standard for national forest planning, NFMA's detailed standards narrow substantially the nearly untrammeled discretion that has characterized multiple use management in the past. In particular, the "diversity provision"—the requirement that planning "provide for diversity of plant and animal communities"—places substantive constraints on agency discretion. Commentators have contended—and courts have agreed—that when read in light of the overall purposes of NFMA and the relevant legislative history, this provision "requires Forest Service planners to treat the wildlife resource as a controlling, co-equal factor in forest management and, in particular, as a substantive limitation on timber production."[83]

NFMA also substantially increases the opportunity for public participation in decisionmaking. In the past, the lack of opportunity for public participation in Forest Service management decisions has been strongly criticized.[84] In contrast, NFMA requires public participation at every significant stage of administrative action. This requirement guarantees to the commercial and noncommercial users of the nation's forests equal access to the decisionmaking process. Finally, NFMA offers still other, more tangible, benefits to wildlife by amending the Knutson-Vandenburg Act[85] to provide that certain receipts from timber sales may be spent for, among other purposes, wildlife habitat management.[86]

The Regulations

The Secretary adopted comprehensive planning regulations in 1979 and revised them in 1982.[87] The central regulation affecting wildlife requires that forest planning "provide for diversity of plant and animal communities and tree species consistent with the overall multiple use objectives of the planning area."[88] Further,

[82] *Id.* § 1604(g)(1).

[83] Seattle Audubon Society v. Moseley, 798 F. Supp. 1484, 1489 (W.D. Wash. 1992), quoting Wilkinson & Anderson, *supra* note 43, at 296. *See also* Sierra Club v. Espy, 822 F. Supp. 356 (E.D. Tex. 1993) (paraphrasing the same language in Wilkinson & Anderson). See Wilkinson & Anderson at 291–96 for a discussion of the legislative history of the diversity provision.

[84] *See, e.g.*, Note, *Managing Federal Lands: Replacing the Multiple Use System*, 82 Yale L.J. 787, 791–92 (1973).

[85] 16 U.S.C. § 576–576b.

[86] *Id.* § 576b.

[87] A Committee of Scientists was appointed to advise the Secretary with respect to the regulations. 43 Fed. Reg. 39056 (1978). The Committee issued its report in 1979. *Final Report of the Committee of Scientists,* 44 Fed. Reg. 26599 (1979).

[88] 36 C.F.R. § 219.26 (1996). Diversity means "[t]he distribution and abundance of differ-

> [m]anagement prescriptions, where appropriate and to the extent practicable, shall preserve and enhance the diversity of plant and animal communities . . . so that it is at least as great as that which would be expected in a natural forest and the diversity of tree species similar to that existing in the planning area.[89]

Other regulations give specific direction as to how the Forest Service should conduct planning activities to preserve diversity. First, and most important, the Forest Service shall manage fish and wildlife habitat to

> maintain viable populations of existing native and desired non-native vertebrate species in the planning area.[90] For planning purposes, a viable population shall be regarded as one which has the estimated numbers and distribution of reproductive individuals to insure its continued existence is well distributed in the planning area.[91]

Next, in order to ensure that viable populations will be maintained, the regulations require that habitat which will support "a minimum number of reproductive individuals . . . be well distributed."[92] Finally, the Forest Service must monitor the population trends of certain "management indicator species" (MIS), which shall be selected "because their population

---

ent plant and animal communities and species within the area covered by a land and resource management plan." *Id.* § 219.3. In 1995, the Forest Service issued a proposal to amend the diversity regulations. 60 Fed. Reg. 18886 (April 13, 1995).

[89] *Id.* § 219.27(g). In order to aid in planning for diversity, "[i]nventories shall include quantitative data making possible the evaluation of diversity in terms of its present and prior condition." *Id* § 219.26.

[90] *Id.* § 219.19. The "planning area" is defined as "the area of the National Forest System covered by a regional guide or forest plan." *Id.* § 219.3. Accordingly, the court in Sharps v. U.S. Forest Service, 823 F. Supp. 668, 679 (D.S.D. 1993), held that neither § 219.19 nor § 219.27 applied to district plans and that the agency was not required to maintain a viable black-tailed prairie dog population at the district level. *But cf.* Sierra Club v. Robertson, 784 F. Supp. 593, 609 (W.D. Ark. 1991) (noting in dictum that § 219.26 requires that "diversity shall be considered throughout the planning process," which suggests that diversity should be considered at lower planning levels). One court has held that although the Forest Service "has the discretion to consider conditions surrounding the forest," it is not required to compensate in the forest plan for lack of diversity in surrounding areas. Sierra Club v. Robertson, 845 F. Supp. 485, 502 (S.D. Ohio 1994).

[91] The Forest Service's proposed amendments to the regulations would eliminate the viability requirement and instead "create[s] a system for protection of habitat capability for sensitive species in order to prevent the need for listing the species as threatened or endangered under [Endangered Species Act] and to preclude extirpation of the sensitive species from the plan area." 60 Fed. Reg. 18886, 18894 (1995). This change, if adopted, would appear to place the burden of proof on challengers to Forest Service actions to demonstrate that management actions would lead to extirpation of a species, where the current regulation places the burden of proof on the agency to demonstrate that its actions will maintain viable populations of species. See discussion of spotted owl cases, *infra* at text accompanying notes 105–166, for judicial interpretation of the current viability regulation.

[92] 36 C.F.R. § 219.19 (1996).

changes are believed to indicate the effects of management activities.''[93] Planners must estimate the effects of various management proposals on the populations and habitat of the MIS.[94]

The regulations give little guidance with regard to implementation of NFMA's requirement that plans protect bodies of water ''where harvests are likely to seriously and adversely affect water conditions or fish habitat.''[95] The only regulations that make specific reference to this statutory requirement are those pertaining to riparian management, which provide for ''special attention'' to land and vegetation ''for approximately 100 feet from the edges of all perennial streams, lakes, and other bodies of water.''[96] The regulation simply repeats the statutory language without elaborating as to how damage to water conditions or habitat is to be measured or prevented. Moreover, the regulations do not appear to recognize that ''timber harvesting on erosive slopes outside riparian areas can also have serious adverse effects on water quality and fish resources.''[97]

It would appear that anadromous fish are obvious candidates for selection as management indicator species, since they have ''special habitat needs that may be influenced significantly by planned management activities,'' they are commonly fished, and their population changes ''indicate the effects of management activities . . . on water quality.''[98] In addition, NFMA requires that plans ensure protection from timber harvesting that will ''seriously and adversely affect water conditions or fish habitat.''[99] However, despite NFMA's clear mandate for protection of fish habitat,

[93]*Id.* § 219.19(a)(1). The regulation provides that the following categories shall be represented where appropriate:

> Endangered and threatened plant and animal species identified on State and Federal lists for the planning area; species with special habitat needs that may be influenced significantly by planned management programs; species commonly hunted, fished, or trapped; [and] non-game species of special interest.

Id. The Forest Service Manual also provides for identification and consideration of ''sensitive'' species.

[94]*Id.* For detailed discussion of the Act's wildlife regulations, see Wilkinson & Anderson, *supra* note 43, at 296–306.

[95]16 U.S.C. § 1604(g)(3)(E)(iii).

[96]36 C.F.R. § 219.27(e) (1996).

[97]Wilkinson & Anderson, *supra* note 43, at 224. The authors suggest that

> the Forest Service should establish specific water quality standards in the forest plans. At a minimum, the water quality standards should include maximum temperature and sediment levels. If timber cannot be harvested in an area without exceeding the water quality standards—due to steep slopes, unstable soils, or other factors—the forest plan should identify the area as unsuitable for timber production.

Id. at 224–25.

[98]36 C.F.R. § 219.19(a)(1) (1996).

[99]16 U.S.C. § 1604(g)(3)(E)(iii).

until recently the Forest Service has taken the position that the Clean Water Act[100] sets the only standards applicable to water quality in the national forests.[101] This position is changing, as the 1994 Record of Decision for management of old-growth species for the Pacific Northwest region and northwestern California—an area that includes many of the nation's major salmon streams—recognizes the Forest Service's responsibility to maintain viable populations of anadromous fish.[102]

Litigation under NFMA

Challenges to Forest Service decisions, including final regional guides, management plans, or projects and activities such as timber sales, must first be pursued through the agency's administrative appeal process.[103] Appeal regulations adopted in 1994 provide that a project or activity decision that has been appealed shall not be implemented until fifteen days after the appeal has been resolved.[104] This stay provision does not apply to programmatic decisions. Final agency actions may be challenged in federal district court under the Administrative Procedure Act.

Spotted Owls and Old-Growth Forests: Setting the Stage for Litigation

In the late 1980s and early 1990s, courts began to apply the Act's diversity provision to specific management plans and activities. A series of cases involving the northern spotted owl[105] and its old-growth forest hab-

[100]33 U.S.C. §§ 1251–1387.

[101]For example, although timber harvesting on erosive slopes outside riparian areas can have serious adverse effects on water quality and fish habitat, Forest Service regulations regarding fish habitat are limited to riparian area management. This practice might satisfy the Clean Water Act's water quality standards, but not NFMA's viability requirement. *See* Wilkinson & Anderson, *supra* note 43, at 224–25. They note that "the NFMA water quality provisions, which are subsequent to and more specific than section 208 of the Clean Water Act of 1972, plainly supplement the Clean Water Act requirements for national forest lands."

[102]See discussion of the Record of Decision, *infra* at 114. *See* Notice of Availability of Environmental Assessment for the Implementation of Interim Strategies for Managing Anadromous Fish-Producing Watersheds in Eastern Oregon and Washington, Idaho, and Portions of California, 59 Fed. Reg. 14356 (1994).

[103]Prior to 1992, the appeal system was based on administrative authority. However, in that year Congress directed the agency to adopt an appeal system with certain provisions. Dep't of the Interior and Related Agencies Appropriations Act for Fiscal Year 1993, Pub. L. No. 102–381, § 322, 106 Stat. 1419, 16 U.S.C. § 1612 note. See 36 C.F.R. Part 217 (1996) for regulations governing appeal of programmatic decisions, such as regional guides and management plans. See 36 C.F.R. Part 215 (1996) for regulations governing appeals of projects and activities, such as timber sales and road construction.

[104]36 C.F.R. § 215.10(b) (1996).

[105]*Strix occidentalis caurina.* The northern spotted owl is a medium-sized forest owl that nests in cavities in large conifers, preys on small mammals, and is highly territorial and site tena-

itat moved this previously obscure provision to the forefront of public land wildlife law.

The old-growth Douglas fir forests of western Washington and Oregon and northwestern California "hold some of the world's most valuable timber, wildlife, water, and recreational resources."[106] Almost all the remaining old growth is on federal lands, with nearly 75 percent in national forests.[107] The Forest Service selected the northern spotted owl as a management indicator species[108] for old-growth habitat in the management plans for western Washington and Oregon forests. Spotted owl populations were declining as the owl's habitat declined.[109]

In 1986, the Forest Service issued its draft Supplemental Environmental Impact Statement (SEIS) to the Regional Guide for Region 6 (Oregon and Washington) on management of spotted owl habitat.[110] The management approach of the Forest Service's preferred alternative was to set aside "spotted owl habitat areas" ("SOHAs"), where no logging would be permitted, for 550 pairs of owls. Each SOHA would include as many as 2,200 acres of old-growth habitat, depending upon the territorial needs of owl pairs in a particular area. Under this alternative, approximately 25 percent of existing spotted owl habitat would be logged after fifteen years, and 60 percent after fifty years.[111]

The draft Spotted Owl SEIS produced a storm of controversy in the Pacific Northwest. It generated over 40,000 public comments, more than any other previous Forest Service proposal except the nationwide Roadless Area Review Evaluation (RARE II) recommendations.[112] The timber industry objected to setting aside additional old-growth forest without further research, and environmentalists recommended a moratorium on logging in *any* spotted owl habitat.[113]

In 1988, the Forest Service released its final SEIS and Record of Decision on spotted owl management, adopting the preferred alternative with

---

cious. Its range is from southwestern British Columbia to northwestern California. Interagency Scientific Committee to Address the Conservation of the Northern Spotted Owl, U.S. Dep't of Agriculture & U.S. Dep't of the Interior, A Conservation Strategy for the Northern Spotted Owl 9 (1990) (hereinafter "A Conservation Strategy").

[106]Wilkinson & Anderson, *supra* note 43, at 378.

[107]U.S. Dep't of Interior, Draft Recovery Plan for the Northern Spotted Owl 31 (1992). Scientists estimate that 80 percent of the original old-growth forest was gone by the 1980s. *Id.*

[108]*See* 36 C.F.R. § 219.19 (1996). See discussion of management indicator species, *supra* at text accompanying notes 93–94.

[109]A Conservation Strategy, *supra* note 105, at 20.

[110]Forest Service, U.S. Dep't of Agriculture, Draft Supplement to the Environmental Impact Statement for an Amendment to the Pacific Northwest Regional Guide (1986).

[111]*Id.* at 4–15.

[112]Wilkinson & Anderson, *supra* note 43, at 379.

[113]*Id.*

few changes. Critics attacked the scientific foundation for the plan.[114] Both the timber industry and environmental groups immediately appealed the Record of Decision to the Chief of the Forest Service, who denied both appeals. This decision set the stage for more than five years of litigation and congressional reaction before a plan that passed muster with the courts was adopted. The controversy remains unsettled to this day, however; implementation of the final plan remains blocked by congressional action.

Phase I of the Spotted Owl Litigation: Seattle Audubon Society v. Robertson

Environmental groups filed the first case, *Seattle Audubon Society v. Robertson,*[115] in February 1989. The plaintiffs claimed the Forest Service had violated the National Forest Management Act's diversity provision and the National Environmental Policy Act.[116] In March, the district court granted a preliminary injunction against imminent timber sales.[117] Before the issue could be resolved on the merits, Congress passed "Section 318," a rider to an appropriations bill that attempted to insulate Forest Service and Bureau of Land Management timber sales in spotted owl habitat from legal challenges based on NFMA or NEPA.[118] The district court dissolved the injunction and timber sales proceeded.[119]

[114]Victor M. Sher, *Travels with Strix: The Spotted Owl's Journey Through the Federal Courts,* 14 Pub. Land L. Rev. 41 (1993) (noting, in particular, criticism of the agency's population viability analysis). *See also* Keith Ervin, Fragile Majesty 214 (1989) ("The shortcomings of the draft SEIS alarmed biologists who had been studying the spotted owl. Not only were spotted owl habitat areas smaller than the home ranges typically used by owls, one-third of all existing SOHAs in Oregon and Washington didn't even have owls.")

[115]No. C89–160WD (W.D. Wash. 1989).

[116]The plaintiffs also claimed that the agency had violated the Migratory Bird Treaty Act. See Chapter 4 for a discussion of the Migratory Bird Treaty Act issues in the spotted owl cases.

[117]The injunction affected half of the agency's timber sales in Oregon and Washington. *See* Ervin, *supra* note 114, at 216.

[118]Dep't of Interior and Related Agencies Appropriations Act for Fiscal Year 1990, Pub. L. No. 101–121, § 318, 103 Stat. 745–50 (1989).

[119]The plaintiffs challenged the constitutionality of this provision. The district court rejected the challenge, but the Ninth Circuit reversed. Seattle Audubon Society v. Robertson, 914 F.2d 1311 (9th Cir. 1990). The court of appeals held that because Congress had attempted to direct a specific result in pending litigation without changing the underlying law, Section 318 was an unconstitutional intrusion into the decisionmaking process of Article III courts. *Id.* at 1316. The Supreme Court reversed, finding that Congress was within its constitutional authority because it had amended the law, rather than directing the courts to reach a particular decision under old law. Robertson v. Seattle Audubon Society, 503 U.S. 429, 438 (1992). When the court of appeals rendered its decision, all but approximately 16 of the Forest Service's fiscal year 1990 timber sales, and all of the BLM's sales, had been sold. By the time the Supreme Court decided the matter, the one-year appropriations law had expired.

Although the Forest Service case was dormant during 1990 because of Section 318, there were other significant developments regarding the spotted owl that would later affect the litigation. In April 1990, a committee of government scientists—appointed as a result of the *Seattle Audubon* injunction and charged with developing a conservation strategy for the owl—issued its report.[120] The Interagency Scientific Committee, chaired by Forest Service biologist Jack Ward Thomas,[121] concluded that the owl was "imperiled" across its range "because of continuing losses of habitat from logging and natural disturbances,"[122] and that existing management strategies had "contributed to a high risk that spotted owls will be extirpated from significant portions of their range."[123] The committee recommended an alternative strategy, in which large areas that included both late successional and second growth forest—able to support multiple pairs of owls in one block—would be protected from logging. In addition, the Committee recommended changes in logging practices on all federal forest lands that would preserve enough structure to allow dispersal of juvenile owls between protected areas.[124]

Another important development was the listing of the spotted owl as a threatened species under the Endangered Species Act (ESA). In late 1987, the Fish and Wildlife Service had determined that listing the owl was "not warranted."[125] Conservation groups challenged this determination. In 1988, the district court found that the agency's decision was "arbitrary and capricious and contrary to law."[126] The court ordered the agency to reconsider, and in 1989 the Fish and Wildlife Service reversed itself and listed the owl as threatened.[127]

Section 318 expired at the end of the fiscal year (September 30, 1990) and was not reenacted. The Forest Service had not prepared a new SEIS or standards and guidelines for spotted owl management, as required by NFMA. Instead, on October 3, the agency published an announcement in the Federal Register, vacating the 1988 Record of Decision and the SEIS.[128] The Forest Service declared its intent to "conduct timber management activities in a manner not inconsistent with" the recommendations of the Interagency Scientific Committee.[129]

[120]A Conservation Strategy, *supra* note 105.

[121]Dr. Thomas was later appointed Chief of the Forest Service by President Bill Clinton.

[122]A Conservation Strategy at 1.

[123]*Id.* at 18.

[124]*Id.* at 2–4.

[125]52 Fed. Reg. 48552 (1987).

[126]Northern Spotted Owl v. Hodel, 716 F. Supp. 479, 483 (W.D. Wash. 1988). See Chapter 7 for further discussion of this case.

[127]55 Fed. Reg. 26114 (1990).

[128]55 Fed. Reg. 40412 (1990).

[129]*Id.* at 40413.

Phase II of the Spotted Owl Litigation: Seattle Audubon Society v. Evans

The plaintiffs in *Seattle Audubon* amended their complaint and moved for summary judgment, charging *inter alia,* that the Forest Service still had not prepared an adequate EIS under NEPA and had failed to adopt standards and guidelines to protect the viability of the owl as required by the NFMA's diversity provision and implementing regulations. The Forest Service argued that the owl's listing under the Endangered Species Act relieved it of any obligation under NFMA to provide for the owl's viability.

In *Seattle Audubon Society v. Evans,* the court granted summary judgment to the plaintiffs on their NFMA claim.[130] The court held: "The duty to maintain viable populations of existing vertebrate species requires planning for the entire biological community—not for one species alone. It is distinct from the duty, under the ESA, to save a listed species from extinction."[131] In support of this important ruling, the court noted:

> NFMA was enacted three years later than ESA, and nothing in its language or legislative history suggests that Congress intended to exclude endangered or threatened species from NFMA's procedural and substantive requirements. The regulations under NFMA explicitly address endangered and threatened species. They do not suggest that ESA alone governs, or imply any conflict between the two statutes.[132]

With regard to the plaintiffs' other claims, the court found that (1) changing the owl management plan by publishing a notice in the Federal Register, without providing for public participation, was not permitted under NFMA, and (2) because NFMA requires an EIS when the Forest Service adopts or amends a Regional Guide, the court did not need to rule on the NEPA claim.

The plaintiffs requested a permanent injunction barring timber sales in spotted owl habitat until the Forest Service issued a legally valid management plan. The timber industry asked for an evidentiary hearing regarding the scope of injunctive relief.[133] Following an extensive hearing at which scientists, economists, and managers testified, the court issued the injunction.[134]

[130]No. C89–160WD, 1991 WL 180099 (W.D. Wash. Mar. 7, 1991) (Order on Motions for Summary Judgment and for Dismissal).

[131]*Id.* at 6. The judge also rejected the agency's argument that its Federal Register notice was adequate as a set of standards and guidelines under NFMA.

[132]*Id.* at 6. The regulations that address endangered and threatened species are found at 36 C.F.R. §§ 219.19(a)(1) & (7), 219.27(a)(8) (1996).

[133]An industry trade association had been granted intervenor status.

[134]Seattle Audubon Society v. Evans, 771 F. Supp. 1081 (W.D. Wash. 1991).

The court first found that the owl was threatened with extinction. It then noted that the Interagency Scientific Committee's conservation strategy, while supported by well-qualified scientists, had been criticized by other equally well-qualified scientists as "over-optimistic and risky," and that the public had not had an opportunity for comment.[135] With regard to the issue of irreparable harm, the court found that continued logging of spotted owl habitat would foreclose options and might "push[] the species beyond a threshold from which it could not recover."[136] Finally, the court found that the economic effects of enjoining timber sales in spotted owl habitat would be "temporary and can be minimized in many ways."[137] The court ordered the Forest Service to complete a new management plan and EIS by March, 1992.

The court concluded its landmark opinion with this observation:

> The argument that the mightiest economy on earth cannot afford to preserve old growth forests for a short time, while it reaches an overdue decision on how to manage them, is not convincing today. It would be even less so a year or a century from now.[138]

The Ninth Circuit affirmed the district court's decision.[139] The court rejected the government's contention that listing a species under the Endangered Species Act relieved the Forest Service of its obligations under NFMA to provide for diversity of species because a threatened or endangered species under the Endangered Species Act is not a "viable" population within the meaning of the Forest Service regulation:

> The effect of the Forest Service's position in this litigation, were it to be adopted, would be to reward the Forest Service for its own failures; the net result would be that the less successful the Forest Service is in maintaining viable populations of species as required under its regulations, the less planning it must do for the diversity of wildlife sought by the statute. This is directly contrary to the legislative purpose of the National Forest Management Act.[140]

Further, the court said, "The ESA list is not a list of animals to be written off. It is a mandate for all agencies involved to take aggressive steps to avoid a species' extinction and preserve its viability."[141]

[135] *Id.* at 1093.
[136] *Id.* at 1096.
[137] *Id.*
[138] *Id.*
[139] Seattle Audubon Society v. Evans, 952 F.2d 297 (9th Cir. 1991).
[140] *Id.* at 301.
[141] *Id.* at 302. Despite this ruling, in Mt. Graham Red Squirrel v. Madigan, 954 F.2d 1441 (9th Cir. 1992), the Forest Service once again argued in the same circuit that it had no duty under the National Forest Management Act to maintain a viable population of an endan-

Phase III of the Spotted Owl Litigation: Seattle Audubon Society v. Moseley

In March 1992, the Forest Service issued a new Record of Decision, officially adopting the Interagency Scientific Committee's strategy as its spotted owl management plan.[142] Once again, conservation groups challenged the plan and EIS, charging that the plan did not adequately provide for either spotted owls or other old-growth species. Once again, the district court found that the plan and EIS were not legally sufficient[143] and issued a new injunction against timber sales in spotted owl habitat.[144] The Ninth Circuit affirmed.[145]

The basis for the district court's decision was the inadequacy of the EIS under NEPA, rather than a violation of the NFMA's diversity provision. The court found a NEPA violation in the EIS's failure to discuss significant scientific information about spotted owl population declines that cast doubt on the Interagency Scientific Committee's population viability analysis. The court was careful to note that it was not rejecting the EIS merely because experts disagreed, "but because the FEIS lack[ed] reasoned discussion of major scientific objections."[146] In addition, the EIS failed to discuss the impact on owl populations of decisions by other agencies—such as the Bureau of Land Management—not to implement the Interagency Scientific Committee strategy.[147]

The court also addressed the impact of the Forest Service's plan on other old-growth species under NEPA, rather than NFMA. The EIS stated

---

gered species because a listed species was no longer "viable." In this case, the endangered Mt. Graham red squirrel was further imperiled by the proposed construction of an international observatory on most of its remaining habitat, which was on Forest Service land. The court declined to rule on the issue, finding that Congress had precluded the court from granting relief under the National Forest Management Act by enacting the Arizona-Idaho Conservation Act, which directed the Forest Service to issue a Special Use Authorization for the project. *Id.* at 1460. The appellate court thus upheld the district court's grant of summary judgment to the Forest Service on the plaintiffs' claim that the agency had violated NFMA's diversity provision, even though the Conservation Act did not specifically exempt the project from the National Forest Management Act.

[142]Forest Service, U.S. Dep't of Agriculture, Record of Decision on Management of the Northern Spotted Owl in the National Forests (1992).

[143]Seattle Audubon Society v. Moseley, 798 F. Supp. 1473 (W.D. Wash. 1992).

[144]798 F. Supp. 1484 (W.D. Wash. 1992). The court ordered the Forest Service to prepare yet another management plan and EIS by August 1993. Seattle Audubon Society v. Moseley, 798 F. Supp. 1494 (W.D. Wash. 1992).

[145]Seattle Audubon Society v. Espy, 998 F.2d 699 (9th Cir. 1993).

[146]798 F. Supp. at 1482. The Ninth Circuit agreed, stating that the EIS did not "address in any meaningful way the various uncertainties surrounding the scientific evidence upon which the [Interagency Scientific Committee] rested." 998 F.2d at 704.

[147]The Committee stated that the success of its strategy depended upon it being implemented on all federal lands. 798 F. Supp. at 1479.

that species other than the owl would have only a "low to medium-low" chance of survival under the plan. However, the Forest Service argued that this was the opinion of others, not the agency's own opinion. Significantly, the court stated that had this opinion been the agency's own view, the court would have granted summary judgment to the plaintiffs under NFMA.[148] Instead, the court found a NEPA violation because the only discussion of other species in the EIS indicated that they would be put at risk by the plan, and the agency failed to explain the magnitude of the risk or to justify "a potential abandonment of conservation duties imposed by law."[149]

Although the court's holding was based on a NEPA violation, the court found the EIS to be inadequate because NFMA requires the agency to plan for viable populations of native species, not just management indicator species. The court declared: "To adopt a plan that would preserve a management indicator species . . . such as the spotted owl, in a way that exterminated other vertebrate species would defeat the purpose of monitoring to assure general wildife viability."[150] Further, the court warned the agency that "[w]hatever plan is adopted, it cannot be one which the agency knows or believes will probably cause the extirpation of other native vertebrate species from the planning areas."[151]

The court's statements might appear to call into question the concept of planning for management indicator species to meet the requirements of NFMA's diversity provision: a plan that is adequate for the owl—an old-growth MIS—might nevertheless be inadequate for other old-growth species. This result, however, is consistent with the use of the MIS approach as one tool—but not the exclusive tool—in planning for diversity.[152] The court's findings underscore the importance of the Forest Service's choice of indicator species.[153] It appears that the agency acts at its peril if it does not choose a representative sampling of indicator species, or if it declines to select a species which it believes to be at risk.

Similarly, under NEPA, an EIS must discuss the viability of other species when there is evidence that the preferred alternative would provide for the MIS but not for other species. For example, public comments in re-

[148]798 F. Supp. at 1490.

[149]*Id.* at 1483.

[150]*Id.* at 1489.

[151]*Id.* at 1490.

[152]The court cited Wilkinson & Anderson, *supra* note 43, at 300: "The use of MIS in no way diminishes the requirement to maintain well-distributed, viable populations of existing vertebrates; in fact, proper use of MIS should help to ensure them." 798 F. Supp. at 1489. On the other hand, "the agency is not required to make a study or develop standards and guidelines as to every species." *Id.* at 1499.

[153]As Wilkinson & Anderson, *supra* note 43, observe at 300: "It is evident that the success or failure of wildlife and fish resources planning will depend largely upon the manner in which individual national forests choose their MIS."

sponse to a draft EIS, noting evidence of detrimental impacts of a proposed plan on non-MIS species, could alert the agency to the need for further investigation and discussion in the EIS.[154] If, however, after reasoned discussion the agency were to conclude that the plan is adequate for other species as well as for the MIS, a court is not likely to overturn this determination simply because outside experts disagree with it.[155]

The controversy over management of Northwest old-growth forests captured national attention. In the 1992 presidential election, candidate Bill Clinton promised to hold a "forest summit" if he were elected, in which interested parties would work out a solution to the seemingly intractable problem. The summit was held in April 1993 in Portland, Oregon, and at the conclusion of the one-day meeting, President Clinton directed his staff to assemble another interagency team, again headed by Dr. Jack Ward Thomas, to recommend a management strategy that would "comply with existing laws and produce the highest contribution to economic and social well being."[156]

The Forest Ecosystem Management Assessment Team, or "FEMAT" as it was called, analyzed ten management options. The team released its report in July 1993,[157] and the report provided the basis for the Forest Service's new management plan and EIS.[158] The Forest Service selected as its preferred alternative "Option 9," which included a system of late-successional forest reserves, with limited management activities permitted in the reserves; protection for riparian areas; and a process of "watershed analysis" for selected watersheds, in order to provide for at-risk species of anadromous fish. In early 1994, the agency issued its Record of Decision, adopting the preferred alternative.[159]

[154] *See* 40 C.F.R. § 1502.9(b) (1996): "The agency shall discuss at appropriate points in the final statement any responsible opposing view which was not adequately discussed in the draft statement and shall indicate the agency's response to the issues raised."

[155] *See, e.g.*, Inland Empire Public Lands Council v. Schultz, 992 F.2d 977, 981 (9th Cir. 1993) ("NEPA does not require that we decide whether an EA is based on the best scientific methodology available, nor does NEPA require us to resolve disagreements among various scientists as to methodology."); Sierra Club v. Robertson, 784 F. Supp. 593, 608–09 (W.D. Ark. 1991).

[156] Forest Ecosystem Management Assessment Team, U.S. Dep't of Agriculture, *et al.* Forest Ecosystem Management: An Ecological, Economic and Social Assessment I-1 (1993).

[157] *Id.* The report was an unprecedented joint project of six government agencies: Forest Service, Bureau of Land Management, Fish and Wildife Service, National Park Service, National Marine Fisheries Service, and Environmental Protection Agency. The report was to form the blueprint for forest management on both Forest Service and Bureau of Land Management lands.

[158] Forest Service, U.S. Dep't of Agriculture and Bureau of Land Management, U.S. Dep't of Interior, Draft Environmental Supplemental Impact Statement on Management of Habitat for Late-Successional and Old-Growth Forest Related Species Within the Range of the Northern Spotted Owl (1993). The Forest Service received over 100,000 public comments on the draft EIS.

[159] Forest Service, U.S. Dep't of Agriculture, and Bureau of Land Management, U.S. Dep't

Both industry and conservation groups challenged the Forest Service's new plan. The court rejected both challenges.[160]

Congress did not allow the matter to rest, however. In 1995, Congress passed a rider to the 1995 Emergency Supplemental Appropriations and Rescissions Act[161] that expedited the preparation and award of timber sales under the new forest plan. The rider directed the Forest Service to release or award in fiscal years 1995 and 1996 "all timber sale contracts offered or awarded before that date in any unit of the National Forest System or district of the Bureau of Land Management subject to section 318" of the 1990 appropriations bill, discussed above.[162] The 1995 rider directed that the sales be awarded "[n]otwithstanding any other provision of law,"[163] thus insulating covered sales from challenge under environmental laws.

Soon after the 1995 rider was passed, the timber industry brought suit, asserting that the rider mandated the release of all timber sale contracts previously awarded in Washington and Oregon, under their original advertised terms. In *Northwest Forest Resource Council v. Glickman*, the district court in Oregon agreed with the industry's interpretation of the rider.[164] The court then entered an injunction ordering the release of all timber sales offered or awarded in Washington or Oregon from October 1, 1990 to July 27, 1995.[165] The Ninth Circuit refused to stay the injunction while the district court's ruling was pending appeal, and the Forest Service began to release sales as directed. Conservationists responded with civil disobedience, leading to hundreds of arrests. The Ninth Circuit later affirmed the district court's decision.[166]

---

of Interior, Record of Decision for Amendments to Forest Service and Bureau of Land Management Planning Documents Within the Range of the Northern Spotted Owl and Standards and Guidelines for Management of Habitat for Late-Successional and Old-Growth Forest Species Within the Range of the Northern Spotted Owl (1994).

[160]Seattle Audubon Society v. Lyons, 871 F. Supp. 1291 (W.D. Wash. 1994). Both sides raised a flurry of claims, attacking both the procedures used to develop the plan and the substantive result. None was successful.

[161]Act of July 27, 1995, Pub. L. No. 104–19, § 2001.

[162]*Id.* § 2001(k)(1).

[163]*Id.*

[164]No. 95–6244 (D. Or. Sept. 13, 1995), *aff'd*, 82 F.3d 825 (9th Cir. 1996).

[165]No. 95–6244 (D. Or. October 17, 1995).

[166]82 F.3d 825 (9th Cir. 1996). The rider resulted in a barrage of cases, both in the Northwest and around the country, regarding whether particular sales were covered by the rider. For example, in the Oregon case, the court ruled that the rider bars challenges to sales offered under the Forest Service's 1994 old-growth forest plan. Oregon Natural Resources Council v. Thomas, No. 95–6272-HO (D. Or. Dec. 5, 1995), *aff'd*, 92 F.3d 792 (9th Cir. 1996). Another ruling invalidated the Forest Service's protocol for determining whether the marbled murrelet, a threatened bird, was nesting in a particular area. Northwest Forest Resource Council v. Glickman, No. 95–6244-HO (D. Or. Jan. 19, 1996). (Congress had provided in section 2001(k) that timber sales could not be awarded if threatened or endangered species were known to be nesting in the sale area.) The Ninth Circuit reversed this ruling and upheld the FWS murrelet protocol, providing one of the few victories for conservationists in the litigation spawned by the rider. 97 F.3d 1161 (9th Cir. 1996).

Other Litigation under the Diversity Provision

Cases subsequent to the *Seattle Audubon* series have dealt with the scientific methods the Forest Service has used in planning to meet the diversity requirement. Conservation groups challenged forest plans for the Chequamegon and the Nicolet National Forests in Wisconsin on the basis that the agency had failed to follow principles of conservation biology in developing the plans.[167] In both cases, the court held that the plaintiffs had not shown that the Forest Service acted arbitrarily or capriciously.

In *Sierra Club v. Marita,* the plaintiffs charged that forest planners for the Chequamegon National Forest had failed to consider the adverse effects of forest fragmentation on population viability, including the isolation of populations of animals separated by unsuitable habitat (the principle of "island biogeography") and the effects on species' suitable habitat of forces from neighboring habitats ("edge effects").[168] Instead, planners had simply

> determined the amount of each indicator species' habitat that would be available under each alternative forest plan and divided that amount by the amount of habitat necessary to support a single member or pair of the species. . . . This calculation . . . resulted in an estimate of the total population that each alternative plan could support.[169]

Plaintiffs attacked this methodology because it did not consider spatial distribution of the habitat.

The plaintiffs had submitted as part of the administrative record evidence from experts on biological diversity attesting to the detrimental effects of forest fragmentation on population viability. They had suggested that in order to minimize the effects of fragmentation, the Forest Service establish two "Diversity Management Areas" (DMAs), each consisting of about 50,000 acres, in which no harvesting or road construction would take place.

The forest supervisor's draft of the final forest plan included provision for two contiguous areas, totalling from 15 to 25 percent of the forest, "to be managed in accordance with the principles of island biogeography."[170] The Regional Forester decided, however, that the forest plan should not establish any DMAs and should make no other provision for

[167]Sierra Club v. Marita, 845 F. Supp. 1317 (E.D. Wis. 1994) ("Marita II"), *aff'd,* 46 F.3d 606 (7th Cir. 1995); and Sierra Club v. Marita, 843 F. Supp. 1526 (E.D. Wis. 1994) ("Marita I"), *aff'd,* 46 F.3d 606 (7th Cir. 1995). The same judge heard both cases, and the opinions are nearly identical.

[168]845 F. Supp. 1317 (E.D. Wis. 1994).

[169]*Id.* at 1324.

[170]*Id.* at 1326.

the study of island biogeography. In support of this determination, the Record of Decision—issued in 1986—stated that

there is conflicting scientific evidence regarding the necessity of providing large areas of old growth habitat. Very little research has been done in the Lake States to provide information about the need, amount and distribution of this type of habitat. More information is needed to make an informed decision.[171]

The court found that it could "safely assume that the principles of conservation biology set forth by plaintiffs represent sound ecological theory."[172] However, the court declined to find that the Forest Service had acted arbitrarily in failing to apply these principles to the forest plan, because

[w]hatever their theoretical validity . . . considerable uncertainty seems to surround the question of how exactly these principles should be applied. Nowhere in plaintiffs' exhaustive briefs and supporting materials does there appear any suggestion of what methodology the Service should have used to incorporate principles of conservation biology into its planning process. . . . It is therefore incumbent upon plaintiffs to explain how, precisely, those principles might have been considered; yet no such explanation has been given.[173]

The court further stated that even if the plaintiffs had identified a specific way to apply the principles in the Chequamegon plan, the court would be "reluctant to conclude that the Service acted irrationally in failing to embrace that science."[174] The court then cited the 1979 report of the Committee of Scientists that had assisted in developing the National Forest Management Act's regulations, which found that "there remains a great deal of room for honest debate" as to how precisely planners should provide for biological diversity in a particular area.[175] The court said that because the Committee had not mentioned the ecological principles advanced by the plaintiffs, the court could not read those principles into the National Forest Management Act.[176]

[171] *Id.*

[172] *Id.* at 1329.

[173] *Id.* In court, the plaintiffs had not maintained that the agency was obligated to adopt their DMA proposal. Rather, they contended only that the agency was "obligated to take into account the scientific principles upon which the proposal was based." *Id.*

[174] *Id.* at 1330.

[175] Report of the Committee of Scientists, *supra*, note 87 at 44 Fed. Reg. 26609 (1979).

[176] 845 F. Supp. at 1330. For other cases in which the court upheld the Forest Service's method of planning for diversity, see Krichbaum v. Kelley, 844 F. Supp. 1107 (W.D. Va. (1994); (planning for even-aged management did not violate the diversity requirement; naturally occurring ecosystems are not the sole measure of diversity); (Sierra Club v. Marita, 843 F. Supp. 1526 (E.D. Wis. 1994); Sierra Club v. Robertson, 845 F. Supp. 485, 502 S.D. Ohio 1994) ("The Secretary's regulations do not require the planners to use any particular

The court's conclusion that the Forest Service did not act arbitrarily or capriciously when it failed to base its diversity analysis on the principles of conservation biology was qualified by a significant footnote:

> Note the use of the past tense in this sentence. The court has been presented with information relating to the state of scientific knowledge as of the early 1980s and therefore knows nothing of what may have developed in this area since then. Thus, the court's conclusions regarding the rationality of defendant's mid-1980s analysis do not necessarily apply to its subsequent analyses.[177]

The holding in this case thus appears to be of limited significance for two reasons. First, as the court noted, it evaluated the Forest Service's actions in light of existing knowledge in the early 1980s, and the principles set forth by the plaintiffs in *Sierra Club. v. Marita* have gained greater support and specificity since that time.[178] Query whether the Committee of Scientists would fail to mention the principles of conservation biology if it were to consider the issue in the 1990s?[179] Second, in the absence of expert opinion applying these principles to the habitat in question, the court deferred to agency expertise. The court emphasized that the plaintiffs' objectives were of a theoretical nature, leaving open the opportunity

---

method in their analysis of diversity"); Oregon Natural Resources Council v. Lowe, 836 F. Supp. 727, 733 (D. Or. 1993); Forest Conservation Council v. Espy, 835 F. Supp. 1202 (D. Idaho 1993) (Forest Service's conclusion that paving a forest road along the South Fork of the Salmon River would be the best way to reduce sediment and protect salmon not arbitrary or capricious); Sierra Club v. Robertson, 810 F. Supp. 1021, 1027–28 (W.D. Ark. 1992) ("The agency's judgment in assessing issues requiring a high level of technical expertise, such as diversity, must therefore be accorded the considerable respect that matters within the agency's expertise deserve.")

[177]845 F. Supp. at 1330 n.8.

[178]The Forest Service may have a duty to supplement a regional guide or forest plan and EIS if, after the decision documents were prepared, significant new information as to the likelihood that the plan will ensure viable populations has come to light. *See* 40 C.F.R. 1502.9(c) (1996) and Portland Audubon Society v. Lujan, 712 F. Supp. 1456 (D. Or. 1989). The new information could be brought to the agency's attention by means of a petition to amend a guide or plan. If the agency declines to do so, however, the petitioner would have a heavy burden to demonstrate that the new information is significant enough to warrant supplementation. *See* Marsh v. Oregon Natural Resources Council, 490 U.S. 360, 373–74, 377 (1989) (agencies may employ a "rule of reason" in deciding whether to supplement; agency decision is overturned only if arbitrary and capricious; where the decision rests upon a high level of expertise, the court should defer to the agency's informed discretion); Forest Conservation Council v. Espy, 835 F. Supp. 1202, 1215 (D. Idaho 1993) (no duty to supplement EIS on decision to pave forest road to protect chinook salmon habitat because species was listed as threatened after decision documents were prepared).

[179]The Court of Appeals in Marita noted that "[t]he Service acknowledged at oral argument that conservation biology was the 'new trend in science,' indicating that the Service may well change its mind when evaluating future forest plans." 46 F.3d 606, 620 (7th Cir. 1995).

for plaintiffs in a future challenge to focus more practically on how the new conservation principles they advocated should be applied in a given instance.

The Clear-Cutting Provision and Red-Cockaded Woodpeckers

In Sierra Club v. Espy,[180] the court granted a preliminary injunction against even-aged management on the national forests in Texas under the provision of the National Forest Management Act that even-aged logging must be "carried out in a manner consistent with the protection of soil, watershed, fish, wildlife, recreation, and esthetic resources."[181] The court stated that "the mandate of this Section is amplified" by the diversity provision. Accordingly, the court found that

> [t]he NFMA clearly requires the Service's Planners to treat the natural resources of our national forests as controlling, co-equal factors in forest management—in particular, as substantive limitations on the particular logging practices that can take place in these forests.[182]

The court then declared that "the monoculture created by clear-cutting and resultant even-aged management techniques is contrary to NFMA-mandated bio-diversity."[183] The magistrate had found that the endangered red-cockaded woodpecker and inner-forest species such as the pileated woodpecker and the fox squirrel decline under even-aged management, and that the Forest Service had failed to consider "ecosystems of old growth forests as an element of diversity."[184]

Conclusion

The *Seattle Audubon* cases established that the National Forest Management Act calls for a new direction in wildlife management on national forests. The appointment of a biologist, who came to prominence in those cases, as Chief of the Forest Service gave hope that the fish and wildlife resource will be given equal consideration with timber production. According to a report prepared in 1993 by Dr. Thomas before his appointment as Chief:

> [I]t appears that a significant objective of land management (particularly that of the National Forests) can now be described as the preservation of biodiversity.
>
> It is difficult not to accept this, if the regulations issued pursuant to [the] National Forest Management Act of 1976 that calls [*sic*] for maintenance of *viable*

[180] 822 F. Supp. 356 (E.D. Tex. 1993).
[181] 16 U.S.C. § 1604(g)(3)(F)(v).
[182] 822 F. Supp. at 364.
[183] *Id.*
[184] *Id.* at 366, 368.

populations of native and desired non-native vertebrates *well distributed* within the planning areas . . . are to be taken seriously. And the Federal courts have said that the Act is to be so considered.[185]

The standard of judicial review of Forest Service actions, however, has not changed. Courts will continue to defer in most cases to the agency's discretion as to how best to meet the Act's requirements. However, the courts have shown an increasing willingness to overturn Forest Service decisions when the agency has failed to conduct adequate scientific analysis or cannot support its conclusions with credible scientific opinion, and where the plaintiffs have made a credible showing of specific harms that are likely to result from the Forest Service's proposed action. In other instances, the court is likely to defer to the agency's expertise.

[185]Forest Service, U.S. Dep't of Agriculture, Viability Assessments and Management Considerations for Species Associated with Late-Successional and Old-Growth Forests of the Pacific Northwest 7 (1993) (emphasis in original).

Chapter **10**

MULTIPLE USE LANDS II: NATIONAL RESOURCE LANDS

INTRODUCTION

The National Resource Lands are the other major category of multiple use federal lands, in addition to the national forests. They comprise the areas of the public domain that have not been reserved or withdrawn for particular uses. These lands are located almost exclusively in the far West and Alaska. Since 1946, they have been managed by the Department of the Interior's Bureau of Land Management (BLM).[1]

The BLM has jurisdiction over the largest area of any land manager, public or private, in the United States. The agency manages more fish and wildlife habitat than any other, and its lands contain more than 3,000 species of wildlife.[2] The vast majority of BLM lands in the lower forty-eight states—a total of 177 million acres—are arid or semi-arid lands in eleven far Western states.[3] Livestock grazing, permitted on over 90 percent of these lands, is the primary commercial activity.[4] Grazing can have signifi-

[1]The Bureau of Land Management came into being in 1946 as a result of the consolidation of Interior's General Land Office and its Grazing Service according to section 403 of the Reorganization Act of 1945, Dec. 20, 1945, ch. 582, 59 Stat. 613.

[2]Bureau of Land Management, U.S. Dep't of the Interior, Public Land Statistics 1991, at 37 (1992) (hereinafter Statistics). Over 100 threatened or endangered species are found on BLM lands, *id.* at 44, and these lands contain 30,000 miles of fishable streams. *Id.* at 38.

[3]*Id.* at 6. The BLM manages another ninety million acres in Alaska.

[4]Bureau of Land Management, U.S. Dep't of the Interior, Rangeland Reform '94 Draft Environmental Impact Statement at 5 (1994). In fact, "[l]ivestock grazing is the most extensive commercial use of public lands in the United States." Joseph M. Feller, *'Til the Cows Come Home: The Fatal Flaw in the Clinton Administration's Public Lands Grazing Policy,* 25 Envt'l L. 703 (1995).

cant impacts on the rangeland environment.[5] Other primary uses of BLM lands affecting wildlife are timber production, mineral exploration and production, water development, and recreation.[6]

The history of management of the National Resource Lands is strikingly similar to that of the National Forest System. Although early statutory authority appeared to favor particular uses, the BLM embraced the principles of multiple use. Congress later confirmed that practice in a law compelling multiple use management. Finally, as with the nation's forest resources, increasing conflicts among competing users led to enactment of major legislation to guide BLM discretion in resolving these conflicts. In recent years, pressure for increased protection of wildlife habitat on BLM lands has paralleled that on national forest lands, with more modest results.

This chapter examines those aspects of the laws and regulations governing National Resource Land management that relate directly to wildlife.[7] The more general analysis of "multiple use" as a legally enforceable standard will not be repeated; the discussion of that issue in connection with the National Forest System is equally applicable here. In fact, there has been relatively little litigation in which the multiple use standard has been put forth as a limitation on BLM management discretion; most such litigation has concerned the Forest Service.

Implementation of wildlife programs on National Resource Lands is carried out cooperatively by BLM and the individual states. This cooperative approach, which was in effect well before the Sikes Act Extension,[8] traditionally has allocated "species management" responsibilities to the states and "habitat management" responsibilities to BLM.[9] Prior to the Sikes Act Extension, BLM formalized its cooperative agreements with various states as "Memoranda of Understanding" that set forth in general terms the division of responsibilities between the states and BLM.[10]

[5] *See* National Research Council, *Rangeland Health: New Methods to Classify, Inventory, and Monitor Rangelands* 18–26 (1994); Thomas L. Fleischner, *The Ecological Costs of Livestock Grazing in Western North America,* 8 Conservation Biology 629 (1994); Joseph M. Feller, *What is Wrong With the BLM's Management of Livestock Grazing on the Public Lands?* 30 Idaho L. Rev. 560–563 (1993–94).

[6] *See* 43 U.S.C. § 1702(*l*) (defining "principal or major uses" of BLM lands within the meaning of the Federal Land and Policy Management Act, 43 U.S.C. §§ 1701–1782).

[7] For an exhaustive discussion of BLM management, including wildlife, through 1984, see the series of articles by Professor George Cameron Coggins in Environmental Law in 1982–84. *See especially* George Cameron Coggins, *The Law of Public Rangeland Management III: A Survey of Creeping Regulation at the Periphery, 1934–1982,* 13 Envt'l L. 295, 325–51 (1983) (federal wildlife laws and public rangeland); George Cameron Coggins, *The Law of Public Rangeland Management IV: FLPMA, PRIA, and the Multiple Use Mandate,* 14 Envt'l L. 1 (1983).

[8] *See* 35 Fed. Reg. 14573 (1970).

[9] *See* remarks of Curt Berklund, Director, BLM, before the Sixty-fifth Convention of the International Association of Game, Fish and Conservation Commissioners 84 (1975).

[10] The actual management programs were contained in individual Habitat Management

THE BLM'S STATUTORY MANDATES

The Taylor Grazing Act and the Classification and Multiple Use Act

Until 1934, no statute governed management of the unreserved and unappropriated public domain. In that year, Congress passed the Taylor Grazing Act.[11] The Act was born of a concern for deterioration of the range as a result of uncontrolled grazing and adverse weather, and for the imminent catastrophe facing the grazing industry as a result of range deterioration and the economic Depression.

The purpose of the Taylor Grazing Act was to bring about a more orderly use and regeneration of the range. The Act authorized the Secretary of the Interior to establish "grazing districts . . . from any part of the public domain . . . which in his opinion are chiefly valuable for grazing and raising forage crops."[12] In these grazing districts, the BLM regulates the privilege of grazing livestock by issuing grazing permits.[13]

The Taylor Grazing Act was directed primarily at rescuing the Western grazing industry, just as the Forest Service's 1897 Organic Act was aimed principally at assuring future timber supplies from the national forests. Both acts, however, had additional purposes as well. Among the purposes specified in the 1897 Organic Act was the "protection" of the forests; the counterpart in the Taylor Act was the directive to the Secretary to "do any and all things necessary to . . . preserve the land and its resources from destruction or unnecessary injury."[14] Whereas the Organic Act made no express mention of wildlife, the Taylor Act apparently contemplated that wildlife was among the resources the Secretary was to preserve, for it required him or her to cooperate with various parties "interested in the use of grazing districts," including "official State agencies engaged in conservation or propagation of wild life."[15]

The Classification and Multiple Use Act of 1964 laid to rest any doubt as to BLM's authority to manage lands under its jurisdiction for wildlife conservation. This law directed the Secretary to administer the National Resource Lands under principles of multiple use and sustained yield.[16]

Plans, or "HMPs," each applicable to a particular biological unit or ecotype. Memoranda of Understanding have been signed with each of the eleven Western states; over 150 HMPs have been developed for those same states. Compliance with the Sikes Act Extension is effected by means of supplemental agreements to the existing Memoranda and HMPs.

[11] 43 U.S.C. §§ 315–315f, 315h-315r.

[12] *Id.* § 315.

[13] *See* 43 U.S.C. § 315b. Regulations regarding grazing administration are found at 43 C.F.R. Part 4100 (1995).

[14] 43 U.S.C. § 315a.

[15] *Id.* § 315h.

[16] *Id.* §§ 1411–1418 (1970) (omitted).

The Act listed the same basic management purposes, including fish and wildlife, as did the Multiple-Use Sustained-Yield Act four years earlier, but it added several other purposes, including industrial development, mineral production, and occupancy.[17] Moreover, just as the 1960 Act had declared itself to be "not in derogation of" the 1897 Organic Act, so too the Classification and Multiple Use Act of 1964 declared itself to be "[c]onsistent with and supplemental to the Taylor Grazing Act."[18] Thus BLM's discretionary management authority was at least as broad as, if not broader than, that of the Forest Service.

The multiple use standard imposes few objective limits on management discretion,[19] and those limits are legislative rather than judicial in origin. The most significant of these for the national forests, the Wilderness Act, initially did not apply to lands under BLM's jurisdiction.[20] Thus BLM multiple use management was, until the 1970s, virtually unfettered.

The Secretary initially promulgated regulations that required an "allocation of vegetation resources among livestock grazing, wild free-roaming horses and burros, wildlife, and other uses."[21] District Managers had almost complete discretion in determining to what extent available forage would be allocated to wildlife. The advisory boards with which they were to consult on such questions were composed primarily or exclusively of livestock operators, thus assuring a pro-grazing public input.[22]

In the last twenty-five years, congressional and presidential actions have narrowed the scope of BLM discretion under the Classification and Mul-

[17] *Id.* § 1411(a).

[18] *Id.*

[19] See discussion in Chapter 9 with regard to the multiple use standard and the National Forest System.

[20] No BLM lands were considered for inclusion in the National Wilderness Preservation System until 1976, when the Federal Land Policy and Management Act, 43 U.S.C. § 1782, directed a study of certain BLM lands to determine their suitability for wilderness designation. See discussion of the wilderness system in Chapter 8.

[21] 43 C.F.R. § 4110.2–2 (1981) (since revised).

[22] 43 U.S.C. § 315o-1. Advisory boards were to be composed of from five to twelve local stockmen and one wildlife representative. *Id.* The Federal Advisory Committee Act, Pub. L. No. 92–463, 86 Stat. 770, codified as an appendix to 5 U.S.C., terminated all existing BLM advisory committees as of January 5, 1975. BLM subsequently established a three-tiered system of advisory boards, the membership of which was to be "balanced in terms of the points of view represented and the functions to be performed." 40 Fed. Reg. 25452, 25454 (1975).

The Federal Land Policy and Management Act of 1976, 43 U.S.C. § 1753, authorized the reestablishment of "grazing advisory boards" for each BLM district office and National Forest headquarters office in the sixteen contiguous Western states having jurisdiction over more than 500,000 acres of land used for commercial livestock grazing. Until 1995, the boards were composed entirely of ranchers. *See* 43 C.F.R. § 1784.2-1(b) (1994) (since revised). In that year, the grazing advisory boards were replaced by Resource Advisory Councils, which have broader advisory duties and more balanced representation. *See* 43 C.F.R. § 1784.6 (1995). See discussion of current regulations *infra*, at text accompanying notes 56–64.

tiple Use Act of 1964. The most significant federal statutes are the National Environmental Policy Act of 1969 and the Federal Land Policy and Management Act of 1976. Another major legislative limitation on BLM discretion is the Wild Free-Roaming Horses and Burros Act.[23] Finally, an Executive Order in 1972 regulating the use of "off-road vehicles" (ORVs) on public lands has restricted management on certain BLM lands, particularly those in the California desert.[24]

The Federal Land Policy and Management Act of 1976

Closer public scrutiny of BLM management decisions has come about as a result of the Federal Land Policy and Management Act of 1976 (hereinafter "FLPMA").[25] Culminating years of efforts to give BLM an "organic act" defining its responsibilities and authorities, FLPMA has many similarities to the contemporaneous National Forest Management Act.

As its most basic mandate, FLPMA directs the Secretary of the Interior to develop and maintain "land use plans which provide by tracts or areas for the use of the public lands."[26] Plans must be based on an inventory of all public lands and their resources that the Secretary maintains on a continuing basis,[27] and they are to be coordinated with land use plans for the national forests.[28] Finally, the Act directs the Secretary, in the development of land use plans, to "use a systematic interdisciplinary approach to achieve integrated consideration of physical, biological, economic, and other sciences."[29]

FLPMA sets no definite timetable for the development of land management plans.[30] It does, however, require that "regulations and plans for the protection of public land areas of critical environmental concern be promptly developed."[31] These are areas where "special management attention is required . . . to protect and prevent irreparable damage to . . . fish and wildlife resources or other natural systems or processes."[32] The

[23]See discussion of the Wild Free-Roaming Horses and Burros Act and multiple-use lands in this chapter.

[24]See discussion of litigation under this Executive Order *infra* at text accompanying notes 123–150.

[25]43 U.S.C. §§ 1701–1784.

[26]*Id.* § 1712(a).

[27]*Id.* § 1711(a).

[28]*Id.* § 1712(b).

[29]*Id.* § 1712(c)(2). Regulations governing the planning process are found at 43 C.F.R. Part 1600 (1995).

[30]*But see* Natural Resources Defense Council v. Jamison, 815 F. Supp. 454 (D.D.C. 1992) (ordered BLM to establish a schedule for completing resource management plans for certain areas, after a delay of 15 years during which plans met only requirements of FLPMA transition period regulations).

[31]43 U.S.C. § 1701(a)(11).

[32]*Id.* § 1702(a).

concept of "areas of critical environmental concern" was an innovative idea offering the opportunity for environmentally sensitive land management without the severe strictures of wilderness areas.[33] Unfortunately, however, the concept has been little used.

At all stages of planning and implementation, the Act requires active public participation. It declares as a statement of policy that in exercising discretionary authority, the Secretary shall "be required to establish comprehensive rules and regulations after considering the views of the general public; and to structure adjudication procedures to assure adequate third party participation, objective administrative review of initial decisions, and expeditious decision making."[34] More specifically, FLPMA requires "public involvement"[35] in the formulation of plans and programs relating to management of the public lands[36] and in the promulgation of rules and regulations, in accordance with the Administrative Procedure Act, "to carry out the purposes of this Act and of other laws applicable to public lands."[37] The Act also requires opportunity for a public hearing in connection with all land withdrawals other than emergency withdrawals.[38]

While the National Forest Management Act contains detailed limitations on the scope of Forest Service management discretion, thus providing judicially enforceable standards, FLPMA has few precise standards. Instead, it substitutes Congress itself as the overseer of certain BLM management decisions. For example, by concurrent resolution of the House and Senate, Congress can veto proposed sales of tracts of the public lands exceeding 2,500 acres,[39] withdrawals of more than 5,000 acres,[40] the termination of certain preexisting withdrawals,[41] and management decisions

[33] *See* 43 C.F.R. § 1610.7–2 (1995).

[34] 43 U.S.C. § 1701(a)(5). See 43 C.F.R. Part 4, Subparts B and E (1995), for regulations governing appeal of BLM decisions.

[35] "Public involvement" is defined as "the opportunity for participation by affected citizens in rulemaking, decisionmaking, and planning with respect to the public lands, including public meetings or hearings held at locations near the affected lands, or advisory mechanisms, or such other procedures as may be necessary to provide public comment in a particular instance." *Id.* § 1702(d). See discussion of recent regulatory changes regarding public involvement *infra* at text accompanying notes 57–64.

[36] *Id.* § 1712(a),(f). Regulations governing public participation in resource management planning are found at 43 C.F.R. § 1610.2 (1995). For a discussion of the notice provision of the regulations, 43 C.F.R. § 1610.4–1 (1995), see National Parks and Conservation Ass'n v. Federal Aviation Administration, 998 F.2d 1523, 1531 (10th Cir. 1993) (*quoting* 43 C.F.R. § 1610.4–1: notice must be provided "at the onset of the planning process").

[37] *Id.* § 1740. For a discussion of public participation in BLM decisionmaking, see Joseph M. Feller, *Grazing Management on the Public Lands: Opening the Process to Public Participation,* 26 Land & Water L. Rev. 571 (1991).

[38] 43 U.S.C. § 1714(h).

[39] *Id.* § 1713(c).

[40] *Id.* § 1714(c)(1).

[41] *Id.* § 1714(*l*).

that totally eliminate one or more of the "principal or major uses"[42] for two or more years with respect to any tract of land of 100,000 acres or more.[43]

Congress amended FLPMA in 1978 to give specific guidance to the BLM on range management. In the Public Rangelands Improvement Act of 1978 (PRIA),[44] Congress declared a "national policy and commitment" to "manage, maintain and improve the condition of the public rangelands so that they become as productive as feasible for all rangeland values."[45] The definition of "range condition" includes wildife habitat,[46] as do "rangeland values."[47]

FLPMA also contains several measures specifically geared to wildlife. First, it directs that half the moneys received by the United States as fees for grazing livestock be put in a special fund to be spent solely for "range betterment," including "fish and wildlife habitat enhancement."[48] Second, it permits the Secretary to exchange public lands for private lands upon a determination that the exchange will serve the public interest, and it requires the Secretary to consider the fish and wildlife aspects of the proposed exchange in making this determination.[49] Finally, although FLPMA declares a general policy that the United States should "receive fair market value of the use of the public lands and their resources,"[50] it expressly does not authorize the Secretary to require "[f]ederal permits to hunt and fish on public lands or on lands in the National Forest System and adjacent waters."[51] It does, however, authorize the Secretaries of the Interior and Agriculture to designate areas under their jurisdiction "where, and establish periods when, no hunting or fishing will be permitted for reasons of public safety, administration, or compliance with provisions of applicable law."[52]

A final provision of FLPMA of substantial indirect significance for wildlife is its directive that the Secretary review all roadless areas of 5,000 acres or more for possible inclusion in the National Wilderness Preservation

[42]One of the six "principal or major uses" specified in the Act is "fish and wildlife development and utilization." *Id.* § 1702(*l*).

[43]*Id.* § 1712(e)(2).

[44]Act of Oct. 25, 1978, Pub. L. No. 95–514, 92 Stat. 1803, codified at 43 U.S.C §§ 1739, 1751–1753, and 1901–1908.

[45]*Id.* § 1901(b).

[46]*Id.* § 1902(d).

[47]*Id.* § 1901(a)(6).

[48]*Id.* § 1751(b)(1).

[49]*Id.* § 1716(a).

[50]*Id.* § 1701(a)(9).

[51]*Id.* § 1732(b).

[52]*Id.* See discussion of litigation regarding hunting and fishing administration, *infra* at text accompanying notes 154–164.

System. That review and the litigation it has spawned are considered in detail in Chapter 8.

The effects of FLPMA on BLM land management are similar to those of the National Forest Management Act of 1976 on the Forest Service, in that while both acts reaffirm the management principle of "multiple use,"[53] they limit the unfettered discretion that BLM and the Forest Service had previously enjoyed under that standard.[54] In FLPMA, more stringent congressional oversight provides the limitations, in contrast to the NFMA's enumeration of detailed standards and guidelines. A critical difference for wildlife is FLPMA's lack of a directive that BLM "provide for diversity of plant and animal communities," a provision of NFMA that has had a major impact on national forest management.[55] The two acts are similar in vastly expanding the opportunity for citizen participation in management decisions, perhaps FLPMA's greatest promise for wildlife conservation on BLM lands.

Meaningful citizen participation in BLM decisionmaking was not encouraged, however, until quite recently. For nearly two decades, regulations effectively limited public involvement to ranchers.[56] In 1995, the BLM revised its rules on public involvement, stating that "increased public participation is essential to achieving lasting improvements in the management of our public lands."[57] The regulations replaced district grazing advisory boards, which advised BLM officials on the development of allotment management plans and consisted entirely of ranchers, with Resource Advisory Councils.[58] Each Resource Advisory Council is to advise local BLM officials regarding "preparation, amendment and implementation of land use plans for public lands and resources within its area."[59]

The rule provides that each council "shall be representative of the interests of" three general groups: commercial interests, including grazing, transportation, developed recreation, and timber; environmental, dis-

[53] *See, e.g.*, 43 U.S.C. §§ 1701(a)(7), 1702(c), 1732(a).

[54] In Utah v. Andrus, 486 F. Supp. 995, 1003 (D. Utah 1979), however, the court strongly implied that FLPMA provides no standards against which the validity of individual management decisions can be gauged:

> If all the competing demands reflected in FLPMA were focused on one particular piece of public land, in many instances only one set of demands could be satisfied. A parcel of land cannot both be preserved in its natural character and mined. Thus, it would be impossible for BLM to carry out the purposes of the Act if each particular management decision were evaluated separately. It is only by looking at the overall use of the public lands that one can accurately assess whether or not BLM is carrying out the broad purposes of the statute.

[55] See discussion of NFMA's diversity provision in Chapter 9.

[56] For example, see note 22 *supra*, regarding grazing advisory boards.

[57] 60 Fed. Reg. 9894, 9895 (1995).

[58] *See* 43 C.F.R. § 1784.6 (1995).

[59] *Id.* § 1784.6–1(b).

persed recreation, archeological and historical, and wild horse and burro groups; and public officials, tribes, academicians, and the public at large.[60] All councils have from ten to fifteen members, but the rule allows flexibility in structuring the groups; three models are set forth.[61]

In addition to creating the new advisory committees, the amendments replaced the term "affected interests" with the term "interested public."[62] The rules now make clear that any party who writes to express concern about grazing on an allotment will be recognized as a member of the "interested public" and will be notified of and consulted on significant management decisions.[63] BLM must consult with the interested public on grazing permit issuance, renewal, or modification, increasing and decreasing permitted use, and development of activity plans and range improvement programs.[64]

The changes in BLM's public participation regulations were part of a comprehensive revision of grazing regulations. Other proposed revisions included strengthening the rangeland health standards for purposes of grazing administration, and an increase in the grazing fee.[65] Interior adopted the standards for determining rangeland health,[66] but not the grazing fee increase.[67] One long-time critic of BLM grazing management asserted that while the new rules were an improvement in grazing administration, both the proposed and final regulations suffered from a "fatal

[60]43 C.F.R. § 1784.6–1(c) (1995). The rule requires the Secretary to "provide for balanced and broad representation from within each category." *Id.* § 1784.6–1(d). Members must have "demonstrated a commitment to collaborate in seeking solutions to resource management issues," in addition to some experience or training with regard to relevant issues and knowledge of the geographical area involved. *Id.* § 1784.2–1(b).

[61]*Id.* § 1784.6–2. The rule also allows councils to form subgroups of "rangeland resource teams" and "technical review teams" to advise them. Technical review teams consist of federal employees and paid consultants and may be created for a particular task. They are to gather and analyze data and make recommendations to their council.

[62]43 C.F.R. § 4100.0–5 (1995).

[63]*Id.* In the previous rule, the authorized officer had discretion to determine whether or not a party was an "affected interest."

[64]60 Fed Reg. 9894, 9897 (1995).

[65]*Id.* at 9898–99.

[66]43 C.F.R. Subpart 4180 (1995). This subpart lists range conditions that must be met. For example, § 4180.1(d) requires that "[h]abitats are, or are making significant progress toward being, restored or maintained for Federal threatened and endangered species, Federal Proposed, Category 1 and 2 Federal candidate and other special status species." If the standards are not met, grazing practices must be modified no later than the start of the next grazing year. Section 4110.3–2 provides that the authorized officer "*shall* reduce permitted grazing use" when monitoring or field observation show grazing use is not consistent with subpart 4180 (emphasis added).

[67]Instead, Interior decided "not to promulgate the fee increase provision of the proposed rule in order to give the Congress the opportunity to hold additional hearings on this subject and to enact legislation addressing appropriate fees for grazing on public lands." 60 Fed. Reg. at 9899 (1995).

flaw'': the lack of "any provision for reviewing the public lands to identify those areas that are ecologically unsuitable for livestock grazing or those areas where the costs and environmental impacts of grazing are disproportionate to its economic benefits."[68]

Litigation

Most litigation against the BLM with significant implications for wildlife has involved the grazing program. Other major subjects of legal action have been forest management, off-road vehicle use, and hunting and fishing regulations.[69]

Grazing Management

Despite Congress's clear expressions of continuing concern about overgrazing and deteriorating range conditions,[70] only two court challenges to the BLM's grazing management have succeeded.[71] The first case—decided before FLPMA was enacted in 1976—involved a challenge under the National Environmental Policy Act. Other significant cases since FLPMA have involved both that statute and NEPA.

In 1974, BLM prepared a draft EIS for its entire livestock grazing program. The agency intended that the programmatic statement serve as the foundation for all subsequent actions implementing the program. The

[68]Feller, *supra* note 4, at 712.

[69]Cases involving the Wild Free-Roaming Horses and Burros Act are discussed in a separate section in this chapter, since the Act applies to both BLM and Forest Service lands. Note, however, that 90 percent of the horses are on BLM's arid lands, and all of the significant litigation to date has been against the BLM.

[70]Forty-four years after the Taylor Grazing Act was passed, Congress found that "vast segments of the public rangelands" were in "an unsatisfactory condition," and that these conditions "threaten[ed] important and frequently critical fish and wildlife habitat" and "prevent[ed] expansion of the forage resource and resulting benefits to livestock and wildlife production." 43 U.S.C. § 1901(a)(1),(3).

[71]The cases are Natural Resources Defense Council v. Hodel, 618 F. Supp. 848 (E.D. Cal. 1985) (regulation permitting cooperative agreements between the agency and grazing permittees was an abuse of discretion), discussed *infra* at text accompanying notes 90–93, and Natural Resources Defense Council v. Morton, 388 F. Supp. 829 (D.D.C. 1974), *aff'd* without opinion, 527 F.2d 1386 (D.C. Cir.), *cert. denied*, 427 U.S. 913 (1976) (EIS for BLM's grazing program inadequate). The court's opinion in Natural Resources Defense Council v. Hodel contains a detailed discussion of BLM grazing management from 1934 to 1985.

Conservation groups successfully challenged BLM grazing management in a recent administrative action. See discussion of National Wildlife Fed'n v. Bureau of Land Management, No. UT-06–91–01 (U.S. Dep't of the Interior, Office of Hearings & Appeals, Hearings Div., Dec. 20, 1993), petition for stay denied, 128 IBLA 231 (Mar. 1, 1994), petitioner's motion to dismiss intervenors' appeal for lack of standing granted in part, denied in part, 129 IBLA 124 (Apr. 13, 1994), *infra* at text accompanying notes 82–86.

Natural Resources Defense Council, however, argued that the EIS was insufficient as a basis for actual "on the ground" actions to be taken in each of BLM's fifty-two grazing districts. In *Natural Resources Defense Council, Inc. v. Morton,* the District Court for the District of Columbia agreed and ordered BLM to prepare individual statements for each of its more than 200 planning units.[72] The court found that the program-wide EIS "does not provide the detailed analysis of local geographic conditions necessary for the decision-maker to determine what course of action is appropriate under the circumstances."[73]

As a result of the 1974 *Natural Resources Defense Council* decision and congressional directive in FLPMA,[74] the BLM determined that it would prepare resource management plans by grazing districts. In 1985, the Natural Resources Defense Council challenged one of the district plans and its accompanying EIS as inadequate under both NEPA and FLPMA. While clearly sympathetic to the plaintiffs' claims, the court in *Natural Resources Defense Council v. Hodel*[75] felt constrained to uphold the BLM's actions.[76]

First, the court rejected the plaintiffs' contention that the law required the plan and EIS to be more specific for each grazing allotment[77] in the district with respect to numbers of livestock, seasons of use, and other components of range management. The court found that neither NEPA nor FLPMA/PRIA required this level of specificity.[78] With regard to FLPMA and PRIA, the court stated that "nowhere in the statutes did Con-

[72]388 F. Supp. 829 (D.D.C. 1974), *aff'd* without opinion, 527 F.2d 1386 (D.C. Cir.), *cert. denied,* 427 U.S. 913 (1976).

[73]*Id.* at 838–39. For subsequent related litigation, see Natural Resources Defense Council v. Andrus, 11 Envt'l Rep. Cas. 1523 (D.D.C. April 14, 1978).

[74]Two years after the district court's decision in *Natural Resources Defense Council v. Morton,* Congress passed FLPMA. As discussed previously, FLPMA requires the BLM to prepare land use plans "by tracts or areas." 43 U.S.C. § 1712(a).

[75]624 F. Supp. 1045 (D. Nev. 1985), *aff'd,* 819 F.2d 927 (9th Cir. 1987).

[76]The court stated: "Many of the complaints raised by plaintiffs have factual merit and suggest the possibility of either bad management or insensitivity to certain environmental concerns on the part of BLM. . . . [But] I am powerless to substitute my judgment for that of the BLM in these matters." 624 F. Supp. at 1047–48.

[77]Districts are further divided into "allotments" for administrative purposes. *See* 43 C.F.R. § 4100.0–5 (1995). The BLM prepares allotment management plans (AMPs) as well as district resource management plans. *See id.,* § 4120.2.

[78]624 F. Supp. at 1051, 1059. The court found that "[a] document addressing the ecological and other impacts for each set of permutations of stocking levels would be a completely unmanageable undertaking" and that Natural Resources Defense Council v. Morton did not require localized EISs. 624 F. Supp. at 1051. *But see* National Wildlife Fed'n v. Bureau of Land Management, No. UT-06–91–01 (U.S. Dep't of the Interior, Office of Hearings & Appeals, Hearings Div., Dec. 20, 1993), petition for stay denied, 128 IBLA 231 (Mar. 1, 1994), petitioner's motion to dismiss intervenors' appeal for lack of standing granted in part, denied in part, 129 IBLA 124 (Apr. 13, 1994), discussed *infra* at text accompanying notes 82–86.

gress describe in detail what sort of information must be included in a land use plan."[79] It was adequate, the court said, that the plan "set overall limits on grazing."[80]

The plaintiffs also charged the BLM with violating substantive mandates of FLPMA and PRIA, including failure to curb overgrazing and failure to assure improvement of the public rangelands. The court deferred to agency discretion with regard to methods of measuring range conditions and choice of remedy if overgrazing is found, stating that FLPMA and PRIA "contain only broad expressions of concern and desire for improvement" rather than standards the court could use to determine whether the plan was arbitrary and capricious under the Administrative Procedure Act.[81] Thus for many years the BLM's grazing program appeared to be effectively insulated from meaningful judicial review, despite the holding in *Natural Resources Defense Council v. Morton* and the passage of FLPMA.

In a landmark administrative decision almost ten years later, however, an administrative law judge (ALJ) in the Department of the Interior found that NEPA and FLPMA do require specificity in allotment management plans and a reasoned decision as to whether grazing is in the public interest in particular areas. The ALJ in *National Wildlife Fed'n v. Bureau of Land Management*[82] first held that the BLM violated NEPA by failing to prepare an EIS that analyzed the specific environmental impacts of livestock grazing in the Comb Wash Allotment in southeastern Utah.[83] He found that the EIS was inadequate because it lacked site-specific information about the allotment.[84] The ALJ further found that the BLM had violated FLPMA by permitting grazing in five environmentally sensitive canyons without making a "reasoned and informed decision that the ben-

[79]624 F. Supp. at 1059.

[80]*Id.*

[81]*Id.* at 1058. The court stated:

> Plaintiffs are understandably upset at what they view to be a lopsided and ecologically insensitive pattern of management of public lands at the hands of the BLM, a subject explored at length by many commentators. Congress attempted to remedy this situation through FLPMA, PRIA and other acts, but it has done so with only the broadest sorts of discretionary language, which does not provide helpful standards by which a court can readily adjudicate agency compliance.

Id. at 1062.

[82]No. UT-06-91-01 (U.S. Dep't of the Interior, Office of Hearings & Appeals, Hearings Div., Dec. 20, 1993), petition for stay denied, 128 IBLA 231 (Mar. 1, 1994), petitioner's motion to dismiss intervenors' appeal for lack of standing granted in part, denied in part, 129 IBLA 124 (Apr. 13, 1994).

[83]*Id. supra* note 78, at 17–22.

[84]*Id.* at 21. The ALJ found that the BLM failed to consider the impacts of grazing on, *inter alia*, riparian areas, vegetation, and wildife habitat; that at least 70 percent of wildlife species in the arid southwest are dependent on riparian areas; and that the health of the riparian areas depends largely upon vegetation. *Id.* at 8, 11.

efits of grazing the canyons outweigh the costs.''[85] He prohibited the BLM from authorizing grazing in the canyons unless and until the agency prepared an adequate EIS and made a reasoned and informed decision that grazing in the canyons is in the public interest.[86]

A comparison of *Natural Resources Defense Council v. Hodel* and *National Wildlife Federation v. Bureau of Land Management* reveals one factor that may explain, in part, the different outcomes. The *Natural Resources Defense Council* case was decided on motions for summary judgment, while there was a full-scale evidentiary hearing in *National Wildlife Federation.* The court in *Natural Resources Defense Council* noted that neither party requested a trial, and that based on the record, the court was ''unable to attempt the fine evaluation and weighing that would even bring me close to deciding that the BLM's decision was irrational.''[87] In contrast, in *National Wildife Federation* ''appellants presented overwhelming evidence that grazing has significantly degraded and may continue to significantly degrade the quality of the human environment.''[88] On the basis of the evidence presented, the ALJ made several specific factual findings as to the detrimental impacts of grazing on riparian areas, vegetation, wildlife habitat, recreation, visual quality, and archeological sites. The BLM's broad discretion in grazing management places a heavy burden of proof on those who ask a court to overturn an agency decision, and it appears that challengers must present specific and credible evidence of abuse of discretion in order to succeed.

In one case, the court had no trouble finding that BLM had abused its discretion and violated its duties under several statutes. In 1984, the BLM revised its grazing regulations to virtually eliminate what little oversight it historically had exercised with regard to grazing practices.[89] In a second case entitled *Natural Resources Defense Council v. Hodel,* the court invalidated several of the regulations.[90] A major focus of the plaintiffs' case was a regu-

[85] *Id.* at 23.

[86] *Id.* at 34. In another significant ruling, the ALJ, contrary to the court in Natural Resources Defense Council v. Hodel, found that the BLM's policy of requiring monitoring data that demonstrate excessive forage utilization or an unsatisfactory trend in vegetation before reducing livestock numbers violates FLPMA. *Id.* at 25.

For a detailed discussion of the BLM's policies on range monitoring and improvement of range conditions, see Feller, *supra* note 5, at 570–86. (The Feller article is an in-depth critique of BLM grazing management.) A 1995 amendment to BLM's grazing regulations provides that changes in permitted grazing levels may be justified by ''monitoring, ecological site inventory or other acceptable methods,'' rather than solely by narrowly-defined range monitoring data. 43 C.F.R. § 4110.3–2 (1995).

[87] 624 F. Supp. at 1062.

[88] National Wildlife Fed'n, *supra* note 78, at 4. For a discussion of the BLM's management of the Comb Wash allotment and the evidence presented at the hearing, see Feller, *supra* note 5 at 586–95.

[89] *See* 49 Fed. Reg. 6440 (1984).

[90] 618 F. Supp. 848 (E.D. Cal. 1985).

lation that allowed the BLM to enter into "Cooperative Management Agreements" (CMAs), in which the BLM allowed selected ranchers a free hand in managing their livestock grazing. The CMAs did not "contain specific performance standards such as numbers of animals or seasons of use," and they listed "no terms or conditions whatsoever which prescribe the manner in or extent to which livestock grazing shall be managed."[91] In an effort to give the permittee "secure tenure," the BLM would not evaluate performance for five years after the agreement was signed, and the permittee was automatically entitled to a renewal for ten years if the agreement's objectives were being met. The court noted that the BLM "forfeits any remedy for the rancher's failure to meet objectives except that of denying renewal of the CMA after ten years of non-compliance. This is secure tenure indeed."[92] The court agreed with the plaintiffs that this regulation was "a naked violation of defendants' affirmative duties under the Taylor Grazing Act, FLPMA, and PRIA."[93] The court concluded that the CMA program was "contrary to Congressional intent and was enacted without proper regard for the possible environmental consequences which may result from overgrazing on the public lands."[94]

The conclusion that follows from the two *Natural Resources Defense Council* cases is that while the BLM has wide discretion in regulating grazing, the agency must exercise at least some oversight to reduce overgrazing and its impacts. Further, the *National Wildlife Federation* case promises, at long last, meaningful review of BLM grazing management under FLPMA and NEPA. The combination of heightened judicial scrutiny and wider public participation as a result of the revised administrative regulations may signal a new era in BLM management. It remains to be seen whether the potential for change will be realized.

One other BLM grazing case with significance for wildlife was decided under an obscure statute passed to protect homesteaders over 100 years ago. In *United States ex rel. Bergen v. Lawrence* the court affirmed a district court order that the defendant remove a cattle fence he had constructed on his private land abutting BLM land.[95] The defendant had constructed a twenty-eight-mile fence enclosing over 20,000 acres of "checkerboard"

[91] *Id.* at 863.
[92] *Id.*
[93] *Id.* at 868. The court found that the CMA program

> disregards defendants' duty to prescribe the manner in and extent to which livestock practices will be conducted on public lands. The program also overlooks defendants' duty of expressly reserving, in all permits, sufficient authority to revise or cancel livestock grazing authorization when necessary.

Id.
[94] *Id.* at 852–53.
[95] 848 F.2d 1502 (10th Cir. 1988).

private, state, and federal lands in Wyoming.[96] The fence obstructed the passage of pronghorn antelope and excluded them from their winter range.[97]

The court found that the defendant had violated the Unlawful Inclosures of Public Lands Act[98] even though the fence was constructed on private land. The Act declares enclosures of federal lands to be unlawful and requires that such enclosures be removed.[99] The court ordered the defendant to remove or modify the fence to conform to BLM standards.[100]

The appellate court agreed with the district court's conclusion that this case was controlled by the 1897 United States Supreme Court case of *Camfield v. United States*,[101] a "virtually identical"[102] situation. The court rejected arguments that the removal order imposed a "servitude" on the plaintiff's lands, or granted the antelope an "easement" across them, which constituted a taking for which the plaintiff must be compensated. The court stated, "In declaring that the fence must be removed, the district court did not grant the antelope any easement across Lawrence's private lands, nor do we."[103] Instead, the court said, the issue was simply whether the fence unlawfully enclosed federal lands.

The defendant further argued that the Unlawful Inclosures of Public Lands Act was intended only to allow passage by people, not wildlife. The court rejected this argument as well. The court found that the Act was intended to prevent the obstruction of free passage for all lawful purposes over public lands, and that access for antelope to winter forage is a lawful purpose under FLPMA.[104]

Forest Management

Although the vast majority of BLM lands are arid lands on which the primary commercial activity is grazing, in western Oregon the BLM man-

[96]The defendant owned the private land and had grazing permits on the federal and state sections. The fence enclosed approximately 9,600 acres of public land, but it was constructed entirely on private lands, except where it crossed the common corners of state and federal sections.

[97]The court noted that the winter of 1983 was unusually severe and antelope collected against the fence and starved in an unsuccessful attempt to reach their winter range. *Id.* at 1504.

[98]43 U.S.C. §§ 1061–1066. The Act, passed in 1885 in response to "range wars," was intended to prevent the exclusion of homesteaders from the public lands. 848 F.2d at 1509.

[99]43 U.S.C. § 1061.

[100]BLM standards pursuant to the Taylor Grazing Act require that fences must be designed to allow antelope to go under and over the fence. *See* 848 F.2d at 1504, n. 4.

[101]167 U.S. 518 (1897).

[102]United States ex rel. Bergen v. Lawrence, 620 F.Supp. 1414, 1416 (D. Wyo. 1985).

[103]848 F.2d. at 1505.

[104]The court cited 43 U.S.C. § 1701(a)(8), which directs that "the public lands be managed in a manner . . . that will provide food and habitat for fish and wildlife and domestic animals."

ages 2.5 million acres of forest land that produced about 1 billion board feet of timber annually in the mid-1980s.[105] These lands were the focus of the first case, filed in 1987, in which conservationists sought to limit logging in the old-growth forest habitat of the threatened northern spotted owl. The cases against the Forest Service, which came to dominate the spotted owl litigation, are discussed in depth in Chapter 9. This section will briefly address the conservationists' NEPA challenge to old-growth forest management on BLM lands.

As previously noted, FLPMA contains no provision similar to NFMA's diversity requirement; thus the BLM has no express statutory or regulatory duty to provide for viable populations of wildlife on its lands.[106] Plaintiffs in the BLM spotted owl cases alleged violations of NEPA, the Migratory Bird Treaty Act,[107] the Oregon and California Lands Act,[108] and FLPMA. The NEPA claims proved to be the most significant in a series of cases paralleling those against the Forest Service. Conservation groups ultimately succeeded in obtaining an injunction against timber sales in spotted owl habitat on BLM lands until the BLM prepared an EIS in which it considered the potential impacts of logging on the spotted owl.[109]

In October 1987, conservation groups charged the BLM with failure to supplement their timber management plans and accompanying EISs, all of which had been prepared between 1979 and 1983, after new scientific information revealed increased threats to the survival of the spotted owl from continued logging of old growth.[110] Within two months of the filing of the lawsuit, however, Congress enacted the first "rider" to an appropriations bill limiting the court's authority to review federal timber management plans under environmental laws.[111] Section 314 of the Department of Interior Appropriations Act for Fiscal Year 1988 prohibited challenges to BLM timber management plans "solely on the basis that the

[105]These lands were granted to the railroads in the late 1800s and taken back by the federal government in 1916. The lands are subject to the Oregon and California Lands Act, 43 U.S.C. § 1181.

[106]The BLM does, of course, have a duty to comply with the Endangered Species Act, which applies to all lands and all "persons." The spotted owl was listed as threatened under the Act in 1990. See Chapter 7 for a discussion of ESA cases involving the spotted owl.

[107]16 U.S.C. §§ 703 *et seq.* See discussion of the Migratory Bird Treaty Act claims in the spotted owl cases in Chapter 4.

[108]43 U.S.C. § 1181.

[109]*See* Portland Audubon Society v. Babbitt, 998 F.2d 705 (9th Cir. 1993).

[110]Although NEPA does not expressly require supplementation of an EIS, implementing regulations require an agency to prepare a supplement if there "are significant new circumstances or information relevant to environmental concerns and bearing on the proposed action or its impacts." 40 C.F.R. § 1502.9(c)(1)(ii) (1996). The United States Supreme Court approved this interpretation of NEPA in Marsh v. Oregon Natural Resources Council, 490 U.S. 360, 372 (1989).

[111]See discussion of a later appropriations rider limiting judicial review of Forest Service and BLM timber management, known as "Section 318," in Chapter 9.

plan does not incorporate information available subsequent to the completion of the existing plan." [112] Section 314 did not, however, bar challenges to "particular activities" carried out under the plans, such as individual timber sales.

In April 1988, the district court judge relied on section 314 to dismiss the plaintiffs' complaint, finding that she had no jurisdiction to hear the case.[113] The Ninth Circuit reversed on the grounds that it was unclear whether Congress in section 314 had intended to bar the suit.[114] On remand, after a lengthy evidentiary hearing, the district court concluded that the BLM had failed to consider "new, significant, and probably accurate information" about the risk of extinction facing the spotted owl, and that the failure was arbitrary and capricious and violated NEPA.[115] However, the court again held that section 314 barred plaintiffs' NEPA claim. This time, the Ninth Circuit affirmed.[116]

Section 314 was reenacted for Fiscal Year 1990 but not for Fiscal Year 1991. In 1991, plaintiffs once again asked the court to find the BLM in violation of NEPA. After additional litigation regarding the continuing effect of section 314,[117] the district court granted the plaintiffs' motion for summary judgment and entered a permanent injunction against further BLM timber sales in spotted owl habitat until the agency complied with NEPA.[118] The Ninth Circuit affirmed.[119]

In 1994, the Forest Service and the BLM adopted a joint management plan for old-growth forests in western Washington and Oregon and northern California.[120] Both industry and conservation groups challenged the

[112]Pub. L. No. 100–202, § 314, 101 Stat. 1329–254 (1987).

[113]Portland Audubon Society v. Hodel, 18 Envt'l L. Rep. (Envt'l L. Inst.) 21210 (D. Or. 1988).

[114]Portland Audubon Society v. Hodel, 866 F.2d 302 (9th Cir.), *cert. denied,* 492 U.S. 911 (1989).

[115]Portland Audubon Society v. Lujan, 712 F. Supp. 1456, 1485 (D. Or. 1989).

[116]Portland Audubon Society v. Lujan, 884 F.2d 1233 (9th Cir. 1989), *cert. denied,* 494 U.S. 1026 (1990).

[117]*See* Seattle Audubon Society v. Evans, 952 F.2d 297, 303–04 (9th Cir. 1991) (finding that Section 314 expired on September 30, 1990).

[118]Portland Audubon Society v. Lujan, 795 F. Supp. 1489 (D. Or. 1992).

[119]Portland Audubon Society v. Babbitt, 998 F.2d 705 (9th Cir. 1993). For a discussion of the BLM spotted owl cases, see Victor M. Sher, *Travels with Strix: The Spotted Owl's Journey Through the Federal Courts,* 14 Pub. Land L. Rev. 41, 58–63 (1993).

In a case in which plaintiffs challenged a single timber sale in spotted owl habitat, the court held that the BLM had not violated NEPA or FLPMA by failing to prepare a site-specific supplemental EIS for the sale and that the BLM's decision to emphasize timber management over wildlife conservation in certain areas was not inconsistent with either FLPMA's multiple use mandate or the Oregon and California Lands Act. Headwaters v. Bureau of Land Management, 914 F.2d 1174 (9th Cir. 1990), *petition for rehearing en banc denied,* 940 F.2d 435 (1991).

[120]See discussion of the old-growth forest plan in Chapter 9.

plan, but the court upheld it.[121] Congress later enacted a rider to a 1995 Emergency Supplemental Appropriations Act that overrode many of the plan's provisions and reopened thousands of acres of old-growth forests to logging.[122]

Off-Road Vehicle Use

In the last twenty-five years the increasing use of off-road vehicles (ORVs) on public lands has spawned an executive order, administrative regulations, and litigation by both conservation groups and ORV enthusiasts over access to public lands by ORV users. Most litigation has focused on the California desert, an area singled out by Congress in FLPMA for special management because of its fragile ecosystem and increasing recreational use.[123]

A 1972 executive order directed various agency heads, including the Secretary of the Interior, to establish procedures for the designation of particular areas of land under their respective jurisdictions where ORVs would be permitted or prohibited.[124] In 1975, conservation groups successfully challenged BLM regulations implementing this executive order in *National Wildlife Federation v. Morton.*[125]

In *Morton,* the plaintiffs persuaded the court that BLM's implementing regulations failed to comply with the terms of the Order in several respects, among them by providing for blanket designations rather than site-specific designations and by failing to provide for adequate public participation. Of most interest for wildlife purposes, however, was the court's conclusion that the BLM regulations "significantly diluted the standards" of the Order.[126]

The court pointed to the Order's requirement that "[a]reas and trails shall be located to minimize harassment of wildlife or significant disruption of wildlife habitat," whereas the regulations provided only for "[c]onsideration of the need to minimize harassment of wildlife or significant disruption of wildlife habitat."[127] While the language might appear to be an innocent attempt to paraphrase the Order, the court believed that it reflected a subtle change of emphasis contrary to the Or-

[121]Seattle Audubon Society v. Lyons, 871 F. Supp. 1291 (W.D. Wash. 1994).

[122]Pub. L. No. 104–19, § 2001, 109 Stat. 240 (1995). See Chapter 9 for a discussion of litigation under this rider.

[123]*See* 43 U.S.C. § 1781.

[124]Exec. Order No. 11644, 37 Fed. Reg. 2877 (1972), as amended by Exec. Order 11989, 3 C.F.R. at 120 (1978 Comp.). BLM regulations provide that ORV use will be administered in accordance with these executive orders. *See* 43 C.F.R. § 420.25(a) (1995).

[125]393 F. Supp. 1286 (D.D.C. 1975).

[126]*Id.* at 1295.

[127]39 Fed. Reg. 13612, 13614 (1974).

der's strict mandate. In response to the suit, BLM revised the regulation to repeat verbatim the standards of the Order.[128]

Amendment of the regulation did not end the matter. In *American Motorcyclist Ass'n v. Watt*, the first of several cases involving ORV use in the California desert, conservation groups successfully challenged the ORV route designation criteria in the California Desert Conservation Area Plan as being contrary to FLPMA, the Order, and the Order's revised implementing regulation.[129]

The California Desert Conservation Area (CDCA) encompasses 25 million acres in southeastern California, 12 million of which are managed by the BLM. In FLPMA, Congress found that "the California desert environment is a total ecosystem that is extremely fragile, easily scarred, and slowly healed,"[130] and that its resources, including "certain rare and endangered species of wildlife, plants, and fishes,"[131] were seriously threatened by air pollution and increased use, especially recreational use. Congress designated the area for special management "to provide for the immediate and future protection and administration" of the area[132] and directed the Secretary to prepare a management plan for it.[133]

Plaintiff conservation groups challenged the ORV route designation criteria in the CDCA Management Plan, which differed from that of the Order and the regulation. The Plan's list of factors to be considered in approving an ORV route included whether the route would cause "considerable adverse impacts." As in *Morton*, the court found that the difference in language was significant. The court believed that "[t]he 'considerable adverse impacts' standard is qualitatively different than the minimization criteria mandated by [43 C.F.R.] § 8342.1 and in practice is almost certain to skew route designation decision-making in favor of ORV use."[134] The court concluded that the Plan was "very likely to result in a route selection process which does not comply in significant respects" with the regulation's express standards.[135] The court enjoined the BLM from approving any ORV route in the CDCA that did not comply with the regulation's criteria.

[128]The regulations, appearing at 43 C.F.R. Part 8340 (1995), add the requirement that "[s]pecial attention . . . be given to protect endangered species and their habitats." *Id.* at § 8342.1(b). For a more extended discussion of this controversy, see Rosenberg, *Regulation of Off-Road Vehicles*, 5 Envt'l Aff. 175 (1976).

[129]543 F. Supp. 789 (C.D. Cal. 1982). The American Motorcyclist Association and conservation groups challenged different aspects of the plan.

[130]43 U.S.C. § 1781(a)(2).

[131]*Id.* § 1781(a)(3).

[132]*Id.* § 1781(b).

[133]*Id.* § 1781(d).

[134]543 F. Supp. at 797.

[135]*Id.*

In the same case, the American Motorcyclist Association moved to enjoin the BLM from implementing the Management Plan, which, despite the route designation criteria, did restrict the use of ORVs in the conservation area. The district court denied a preliminary injunction request even though it found that the Association was likely to prevail on its claim that the BLM committed procedural violations of FLPMA in adopting the Plan.[136] The Ninth Circuit affirmed denial of the injunction. Citing Congress's concern for the fragile desert environment, the court found that "the public interest in protecting and managing the CDCA would be severely disserved by enjoining the Plan."[137]

The next round in the California desert litigation involved a challenge by conservation groups to the BLM's decision to permit ORV use in a particular area. In *Sierra Club v. Clark*,[138] plaintiffs sued to compel the BLM to close the 5,500-acre Dove Springs Canyon area to ORV use, an area of "rich and varied biota."[139] Prior to adoption of the CDCA Management Plan, the entire area had been open to ORV use. This use had steadily increased and had "been accompanied by severe environmental damage."[140] However, the Plan maintained unrestricted use of ORVs in 3,000 of the canyon's 5,500 acres.

Plaintiffs alleged that the BLM's failure to close the Dove Springs area to ORV use violated Executive Order 11644 and its implementing regulation, as well as FLPMA's requirements that the Secretary prevent "unnecessary or undue degradation" of public lands[141] and that he or she maintain and conserve resources of the CDCA.[142] The court rejected these arguments and held that the management plan's accommodation of ORV use was not unreasonable, given Congress's express statement that ORV use was to be permitted "where appropriate."[143] The plaintiffs had argued that it is unreasonable for the Secretary to conclude that ORV use is "appropriate" when that use "substantially impairs productivity of renewable resources and is inconsistent with maintenance of environmental quality."[144] The court stated:

> We can appreciate the earnestness and force of Sierra Club's position, and if we could write on a clean slate, would prefer a view which would disallow the

[136]American Motorcyclist Ass'n v. Watt, 534 F. Supp. 923 (C.D. Cal. 1981), *aff'd*, 714 F.2d 962 (9th Cir. 1983).

[137]American Motorcyclist Ass'n v. Watt, 714 F.2d 962, 967 (9th Cir. 1983).

[138]756 F.2d 686 (9th Cir. 1985).

[139]The evidence showed that "Dove Springs Canyon possesses abundant and diverse flora and fauna. Over 250 species of plants, 24 species of reptiles, and 30 species of birds are found there. It also offers good habitat for the Mojave ground squirrel, the desert kit fox, and the burrowing owl." *Id.* at 688.

[140]*Id.*

[141]43 U.S.C. § 1732(b).

[142]*Id.* §§ 1781(b) and (d).

[143]*Id.* § 1781(a)(4).

[144]756 F.2d at 691.

virtual sacrifice of a priceless natural area in order to accommodate a special recreational activity. But we are not free to ignore the mandate which Congress wrote into the Act.[145]

The court held that the Secretary had not abused "the broad discretion committed to him by an obliging Congress."[146]

The Ninth Circuit also rejected a challenge to an amendment to the CDCA Management Plan that permitted an ORV race from Barstow, California, to Las Vegas, Nevada, through the Conservation Area.[147] The BLM had permitted the race annually from 1967 to 1974 but refused to issue a permit starting in 1975, on the basis of adverse impacts to desert resources from previous races. Each year the permit was denied, ORV users engaged in " 'protest' rides" across the old race routes, with "considerable impacts on the biological, geological, cultural and other resources of the desert environment."[148]

In 1983, the BLM amended the Management Plan to permit the race. The agency had prepared an EIS and included specific mitigation measures to reduce environmental impacts. Plaintiffs challenged the Plan amendment as a violation of FLPMA and the Executive Order. Citing the earlier *Sierra Club* decision, the court held that the amendment was "a proper exercise of the BLM's discretion in providing for combined use of the desert."[149] The court noted that the BLM had attempted to "balance desired use and ecological concerns through the permit and mitigation requirements."[150]

In 1994, Congress finally resolved most of the long-standing controversy over the California desert by passing the California Desert Protection Act.[151] The Act classified more than 7.5 million acres of the California desert as wilderness, national parks and other protected areas.[152] Over 3.5 million acres of BLM land received wilderness designation.[153]

[145]*Id.* The court noted that the plaintiffs' interpretation of the language "would inevitably result in the total prohibition of ORV use because it is doubtful that any discrete area could withstand unrestricted ORV use without considerable adverse effects." *Id.*

[146]*Id.*

[147]Sierra Club. v. Clark, 774 F.2d 1406 (9th Cir. 1985).

[148]*Id.* at 1408.

[149]*Id.* at 1409.

[150]*Id.* at 1409–10. It appears that the protest rides were a significant factor in the court's—and perhaps the BLM's—decision. The court observed that "[w]hile there is little doubt that negative impacts resulted from the 1983 race, so is there little doubt that harm would result if uncontrolled 'protest rides' were to continue." *Id.* at 1410.

[151]Pub. L. No. 103–433, 108 Stat. 4471 (1994).

[152]The Act constitutes the largest protective land withdrawal since the passage of the Alaska National Interest Conservation Lands Act of 1980.

[153]Pub. L. No. 103–433, § 102, 108 Stat. at 4472–81 (1994). Motorized transport is prohibited in wilderness. See discussion of the Wilderness Act in Chapter 8.

Hunting and Fishing Regulations

As previously noted, section 302(b) of FLPMA[154] authorizes the Secretaries of Agriculture and the Interior to prohibit hunting and fishing in designated areas "for reasons of public safety, administration, or compliance with provisions of applicable law." According to the Conference Report, "[t]he word 'administration' authorizes exclusion of hunting and fishing from an area in order to maintain supervision. It does not authorize exclusions simply because hunting and fishing would interfere with resource-management goals."[155] However, in introducing the conferees' bill to the Senate, Senator Metcalf disputed this language and insisted that the BLM and the Forest Service had the authority to close an area to hunting and fishing to protect declining wildlife populations.[156]

Congressional intent with respect to section 302(b) figured prominently in a complex series of cases involving a state program of wolf reduction on federal lands in Alaska. In *Defenders of Wildlife v. Andrus,* the plaintiffs

[154]43 U.S.C. § 1732(b).

[155]H.R. Rep. No. 94–1724, 94th Cong., 2d Sess. 60 (1976), *reprinted in* 1976 U.S.C.C.A.N. at 6231.

[156]Senator Metcalf's precise statement was as follows:

> Unfortunately, in attempting to define the term "administration," the statement of managers confuses the issue and could be wrongly interpreted to prevent the Secretary from protecting the public lands.
>
> Traditionally, the States have regulated fishing and hunting of resident species of wildlife. The BLM and the Forest Service have not attempted to manage resident species of wildlife, but have focused on management of their habitat. This bill does nothing to change that. However, as a property owner the Federal Government has certain rights, and those rights have been upheld by the Supreme Court, most recently in Kleppe against New Mexico. . . .
>
> The conference report does not in any way surrender Congress' power.
>
> The language of the statement of the managers could be interpreted as so narrowing the definition of 'administration' that the agency would be unable to close an area to hunting even where the number of a species is drastically reduced. Carried further this language could be interpreted to mean that an area which was used for habitat research could not be closed to hunting or fishing "simply because hunting and fishing would interfere with resource management goals."
>
> In this legislation for the first time we are giving BLM basic statutory authority to manage the public lands on a multiple-use basis. Two of those uses are hunting and fishing, but they should not take precedence over all other uses. Further, it makes no sense to give an agency authority and then to tie its hands.
>
> When this matter was discussed by the conferees, the right—indeed the responsibility—of BLM and the Forest Service to manage wildlife habitat was agreed to by all. I believe the language in the statement of managers could be interpreted differently and thus does not accurately reflect the conferees' agreement on this issue.

122 Cong. Reg. 34511 (1976).

alleged that under that section the Secretary of the Interior had the authority to halt the state program and, having that authority, the duty under the National Environmental Policy Act to prepare an environmental impact statement before permitting it through acquiescence.[157] The district court, in ruling on plaintiffs' motion for preliminary injunction, agreed with both contentions.[158]

The state of Alaska, which had elected not to join the earlier litigation, later instituted its own action in Alaska against the Secretary of the Interior, challenging the lawfulness of actions taken by the Secretary to carry out the former court's order.[159] The Alaska court, though it initially concluded on the basis of legislative history that the Secretary lacked authority to halt the wolf kill, ultimately decided that the statute was clear on its face that such authority existed.[160] On the question of NEPA's applicability, however, the Alaska court disagreed with the former court, reasoning that more than mere federal acquiescence in the state program was necessary to trigger NEPA's requirements.

These issues reached the courts of appeals for both the Ninth and District of Columbia Circuits. Although the federal government had by then reversed its position and concluded that it did, indeed, have authority to halt the wolf kill, neither court ruled on that issue. Instead, both courts agreed that whether the authority existed or not, the Secretary had no duty under NEPA to prepare an environmental impact statement in the absence of an affirmative federal action authorizing the wolf kill.[161] The Ninth Circuit based its conclusion on an analysis of NEPA alone. The District of Columbia Circuit reasoned more broadly from the relationship of NEPA and FLPMA that the "cautious and limited permission to intervene in an area of state responsibility and authority" provided by section 302(b) of FLPMA did not impose "such supervisory duties on the Secretary that each state action he fails to prevent becomes a 'Federal action.' "[162]

In the final footnote to its opinion, the District of Columbia Circuit suggested the possibility of a violation of FLPMA independent of its relationship with NEPA.[163] Though the footnote did not elaborate upon the matter, the court appeared to refer to the wolf kill as a possible violation of FLPMA's substantive duty to "manage the public lands under principles

[157]7 Envt'l L. Rep. (Envt'l L. Inst.) 20225 (D.D.C. Feb. 14, 1977).

[158]*Id.* at 20228–29. The court relied heavily on the legislative history quoted in note 156 *supra.*

[159]Alaska v. Andrus, 429 F. Supp. 958 (D. Alas. 1977).

[160]*Id.* at 962.

[161]Alaska v. Andrus, 591 F.2d 537 (9th Cir. 1979); Defenders of Wildlife v. Andrus, 627 F.2d 1238 (D.C. Cir. 1980). The latter case was the appeal from an action filed after dismissal of the earlier case by the same name cited in note 159 *supra.*

[162]627 F.2d at 1250.

[163]*Id.* at 1250 n. 10.

of multiple use and sustained yield.''[164] The court declined to express any view on that but clearly raised the possibility of substantive review under FLPMA of federal land management actions.

THE WILD FREE-ROAMING HORSES AND BURROS ACT AND FEDERAL MULTIPLE USE LANDS

Introduction

One of the most significant federal statutes that place additional duties on federal multiple use land managers is the Wild Free-Roaming Horses and Burros Act (Wild Horses Act).[165] The Act applies to all lands under the jurisdiction of the Forest Service and the BLM. Over 95 percent of wild horses and burros live on BLM lands, with the remainder on Forest Service lands.[166]

Congress passed the Wild Free-Roaming Horses and Burros Act in 1971 to protect these ''living symbols of the historic and pioneer spirit of the West,'' which were thought to be disappearing.[167] In the early 1900s, the West was home to an estimated 2 million wild horses. By the 1950s, their numbers had dropped to fewer than 20,000. Under the Act's protection,[168] their population had grown to an estimated 55,000 in 1992.[169]

The success of the Act in increasing the population of wild horses has, not surprisingly, resulted in greater competition for forage and water between wild horses and livestock. Much of the later litigation under the Act has dealt with the problem of ''excess'' horses and has been based on 1978 amendments that increased the Secretary's authority to remove such horses.

[164]43 U.S.C. § 1732(a).

[165]16 U.S.C. §§ 1331–1340. The constitutionality of the Wild Horses Act, upheld in Kleppe v. New Mexico, is discussed in Chapter 2.

[166]Betsy A. Cody, Congressional Research Service, Wild Horse and Burro Management CRS-2 (Report No. 93–521 ENR, May 24, 1993).

[167]16 U.S.C. § 1331. Interestingly, wild horses are not native to the Western range. They are descended from horses brought to North America by Spanish explorers. They gradually mixed with domestic horses that had escaped from or were released by settlers and native Americans. Cody, *supra* note 166, at CRS-1.

Prior to the law's enactment in 1971, the only protection that wild horses and burros received from federal land managers was the limited protection afforded by BLM's ''wild horse policy'' initiated in 1967. That policy was set forth in Instruction Memo 67–361 (Sept. 5, 1967). Among other things, it required that consideration ''be given to reservation of forage for wild horses or burros where it is needed and is definitely in the public interest.''

[168]In addition to directing the Secretaries of Agriculture and Interior to manage and protect wild horses and burros, the Act provides criminal penalties for any person who maliciously kills or harasses a wild horse or burro, among other acts. *See* 16 U.S.C. § 1338.

[169]Cody, *supra* note 166, at CRS-1,4.

The 1971 Act and Early Litigation

The initial litigation under the Act concerned the obscure and arguably conflicting management standards in the original legislation.[170] On the one hand, the Act's directive that protected horses and burros "be considered . . . an integral part of the natural system of the public lands" suggested simply that they were to be given the same "due consideration" as other types of resources under traditional multiple use management. This interpretation was consistent with at least one reading of the Act's legislative history, which is that the Act was intended to give a previously lacking "status" to horses and burros.[171] The only time the original Act specifically referred to multiple use management, however, was in connection with those ranges that the Secretary was authorized to designate as sanctuaries for wild horses and burros[172] and which were to be "devoted principally but not necessarily exclusively to their welfare in keeping with the multiple-use management concept for the public lands."[173]

On the other hand, parts of the 1971 Act implied that protected horses and burros were to have a "first among equals" status. For example, the Act directs that "[a]ll management activities shall be at the minimal feasible level."[174] Moreover, in a provision substantially amended in 1978, the Act authorized the Secretary to destroy protected horses and burros only when "such action is necessary to preserve and maintain the habitat in a suitable condition for continued use."[175] When an area was found to be overpopulated, horses and burros could not be destroyed for this reason unless destruction was "the only practical way to remove excess animals from the area."[176]

Even if Congress intended to afford wild horses and burros a first among equals status, it clearly did not contemplate that the protected animals would displace all other types of range users over a continually expanding area. The Act's directive that they be considered an integral part of the natural system of the public lands applies only "in the area

[170]To some extent, 1978 amendments clarified the standards. *See* Pub. L. No. 95–514, § 14, 92 Stat. 1810, amending 16 U.S.C. §§ 1332(f), 1333(b).

[171]*See* H.R. Rep. No. 92–480, 92d Cong., 1st Sess. at 4 (1971). The Senate report suggests, however, that the purpose of the Act, in part, was to codify the preexisting practice of considering the interests of wild horses and burros in making multiple use decisions. *See* Sen. Rep. No. 92–242, 92d Cong., 1st Sess. at 3 (1971).

[172]16 U.S.C. § 1333(a).

[173]*Id.* § 1332(c). The authority to designate wild horse ranges is "purely discretionary with the Secretary," Dahl v. Clark, 600 F. Supp. 585, 595 (D. Nev. 1984), who has little utilized the authority.

[174]*Id.* § 1333(a).

[175]16 U.S.C. § 1333(c) (1976) (current version at 16 U.S.C. § 1333(b)).

[176]*Id.*

where presently [1971] found."[177] Moreover, the Act carefully prescribes that designated ranges include "the amount of land necessary to sustain an *existing herd* or herds of wild free-roaming horses and burros, *which does not exceed their known territorial limits.*"[178] Finally, the Act contains an express denial of any authority "to relocate wild free-roaming horses or burros to areas of public lands where they do not presently exist."[179]

The first decisions under the original Act reflected its dichotomous nature. In *American Horse Protection Association, Inc. v. Frizzell,* the plaintiff challenged the proposed roundup of 400 wild horses the BLM claimed were in excess of the grazing capacity of Stone Cabin Valley, Nevada.[180] Licensed livestock in the Valley accounted for more than 75 percent of its estimated grazing capacity. Of the remainder, BLM had reserved nearly half for wildlife. The other half was deemed insufficient to support the 900 to 1200 wild horses living in the Valley.

The plaintiff contended that the proposed roundup violated both the Wild Horses Act and the National Environmental Policy Act. The NEPA claim was confused somewhat by the fact that as a result of the 1974 *Natural Resources Defense Council* case previously discussed, BLM was preparing an EIS with regard to the overall management of an area that encompassed Stone Cabin Valley.[181] The court assumed that when the BLM completed the EIS it would contain a long-term solution to the overgrazing problem, and thus the court was willing to view the proposed roundup "as an interim measure to preserve the range" pending the EIS's completion.[182] Focusing narrowly on the roundup alone, the court concluded that NEPA was not applicable because the roundup would not have a significant impact on the Stone Cabin Valley environment. The court suggested it would rule otherwise only if it were satisfied that "the proposed roundup would extinguish the wild horse population" there.[183]

As to the claimed violation of the Wild Horses Act, the court emphasized the broad discretion the Act vests in the Secretary and rejected plaintiff's assertion that under the Act "wild horses were given a higher priority on the public lands than other grazers."[184] The court, by stating that the only statutory or regulatory support for plaintiff's assertion was an administrative regulation authorizing the closure of certain areas to livestock grazing when it was necessary to allocate all available forage to horses and

[177] *Id.* § 1331.
[178] *Id.* § 1332(c) (emphasis added).
[179] *Id.* § 1339.
[180] 403 F. Supp. 1206 (D. Nev. 1975).
[181] *See* text accompanying notes 72–73 *supra.*
[182] 403 F. Supp. at 1219 n. 10.
[183] *Id.* at 1219.
[184] *Id.* at 1220.

burros, refused to read into the Act the directive that wild horses and burros be given special status generally.[185] Moreover, the court believed that the regulation to which it referred applied only "to a situation where wild horses and burros were in danger of extinction."[186]

In short, the court held that under the principle of multiple use, "neither wild horses nor cattle possess any higher status than the other on the public lands."[187] Underscoring how little room for judicial supervision exists under that standard, the court stated that it probably would not have overturned the Secretary's decision if the decision had been to remove only cattle and no horses.[188]

The second case under the Act reached a strikingly different result. That action, *American Horse Protection Ass'n, Inc. v. Kleppe,* involved essentially similar facts.[189] The BLM proposed a roundup of wild horses near Challis, Idaho, because, the agency claimed, the range was being overutilized by cattle, sheep, wild horses, and other wildlife. This time, however, District Court Judge Richey was unwilling to rest his decision on broad generalizations about the extent of BLM's discretion under the Act. Instead, he found that BLM's estimates of horse populations in the affected area were "unreliable" and "meaningless" and that the proposed roundup was for that reason alone arbitrary and capricious.[190] In addition, however, he found that the Act's requirement that "[a]ll management activities shall be at the minimal feasible level" meant that "careful and detailed consideration must be given to *all* alternative courses of action that would have a less severe impact on the wild horse population."[191] The only alternative involving reduced livestock grazing the BLM had considered was the elimination of all grazing in the region. Judge Richey insisted that the BLM consider other, less drastic, alternatives, including eliminating livestock grazing in the horses' critical winter range.[192]

Judge Richey pointed to still other reasons to void the proposed roundup. One was that the failure to have "professional veterinary assistance on-site at all times during the roundup" or "to develop contingency plans for veterinary assistance in case of unforeseen circumstances during the roundup" violated the Act's requirement that horse removals be con-

[185] *Id.* The current version of the regulation cited by the court appears at 43 C.F.R. § 4710.5 (1995).

[186] 403 F. Supp. at 1220.

[187] *Id.* at 1221.

[188] *Id.* The court noted that in fact the cattle ranchers using the area had "voluntarily" decided to remove both cattle and horses from the Valley. *Id.* n. 12. It is impossible to tell to what extent, if at all, this factor influenced the court's decision.

[189] 6 Envt'l L. Rep. (Envt'l L. Inst.) 20802 (D.D.C. Sept. 9, 1976).

[190] *Id.* at 20803, 20804.

[191] *Id.* at 20804 (emphasis in original).

[192] *Id.*

ducted under "humane conditions and care."[193] Finally, as in the Nevada case the year before, the BLM was preparing an EIS concerning the entire domestic livestock grazing program in the Challis areas. Rather than permit the roundup to proceed as an "interim measure" pending completion of the EIS, however, Judge Richey insisted that to do so "would eliminate one major alternative to the grazing allotments plan proposed" in the BLM's draft EIS, thus violating NEPA.[194]

The 1978 Amendments and Subsequent Litigation

As a result of amendments in 1978, the Secretaries of the Interior and Agriculture now enjoy arguably clearer authority to remove and destroy excess animals. Amended section 3(b)(2) of the Act directs that the Secretary "immediately remove excess animals" whenever he or she makes two determinations: (1) that an overpopulation exists on a given area, and (2) that action is necessary to remove excess animals.[195] This provision imposes two separate duties on the Secretary and suggests that he or she may find an overpopulation to exist, yet find that action is not necessary to remove excess animals. On the other hand, the Act now defines "excess animals" to mean those that "must be removed . . . in order to preserve and maintain a thriving natural ecological balance and multiple-use relationship" in a particular area.[196] Thus, if animals meet this definition, then the Secretary would seem obliged always to order their removal, notwithstanding the wording of section 3(b)(2).[197]

The definition of "excess animals" has further internal ambiguities of its own. Since wild horses and burros are non-native, feral animals, it is unclear how a "natural" ecological balance is to be preserved and maintained; it is similarly interesting to contemplate the potential hurdles that confront any effort to maintain both a "thriving natural ecological balance" and "multiple-use relationship."[198]

Following passage of the 1978 amendments, BLM asked Judge Richey to lift his injunction entered two years earlier. In an unreported opinion,

[193] *Id.*

[194] *Id.*

[195] 16 U.S.C. § 1333(b)(2).

[196] *Id.* § 1332(f).

[197] The current regulation assumes the latter interpretation: "Upon examination of current information and a determination by the authorized officer that an excess of wild horses or burros exists, the authorized officer *shall remove* the excess animals immediately." 43 C.F.R. § 4720.1 (1995) (emphasis added).

[198] The Secretaries were to receive independent advice on how to determine when excess animals exist from a special report required by amended section 4(b)(3), 16 U.S.C. § 1333(b)(3). The report was submitted in 1980. *See* National Science Foundation, Wild and Free-Roaming Horses and Burros: Current Knowledge and Recommended Research (National Academy Press 1980).

he refused because BLM had failed to comply with his order to give serious consideration to protecting the horses' winter range by restricting cattle grazing. The court of appeals agreed that BLM's efforts to fulfill the requirements of the injunction had been "at best, halfhearted" and inadequate.[199] However, the appellate court concluded that Judge Richey had not properly interpreted the 1978 amendments, which, in the appellate court's view, reflected a congressional purpose to "cut back on the protection the Act affords wild horses"[200] by giving the Secretary broad discretion to determine when a wild horse overpopulation exists and directing him to remove excess animals promptly. Rather than lift the district court's injunction, however, the court of appeals remanded for a decision based on the following holding:

> Today we hold only that further consideration of the "winter range" alternative, on which the district court conditioned removal of horses in its 1976 injunction, is, in light of 1978 legislation, not required. It remains open to the district court to determine on remand whether, in light of the goals of the Act as it now stands, and on the basis of the information the Secretary now has, the Agency's current plan to reduce the size of the wild horse herd well below the 340 animals the winter range can support is rationally grounded.[201]

Thus, though circumscribed, the district court was left with limited authority to continue the injunction.

None of the potential ambiguities stemming from the 1978 amendments was addressed in the only other subsequent case involving a challenge to an excess animal removal program. Instead, in *American Horse Protection Ass'n v. Andrus,* the only question was whether the National Environmental Policy Act required that an EIS be prepared for the challenged program.[202] The trial court held that it did not, apparently because of its view that the grazing district EISs being prepared pursuant to the *Natural Resources Defense Council* case totally satisfied the agency's obligation under NEPA.[203] In reversing, the court of appeals did not determine whether an EIS was in fact required, but, at least implicitly, held that the

[199] American Horse Protection Ass'n, Inc. v. Watt, 694 F.2d 1310, 1315 (D.C. Cir. 1982).

[200] *Id.* at 1316.

[201] *Id.* at 1319 (footnote omitted). In a well-reasoned dissenting opinion, Judge Robinson attacked the majority's holding as creating a meaningless semantic distinction. He argued that the district court "could easily have found that the Bureau's intended course of action was not 'rationally grounded' upon the data it had because its deficient consideration of the winter range alternative represented an 'irrational' evaluation of that data," and thus have satisfied the standard articulated by the majority. *Id.* at 1322.

[202] 608 F.2d 811 (9th Cir. 1979).

[203] 460 F. Supp. 880 (D. Nev. 1978).

duty to remove excess animals imposed by the 1978 amendments did not function as an exemption from NEPA.[204]

In a case that was the flip side to the horse protection groups' challenges to BLM horse removal programs, cattle ranchers unsuccessfully sought mandamus directing the BLM to reduce the wild horse population in three grazing allotments in Nevada to 1971 levels. In *Dahl v. Clark*, evidence indicated that the wild horse population in the three allotments had increased tenfold, from 62 in 1971 to 655 in 1984.[205] Almost all of BLM's surveys and studies of range conditions from 1971 to 1982 concluded that the range condition's trend was downward, "mandating a reduction in both livestock and wild horse use."[206]

The case centered on a 1981 directive from then-Secretary of the Interior James Watt that rejected as inaccurate the methods of determining range conditions that the BLM had used for the previous ten years. Secretary Watt ordered BLM officials to maintain numbers of livestock and wild horses at 1981 levels and to start new monitoring studies to determine range utilization.[207] The court observed:

> Although the reason for Secretary Watt's directive is not clear, it appears to the Court that it most likely resulted from the fact that the previous studies indicated that use of the public domain by livestock and by wild horses would have to be drastically reduced due to damage to the range caused by overutilization.[208]

In an ironic turn, the ranchers argued that the previous studies were accurate and that their conclusion of deteriorating range conditions was borne out by plaintiffs' witnesses.[209] Based on the evidence at trial, the court agreed with the ranchers and concluded that "the ranges in question are substantially overused and . . . the environment on the allotments

[204]The Court refused to address the claims of appellants that, even if no EIS were required, the challenged action would represent an abuse of the Secretary's discretion under the Wild Horses Act, or of appellees that the Secretary's discretion was absolute. *See* 608 F.2d at 815 n. 2. In light of this refusal, the trial court's conclusion that in an area that constituted a "checkerboard" of unfenced private and public lands and in which most of the private landowners had requested removal of wild horses, removal of all wild horses in the area was the only practical way in which compliance with section 4 of the Act could be achieved, has little significance. *But see* Mountain States Legal Foundation v. Andrus, 16 Envt'l Rep. Cas. (BNA) 1351 (D. Wyo. April 21, 1981).

[205]600 F. Supp. 585, 586 (D. Nev. 1984).

[206]*Id.*

[207]*Id.*

[208]*Id.* at n. 1.

[209]The court noted that when an EIS prepared in response to the Natural Resources Defense Council case mandated reductions in livestock use in one of the allotments, one of the plaintiffs' expert witnesses was able to persuade the BLM not to make the planned reductions because range conditions were acceptable. *Id.* at 590.

has been severely damaged, because of wild horse and livestock use."[210] The court found that the areas in question were "not in a thriving, ecological condition" and thus the BLM was not carrying out its mandate under the Wild Horses Act. The court held that the decision to maintain wild horses at 1981 levels was arbitrary.[211]

The court was, however, unable to grant the remedy the ranchers sought. The plaintiffs made it "absolutely clear" that they sought to have the wild horse populations "reduced to 1971 levels, and to no other levels."[212] The court held that mandamus would not lie because

> the test as to appropriate wild horse population levels is whether such levels will achieve and maintain a thriving, ecological balance on the public lands. Nowhere in the law or regulations is the BLM required to maintain any specific numbers of animals or to maintain populations in the numbers of animals existing at any particular time.[213]

In another case brought by ranchers, plaintiffs sought to compel the BLM to take measures to prevent wild horses from straying onto private land. In *Fallini v. Hodel*,[214] the court held that while the Wild Horses Act creates a duty on the part of BLM to remove horses from private land after notification that they have strayed,[215] it does not imply a duty to take measures—such as erecting fences—to prevent their straying in the first place. The court found that the legislative history of the Act reveals Congress's desire "not to rely on fenced ranges and to keep management at a minimum."[216]

In the same case, the Fallinis sought to overturn a BLM decision that they violated their 1967 range improvement permit by fencing out wild horses from wells they had installed in 1983 on BLM land.[217] The issue was whether wild horses were "wildlife" within the meaning of the Fallinis' permit, which authorized them to install wells on public land but required that the wells be available for wildlife use. The Fallinis had in-

[210] *Id.* at 592.

[211] *Id.*

[212] *Id.* at 593.

[213] *Id.* at 595. The court stated, however, that "[i]f plaintiffs were looking for a mandamus requiring reduction of wild horse levels to levels other than to 1971, this case might come out differently as to this issue." *Id.* at 595.

[214] 783 F.2d 1343 (9th Cir. 1986).

[215] 16 U.S.C. § 1334.

[216] 783 F.2d at 1346, citing 117 Cong. Rec. 22669–72, 34771–75 (1971). The court found that "[p]revention of straying is subservient to the fundamental goal of protecting the animals with minimal management effort." 783 F.2d at 1346.

[217] Fallini v. Hodel, 963 F.2d 275 (9th Cir. 1992). It is noteworthy that the challenged BLM decision was made in 1983, nine years before the Ninth Circuit finally resolved the case.

stalled guard rails on several of their wells to prevent access by horses.[218] BLM managers concluded that the Fallinis had violated their range improvement permit by erecting guard rails without BLM approval and ordered the ranchers to remove the rails.[219]

The Fallinis appealed this decision to an administrative law judge, who held that the ranchers had not violated either the regulations or their permit. The BLM appealed, and the Interior Board of Land Appeals reversed the ALJ.[220] The Fallinis then appealed to the district court, which reversed the IBLA and also found that the BLM's decision effected a regulatory taking of water rights granted to the Fallinis under state law.[221]

Both the trial court and the appellate court regarded the intent of the BLM and the Fallinis in drafting the 1967 permit as controlling. The Ninth Circuit agreed with the trial court that the parties did not intend the term "wildlife" to include feral horses.[222] The court of appeals reasoned:

> the language should mean what it meant to the parties in 1967 when the permit was issued, and not what it might mean 20 years later after Congress had dramatically changed the legal environment in which range management would occur in the future.[223]

The appellate court found that in 1967, "wild horses were not considered 'wildlife' for grazing permit purposes,"[224] and that the BLM could not claim that a purpose of the range permit was to provide water for wild horses because no wild horses ranged in the area of the Fallinis' well in

[218]The guard rails did not prevent access to the wells by native wildife. The Fallinis' concern was solely with the large number of wild horses: while in 1971 only about 130 wild horses roamed within the BLM allotment in which the Fallinis grazed their cattle, by 1984 about 1,800 wild horses inhabited the allotment. 963 F.2d at 277.

[219]BLM regulations require BLM authorization to modify range improvements. 43 C.F.R. § 4140.1(b)(2) (1995).

[220]Fallini v. BLM, 92 IBLA 200 (1986).

[221]Fallini v. Hodel, 725 F. Supp. 1113 (D. Nev. 1989).

[222]963 F.2d at 278–79. The Ninth Circuit did not reach the question of whether the BLM decision was a regulatory taking.

[223]*Id.* at 278.

[224]*Id.* at 279. The court observed that for range permits in 1967 "wildife" included "the occasional mountain sheep, mule deer, antelope, coyote, kit fox and the birds and rodents that make up the fauna that have evolved in an almost waterless desert." *Id.* The court examined the purposes of the permit in light of the purpose of the Taylor Grazing Act. It cited a previous decision in which it had stated that " '[t]he purpose of the Taylor Grazing Act is to stabilize the livestock industry and protect the rights of sheep and cattle growers from interference.' " Kidd v. United States Dep't of the Interior, Bureau of Land Management, 756 F.2d 1410, 1411 (9th Cir. 1985).

1967. Therefore, the court held, the Fallinis did not violate their permit by failing to get BLM approval for the guard rails.[225]

In a well-reasoned dissenting opinion to the Ninth Circuit decision, Judge Taylor argued that congressional intent, rather that of the parties, should control, and that Congress intended that wild horses be given the status of "wildlife." Judge Taylor observed that although Congress's intent in the 1934 Taylor Grazing Act was to stabilize the livestock industry, that intent must be harmonized with the intent to protect wild horses expressed in the Wild Horses Act. He stated that in the latter act "Congress acknowledged that wild free-roaming horses contribute to the diversity of life forms and are an integral part of the natural system of public lands," and that therefore wild horses are "wildlife" in the intent of Congress and in permits granted under the Taylor Grazing Act.[226]

Conclusion

Conflicts between wild horses and livestock continue, as does the controversy over the BLM's horse removal program.[227] Horse protection groups and some conservation groups argue against removal on the basis that range impacts from livestock far outweigh any damage from wild horses.[228] Other conservation groups and livestock operators support the removal program because they believe that in the absence of predators, the numbers of wild horses and burros will grow out of control, damaging the range for other wildlife as well as for livestock.[229]

[225]After the BLM order in 1983, the Fallinis had removed the guard rails on all but one of their wells. The remaining well, Deep Well, was the subject of this protracted litigation. No horses had inhabited the Deep Well area before the Fallinis provided a source of water. The court's reasoning suggests that had the subject well instead been in an area inhabited by wild horses before installation of the well, the plaintiffs' fencing them out would have violated the permit. This distinction appears to be a tenuous basis for the court's decision.

[226]963 F.2d at 279–80.

[227]*See, e.g.*, U.S. General Accounting Office, Improvements Needed in Federal Wild Horse Program (GAO/RCED-90-110 1990); Animal Protection Institute of America, 109 IBLA 112 (June 7, 1989) (BLM must demonstrate that the animals it removes are in excess of appropriate management levels and that it is removing the animals to prevent range deterioration). In 1992, the BLM released its management plan for wild horses and burros. *See* U.S. Dep't of the Interior, Bureau of Land Management, Strategic Plan for Management of Wild Horses and Burros on Public Lands (1992).

[228]Cody, *supra* note 166, at CRS-5. The Government Accounting Office found in fiscal year 1988, livestock on grazing allotments outnumbered wild horses nearly 100 to 1, and that livestock required 20 times the forage of the wild horses. GAO Report, *supra* note 227, at 24.

[229]Cody, *supra* note 166, at CRS-6.

Chapter **11**

WILDLIFE AND WATER RESOURCE DEVELOPMENT

Early in this century, Supreme Court Justice Oliver Wendell Holmes wrote that "[a] river is more than an amenity, it is a treasure."[1] Conservationists see the fish and wildlife resources of the nation's rivers, lakes, and streams as some of the most valuable parts of that treasure. To others, however, the bounty of waters looks different. They see a source of irrigation water, hydroelectric energy, commercial navigation, municipal and industrial water supplies, and the like.

Fish and wildlife values often have been sacrificed for commercial benefits as the nation's water resources have been developed. Congress has recognized the inherent tension among the values water provides and has repeatedly tried to ensure that fish and wildlife are not altogether forgotten. This chapter examines some of the most important of Congress's efforts, including the Fish and Wildlife Coordination Act, the Federal Power Act and its various amendments, and the Northwest Power Act.

THE FISH AND WILDLIFE COORDINATION ACT

Early History

In the midst of the Great Depression, Congress enacted a remarkably forward-looking statute, the Fish and Wildlife Coordination Act (hereinafter "the Coordination Act").[2] The Coordination Act authorized "investigations . . . to determine the effects of domestic sewage, trade

[1]State of New Jersey v. State of New York, 283 U.S. 336, 342 (1931).
[2]16 U.S.C. §§ 661–667e.

wastes, and other polluting substances on wild life,''[3] encouraged the "development of a program for the maintenance of an adequate supply of wild life" on federal lands,[4] and called for state and federal cooperation in "developing a Nation-wide program of wild-life conservation and rehabilitation."[5] For the most part, however, this 1934 statute was purely hortatory.

Only two provisions of the original Coordination Act appeared mandatory. One required the federal government to consult with the Bureau of Fisheries prior to constructing any dam or issuing a permit authorizing the construction of a dam.[6] Consultation had a single, narrowly limited objective: to determine if fish ladders or other aids to migration were "necessary . . . [and] economically practicable."[7] The second provision required that the Bureau be given opportunity to use the waters of any federal impoundment "for fish-culture stations and migratory-bird resting and nesting areas . . . not inconsistent with the primary use of the waters."[8] These provisions appeared mandatory on their face, but the 1934 Act's legislative history suggested otherwise. The bill's proponents explained that "there is nothing but a spirit of cooperation which is insisted on in this bill. There is nothing mandatory about the bill."[9]

The 1946 and 1958 Amendments

After only twelve years, Congress concluded that the Coordination Act had "proved to be inadequate in many respects."[10] The result was a major legislative overhaul in 1946.[11] On one hand, the 1946 revisions represented a retreat from some of the more ambitious goals of 1934. Deleted were the goals of establishing a nationwide program of wildlife conservation and of maintaining an "adequate supply" of wildlife on the federal lands.

On the other hand, the 1946 amendments also expanded the scope of some of the original Act's specific directives. For example, Congress required consultation with the U.S. Fish and Wildlife Service (successor to the Bureau of Fisheries) and with the appropriate state wildlife agency not only in the case of dam construction but also "[w]henever the waters of any stream or other body of water are authorized to be impounded,

[3]Act of March 10, 1934, ch. 55, § 2, 48 Stat. 401.
[4]*Id.* § 5.
[5]*Id.* § 1.
[6]*Id.* § 3(b).
[7]*Id.*
[8]*Id.* § 3(a).
[9]85 Cong. Rec. at S 2011 (Feb. 6, 1934).
[10]H.R. Rep. No. 79–1944, 79th Cong., 2d Sess. 1 (1946).
[11]Act of August 14, 1946, ch. 965, 60 Stat. 1080.

diverted, or otherwise controlled for any purpose whatever by any department or agency of the United States, or by any public or private agency under Federal permit."[12] The resulting conservation recommendations were required to "be made an integral part of any report submitted by any agency of the Federal Government responsible for engineering surveys and construction of such projects."[13] The object of consultation was not merely to aid fish migration, but to prevent "loss of and damage to wildlife resources,"[14] a goal of enormous potential scope because of the Act's broad (and circular) definition of "wildlife" as "birds, fishes, mammals, and all other classes of wild animals and all types of aquatic and land vegetation upon which wildlife is dependent."[15]

The 1946 amendments also expanded the Coordination Act's reach by requiring that, whenever the duty to consult applied to any federal project, so too did the requirement of making "adequate provision consistent with the primary purposes of such impoundment, diversion, or other control . . . for the conservation, maintenance, and management of wildlife."[16] The Act directed the appropriate state wildlife agency or the Secretary of the Interior to make available the waters and lands thus utilized if the area had "particular value" for migratory bird management.[17] A directive in the original House bill that the Corps of Engineers consider impacts on wildlife in its operation of impoundment facilities would have given the "adequate provision" requirement further specific content. However, the War Department successfully lobbied against that restriction.[18]

The only significant reported case to consider the 1946 amendments is *Rank v. Krug*, in which the court found "great cogency" in plaintiff landowners' argument that the statutorily required "adequate provision" for wildlife had not been made with respect to the Bureau of Reclamation's Friant Dam on the San Joaquin River.[19] Nonetheless, viewing the state as the proper party to insist upon compliance with the Act, the court refused to hear plaintiffs' claim. Resting on a principle now largely abandoned, the court held:

> It is too plain to need argument that a citizen cannot compel compliance where that duty is lodged with regularly selected officials whose duties are clearly defined

[12] *Id.* § 2.

[13] *Id.* For a discussion of how limited this obligation was, see Sierra Club v. Sigler, 532 F. Supp. 1222, 1242–43 (S.D. Tex. 1982), *rev'd on other grounds,* 695 F.2d 957 (5th Cir. 1983).

[14] Act of August 14, 1946, ch. 965, § 2, 60 Stat. 1080.

[15] *Id.* § 8.

[16] *Id.* § 3.

[17] *Id.*

[18] *See* Shipley, *The Fish and Wildlife Coordination Act's Application to Wetlands,* in A. Reitze, Environmental Planning: Law of Land and Resources, ch. 2 at 51 (1974).

[19] 90 F. Supp. 773, 801 (S.D. Cal. 1950).

by statute, any more than a private citizen could step in and assume the duties of a prosecuting attorney or governor.[20]

The court did not amplify its reference to the state's "duty" to compel compliance, and the court expressly refrained from deciding whether that duty could be enforced through a private mandamus action.

After another twelve years, Congress again concluded that the results of the amended Coordination Act had "fallen far short of the results anticipated."[21] The result was another major overhaul in 1958,[22] principally to require that wildlife conservation be given "equal consideration" with other features of water resource development.[23] The object of earlier versions of the Act had been preventing wildlife losses, and the 1958 amendments introduced the further goal of wildlife enhancement.[24] In addition, Congress expanded the list of water-related activities to which the Act's consultation requirement applied to include channel deepenings and all modifications of any body of water.[25] Finally, in addition to the existing requirement that the resource agencies' recommendations be made an "integral part" of any federal construction agency's report, the 1958 amendments required that "the project plan . . . include such justifiable means and measures for wildlife purposes as the reporting agency finds should be adopted to obtain maximum overall project benefits."[26]

Two distinctly different types of federal actions are subject to the Coordination Act as a result of the 1946 and 1958 amendments. First, for major federal water development projects, such as dams, reclamation efforts, and channelization projects, the reports and recommendations of the Fish and Wildlife Service (or, since 1970, the National Marine Fisheries Service) and of the state wildlife agency must be given "full consideration" by the federal project agency and must be made "an integral part of any report prepared or submitted by" it to Congress or any other entity having authority to authorize the project.[27]

The second category of projects subject to the Act includes water-related

[20]*Id.* at 801.

[21]Sen. Rep. No. 85–1981, 85th Cong., 2d Sess. 4 (1958), *reprinted in* 1958 U.S.C.C.A.N. at 3449.

[22]Act of August 12, 1958, Pub. L. No. 85–624, 72 Stat. 563.

[23]16 U.S.C. § 661.

[24]*Id.* § 662(a).

[25]*Id.* Federal activities exempt from the requirements are described at 16 U.S.C. § 662(h).

[26]*Id.* § 662(b).

[27]*Id.* Typically, the recommendations of the wildlife agencies take the form of suggested modifications in project design or acquisition of land valuable as wildlife habitat to replace what will be lost as a result of the project. By requiring that "mitigation plans" be included in reports to Congress, that body ultimately decides how much money will be appropriated to carry out fish and wildlife mitigation efforts and thus how effective the Coordination Act will be with respect to any particular water development project.

activities for which federal permits are required, most notably Corps of Engineers permits issued under section 404 of the Clean Water Act.[28] The wildlife agencies can recommend that the permitting agency (1) deny a permit, an option that is seldom available as a practical matter with respect to major federal water projects; or (2) condition the permit to reduce adverse impacts on wildlife. The statute does not specify the degree of deference the permitting agency should pay to the wildlife agencies' recommendations beyond the generally applicable goals of wildlife conservation and enhancement.[29]

Early judicial decisions construing the 1958 amendments led some legal commentators to think that the Coordination Act finally would be elevated to a prominent position in federal wildlife law. For example, in *Udall v. Federal Power Commission,* Justice Douglas invoked the Coordination Act in support of his conclusion that the Federal Power Commission had failed to explore and evaluate adequately the likely effects of a proposed dam on the Snake River.[30] While Douglas' opinion rested primarily upon the Federal Power Act[31] and was far from clear as to the nature of the duties the Coordination Act imposed, it did establish that the latter Act required more than perfunctory compliance.[32]

The Coordination Act and NEPA

Congress enacted the National Environmental Policy Act (NEPA) two years after the *Udall* decision.[33] Originally proposed as an amendment to

[28]33 U.S.C. § 1344. Prior to 1972, the Corps of Engineers was responsible for issuing similar permits under section 10 of the Rivers and Harbors Act of 1899, 33 U.S.C. § 403. *But see* text accompanying notes 62–66 *infra.*

[29]The court in Sierra Club v. Sigler, *supra* note 13, held that the Act's requirement in section § 662(b) that the Fish and Wildlife Service's report and recommendations be made an "integral part" of any federal agency project report does not apply to federally licensed or permitted private development projects.

[30]387 U.S. 428 (1967).

[31]16 U.S.C. § 800(b). For further discussion of the Udall case, see text accompanying notes 84–85 *infra.* Douglas' opinion similarly invoked the Anadromous Fish Conservation Act of 1965, Pub. L. 89–304, § 2, 79 Stat. 1125 (current version at 16 U.S.C. § 757b), which merely directed the Secretary of the Interior to make studies and recommendations "for the conservation and enhancement of anadromous fishery resources," as evidencing a special congressional concern for anadromous fish.

[32]In fact, Justice Douglas' opinion directed that the FPC consider whether the dam ought to be constructed at all. On remand to the FPC, the Presiding Examiner concluded that construction should be deferred and the status quo preserved until 1975, by which time the Departments of the Interior and Agriculture and the Congress could determine the Snake River's suitability for inclusion in the National Wild and Scenic Rivers System. Pacific Northwest Power, Project No. 2243, 1 Envt'l L. Rep. (Envt'l L. Inst.) 30017 (FPC, Feb. 23, 1971). Douglas' vision of a free-flowing river was realized with the creation of Hell's Canyon National Recreation area in 1975. *See* 16 U.S.C. §§ 460gg to 460gg-13, 1274(a)(12).

[33]42 U.S.C. §§ 4321–4361 (hereinafter NEPA).

the Coordination Act, NEPA became instead an independent directive to all federal agencies to consider and evaluate, through the preparation of detailed environmental impact statements, the impacts of all major actions "significantly affecting the quality of the human environment."[34] NEPA eventually came to cast a long shadow over the Coordination Act, diminishing its visibility and significance.

The first case to arise under the Coordination Act after NEPA was enacted was *Zabel v. Tabb.*[35] The Corps of Engineers had refused to issue a permit under section 10 of the Rivers and Harbors Act of 1899[36] because of the adverse ecological effects of the proposed dredging. The permit applicants sued, charging that despite the Coordination Act the Corps could consider only the effects of the proposed action on navigation. The district court agreed with the developers, but the appellate court reversed because of "the government-wide policy of environmental conservation [that] is spectacularly revealed" by the Coordination Act and NEPA.[37] The court in *Akers v. Resor* applied similar reasoning to hold in a pretrial decision that the Coordination Act, construed in light of NEPA, required the preparation of a new wildlife mitigation plan.[38]

On the basis of the *Zabel* and *Akers* decisions, early commentators optimistically described the relationship between the Coordination Act and NEPA as synergistic[39] or symbiotic.[40] However, even before the *Akers* decision, at least one cloud had already appeared on the horizon of NEPA-Coordination Act relations in the form of the *Gillham Dam* decision, *Environmental Defense Fund v. Corps of Engineers.*[41] Plaintiffs challenged a major Corps of Engineers project already two-thirds completed, and the court resurrected the two-decade-old *Rank v. Krug* decision by expressing doubt whether private parties could maintain actions for noncompliance with the Coordination Act. In addition, the court found it unreasonable to order the Corps to engage in consultation under the Coordination Act thirteen years after the project plans had been made. The court concluded that "if defendants comply with the provisions of [NEPA] in good faith, they will automatically take into consideration all of the factors required by the Fish and Wildlife Act."[42]

[34]The legislative history of NEPA is described in R. Liroff, A National Policy for the Environment: NEPA and its Aftermath 10–35 (1976).

[35]430 F.2d 199 (5th Cir. 1970), *cert. denied,* 401 U.S. 910 (1971).

[36]33 U.S.C. § 403.

[37]430 F.2d at 209.

[38]339 F. Supp. 1375 (W.D. Tenn. 1972).

[39]Shipley, *supra* note 18, at 49.

[40]Guilbert, Wildlife Preservation Under Federal Law, in Federal Environmental Law at 557 (E. Dolgin & T. Guilbert eds. 1974).

[41]325 F. Supp. 749 (E.D. Ark. 1971), *injunction dissolved,* 342 F. Supp. 1211, *aff'd,* 470 F.2d 289 (8th Cir. 1972).

[42]325 F. Supp. at 754.

The court's language in the *Gillham Dam* decision raised some major conceptual questions that were further confused in other parts of the opinion. For example, the quoted language can be understood to imply that the Coordination Act's requirements are purely procedural rather than substantive. On the other hand, the court stated that plaintiffs could attempt to prove noncompliance with the Coordination Act as an element of their claim arising under NEPA, "since departures from the congressional policies set forth in [the Coordination Act] . . . should be acknowledged in any" impact statement.[43] In fact, however, the *Gillham Dam* impact statement did not acknowledge any such "departures" but actually claimed benefits for enhancement of fish and wildlife resources. Nonetheless, while professing to have no power of substantive review under NEPA, the court proceeded to find the impact statement inadequate in part because "the evidence indicates that there are a number of serious possibilities for injury to fish and wildlife which might occur if the dam is constructed."[44]

The *Gillham Dam* decision's statement that compliance with NEPA constitutes compliance with the Coordination Act is, accordingly, both confused and confusing. Many courts nonetheless have uncritically accepted it as a basis for dismissing Coordination Act claims when coupled with NEPA claims.[45] Few of the later decisions attempt to clarify the precise meaning of NEPA-Coordination Act equivalence. Several commentators, however, have suggested that equating the Coordination Act with NEPA may not withstand critical scrutiny.[46]

The question whether NEPA imposed substantive as well as procedural duties was unresolved at the time of the *Gillham Dam* decision. Many commentators, and at least a few courts, asserted that NEPA imposed substantive, judicially enforceable standards of environmental quality.[47] Seven

[43] *Id.*

[44] *Id.* at 760.

[45] *See, e.g.*, Environmental Defense Fund v. Froehlke, 473 F.2d 346 (8th Cir. 1972); County of Trinity v. Andrus, 438 F. Supp. 1368, 1383 (E.D. Cal. 1977); Save Our Sound Fisheries Ass'n v. Callaway, 387 F. Supp. 292 (D.R.I. 1974); and Cape Henry Bird Club v. Laird, 359 F. Supp. 404 (W.D. Va.), *aff'd*, 484 F.2d 453 (4th Cir. 1973).

[46] *See* Parenteau, *Unfulfilled Mitigation Requirements of the Fish and Wildlife Coordination Act*, 42 N. Am. Wildlife Conf. Proc. 179 (1977), and Blumm, *Hydropower vs. Salmon: The Struggle of the Pacific Northwest's Anadromous Fish Resources for a Peaceful Coexistence with the Federal Columbia River Power System*, 11 Envt'l L. 211, 268–76 (1981).

[47] For a sampling of some of the commentary, including citations to the relevant cases, *see* Yarrington, *Judicial Review of Substantive Agency Decisions: A Second Generation of Cases Under the National Environmental Policy Act*, 19 S.D. L. Rev. 279 (1974); Kumin, Substantive Review Under the National Environmental Policy Act: EDF v. Corps of Engineers, 3 Ecology L.Q. 173 (1973); Arnold, *The Substantive Right to Environmental Quality Under the National Environmental Policy Act*, 3 Envt'l L. Rep. (Envt'l L. Inst.) 50028 (1973); and Note, *The Least Adverse Alternative Approach to Substantive Review Under NEPA*, 88 Harv. L. Rev. 735 (1975).

years after *Gillham Dam*, however, the Supreme Court concluded that although NEPA established "significant substantive goals for the Nation," its obligations were "essentially procedural."[48]

The Supreme Court's holding that NEPA's obligations are only procedural might have provided the impetus for courts to reexamine whether compliance with NEPA necessarily constitutes compliance with the Coordination Act as well. Some of the latter act's requirements suggest a substantive component. For example, unless the Coordination Act's mandate that project plans "include such justifiable means and measures for wildlife purposes as the reporting agency finds should be adopted to obtain maximum overall project benefits" is intended to vest absolute discretion in such agencies, this mandate would appear to provide a substantive standard against which reviewing courts can measure an agency's compliance.[49] The Act's central objective, that wildlife conservation receive "equal consideration" with other features of water resource development programs, also implies that Congress did not intend absolute discretion.[50]

Despite these considerations and despite the opportunity to reexamine the *Gillham Dam* notion of NEPA and Coordination Act equivalence that

[48]Vermont Yankee Nuclear Power Corp. v. Natural Resources Defense Council, 435 U.S. 519, 558 (1978). Two years later, this decision was reinforced in Stryker's Bay Neighborhood Council, Inc. v. Karlen, in which the Court declared that NEPA required only "consider[ation of] . . . environmental consequences. . . . NEPA requires no more." 444 U.S. 223, 227–228 (1980).

[49]16 U.S.C. § 662(b). The court in County of Trinity v. Andrus, *supra* note 45, held that a statutory directive to the Secretary of the Interior "to adopt appropriate measures to insure the preservation and propagation of fish and wildlife" in connection with a water diversion project constitutes a judicially reviewable substantive standard.

[50]16 U.S.C. § 661. In Sierra Club v. Alexander, a case involving a challenge to the Corps of Engineers' issuance of a section 404 "dredge and fill" permit in the face of unresolved objections by the Fish and Wildlife Service, the court held that the Coordination Act required the Corps to give serious consideration to the Service's recommendations but need not follow them. 484 F. Supp. 455, 470 (N.D.N.Y. 1980), *aff'd* 633 F.2d 206 (2d Cir.). *See also* Sierra Club v. Sigler, *supra* note 13, at 1243, holding that "Congress intended only that the Corps consult [the Fish and Wildlife Service] before issuing a permit for private dredge and fill operation." That view finds support in the legislative history of the Coordination Act's 1958 amendments, where it is emphasized that wildlife agencies are not intended to exercise a "veto power" over agencies required to consult with them. *See* Hearings on H.R. 13138 Before the Senate Comm. on Interstate and Foreign Commerce, 85th Cong., 2d Sess. 4 (1958). For an interpretation of another provision mandating consultation but disclaiming any grant of veto power to the agency consulted, see the discussion of section 7 of the Endangered Species Act in Chapter 7. *See also* the statement in Sierra Club v. Butz, 3 Envt'l L. Rep. (Envt'l L. Inst.) 20292, 20293 (9th Cir. March 16, 1973), that the requirement of "due consideration" under the Multiple Use-Sustained Yield Act "can hardly be satisfied by a showing of knowledge of the consequences and a decision to ignore them." The Coordination Act's standard of "full consideration" would seem to demand no less.

the Supreme Court's later NEPA decisions provided, courts have continued to uncritically embrace the equivalence view.[51] In *Texas Comm. on Natural Resources v. Marsh,* the Fifth Circuit Court of Appeals underscored in a few short sentences the Coordination Act's ineffectiveness:

> NEPA and [the Coordination Act] do not require the Corps to adopt a mitigation plan that addresses all adverse impacts of its projects. Nor do they require the Corps to adopt all suggestions made by USFWS. Rather, the Corps meets the statutory requirements if it gives serious consideration to the views expressed by USFWS. . . . The Supreme Court has made it clear that to "consider" environmental factors does not mean to give them special weight.[52]

Duties of Fish and Wildlife Agencies under the Coordination Act

Just as the courts have been unwilling to recognize that the Coordination Act imposes any substantive obligations on development and licensing agencies, they also have not clearly defined the requirements the Act imposes on the Fish and Wildlife Service and the National Marine Fisheries Service. Only one case, *Sun Enterprises, Ltd. v. Train,* shows an earnest attempt to define those requirements.[53]

In *Sun Enterprises,* the Environmental Protection Agency Administrator issued a National Pollution Discharge Elimination System (NPDES) permit pursuant to section 402 of the Federal Water Pollution Control Act.[54] At the time he received the permit application, the Administrator had notified the Fish and Wildlife Service and invited its comments. Rather than respond substantively, the Service advised that, due to lack of personnel, the agency would make no recommendation on the application. The Administrator later issued the permit, and plaintiffs challenged it on

[51] *See, e.g.,* County of Bergen v. Dole, 620 F. Supp. 1009, 1064 (D. N.J. 1985), and Environmental Defense Fund v. Alexander, 501 F. Supp. 742, 766–767 (N.D. Miss. 1980), *aff'd in part and rev'd in part on other grounds sub nom.* Environmental Defense Fund v. Marsh, 651 F.2d 983 (5th Cir. 1981). One of the few cases that explicitly distinguishes Gillham Dam is National Wildlife Federation v. Andrus, 440 F. Supp. 1245 (D.D.C. 1977), where action was brought under both NEPA and the Coordination Act to enjoin the Bureau of Reclamation from constructing a power plant. In granting the injunction, the court held that the Bureau had failed to provide Congress with the Fish and Wildlife Service's consultation report regarding the project, as required under the Coordination Act. Significantly, the court refused to follow defendant's urging that Gillham Dam controlled, ruling that "plaintiffs have identified a FWCA policy, that of informing Congress of environmental effects, which may not be duplicated by NEPA. In such circumstances, strict compliance with FWCA should be required." *Id.* at 1255. The court believed this result was "especially true in view of the doubtful authorization by Congress of the [challenged] project in the first place." *Id.*

[52] 736 F.2d 262, 268 (5th Cir. 1984) (citations omitted).

[53] 532 F.2d 280 (2d Cir. 1976).

[54] 33 U.S.C. § 1342.

the grounds that consultation had not occurred. The district court rejected the claim, holding that the Service had no duty to make a substantive reply to the Administrator's consultation request.[55]

On appeal the court held that since section 509 of the Clean Water Act conferred exclusive jurisdiction on the courts of appeal to consider challenges to the issuance of permits not involving alleged violations of effluent standards, the district court had been without jurisdiction to consider the plaintiff's suit. Moreover, the court held that the plaintiff was time-barred from filing under section 509 now, since that provision limited review to petitions filed within ninety days of a permit's issuance.[56] Significantly, the court expressly refrained from deciding whether the plaintiff had a private right of action to enforce the Coordination Act that could confer upon the district court independent jurisdiction to consider the suit. However, the court noted that "compliance with the Coordination Act might be a consideration in connection with a section 509 petition for review of an NPDES permit."[57] In explaining its ruling, the court stated:

> Certainly it would be an unsatisfactory result if the otherwise exclusive mode of review of an NPDES permit's issuance, a §509 petition to the court of appeals, could be circumvented by an action in the district court against Interior. If the failure of Interior to respond substantively to the EPA is reviewable, review must be sought by a §509 petition challenging the issuance of a permit in the absence of either any comment or exercise of discretion not to comment by Interior.[58]

What the court meant by the phrase "in the absence of either any comment or exercise of discretion not to comment by Interior" is difficult to discern. However, the court's discussion of Interior's consultation obligations under the Coordination Act may shed light on the phrase. Rejecting the district court's suggestion that "the lack of any express requirement [in the Act] that Interior respond to EPA is an implicit grant of discretion to Interior allowing it to utilize its resources as it sees fit," the court reasoned that "[t]o employ a narrow construction of the word 'consult,' as urged, would be to emasculate the Coordination Act."[59] Indeed, the court asserted that "the consultation requirement of the Coordination Act [is] especially important in the context of NPDES permits."[60] Thus

[55]394 F. Supp. 211, 218 (S.D.N.Y. 1975).

[56]532 F.2d at 289.

[57]*Id.* at n. 10.

[58]*Id.* at 289 (footnote omitted).

[59]*Id.* at 290.

[60]*Id.* The court thus answered the question whether the Coordination Act applies to NPDES permits with an emphatic "yes." *See id.* at 288 n. 9. The court was unwilling, however,

Interior's position that funding and personnel are inadequate to meet the burdensome demands of reviewing NPDES permits is entitled to little weight. . . . Whatever the reason, while we appreciate the difficulties involved in reviewing the large number of applications forwarded by EPA to Interior, we cannot condone what amounts to administrative or executive repeal of an act of Congress.[61]

The court appears to say that the Coordination Act imposes upon Interior either a nondiscretionary duty to consult, or a duty to exercise a cognizable standard in declining to consult. In either case, Interior's failure to respond to a consultation request should be reviewable.

The problem underscored in *Sun Enterprises*, insufficient funds and personnel for the Fish and Wildlife Service to carry out the Coordination Act's mandate, emerged in the mid-1970s. The problem did not surface before then largely due to the Corps of Engineers' practice prior to 1970 of not requiring permits under section 10 of the Rivers and Harbors Act of 1899[62] to erect structures or fill wetlands shoreward of established "harbor lines," or under section 13[63] to discharge refuse into navigable waters. The 1970 reversal of the Corps' harbor line policy at the urging of the House Committee on Government Operations,[64] and the passage of the Federal Water Pollution Control Act amendments of 1972, which substituted the NPDES permit system for the permit requirements of section 13 of the 1899 Act, brought about a quantum leap in the number of permits subject to Fish and Wildlife Service review under the Coordination Act.[65] The workload increased still further as a result of the decision in

---

to express a view as to whether the Act continued to apply once authority for issuance of NPDES permits was transferred to a state. *See id.* at 289 n. 10. A fascinating history of the persistent and ultimately successful efforts of Congressmen Reuss, Dingell, and Vander Jagt to persuade EPA, the Bureau of Sports Fisheries and Wildlife, and the National Oceanic and Atmospheric Administration that the Coordination Act did apply to section 402 permits is contained in Protecting America's Estuaries: Florida, Hearings Before the Subcomm. on Conservation and Natural Resources of the House Comm. on Government Operations, 93rd Cong., 1st Sess., app. 4, pt. C, at 1261–1342 (1973).

[61] 532 F.2d at 290.

[62] 33 U.S.C. § 403.

[63] *Id.* § 407.

[64] *See* House Comm. on Government Operations, Our Waters and Wetlands: How the Corps of Engineers Can Help Prevent Their Destruction and Pollution, H.R. Rep. No. 91–917, 91st Cong., 2d Sess. (1970). The Corps' harbor line policy declares these lines to be guidelines only for determining the impact on navigation of shoreward structures or fills. *See* 47 Fed. Reg. 31793, 31806 (1982).

[65] The total number of permit applications and projects to which the Coordination Act applied increased nearly eightfold between 1970 and 1974, while the total number of federal wildlife personnel investigating and reporting on the same increased from 258 to only 411. *See* General Accounting Office, Improved Federal Efforts Needed to Equally Consider Wildlife Conservation with Other Features of Water Resource Development, app. IV, at 57, 59 (1974), and Department of the Interior, Fish and Wildlife Service, Conserving Our Fish and Wildlife Heritage, Annual Report–FY 1975.

Natural Resources Defense Council v. Callaway substantially expanding the Corps of Engineers' jurisdiction to issue dredge and fill permits under section 404 of the Clean Water Act.[66]

Private Rights of Action under the Coordination Act

While the practical problem posed in *Sun Enterprises* is significant, solving it will not enhance the Coordination Act's utility to environmental litigants if the legal problem presented by the first case under the Coordination Act persists. Courts in several cases in the 1970s and early 1980s brought the Coordination Act back full circle to where it had been decades earlier in *Rank v. Krug* by holding that private plaintiffs cannot assert causes of action under it.[67]

In *Sierra Club v. Morton*[68] the district court held that plaintiffs could not assert a private right of action under the Coordination Act, though private plaintiffs could assert claims under sections 9 and 10 of the Rivers and Harbors Act.[69] Ultimately, that case found its way to the Supreme Court, although by then only the Rivers and Harbors Act questions remained. The court ruled that there was no private right of action under that statute.[70] Though the Court's opinion never mentioned the Coordination Act, it raises serious doubt as to whether a different result would apply to the latter statute.[71] However, neither does the Court's opinion address the Administrative Procedure Act (APA), which provides generally for a right of action to any "person suffering legal wrong because of agency action, or adversely affected or aggrieved by agency action within the meaning of a relevant statute."[72] The Supreme Court's opinion may, therefore, elevate form over substance since a properly framed complaint under the APA would seem to permit judicial review of agency action alleged to violate the Coordination Act.[73]

[66]392 F. Supp. 685 (D.D.C. 1975).

[67]*See, e.g.*, Environmental Defense Fund v. Alexander, *supra* note 51, and County of Trinity v. Andrus, *supra* note 45.

[68]400 F. Supp. 610 (N.D. Cal. 1975), *aff'd in part, rev'd in part on other grounds sub nom.* Sierra Club v. Andrus, 610 F.2d 581 (9th Cir. 1979), *rev'd on other grounds sub nom.* California v. Sierra Club, 451 U.S. 287 (1981).

[69]33 U.S.C. §§ 401, 403.

[70]California v. Sierra Club, *supra* note 68.

[71]Since the Coordination Act, unlike the Rivers and Harbors Act, contains no provision authorizing the Department of Justice to enforce it, the argument in support of a private right of action would appear stronger for the Coordination Act.

[72]5 U.S.C. § 702.

[73]In Chrysler Corp. v. Brown, 441 U.S. 281 (1979), the Court held that the absence of a private right of action under a specific statute is no bar to a litigant obtaining APA review of alleged agency violations of that statute. This holding follows from 5 U.S.C. § 704: "[F]inal agency action *for which there is no other adequate remedy in a court* [is] subject to judicial review."

Even if the Coordination Act has been an ineffective tool for environmental litigants, it does not necessarily follow that the Act has failed to achieve its purposes, for it is possible that the affected agencies have diligently implemented the statutory policies without the stimulus of judicial intervention. The evidence on this, however, is discouraging.[74]

In an effort to improve the law's effectiveness, in 1979 the Departments of the Interior and Commerce jointly published proposed regulations to implement the Coordination Act.[75] Following the public comment period, however, they were withdrawn for reconsideration. Modified regulations were proposed in 1980.[76] In the Reagan administration, however, the proposed regulations were again reconsidered and ultimately abandoned.[77] The Fish and Wildlife Coordination Act is thus without implementing regulations, largely overshadowed by NEPA, and undercut by a series of disabling judicial interpretations. Its promise, once viewed with considerable optimism, remains largely unfulfilled.

THE FEDERAL POWER ACT

The federal government has played at least two roles in water resource development. First, it has been directly responsible for harnessing the nation's rivers through its own development projects, such as hydroelectric dams, channelization, and "reclamation" projects. A major part of the missions of federal agencies such as the Bureau of Reclamation, Army Corps of Engineers, and Tennessee Valley Authority has been to carry out these activities.

In its other role, the federal government has regulated nonfederal water resource development. A variety of nonfederal interests, including municipalities, public utilities, and various private parties, have been important players in this arena. The federal government's responsibilities for federally owned land and for maintaining the navigability of navigable waters have provided the basis for long-standing federal oversight and regulation of nonfederal dam building. Prior to 1920, federal regulatory responsibilities were dispersed among the Secretaries of Agriculture, Interior, and War, and efforts were largely uncoordinated.

(Emphasis added.) Earlier, the Ninth Circuit recognized that the APA provided a basis for reviewing claims of Coordination Act violations in Ass'n of N.W. Steelheaders v. United States Army Corps of Engineers, 485 F.2d 67, 70 (9th Cir. 1973). For a similar case in another context subsequent to the Supreme Court's opinion, see California v. Watt, 683 F.2d 1253, 1270 (9th Cir. 1982).

[74] *See, e.g.,* General Accounting Office, *supra* note 65, at 43; Houck, *Promises, Promises: Has Mitigation Failed?* 10 Water Spectrum 31 (Spring 1978).

[75] 44 Fed. Reg. 29300 (1979).

[76] 45 Fed. Reg. 83412 (1980).

[77] 47 Fed. Reg. 31299 (1982).

The 1920 Act

Congress perceived the lack of coordination among the various federal agencies as an impediment to developing the nation's hydropower resources, and in 1920 it enacted the Federal Water Power Act[78] to ensure more orderly hydropower development. The Act's purposes were "to establish a national policy to promote the comprehensive development of water power on government lands and navigable waters other than by the government itself" and to administer that policy "by abolishing the piecemeal authorities of the Secretaries of the Interior, Agriculture and War over the nation's hydroelectric resources and centralizing them in the Federal Power Commission."[79] Fifteen years later, as part of the Public Utility Act of 1935, Congress conferred additional powers upon the Commission and renamed the 1920 legislation the Federal Power Act.[80]

The original Federal Power Act never explicitly mentioned fish or other wildlife. Section 4(e) of the Act authorized the Commission to "issue licenses . . . for the purpose of constructing, operating, and maintaining dams" and related facilities on waters subject to congressional jurisdiction.[81] Only two requirements constrained this authority. First, the Commission had to find that the "contemplated improvement" was "desirable and justified in the public interest."[82] Second, under Section 10(a) the Commission had to find that the proposed project "will be best adapted to a comprehensive plan for improving or developing a waterway . . . for the use or benefit of interstate or foreign commerce, for the improvement and utilization of waterpower development, and for other beneficial public uses, including recreational purposes."[83]

The Fish and Wildlife Coordination Act's duties apply broadly to all manner of activities affecting the nation's waters and thus overlap to a considerable degree the Federal Power Act's separate duties. Although, as discussed above, judicial interpretation has rendered the Coordination Act largely toothless, amendments to the Federal Power Act in the 1980s and court decisions provide hope that it may escape a similar fate.

A Supreme Court case in the 1960s gave little guidance as to the meaning of the Federal Power Act's requirement that a project be in the "public interest." *Udall v. Federal Power Commission* involved a challenge to a license to construct a hydroelectric dam on the Snake River, the proposed

[78]Act of June 10, 1920, ch. 285, 41 Stat. 1063.

[79]State of California ex rel. State Water Resources Control Board v. Federal Energy Regulatory Commission, 966 F.2d 1541, 1556 (9th Cir. 1992) (*quoting* Escondido Mutual Water Company, 9 FERC 61,241, at 61,519–20 (1979)).

[80]Act of Aug. 26, 1935, ch. 687, 49 Stat. 803.

[81]16 U.S.C. § 797(e).

[82]*Id.*

[83]*Id.* § 803(a).

High Mountain Sheep Dam.[84] Concern for the proposed dam's impacts on anadromous fish prompted the Secretary of Interior to challenge the license. Justice Douglas, in concluding that the Commission had not properly discharged its duties, held that

> [t]he test is whether the project will be in the public interest. And that determination can only be made after an exploration of all issues relevant to the "public interest," including . . . the preservation of anadromous fish for commercial and recreational purposes, and the protection of wildlife.[85]

While clarifying that the Commission had to consider wildlife impacts in determining the public interest, Justice Douglas's opinion gave little guidance as to how the Commission was to determine those impacts and weigh them against other significant factors.

The Ninth Circuit in *Confederated Tribes and Bands of the Yakima Indian Nation v. FERC* held that the licensing body must consider fish and other wildlife impacts prior to licensing, not afterwards.[86] This case involved the relicensing of a dam originally licensed in 1930.[87] There the Federal Energy Regulatory Commission ("FERC," the successor agency to the Federal Power Commission as of 1977) issued a renewal license with a condition that allowed it later to modify the license to impose fish conservation requirements after considering the results of an ongoing comprehensive study of the effects of several dams on the same river. FERC argued that delaying imposition of these requirements until after the study was completed comported with its duty to ensure that the project was best adapted to a comprehensive plan for the waterway.

The court disagreed, noting that FERC could not modify or amend the license without the licensee's consent,[88] and that issuance of the permit removed any incentive to complete the fish study quickly. The court ruled that FERC had to fully consider fishery impacts prior to relicensing. In the court's view, FERC should have deferred renewing the license and instead issued only annual licenses until the study was completed.[89]

[84]387 U.S. 428 (1967).

[85]*Id.* at 450.

[86]746 F.2d 466 (9th Cir. 1984).

[87]Federal Power Act licenses are issued for limited terms, not to exceed 50 years. Section 15 of the Act authorizes renewal of an expiring license (generally called "relicensing") upon undertaking the same inquiry into the public interest that is required of original licensing decisions. The relicensing process often is protracted. Even though an applicant can file for renewal as much as five years prior to expiration of the original license, it is not unusual for the original license to expire before the relicensing process is completed. In that event, annual licenses on the same terms and conditions as the original license are issued until a final determination on the relicensing application is made. For details, *see* 16 U.S.C. § 808(a).

[88]*See* 16 U.S.C. § 799.

[89]Two years later, however, the same court in National Wildlife Federation v. FERC, 801

The *Yakima* court indicated that FERC had discretion to impose conservation conditions in any annual license, pending completion of the relicensing process. Whether FERC had a *duty* to impose such conditions in annual licenses was presented to the court in *Platte River Whooping Crane Critical Habitat Maintenance Trust v. FERC.*[90]

At issue in that case were two expiring licenses for projects on the Platte River. One license allowed FERC unilaterally to impose additional conservation conditions; both licenses allowed it to impose additional conditions with the consent of the licensee. FERC denied the Trust's request that the Commission either use its unilateral authority to impose conditions or pursue consensual conditions with the licensees. The Trust challenged FERC's refusal as an abuse of its discretion.

The court rejected FERC's contention that because the issuance of an annual license is a nondiscretionary act, it had no obligation to review the projects' ongoing environmental impacts under annual licenses. Finding that FERC had been apprised for nearly a decade that the projects' continued operation posed a potential threat to the endangered whooping crane, the court concluded that "it was irresponsible for FERC to . . . refuse even to conduct a preliminary investigation into this threat and the availability of interim measures to combat it."[91]

The 1986 Amendments

The *Platte River* court based its decision at least partly on 1986 amendments to the Federal Power Act made by the Electric Consumers Protection Act ("ECPA").[92] The ECPA directed that

> in addition to the power and development purposes for which licenses are issued, [FERC] shall give equal consideration to the purposes of energy conservation, the protection, mitigation of damage to, and enhancement of, fish and wildlife (including related spawning grounds and habitat), the protection of recreational opportunities, and the preservation of other aspects of environmental quality.[93]

This requirement essentially restated the largely ineffectual "equal consideration" requirement that had been a part of the Fish and Wildlife

F.2d 1505 (9th Cir. 1986), was unwilling to hold that an applicant must prepare a comprehensive plan prior to the issuance of a "preliminary permit," a permit that gives the permittee a "priority of application for a license," thus encouraging it to invest the time and money necessary to develop its project proposal fully. *See* 16 U.S.C. § 798. The court acknowledged that a comprehensive plan might be needed at that stage, but only if a permit applicant could not gather the information needed for proper consideration of the application without such a plan.

[90]876 F.2d 109 (D.C. Cir. 1989).

[91]*Id.* at 117.

[92]Act of October 16, 1986, Pub. L. No. 99–495, 100 Stat. 1243.

[93]16 U.S.C. § 797(e).

Coordination Act since 1958. The ECPA's legislative history, however, indicates that Congress intended much more. The conference report explained that "equal consideration must be viewed as a standard, *both procedural and substantive,* that cannot be satisfied by mere consultation."[94]

The ECPA amendments made other important changes. They expanded the elements of the comprehensive plan to which licensed projects must be adapted. All licenses were to be subject to the condition that the project be "best adapted to a comprehensive plan" for "the adequate protection, mitigation, and enhancement of fish and wildlife (including related spawning grounds and habitat)."[95] To carry out these obligations, new section 10(j) of the Federal Power Act required FERC to include in each project license conditions for the protection, mitigation, and enhancement of fish, wildlife, and related spawning grounds and habitat, based on recommendations of various federal and state fish and wildlife agencies.[96] While not strictly bound to incorporate all recommendations as license conditions, FERC can reject them only upon finding that they are "inconsistent with the purposes and requirements" of the Federal Power Act or other laws and that the conditions imposed by FERC protect and mitigate damage to fish and wildlife.[97]

These changes, though codifying a significant role for fish and wildlife agencies in the FERC licensing process, left the ultimate decisionmaking authority in FERC's hands. That was the clear holding of both *National Wildlife Federation v. FERC*[98] and *State of California ex rel. State Water Resources Control Board v. FERC.*[99] The Ninth Circuit Court of Appeals in the latter case took no note of the language of the conference report quoted above. Rather, it emphasized a different aspect of ECPA's legislative history, noting that while the House of Representatives had sought to require equal "treatment" for fish and wildlife, the Senate insisted on equal "consideration." The court attached pivotal significance to this one word difference:

> "Equal consideration" is not the same as "equal treatment," and equal consideration does not dictate FERC's acceptance of the result proposed by the fish and wildlife agencies. FERC must balance the public interest in all of its stated dimensions, give equal consideration to conflicting interests, and reach a reasoned factual decision.[100]

[94] H. Conf. Rep. No. 99–934, 99th Cong., 2d Sess. at 22, *reprinted in* 1986 U.S.C.C.A.N. at 2538 (emphasis added).

[95] 16 U.S.C. § 803(a).

[96] *Id.* § 803(j).

[97] *Id.* § 803(j)(2).

[98] 912 F.2d 1471, 1480 (D.C. Cir. 1990).

[99] 966 F.2d 1541 (9th Cir. 1992).

[100] *Id.* at 1550.

In *U.S. Department of the Interior v. FERC,* the District of Columbia Circuit Court of Appeals further limited the advisory role of fish and wildlife agencies in the FERC licensing process. There the agencies had responded to FERC's request for recommendations pursuant to section 10(j) by asserting that further studies were needed before the agencies could give any recommendations. FERC declined to carry out or require the studies, but instead issued the licenses. The Department of the Interior challenged FERC's rejection of its request for further studies. The court held that "the requests for additional studies were not themselves section 10(j) recommendations" and "[t]herefore, whatever heightened burden FERC may bear in rejecting a recommendation from a wildlife agency was not triggered here."[101] The court concluded that if additional studies were needed, the wildlife agencies themselves, not FERC, must conduct them. This ruling is quite significant in light of the limited research budgets of most fish and wildlife agencies.

In at least two situations, however, agencies other than FERC do have the final word regarding license conditions. First, section 4(e) of the Federal Power Act provides that when FERC issues a license for a project with facilities on reserved federal land, the license "shall be subject to and contain such conditions as the Secretary of the department under whose supervision such reservation falls shall deem necessary for the adequate protection and utilization of such reservation."[102] The Supreme Court interpreted this language literally in *Escondido Mutual Water Co. v. La Jolla Band of Mission Indians,* dismissing FERC's contention that it could reject the Secretary of Interior's conditions.[103] Thus, for projects situated wholly or partly on Indian reservations, most national forests, and those national wildlife refuges withdrawn from the public domain, the Secretary with jurisdiction over the reservation can specify the conditions of any FERC license.[104]

The second situation where other agencies also have authority to impose license conditions involves some small hydropower projects that FERC can exempt from normal licensing procedures.[105] Section 30(c) of the Federal Power Act, added in 1978, originally provided that in exempting any project from normal licensing procedures, FERC must consult with the U.S. Fish and Wildlife Service and the state fish and wildlife agency and must include in any exemption decision such terms as those agencies determine are appropriate to prevent loss of fish or other wildlife

[101]952 F.2d at 538, 544.

[102]16 U.S.C. § 797(e).

[103]466 U.S. 765 (1984), *rehearing denied,* 467 U.S. 1267.

[104]The Federal Power Act's definition of "reservations" is found at 16 U.S.C. § 796(2). Construction of hydroelectric projects in national parks is prohibited by the Act of March 3, 1921, ch. 129, 41 Stat. 1353.

[105]*See* 16 U.S.C. §§ 823a(a) and 2705(d).

resources.[106] In *The Steamboaters v. FERC*, the Ninth Circuit Court of Appeals held that because Congress had failed to reference the National Marine Fisheries Service in this provision, its recommendations were not binding in exemption decisions.[107] Congress promptly remedied this oversight by amending the Act in 1986 to include the National Marine Fisheries Service among the agencies whose recommendations are binding.[108]

THE PACIFIC NORTHWEST ELECTRIC POWER PLANNING AND CONSERVATION ACT

Construction of the first of more than thirty hydroelectric dams on the Columbia River and its tributaries began in 1933, the year before the original Fish and Wildlife Coordination Act was passed. Despite the contemporaneous development of the Coordination Act and of hydroelectric power on the Columbia River, dams have been the principal cause of a precipitous decline in the river's anadromous salmon and steelhead trout fisheries.[109]

Congress concluded in 1980 that existing legislation, including the Coordination Act and the Federal Power Act, had not been "adequate to offset the cumulative impact of the hydroelectric dams of the Columbia and its tributaries on fish and wildlife."[110] In that year Congress enacted the Pacific Northwest Electric Power Planning and Conservation Act, commonly called the Northwest Power Act.[111]

The Northwest Power Act was singularly ambitious. Commentators have characterized its goal as the world's largest program of biological restoration.[112] Where the Coordination Act had focused on individual projects and relied heavily on hatcheries to compensate for the loss of fish due to dams, the Northwest Power Act "called for a systemwide remedial program to cover the entire Basin," and "emphasiz[ed] changes in hydro project operations."[113]

[106]16 U.S.C. § 823a(c).

[107]759 F.2d 1382 (9th Cir. 1985).

[108]Pub. L. No. 99-495, § 7(b), 100 Stat. 1248 (codified at 16 U.S.C. § 823a(c)).

[109]For a well-documented account of the history of hydroelectric development of the Columbia and of its impact upon the river's fisheries, see Blumm, *supra* note 46. *See also*, National Research Council, Upstream: Salmon and Society in the Pacific Northwest (1996).

[110]H.R. Rep. No. 96–976, pt. I, 96th Cong. 2d Sess. at 48, *reprinted in* 1980 U.S.C.C.A.N. at 6015.

[111]16 U.S.C. §§ 839–839h. A very useful discussion of the Act and its legislative history is found in *The Pacific Northwest Electric Power Planning and Conservation Act: A Preliminary Analysis*, Anadromous Fish Law Memo, Issue 11 at 1 (Natural Resources Institute, 1981).

[112]Kai N. Lee and Jody Lawrence, *Adaptive Management: Learning from the Columbia River Basin Fish and Wildlife Program*, 16 Envt'l L. 431, 441 (1986).

[113]Northwest Resource Information Center, Inc. v. Northwest Power Planning Council, 35 F.3d 1371, 1378 (9th Cir. 1994).

The 1980 law's principal requirement was that a state-appointed regional planning council[114] develop a regional conservation and electric power plan, including a "program to protect, mitigate, and enhance fish and wildlife, including related spawning grounds and habitat, on the Columbia River and its tributaries."[115] The purpose of the plan and program is to guide the future acquisition of power resources by the Bonneville Power Administration.[116] In addition, Bonneville Power, as well as all federal agencies responsible for managing, operating, or regulating hydroelectric facilities on the Columbia River system, must take into account "to the fullest extent practicable" the council's fish and wildlife program "at each relevant stage of decisionmaking processes" in the operation or regulation of such facilities.[117]

Perhaps most important, the Northwest Power Act directed the federal agencies responsible for managing hydro projects on the Columbia River system to "protect, mitigate, and enhance fish and wildlife, including related spawning grounds and habitat, affected by such projects . . . in a manner that provides equitable treatment for such fish and wildlife."[118] As discussed earlier in this chapter, "equitable treatment," in contrast to "equal consideration," implies more than just a procedural duty to take wildlife impacts into account. It does not, however, mean that fish and wildlife are the overriding considerations, for the Act also provides that fish and wildlife conservation measures must not jeopardize "an adequate, efficient, economical, and reliable power supply."[119]

In *Confederated Tribes and Bands of the Yakima Indian Nation*, the court recognized that "unlike the [Fish and Wildlife Coordination Act, the Northwest Power Act] imposes substantive as well as procedural obligations on FERC."[120] Because the court had already concluded, however, that FERC had failed to carry out its Federal Power Act procedural duty to consider fishery issues prior to licensing, the court did not elaborate further on what the precise nature of those substantives duties is.

In *National Wildlife Federation v. FERC*,[121] the court again declined to examine the nature of the new duties by holding that FERC had failed altogether to consider the Power Planning Council's fish and wildlife program when it issued interim permits to several license applicants. Al-

[114] *See* 16 U.S.C. § 839b(a). The Act prescribes an alternative mechanism for appointment of council members by the Secretary of Energy in the event the state appointment mechanism fails. *Id.* § 839(b).

[115] *Id.* §§ 839b(e)(3)(F), 839b(h)(1)(A).

[116] *Id.* §§ 839b(d)(2), 839d.

[117] *Id.* § 839b(h)(11)(A)(ii).

[118] *Id.* § 839b(h)(11)(A)(i).

[119] *Id.* § 839b(h)(5).

[120] *Supra* note 86 at 473.

[121] *Supra* note 98.

though the Northwest Power Act required FERC to consider the program at "each relevant stage of [its] decisionmaking processes," FERC argued that only licensing, not the issuance of interim permits, was a "relevant stage." The court rejected this claim.

The Ninth Circuit Court of Appeals provided perhaps the most thorough judicial exploration of the Northwest Power Act's requirements in *Northwest Resource Information Center, Inc. v. Northwest Power Planning Council*,[122] in which plaintiffs challenged the Northwest Power Planning Council's Columbia Basin Fish and Wildlife Program. The Council originally promulgated the Program in 1982 and later amended it several times. In December 1992, as part of a continuing process of Program revision, the Council adopted a "Strategy for Salmon" that addressed the important and highly contentious issue of water flows.[123] Indian tribes, environmental interests, and industrial energy users all sought review of the Council's decision.

In developing and revising its Fish and Wildlife Program, the Council must request from state and federal fish and wildlife agencies and from appropriate Indian tribes recommendations for measures to be included in the Program.[124] The Council must give "due weight" to those recommendations.[125] How much deference the Council must pay to the recommendations was one of the key issues raised in *Northwest Resource Information Center.*

The Council had rejected the agencies' and tribes' consensus recommendations concerning flows and biological objectives when it adopted the Strategy for Salmon. The Council defended its action by arguing that the statute vested in it a high degree of discretion, and that it need accord only a nominal amount of deference to the recommendations. The court emphatically disagreed, concluding that the Council must give a high degree of deference not only to the recommendations of the agencies and tribes, but also to their interpretation of the fish and wildlife provisions of the Northwest Power Act itself:

> We conclude that [section 4(h)(7) of the Northwest Power Act] binds, more than unleashes, the Council's discretion with respect to fish and wildlife issues. Indeed, we are convinced that the fish and wildlife provisions of the NPA and their legislative history require that a high degree of deference be given to fishery managers' interpretations of such provisions and their recommendations for program measures.[126]

[122] 35 F.3d 1371 (9th Cir. 1994).
[123] *See* 56 Fed. Reg. 56935 (1992).
[124] 16 U.S.C. § 839b(h)(2).
[125] *Id.* § 839b(h)(7).
[126] 35 F.3d at 1388.

On the basis of the agencies' and tribes' recommendations and public comments on them, the Council must include in the Program measures that meet criteria set forth in section 4(h)(6) of the Northwest Power Act.[127] The Council characterized these criteria as inherently involving a great deal of judgment, and therefore not suitable standards for judicial review. Again the court strongly disagreed, characterizing the criteria as substantive standards against which to measure proposed conservation measures.

Among the criteria by which the Program's measures must be evaluated are that they achieve "sound biological objective[s]" at "minimum economic cost."[128] Although the Strategy for Salmon purported to aim at doubling the salmon population, the court concluded that without a time frame for reaching this goal, it was an "untelling policy statement" and "a far cry from the specificity of a discrete biological objective."[129] The court, however, rejected the industrial energy users' contention that the minimum economic cost criterion required the Council to conduct a cost-benefit analysis of each measure included in the Program. The court found instead that the Council's duty was to conduct an assessment of the economic and power impacts of the Program as a whole to ensure that it did not impair an adequate, efficient, economical, and reliable power supply. The mere fact of increased costs and decreased profits was not significant, for "Congress expected increased costs and lost profits to the hydropower system to the extent the system was responsible for damaging fish and wildlife in the region."[130]

Except for finding that the Council failed to articulate a sound biological objective, the court declined to pass final judgment on whether the Program met the criteria set forth in the Northwest Power Act. The court held that the Council had violated its obligation to explain in the Program the basis for its rejection of the measures unanimously recommended by the fish agencies and tribes. Thus, the court remanded the matter to the Council to explain its action. The tenor of the court's opinion, however, was one of considerable skepticism that the Council had done its job properly. It concluded with a stinging rebuke of the Council's timidity:

> The Council's approach seems largely to have been from the premise that only small steps are possible, in light of entrenched river user claims of economic hardship. Rather than asserting its role as a regional leader, the Council has assumed the role of a consensus builder, sometimes sacrificing the Act's fish and wildlife goals for what is, in essence, the lowest common denominator acceptable to power

[127] 16 U.S.C. § 839b(h)(6).
[128] *Id.* § 839b(h)(6)(C).
[129] 35 F.3d at 1392.
[130] *Id.* at 1395.

interests and [the industries purchasing power from the Bonneville Power Administration].[131]

Like the Fish and Wildlife Coordination Act and the Federal Power Act before it, the Northwest Power Act's promise has gone unrealized. There is a significant difference, however, between the ineffectiveness of the earlier acts and that of the Northwest Power Act. The latter has defined goals, strong language, and court interpretations favorable to fish and wildlife conservation. What is lacking is the political will to implement the Northwest Power Act to achieve its conservation objectives.

[131] *Id.*

Chapter **12**

WILDLIFE CONSERVATION ON PRIVATE LAND

INTRODUCTION

Most of the land in the United States is privately owned. Because federal land is primarily located in the West, states elsewhere typically have a very high proportion of their land base under private control; in Texas, for example, only 2 percent of land is federally owned. Private land is devoted to a wide variety of uses, but agriculture and forestry predominate. Indeed, of the nation's total land base (both public and private lands), an estimated 64 percent is in agricultural use, including grazing. Meanwhile farm residents comprise less than 2 percent of the American population as of 1990, a dramatic decline from 15 percent only four decades earlier.[1]

Private land use, including farming and forestry, has had profound wildlife impacts. A long-standing and difficult challenge for wildlife conservation has been to try to influence the use of private land to advance conservation goals. The federal government has used two main approaches. One is acquisition, converting private land into public land through purchase or condemnation.[2] For the most part, the land acquired is added to the National Park, National Wildlife Refuge, or other conservation systems. At the other end of the spectrum is direct government regulation of land use, in which the government prohibits or restricts uses deemed harmful to the public interest.

[1]Agriculture No Longer Counts, The Washington Post, Oct. 9, 1993, p. A1.

[2]Land acquisition for conservation purposes by private organizations like The Nature Conservancy and numerous smaller land trusts has played an increasingly important role in overall wildlife conservation efforts. *See* National Research Council, Setting Priorities for Land Conservation (1993) at 139–156.

Each approach has significant limitations. The amount of land important to wildlife conservation objectives far exceeds the government's ability to acquire it now or, in all likelihood, anytime in the foreseeable future. Moreover, conservation theory has recently called into question the strategy of acquiring and protecting key wildlife habitat areas as "preserves" that eventually become isolated "habitat islands" surrounded by a sea of incompatible land uses.[3] Thus, even if all the most important wildlife habitat areas were publicly acquired, the very purposes that those acquisitions were intended to achieve could be undermined by activities taking place on the remaining private lands.

The alternative of government regulation of land use also has considerable limitations. First, despite a wide variety of permissible governmental land use controls, the United States Constitution draws a line beyond which the government cannot go. The Fifth Amendment provides that private property may not "be taken for public use, without just compensation." When the government crosses that line, and "takes" private property for a public use, the government must pay for it. Exactly where that line may be in any given situation has been the subject of numerous United States Supreme Court opinions. The common denominator of these many opinions is that no simple mechanical formula can be easily applied to determine when government regulation has gone "too far" and when it has not; a court must decide each case on a unique set of facts.[4] As a result, considerable uncertainty clouds the issue of how much land use regulation the Constitution permits.

Even within permissible constitutional limits, the long-standing history of land use control in the United States has been that local government is the primary unit to exercise these controls. Neither the states nor the federal government have traditionally had much of a role, and efforts to increase state or federal land regulation are almost always met with fierce resistance.

Vesting primary responsibility for land use regulation with local government serves many important public purposes. Effectively conserving wildlife, however, is not one of them. Whether one's focus of concern is a particular species, a particular natural community, or a particular ecosystem, its distribution across the landscape is almost never limited to the area within the jurisdiction of a local government land use agency. Rarely, in fact, have local governments undertaken to regulate land use for the

[3]A good review of this theory and its critics can by found in Amy McEuen, *The Wildlife Corridor Controversy: A Review*, Endangered Species Update, 10(11–12) at 1 (Sept./Oct. 1993). *See also* Brian Dayton and Richard B. Primack, *Plant Species Lost in an Isolated Conservation Area in Metropolitan Boston from 1894 to 1993*, Conservation Biology 10(1) at 30 (1996).

[4]The singularly unhelpful formulation that a government regulation constitutes a taking if it goes "too far" actually can be found in one of the Supreme Court's early takings cases. Pennsylvania Coal Co. v. Mahon, 260 U.S. 393 (1922).

purpose of conserving wildlife or related resources. Were they to try to do so, moreover, they would find themselves in a similar situation to that of the states with respect to migratory birds prior to enactment of the Migratory Bird Treaty Act: The positive efforts of one could be negated by the action or inaction of another.

Wildlife conservation goals are not likely to be achieved without strategies that go beyond acquiring or regulating private land. This chapter examines the considerable range of other strategies reflected in federal law. For example, among the most basic is simply to recognize and acclaim those landowners whose lands harbor important wildlife or other ecological resources. The National Natural Landmarks Program of the National Park Service relies upon this strategy and the expectation that many private landowners will be motivated to preserve what they have if they understand its uniqueness and value. On other lands, potential wildlife values can be achieved through active management. Providing financial assistance in the form of cost-sharing for conservation measures is another way of inducing, rather than compelling, desired action. The Forest Stewardship and Water Quality Incentives Programs administered by the Agriculture Department use this strategy.

American agriculture has long benefitted from a variety of taxpayer-financed programs to stabilize prices, insure against adverse weather, provide access to distant markets, and achieve other goals. The "Swampbuster" program under the 1985 and 1990 Farm Bills conditions eligibility for some of these benefits upon wildlife-related environmental performance. The Farm Bills also initiated important new federal programs to secure wildlife conservation and other conservation benefits through contractual agreements or acquisition of less-than-fee interests in land. The Conservation Reserve and Wetlands Reserve Programs rely upon these approaches. Conceptually, the acquisition of less-than-fee interests is only a variant of the traditional public acquisition approach, of which there are a number of other examples.

The discussion that follows is not intended to be an exhaustive analysis of any of these federal programs, most of which have broader objectives than just wildlife conservation. Rather, its purpose is to give an overview that highlights those aspects most important to wildlife conservation.

RECOGNITION AS ITS OWN REWARD: THE NATIONAL NATURAL LANDMARKS PROGRAM

The premise of the National Natural Landmarks Program is that landowners who are told that their land harbors particularly valuable wildlife or other environmental resources often will seek to maintain those special values out of a sense of civic duty or responsible stewardship. This program, administered by the National Park Service, seeks to identify and encourage the voluntary preservation of "nationally significant examples of the full range of ecological and geological features that constitute the nation's natural heritage."[5] The Secretary of the Interior places sites harboring such features on a National Register of Natural Landmarks and invites the owners of registered sites to enter into voluntary agreements to protect the sites' values. A landowner need not enter into an agreement in order for a site to be placed on the register. Those who do, however, receive a certificate and a bronze plaque. Landowner agreements are revocable and do not convey any property interest to the Secretary.

The National Natural Landmarks Program originated administratively as a result of a creative interpretation of the Historic Sites, Buildings, and Antiquities Act of 1935.[6] That statute declared a "national policy to preserve for public use historic sites, buildings, and objects of national significance for the inspiration and benefit of the people of the United States."[7] Reasoning that significant ecological and geological areas or features were "objects of national significance," Secretary of the Interior Stewart Udall created the program in 1962.[8] By early 1993, some 587 sites had been designated as National Natural Landmarks in every state except Louisiana and Delaware.

It was not until fourteen years after the program's creation that federal legislation first made reference to it. In the 1976 amendments to the National Park Service Act, Congress directed the Secretary to submit annually a complete list of all National Natural Landmarks that "exhibit known or anticipated damage or threats to the integrity of their resources, along with notations as to the nature and severity of such damage or threats."[9] Because this requirement is included in a larger section devoted to the study of areas potentially suitable for inclusion in the National Park System, there is at least an implicit suggestion that landmarks under threat might become priorities for Park Service acquisition. In practice, however, this is unlikely, given the very small size of most National Natural Landmarks; nearly two-thirds are less than 5,000 acres.[10]

As originally envisaged, the National Natural Landmarks Program encompassed sites with qualifying geological, paleontological, or ecological characteristics. A 1973 *Federal Register* notice describing the program listed ten examples of the kinds of areas that could qualify for designation; six of these referred to ecological communities, habitats, or species. Among the examples given were "a habitat supporting a vanishing, rare, or re-

[5]36 C.F.R. § 62.1 (1996).
[6]16 U.S.C. § 461 *et seq.*
[7]16 U.S.C. § 461.
[8]A very useful discussion of the program and its history can be found in National Park Service, Preserving Our Natural Heritage, Volume I: Federal Activities at 285–98 (1977).
[9]16 U.S.C. § 1a-5.
[10]*See* 36 C.F.R. § 62.5(a) (1996).

stricted species," "a relict flora or fauna persisting from an earlier period," and "a seasonal haven for concentrations of native animals, or a vantage point for observing concentrated populations, such as a constricted migration route."[11]

It was not until 1979 that any formal regulations governing the program were established. Interim regulations published that year were replaced the following year with permanent regulations that appeared to give less emphasis to wildlife. The criteria governing landmark designation were intended to recognize "areas which best represent the ecological and geological character of the United States, including, for example, terrestrial communities, aquatic communities, landforms, geological features, habitats of native plant and animal species, and fossil evidence of the development of life on earth."[12] Earlier references to rare, relict, or vanishing species were dropped. Further regulatory revisions in 1987 appeared to relegate wildlife values to an even less prominent status. Primary emphasis is to be given to sites that exemplify a natural region's characteristic biota or geologic features; the presence of high quality habitat for rare, threatened, or endangered species is a secondary criterion for designation.[13]

Lands eligible for designation as National Natural Landmarks can be owned by any private or public owner, other than the National Park Service itself. The first landmarks designated were either privately or municipally owned.[14] Of the 587 sites designated as of mid-1994, well under half were privately owned. Of these, voluntary conservation agreements are in place for no more than half. Federal and state agencies manage a majority of the registered sites, most of which are subject to voluntary conservation agreements.

Interest in the National Natural Landmarks Program has waned since the Carter administration. Since the end of that administration in January 1981, the National Register of Natural Landmarks has added fewer than fifty sites, an annual growth of about four sites.[15] The Register has grown by only one site since 1987. A further indication of waning interest is the fact that the required reports to Congress concerning damaged or threatened landmarks were never submitted for 1990 and 1991.

It is difficult to assess the effectiveness of the National Natural Landmarks Program. According to the most recent report to Congress in May 1995, 58 of the 295 sites for which information was available (20 percent)

[11]38 Fed. Reg. 23982 (1973).

[12]45 Fed. Reg. 81192 (1980).

[13]36 C.F.R. § 62.5 (1996).

[14]The first National Natural Landmark designated was privately owned land within the Mianus River Gorge in Westchester County, New York. It offers an exceptional illustration of piedmont physiography and geomorphology and contains an excellent climax hemlock forest.

[15]*See* National Park Service, National Registry of Natural Landmarks (1994).

were judged to be damaged or under threat.[16] At least some of these sites are threatened by development activities planned by their private owners. Thus, not surprisingly, mere "recognition" of outstanding values on a site will not necessarily induce the owner of the site to preserve those values. The observation of the former Bureau of Outdoor Recreation (since subsumed within the National Park Service) in 1973 seems as true today as when it was originally made:

> While the program represents an important step forward in preserving natural areas, the protection it affords frequently is inadequate and the incentives it provides insufficient. The greatest drawback is lack of funds for acquisition and for payment to landowners for maintenance and management.[17]

Other programs that seek to overcome these drawbacks have since been developed. They are examined in detail in the pages that follow.

CONTRACTING FOR CONSERVATION BENEFITS: THE CONSERVATION RESERVE PROGRAM

American farmers historically have been given strong incentives to grow a variety of key commodities. During the period 1985 through 1992, farmers received nearly $120 billion in taxpayer-funded commodity price supports, disaster relief, and other subsidies for agricultural production.[18] Taking land out of agricultural production to further conservation objectives not only offers no reward to farmers, but threatens their eligibility to continue to receive the subsidies to which they have become accustomed. Keeping land in production, particularly marginal land subject to relatively high erosion or needing relatively large inputs of fertilizer and pesticides, reduces wildlife habitat and contributes to a variety of other environmental problems affecting wildlife.[19] In 1985, Congress recognized that economic incentives for farmers to consider alternatives to intensive crop production would increase opportunities to restore wildlife and its habitat and took significant steps to create these incentives.

On a more or less regular five-year cycle, Congress comprehensively reviews and revises the statutory authority for a host of agricultural programs. The 1985 review resulted in the Food Security Act. One of the Act's eighteen different titles (Title XII) created several new programs aimed at achieving environmental goals rather than traditional agricul-

[16] National Park Service, Damaged and Threatened National Natural Landmarks, 1994 (1995) at 2.

[17] Bureau of Outdoor Recreation, Outdoor Recreation—A Legacy for America at 39 (1973).

[18] Government Support of U.S. Agriculture, San Francisco Chronicle, July 30, 1992.

[19] The power of agricultural subsidy programs to influence farmer behavior is described well in National Research Council, Alternative Agriculture, at 65–85 (1989).

tural goals. Among these was the creation of a "Conservation Reserve Program"[20] that has since been hailed as "one of the greatest contributions to soil, water and wildlife conservation that has ever been seen in North America."[21]

The purpose of the Conservation Reserve Program was to remove up to 45 million acres of farmland from crop production for at least a decade and to require that this land be planted in permanent cover, *i.e.*, grasses, trees, or other perennial forbs. Prior to the Conservation Reserve Program, up to 78 million acres of cropland were withheld from production each year as a result of annual "set-aside" programs that sought to control commodity production by limiting planted acreage. Annual set-asides, however, offered few environmental benefits. Set-aside lands often were left fallow and were highly vulnerable to erosion. Even if planted, the fact that they could be returned to crop production the following year severely limited their value as wildlife habitat. The Conservation Reserve Program, by providing for relatively long-term land retirement, and by requiring planting in permanent cover, sought to overcome the shortcomings of the annual set-aside program.[22]

Participation in the Conservation Reserve Program is voluntary. Farmers with eligible lands submit contract offers to participate in the program and a plan to establish approved vegetative or water cover to enhance wildlife habitat and to carry out other conservation or water quality measures on those lands. The Secretary of Agriculture, through the Natural Resources Conservation Service, determines which offers to accept, taking into account "the extent to which enrollment of the land that is the subject of the contract offer would improve soil resources, water quality, wildlife habitat, or provide other environmental benefits."[23] Contracts run for a period of not less than ten, nor more than fifteen, years.[24]

Farmers who participate in the program receive annual rental payments in cash or commodities. In addition, for those water quality or conserva-

[20]The Conservation Reserve Program is codified at 16 U.S.C. §§ 3831–36.

[21]Mark J. Reefe, Conservation Reserve Program: Ten Years of Providing Habitat, in Linking Agriculture and Resource Conservation Programs, Dec. 17, 1993 (Wildlife Management Institute).

[22]The Conservation Reserve Program has many features in common with the earlier Water Bank Program for Wetlands Preservation authorized by the Water Bank Act of 1970, 16 U.S.C. §§ 1301–11. Under that program, the Secretary of Agriculture entered into 10-year renewable agreements with private landowners and operators in important migratory waterfowl nesting and breeding areas. In return for payment of an agreed upon annual fee, the participating landowner or operator agreed "not to drain, burn, fill, or otherwise destroy the wetland character" of areas included in the program, or "to use such areas for agricultural purposes." *Id.* § 1303. Violations could result in termination of the agreement, forfeiture of future payments and a requirement to refund past payments. *Id.*

[23]*Id.* § 3834(c)(3)(A).

[24]*Id.* § 3831(e).

tion measures and practices determined by the Secretary to be appropriate for cost-sharing, the Secretary pays 50 percent of the cost of establishing them.[25] Cost-sharing is statutorily required for certain practices, including the establishment of wildlife corridors.[26] A participating landowner may receive cost-sharing for the same measures and practices from other nonfederal sources, but the total of such receipts may not exceed 100 percent of the cost.[27]

The program's original emphasis in 1985 was on "highly erodible croplands." By giving these lands an extended rest from production, several goals could be achieved, among them reducing erosion and improving water quality. The Food, Agriculture, Conservation and Trade Act of 1990 broadened the scope of lands eligible for participation in the Conservation Reserve Program. Today, the list of eligible lands includes highly erodible cropland, certain marginal pasture lands, cropland that will be devoted to "permanent wildlife habitat" or other specified purposes, and other cropland that contributes to the degradation of water quality or poses other environmental threats.[28]

Contracts entered into prior to the 1990 amendments typically provided that enrolled lands would be devoted to vegetative cover. The 1990 amendments authorized the conversion of lands previously entered into the reserve from vegetative cover to hardwood trees, windbreaks, shelterbelts, wildlife corridors, or wetlands. In the case of conversion to wetlands, the farmer must convey a long-term or permanent easement to the Secretary; for conversion to any of the other specified purposes, the original contract, if for a term of less than fifteen years, can be extended to fifteen years.[29]

While enjoying substantial discretion to decide which lands are most appropriate for inclusion in the Conservation Reserve Program, the Secretary's authority is not unbounded. The statute provides that he shall designate as "conservation priority areas" those watersheds with "significant adverse water quality or habitat impacts related to agricultural production activities," upon the application of the appropriate state agency.[30] With respect to designated areas, the Secretary must "promot[e] a significant level of enrollment of lands within such watersheds in the program."[31] Total rental payments to any one owner or operator, however, may not exceed $50,000 in any fiscal year.[32] Further, the 1990 amend-

[25] *Id.* § 3834(b).
[26] *Id.* § 3834(b)(3).
[27] *Id.* § 3834(b)(2).
[28] Eligible lands are described in more detail at *id.* § 3831(b).
[29] *See id.* § 3835a.
[30] *Id.* § 3831(f).
[31] *Id.* § 3831(f)(4).
[32] *Id.* § 3834(f)(1).

ments fixed targets or limits for certain categories of lands to be added to the reserve thereafter. Under those amendments, the Secretary is to seek to ensure that at least one-eighth of the land added to the reserve during the period 1991–95 be "devoted to trees, or devoted to shrubs or other non-crop vegetation or water that may provide a permanent habitat for wildlife."[33] On the other hand, no more than a tenth of the acreage added during that period may be marginal pasture land devoted to trees in or near riparian areas.[34]

Farmer participation in the Conservation Reserve Program has been impressive. In its first five years, the program enrolled nearly 34 million acres. By 1992, some 36.5 million acres had been enrolled, nearly 10 percent of all cropland in the United States. Many of those lands, however, are nearing the expiration of their contract terms. What will become of them is a key question with which Congress will soon grapple. That question was actually anticipated by Congress in 1990, when it directed the Secretary of Agriculture to complete a report, by December 31, 1993, of cropland subject to expiring conservation reserve contracts.[35] That study was to examine a wide range of possible alternatives, from acquisition of the land to letting the contracts simply expire. As 1995 drew to a close, however, the future of the Conservation Reserve Program was very much in doubt. Caught up in the political impasse over reducing the budget deficit, legislation to renew the Conservation Reserve Program and other farm programs had not passed either the House or Senate as of the end of 1995.

EASEMENTS FOR ATTAINING ENVIRONMENTAL GOALS ON PRIVATE LANDS: THE WETLANDS RESERVE, ENVIRONMENTAL EASEMENT, AND FOREST LEGACY PROGRAMS

An easement is an interest in land, a "property right" held by an owner other than the owner of the land itself. The death of the landowner or the sale of his land does not extinguish the easement. The rights acquired by subsequent landowners remain subject to the easement. Easements can serve a wide variety of purposes, including guaranteeing public or private access, maintaining open space, and otherwise limiting development on a parcel of land. Easements can be acquired by the government or by private interests. Considered here are three recent federal programs that seek to conserve wildlife habitat and other environmental benefits primarily through the acquisition of easements.

[33] *Id.* § 3832(c).
[34] *Id.* § 3831(b)(3).
[35] Pub. L. 101–624, § 1437, 104 Stat. 3584 (1990).

The Wetlands Reserve Program

Five years after creating the Conservation Reserve Program, Congress created a similar "Wetlands Reserve Program" as part of the Food, Agriculture, Conservation and Trade Act of 1990. Unlike the Conservation Reserve Program, however, which operates through ten-year contracts, the Wetlands Reserve Program operates through the federal purchase of long-term or permanent conservation easements, interests in land that bind successive owners of the land.

Congress's goal was to enroll nearly a million acres of land in the Wetlands Reserve Program by the end of the year 2000.[36] Eligible lands included farmed wetlands, wetlands that had been drained, dredged, filled, or otherwise altered to produce agricultural commodities (*i.e.*, "converted" wetlands), certain other wetlands, and riparian areas linking protected wetlands.[37] Farmers with eligible lands can place their lands into the reserve by entering into an agreement with the Secretary of Agriculture to grant an easement to him and to implement a "wetland easement conservation plan."[38] Under such a plan, the owner must "provide for the restoration and protection of the functional values" of the wetlands that are the subject of the plan.[39] The cost of carrying out the measures and practices required by a plan can be shared by the Secretary. The fraction of eligible costs borne by the Secretary depends upon the length of the easement. For nonpermanent easements, the Secretary's share of the costs must be at least 50 percent, but may not exceed 75 percent; for permanent easements, the Secretary's share must be at least 75 percent, but may not exceed 100 percent.[40] In addition, the Secretary pays the landowner for the easement itself; the amount of the payment is negotiated, but may not exceed the fair market value of the easement.[41]

Although many states recognize permanent easements, some limit their term. The Wetlands Reserve Program authorizes the purchase of permanent, thirty-year, or maximum duration easements allowed under applicable state law. The Secretary must, however, give priority to obtaining permanent easements. In addition, in consultation with the Secretary of the Interior, he is to "place priority on acquiring easements based on the value of the easement for protecting and enhancing habitat for migratory birds and other wildlife."[42]

The Wetlands Reserve Program is, in effect, another federal land ac-

[36] 16 U.S.C. § 3837(b).
[37] *Id.* § 3837(c) and (d).
[38] *Id.* § 3837a(a).
[39] *Id.* § 3837a(b).
[40] *Id.* § 3837c(b).
[41] *Id.* § 3837a(f).
[42] *Id.* § 3837c(d).

quisition program. By focusing on the purchase of easements rather than fee title, however, the program seeks to stretch acquisition dollars over a much larger acreage and to keep lands in private ownership. Landowners retain the right to exclude others from access to their land and to use their land in any manner consistent with the terms of the easement. Thus, the Wetlands Reserve Program is a major new experiment aimed at achieving environmental objectives on private land.

The Environmental Easement Program

At the same time that Congress established the Wetlands Reserve Program, it created a similar "Environmental Easement Program."[43] This program seeks to "ensure the continued long-term protection of environmentally sensitive lands . . . or reduction in the degradation of water quality" by acquiring permanent or maximum legal duration easements from the owners of eligible farms and ranches.[44] In part, the Environmental Easement Program was intended to provide a means of securing longer-term protection of lands entered into the Conservation Reserve Program than the latter program alone provided. Lands entered into the Conservation Reserve Program, as well as other cropland that contains riparian corridors or other environmentally sensitive areas, or that is "an area of critical habitat for wildlife, especially threatened or endangered species," are all eligible for inclusion in the program.[45]

The Environmental Easement Program is procedurally similar to the Wetlands Reserve Program. A landowner conveys an easement to the Secretary of Agriculture and agrees to implement a "natural resource conservation management plan."[46] In addition to being paid for the easement itself, the landowner is entitled to receive cost-share payments for "carrying out the establishment of conservation measures and practices set forth in the plan."[47]

Although conservation easements leave land in private ownership, they do reduce the potential uses of the land for economic activities. Thus, aggressive conservation easement acquisition programs can have local tax and other impacts similar to federal land acquisition programs. For that reason, Congress has provided that, except in special circumstances, no more than 25 percent of the cropland in any county can be enrolled in the Conservation Reserve and Environmental Easement Programs and no

[43] *Id.* § 3839–3839d.
[44] *Id.* § 3839(a).
[45] *Id.* § 3839(b).
[46] *Id.* § 3839a(a)(1).
[47] *Id.* § 3839b.

more than 10 percent may be subject to an easement under those programs.[48]

The Forest Legacy Program

Paralleling the Environmental Easement Program for farms and ranches is a "Forest Legacy Program" aimed at "protecting environmentally important forest areas that are threatened by conversion to nonforest uses."[49] Through conservation easements and other mechanisms, the program seeks to promote forest land protection as well as "the protection of important scenic, cultural, fish, wildlife, and recreational resources, riparian areas, and other ecological values."[50] The Secretary of Agriculture pays fair market value for the property interest acquired, and the participating landowner is required to manage his property in a "manner that is consistent with the purposes for which the land was entered" into the program.[51] Hunting, fishing, hiking, and similar recreational uses are consistent with the program; timber management activities may also be deemed consistent if the Secretary so determines.[52]

COST-SHARING AND TECHNICAL ASSISTANCE: THE FOREST STEWARDSHIP AND STEWARDSHIP INCENTIVES PROGRAMS

Most of the forest land of the United States is owned not by the U.S. Forest Service and other federal agencies, but by private, state, and local government owners. Assisting in the management of nonfederal forest land through technical assistance and similar means has been a long-standing federal objective. In 1978, as part of the Cooperative Forestry Assistance Act, Congress created a "Forestry Incentives Program" for the purpose of encouraging the development, management, and protection of forest land owned by small landowners, the so-called "nonindustrial private forest" landowners.[53] The rather narrow focus of that program was on "land capable of producing crops of industrial wood."[54] Eligible landowners who developed forest management plans in cooperation with the state forester or equivalent official were entitled to receive from the Sec-

[48] *Id.* § 3843(f).

[49] *Id.* § 2103c(a). The Forest Legacy Program was created by Pub. L. 101–624, § 1217, 104 Stat. 3528 (1990).

[50] *Id.*

[51] *Id.* § 2103c(i)-(j).

[52] *Id.* § 2103c(i).

[53] The Cooperative Forestry Assistance Act of 1978, Pub. L. 95–313, 92 Stat. 365, is codified at 16 U.S.C. § 2101 *et seq.* The Forestry Incentives Program, created by Section 4 of the 1978 Act, is codified at 16 U.S.C. § 2103.

[54] 16 U.S.C. § 2103(b).

retary of Agriculture up to 75 percent of the cost of implementing the plan.

A much more broadly focused program was authorized in 1990 to take the place of the original program, which was to be phased out by the end of 1995. The Forest Stewardship Program, though again limited to small landowners, is aimed at all rural lands with existing tree cover or suitable for growing trees.[55] The objective of the Forest Stewardship Program is to provide technical and other assistance to enable small landowners to develop "forest stewardship plans" that encompass a broader set of objectives than just timber production. Specifically, such plans may include measures to "protect soil, water, range, aesthetic quality, recreation, timber, water, and fish and wildlife resources" that are compatible with the landowner's objectives.[56] In addition to the assistance provided in preparing plans, the Secretary is to bestow appropriate recognition on participating landowners, much like the National Natural Landmarks Program.[57]

Assistance in developing a land management plan and recognition for having done so are not the only inducements for landowner participation, however. In addition, under a "Stewardship Incentives Program" created as part of the same 1990 legislation, landowners with a stewardship plan approved by the state forester and who agree to implement the activities in that plan for at least ten years are entitled to receive cost-sharing assistance of up to 75 percent of the cost for approved practices. Approved practices are those that serve any of several specified purposes, of which timber production is but one. Others include "the management and maintenance of fish and wildlife habitat," the establishment and management of forests for aesthetic quality, the management of outdoor recreational opportunities, and others relating to wetlands, water quality, and energy conservation.[58] Through the Forest Stewardship Program, owners of tracts of forest land of up to 5,000 acres in size who wish to manage their lands to achieve wildlife conservation and other environmental benefits do so and have part of the cost borne by the federal government.

ADDING INCENTIVE PAYMENTS TO COST-SHARING: THE AGRICULTURAL WATER QUALITY INCENTIVES PROGRAM

A substantial regulatory framework exists under the Clean Water Act to control water pollution from factories, sewage treatment plants, and similar facilities that discharge waste directly into the nation's waters. This regulatory scheme has achieved significant reductions in water pollution

[55] *Id.* § 2103a(c).
[56] *Id.* § 2103a(f).
[57] *Id.* § 2103a(g).
[58] *Id.* § 2103b(b)(4).

from these "point source" dischargers. No comparable framework exists to control the more diffuse sources of water pollution from urban and agricultural runoff, which carries fertilizers, pesticides, animal wastes, oil, and other pollutants into the nation's waters. These "nonpoint source" discharges have proven highly intractable.

The Food, Agriculture, Conservation and Trade Act of 1990 sought to encourage efforts to minimize the generation and release of agricultural pollutants by offering annual incentive payments and technical assistance to farmers who agree to implement a water quality protection plan approved by the Secretary of Agriculture.[59] Congress set a goal of enrolling 10 million acres of farmland into this "Agricultural Water Quality Incentives Program" through the mechanism of three- to five-year agreements with landowners.[60] Annual payments to participating farmers are based upon the costs they incur, production values foregone, and the amount necessary to encourage participation.[61]

Although the primary focus of the Agricultural Water Quality Incentives Program is on controlling water pollution, wildlife conservation considerations enter explicitly into the program in at least two ways. First, only certain lands are eligible for enrollment in the program. These include "areas where agricultural nonpoint sources have been determined to pose a significant threat to habitat utilized by threatened and endangered species."[62] Thus, the program can be used to further endangered species conservation goals.

In addition, the program offers cost-sharing assistance to participants who voluntarily agree to develop and implement practices "that preserve and enhance wetland or wildlife habitat."[63] For wildlife conservation, such practices include "the establishment of perennial cover, the protection of riparian areas, wildlife corridors, and areas of critical habitat for endangered species."[64] Cost-share assistance may not exceed 50 percent of the cost of the eligible practice and $1,500 per person per contract.

LIMITING ELIGIBILITY FOR GOVERNMENT BENEFITS AS A MEANS OF PROTECTING THE ENVIRONMENT: THE SWAMPBUSTER PROGRAM AND THE COASTAL BARRIER RESOURCES ACT

It is a common observation, indeed, a common complaint, that governmental regulation of private activity has become pervasive. Less commonly

[59] *Id.* § 3838–3838f.
[60] *Id.* § 3838b(a)(11).
[61] *Id.* § 3838b(a)(6)(B)(i).
[62] *Id.* § 3838c(a)(4).
[63] *Id.* § 3838b(a)(4)(A).
[64] *Id.* § 3838b(a)(4)(B)(ii).

complained of, except by taxpayer interests, is the pervasiveness of government largesse in the form of subsidies and other inducements favoring particular industries or economic activities. Often these governmental programs contribute to the destruction of wildlife habitat or other environmental degradation. The "strings" attached to these governmental benefits can be powerful inducements affecting private behavior. This section examines two recent efforts to secure wildlife and other environmental benefits by attaching appropriate strings.

Swampbuster

Whereas the Conservation Reserve and Agricultural Water Quality Incentives Programs offer financial incentives to farmers who agree to carry out environmentally beneficial practices, the "Swampbuster" provisions of the 1985 and 1990 farm bills take the opposite approach. They withhold highly valuable benefits from farmers who carry out harmful practices. Specifically, any person who, after December 23, 1985, drains, dredges, fills, or otherwise alters a wetland for the purpose of producing an agricultural commodity thereon, forfeits his eligibility for price supports, federal crop insurance, Conservation Reserve Program benefits, various agricultural loans, and other benefits to which he would otherwise be entitled.[65] Significantly, this ineligibility for benefits applies not just to the "converted wetland" and the commodities grown on it, but to all commodities on all lands owned by the responsible person. Thus, it serves as a powerful disincentive against the conversion of wetlands to crop production.

Under the original Swampbuster provisions in 1985, a farmer suffered forfeiture of benefits only in a crop year in which he produced agricultural commodities on the converted wetlands. Thus, in subsequent years, if he refrained from producing commodities on the converted wetlands, he would not forfeit any benefits.[66] In 1990, however, amendments greatly broadened the forfeiture provisions. Now, any person who thereafter converts a wetland for the purpose of producing an agricultural commodity forfeits eligibility for the specified benefits "for that crop year and all subsequent crop years."[67] At the same time that Congress was, with one hand, waving this heavy stick against potential malefactors, it was, with the other, extending a carrot to encourage restoration of converted wetlands. A person determined to be ineligible for benefits can regain eligibility "by

[65]The "Swampbuster" provisions described here are roughly paralleled by "Sodbuster" provisions that seek to discourage agricultural production on highly erodible land. The Sodbuster provisions are found at *id.* §§ 3811–13.

[66]This original scope of the Swampbuster requirement is codified at *id.* § 3821(a).

[67]*Id.* § 3821(b).

fully restor[ing] the characteristics of the converted wetland to its prior wetland state.''[68] In addition, the 1990 amendments authorize the Secretary of Agriculture to exempt a person from the ineligibility provisions if he restores a different converted wetland in the same general area of the local watershed at nonfederal expense and makes the restored wetland subject to an easement that will maintain its functions and values.[69]

The 1990 amendments also introduced the possibility of ''graduated sanctions'' as an alternative to a complete prohibition of eligibility for benefits. If the Secretary of Agriculture determines that a person converted a wetland or produced a commodity on a converted wetland ''in good faith and without the intent to violate'' the law's requirements, and if the person has restored the wetland or is in the process of doing so, the Secretary may reduce the benefits due by up to $10,000.[70] The Secretary's decision to apply this provision retroactively, to wetlands converted prior to the 1990 amendments, was upheld in *National Wildlife Federation v. Agricultural Stabilization and Conservation Service.*[71] Notwithstanding these apparent relaxations, the potential loss of vital benefits is an effective deterrent aimed at reducing the loss of wetlands to agricultural expansion.

The Coastal Barrier Resources Act

Swampbuster's approach is a variant on the approach used by the Coastal Barrier Resources Act of 1982 to discourage certain coastal development.[72] The latter Act sought to discourage development on relatively undeveloped coastal barrier islands on the Gulf and Atlantic coasts that were included in a ''Coastal Barrier Resources System.''[73] Like Swampbuster, the Coastal Barrier Resources Act does not prohibit outright any development on areas within the System. Rather, it proscribes any new federal expenditures or financial assistance to construct or purchase roads, airports, bridges, causeways, or other infrastructure, as well as to carry out most erosion control projects and, in general, any other form of expenditure or assistance that is in any way related to development of such areas.[74] In this way, private land owners and others interested in

[68] *Id.* § 3822(i). Wetland restoration plans are, in most instances, to be agreed upon by the Soil Conservation Service and the U.S. Fish and Wildlife Service. *See id.* § 3822(j).

[69] *Id.* § 3822(f)(2) and (3).

[70] *Id.* § 3822(h)(1)-(2).

[71] 941 F.2d 667 (8th Cir. 1991).

[72] 16 U.S.C. §§ 3501–10.

[73] In 1988, the System was expanded to include certain areas along the shore of the Great Lakes.

[74] 16 U.S.C. § 3504. A variety of exceptions related to energy development, military activities, management and protection of fish and wildlife, and other specified purposes are set forth at *id.* § 3505. *See also id.* § 3502(3).

development are free to develop, but without the federal government's direct or indirect assistance.

The mandates of the Coastal Barrier Resources Act extend to all federal agencies. To ensure compliance with its requirements, the Director of the Office of Management and Budget, on behalf of each federal agency concerned, must annually certify to Congress that such agency is in compliance.[75]

The Coastal Barrier Resources System is unlike other "systems," such as the National Park System and the National Wildlife Refuge System, in that it does not imply any federal ownership of or management authority over areas included within the System. Inclusion of areas within the System does not affect their ownership and gives the government no new authority to regulate or control activities in such areas. It may, however, be no less effective in protecting the environment than outright regulation would be.

[75] *Id.* § 3506.

Part **IV**

INTERGOVERNMENTAL WILDLIFE CONSERVATION

The final part of this book examines the complexities that arise with respect to wildlife conservation when the United States deals with other sovereign entities. Chapter 13 focuses on federal and state relations with Native American tribes, while Chapter 14 examines several major international wildlife treaties.

There are important similarities between Native American tribes and foreign nations in their relations with the federal government. Issues of sovereignty are paramount in transactions between independent governments. Native American tribes deal government-to-government with the United States, as foreign nations do. Moreover, the United States Constitution deems treaties the "supreme Law of the Land," along with the Constitution itself and federal laws. This applies to both tribal and international treaties. State laws do not enjoy this status and often are preempted by federal laws.

There also are significant differences between federal-tribe relations and foreign relations. One is that Indian treaties are not, in fact, agreements forged between equal parties. Native Americans ceded much of their lands and waters to the United States in the nineteenth century, after violent battles or under threat of war, as European settlers moved westward and displaced tribes from their lands. Agreements were drafted by federal officials, and tribal representatives rarely were given accurate or complete information as to what they were signing. Further, of critical importance in the field of wildlife law, Congress unilaterally may abrogate tribal treaty rights. In some ways this principle is akin to federal preemption of state law. Finally, while foreign nations have sole jurisdiction over their lands, Native Americans do not; the federal government manages many tribal lands as a "trustee" for tribal interests.

The treaties under which tribes ceded their lands reserved to the tribes certain rights, both on the lands they retained and the lands they ceded. The reserved rights most relevant to wildlife conservation are the rights to hunt and fish. These are central to the Native American culture and economy. Tribes depended upon hunting, fishing, and gathering to provide food, clothing, and shelter. The United States Supreme Court in 1905 characterized the ability to hunt and fish as "not much less necessary to the existence of the Indians than the atmosphere they breathed" (*United States v. Winans*). For many tribes fish remain a primary food source. For most tribes fish and wildlife are still focal points of their culture and religion.

Chapter 13 examines the relationship between Native American rights to hunt and fish, and wildlife regulation by the state and federal governments. The chapter covers state and federal regulation of tribal hunting and fishing (both on- and off-reservation), on-reservation tribal regulation of hunting and fishing, and Native American religious practices and wildlife conservation laws. A related topic, Alaskan native subsistence rights to hunt protected species, is covered in several other chapters.

Today Native Americans are caught in an unenviable bind between a need and desire to exercise their fishing and hunting rights, and knowledge that populations of the animals they fish and hunt are in steep decline. The tribes relied for thousands of years on resources that have faded from abundance to scarcity over the span of a century, or in some cases, mere decades. The causes of decline are many, and Native Americans generally believe it is unfair for them to bear a heavy burden for recovery efforts since they do not bear primary responsibility for the state of the resources. There are no easy resolutions to either the practical dilemmas or the resulting legal conflicts.

Chapter 14 explores the different set of challenges presented in international wildlife conservation. Many species of fish and wildlife cannot be conserved without international cooperation because of their migratory nature. Others are threatened with extinction by the huge global market for wildlife and wildlife products. Still others are found in countries that are too poor to afford the luxury of conserving their living resources without aid from richer nations.

Forging legal instruments to conserve wildlife on a global scale is a delicate and often frustrating process. Unlike in the domestic arena, where the Constitution is acknowledged by all to be the "supreme law," there is no global constitution to which nations have agreed and under which laws that bind sovereign nations may be developed. Instead, relations between countries are governed by contracts, called treaties or conventions, freely negotiated by wholly independent governments.

Many wildlife treaties must be agreed to by a large number of nations if all those that can affect the species' status are to be bound. Often a choice must be made between excluding nations that object to strong conservation measures, but which can thwart the agreement's purpose if not bound by it, and accommodating the desires of the necessary parties by diluting the treaty's force. Even assuming that all requisite countries agree to the model conservation treaty, enforcement can be lax due to the signatories' lack of trust, will, or resources.

Despite these hurdles, the United States has committed to a substantial number of ambitious international agreements to conserve wildlife in this century. Many are currently in force. Chapter 14 examines the evolution of selected wildlife treaties. These include agreements pertaining to particular categories of wildlife (fur seals, whales, polar bears, and endangered species), and two significant international efforts to conserve ecosystems and biodiversity on a broad scale: the legal regime for conservation of the Antarctic ecosystem, and the 1992 Convention on the Conservation of Biological Diversity. Agreements regarding migratory birds and ocean fish are covered in earlier chapters. The United States is a party to all of these agreements except the 1992 Biological Diversity Convention and the Law of the Sea Convention, discussed in Chapter 6.

One can discern in the evolution of international wildlife law a pattern similar to that of domestic law. The initial impetus for many international wildlife treaties was the need to allocate common economic resources like whales, driven by fear that other nations would take more than their share. Then nations realized that the coveted resource was dwindling and that no one would be able to use it in the future if steps were not taken to reduce the rate of exploitation. More recently, leaders have become aware of the global value of healthy ecosystems and the threats they face. Nations are attempting to preserve whole ecosystems and life forms that have no obvious economic value in and of themselves.

The future of wildlife conservation depends in part upon the ability of nations to structure, finance, and enforce creative legal instruments that will curb the impacts on wildlife of a swelling human population and dramatic growth in per capita consumption. International laws must incorporate the full range of our social, ethical, scientific, economic, and technological insights if the other life forms with which we share the earth are to survive for the long term.

Chapter 13

WILDLIFE AND NATIVE AMERICANS

INTRODUCTION

The unique legal status of Native American tribes raises questions as to how wildlife laws apply to tribes and their land. One aspect of this status derives from treaties under which the United States acquired much of its land. In the nineteenth century, tribes ceded lands and waters to the United States in return for guarantees of rights on lands and waters reserved to the Indians and sometimes on the ceded areas as well. These formal treaties are in many ways analogous to international treaties between the United States and foreign nations. As such, they are part of the "supreme law of the land" under the Constitution.

Treaties with Indian tribes are, however, different from other treaties. First, the United States is a "trustee" for the interests of its Native American citizens. Moreover, the Indian treaties were rarely, if ever, agreements between equal parties. The tribes often were at enormous practical disadvantages, effectively forced to sign documents that were drafted by United States government officials and that were not thoroughly or accurately explained to tribal representatives. Courts have sought to rectify these inequities by using rules of construction that favor Native American interests when they interpret treaties.[1]

Not all relations with Native Americans are governed by treaties. Other legal instruments such as executive orders, which established reservations for some tribes, confer special status on these tribes. The government also has a special trust responsibility toward Alaskan natives and recognizes

[1] *See* Choctaw Nation v. United States, 119 U.S. 1, 28 (1886).

certain "aboriginal rights" despite the lack of formal treaties between the government and Alaskan natives. Like treaties, executive orders and aboriginal rights affect application of wildlife laws to Native Americans.

Native Americans present unique issues in wildlife regulation for yet another reason. Many of their religious practices use wildlife or wildlife parts. Regulations aimed at conserving wildlife may affect the exercise of Native American religious practices in ways that do not affect religious practices of other citizens. Native American claims of constitutionally protected free exercise of religion have tested the limits of wildlife regulation as applied to these practices.

STATE AND FEDERAL REGULATION OF NATIVE AMERICAN HUNTING AND FISHING

Many treaties explicitly reserved to the Indians the right to hunt or fish on certain ceded lands or waters, as well as on their reservations. Others were silent on this issue, though courts generally have held that reserving lands for a tribe's exclusive use implies the right to hunt and fish on the reserved lands.[2] Several cases have addressed the issue of how state or federal wildlife conservation laws affect Native American hunting and fishing treaty rights.

State Authority over Off-Reservation Hunting and Fishing

States may regulate taking of wildlife within their borders, in the absence of federal legislation preempting them from exercising authority over particular wildlife. Native Americans have challenged the exercise of that authority when it clashes with their treaty rights. The precise language and historical context of the treaty involved often determine the extent to which state regulation is valid.

One of the first opportunities for the Supreme Court to consider the relationship between a state's authority to regulate taking of wildlife and treaty hunting rights was in 1896, in *Ward v. Race Horse.*[3] The author of the Court's opinion was Justice White, who had written the majority opinion in *Geer v. Connecticut* earlier the same year.[4] Not surprisingly, Justice White was extremely solicitous of Wyoming's claims to far-reaching authority over all wildlife within its borders.

Race Horse, a member of the Bannock tribe, was convicted of violating Wyoming law by killing elk. In a habeas corpus proceeding, he contended

[2]*See* Menominee Tribe of Indians v. United States, 391 U.S. 404 (1968).

[3]163 U.S. 504 (1896).

[4]161 U.S. 519 (1896). The Geer case articulated the doctrine of state "ownership" of wildlife. For further discussion of this case, see Chapter 2 at text accompanying notes 31–37.

that his conviction contravened rights guaranteed to him by an 1869 treaty reserving to the Bannocks "the right to hunt on the unoccupied lands of the United States, so long as game may be found thereon." In 1890, however, Wyoming was admitted to the Union, and soon it enacted game laws that it asserted were applicable even on federal lands within the state. Race Horse committed his offense on federal lands within Wyoming. The case thus squarely presented the issue of treaty rights versus state game laws.

Justice White reasoned that the treaty could not have been intended to create a perpetual right to hunt on federal land. Whatever rights the treaty created could be extinguished simply by transferring the land to private owners, whereupon it would cease to be "unoccupied." Would the mere fact of establishing a state, an action with legal significance that few Indians at the time could possibly have understood, also extinguish treaty rights? Justice White held that it would, since the "complete power to regulate the killing of game within its borders" was a "necessary incident" of a state's sovereign authority.[5]

Despite the treaty, Race Horse and his fellow Bannocks thus could hunt only in accordance with Wyoming's requirements, even on unoccupied federal land. Justice White wrote in the heyday of the state ownership era, which his own opinion in *Geer* had reinforced. He could not have foreseen the day when a state's "complete power" over wildlife would give way to federal authority under the Treaty power or under the Property Clause. Had he been able to do so, he could not have so easily dismissed Race Horse's arguments.

Other cases addressing the question whether states can regulate off-reservation Native American hunting and fishing have focused more specifically on the terms of the particular treaty at issue than on general principles of inherent state authority. For example, in *Oregon Dep't of Fish v. Klamath Indian Tribe*, the Supreme Court held that a 1901 treaty extinguished treaty hunting and fishing rights on lands that had been part of the original Klamath Reservation under an 1864 treaty.[6] To settle a boundary dispute that arose after the original treaty was signed, the 1901 treaty paid the tribe for ceding certain lands that had been erroneously excluded from the original reservation. The question, therefore, was whether the treaty hunting and fishing rights that had attached to the ceded lands prior to the 1901 cession survived thereafter. The Court held, over a vigorous dissent from Justices Marshall and Brennan, that

> no language in the 1901 Agreement evidences any intent to preserve special off-reservation hunting or fishing rights for the Tribe. Indeed, in light of the 1901

[5] 163 U.S. at 510.
[6] 473 U.S. 753 (1985).

diminution, a silent preservation of off-reservation rights would have been inconsistent with the broad language of cession.[7]

The Court's holding, however, in contrast to the premise of the *Race Horse* decision, implicitly recognized that treaty negotiators could have agreed to preserve off-reservation hunting and fishing rights despite the states' general authority over taking of wildlife.

A series of cases in the Pacific Northwest has explored the nature of state authority over Indian fishing rights under a group of treaties called the "Stevens treaties."[8] The first of these cases, *United States v. Winans,* was in 1905.[9] The Supreme Court held that the treaty entitled the Indians to fish on private land even though the state had granted the landowner a license to use a fish wheel, the practical effect of which was to exclude all others from fishing there. The Court based its holding on the treaty's reserving to the Yakima Indians "the right of taking fish at all usual and accustomed places, in common with the citizens of the Territory."

In 1919, in *Seufert Brothers Co. v. United States,* the Court extended the *Winans* decision to apply to customary Indian fishing grounds located outside the area ceded by them.[10] *Winans* and *Seufert* thus established the supremacy of Indian off-reservation treaty rights with respect to the rights of private landowners, while *Race Horse* seemed to subordinate treaty rights to the states' authority to regulate wildlife.

The first major break in the *Race Horse* rule came in 1942 in *Tulee v. Washington,* in which the Supreme Court held that a state requirement that Native Americans purchase fishing licenses was contrary to yet another Stevens treaty.[11] The state had argued that since the treaty guaranteed only the right to fish "in common with" other citizens, the state had only to insure that its regulations not discriminate against Indians. In rejecting that argument and requiring that the state's regulation also be "necessary for the conservation of fish," the Supreme Court planted the seed for a storm of controversy that would long rage in the Pacific Northwest.[12]

The Court's next opportunity to refine its "necessary for the conservation of fish" standard came in 1968 in *Puyallup Tribe v. Department of*

[7] *Id.* at 770.

[8] The Stevens treaties include six treaties negotiated with Indian tribes in the Pacific Northwest by Oregon Territorial Governor Isaac Stevens in 1854 and 1855. Each of these contains similar language regarding fishing rights. Citations to the six treaties appear in Washington v. Washington State Commercial Passenger Fishing Vessel Ass'n, 443 U.S. 658, 662 n. 2 (1979).

[9] 198 U.S. 371 (1905).

[10] 249 U.S. 194 (1919).

[11] 315 U.S. 681 (1942).

[12] *Id.* at 684. *See* text accompanying notes 38–55, *infra.*

Game.[13] Washington state sued in state court to enjoin certain Native Americans from violating state regulations restricting the use of nets for salmon and steelhead fishing. The trial court entered an injunction requiring the Indians to abide by all state fishing regulations. The state supreme court modified the injunction to require adherence only to those regulations that were "reasonable and necessary." The supreme court remanded the case to the trial court for a determination of which regulations met that test. Before the remand, however, the United States Supreme Court reviewed and affirmed the state supreme court's decision, announcing that, although "[t]he right to fish 'at all usual and accustomed places' may . . . not be qualified," nevertheless,

> the manner of fishing, the size of the take, the restriction of commercial fishing, and the like may be regulated by the State in the interest of conservation, provided the regulation meets appropriate standards and does not discriminate against the Indians.[14]

The "appropriate standards" referred to in the Court's "*Puyallup I*" decision were presumably the standards of reasonableness and necessity the state court had established.

The Supreme Court's efforts to articulate standards for state regulation of off-reservation fishing rights thus failed to result in clear guidance to states. With the legal situation in this rather confused posture, Charles A. Hobbs, a former chairman of the Indian Affairs Committee of the American Bar Association, wrote an article in which he proposed a simple, but in retrospect, shattering, suggestion.[15] Hobbs argued that the heart of the problem was the guarantee of a "fair share" of the fish to the Indians and that, since state fisheries agencies had failed to resolve the problem, it became the duty of the courts to "consider the question of 'necessity' from the perspective of what is a fair share of the fish for the Indians."[16]

The Supreme Court was forced to face the "fair share" issue when the *Puyallup* case came back to it in 1973.[17] After the Court's first decision, the Washington Department of Fisheries relaxed its restrictions on commercial net fishing for salmon. However, the Department of Game, which regulated sport fisheries, continued its total prohibition of net fishing for steelhead trout. The state supreme court upheld the ban because "the catch of the steelhead sports fishery alone in the Puyallup River leaves no more than a sufficient number of steelhead for escapement necessary for

[13] 391 U.S. 392 (1968).
[14] *Id.* at 398.
[15] Hobbs, *Indian Hunting and Fishing Rights II*, 37 Geo. Wash. L. Rev. 1251 (1969).
[16] *Id.* at 1260.
[17] 414 U.S. 44 (1973).

the conservation of the steelhead fishery in that river."[18] The effect of state regulation was, therefore, to allocate the entire harvestable steelhead run to the non-Indian sports fishery, leaving none for the Indians.

The Supreme Court, through Justice Douglas, held that the regulation discriminated unlawfully against the Indians and declared that the harvestable run of steelhead "must in some manner be fairly apportioned between Indian net fishing and non-Indian sports fishing."[19] The Court's "*Puyallup II*" decision did not prescribe how the run should be apportioned, but left that critical question for the lower courts to decide.

United States v. Michigan illustrates the potential for recognizing far-reaching off-reservation rights.[20] A federal district court found that Michigan had no authority to regulate off-reservation gill net fishing in Lake Michigan by Chippewa and Ottawa Indians. The Sixth Circuit Court of Appeals, while strongly intimating that the state could regulate this fishing if necessary to prevent extinction or irreparable damage to fisheries, remanded to the district court to determine whether federal regulations preempted state authority or jurisdiction.[21]

When Secretary of the Interior James Watt allowed the regulations to expire in 1981, the state again urged the court of appeals to review the district court's ruling.[22] This time, the court of appeals held that "[t]he state bears the burden of persuasion to show by clear and convincing evidence that it is highly probable that irreparable harm will occur and that the need for regulation exists."[23] Otherwise, the state may not restrict Indian treaty fishing. The state ultimately abandoned its effort to show that its regulation could meet this stringent standard.[24]

State Authority over On-Reservation Hunting and Fishing

In General

The Supreme Court did not directly consider, in either of the first two *Puyallup* decisions or in any other previous case, whether on-reservation Native American hunting or fishing was ever subject to state or federal regulation. The states had long been assumed to be without regulatory

[18]Dep't of Game v. The Puyallup Tribe, 80 Wash. 2d 561, 573, 497 P.2d 171, 171–178 (Wash. 1970).

[19]414 U.S. at 48.

[20]471 F. Supp. 192 (W.D. Mich. 1979), *remanded*, 623 F.2d 448 (6th Cir. 1980).

[21]The federal regulations were adopted under 25 U.S.C. §§ 2 and 9, provisions that confer broad, general authority upon the President and the Secretary of the Interior to manage Indian affairs. It is unclear, however, whether this general authority enables the federal government to act in ways barred to the states.

[22]United States v. Michigan, 653 F.2d 277 (6th Cir. 1981).

[23]*Id.* at 279.

[24]520 F. Supp. 207, 211 (W.D. Mich. 1981).

authority, though a clear basis for this assumption was lacking.[25] The assumption can be traced to the early case of *Worcester v. Georgia*, in which Chief Justice Marshall held that state laws have no force within Indian reservations because reservations are "distinct political communities, having territorial boundaries, within which their authority is exclusive."[26] It was further reinforced by federal legislation giving states authority to assume general criminal or civil jurisdiction over on-reservation Indian activities, but withholding from the grant of criminal jurisdiction any authority over Indian "hunting, trapping, or fishing or the control, licensing, or regulation thereof."[27]

The "fair apportionment" standard of the Supreme Court's *Puyallup II* decision inevitably forced the courts to consider whether in fact the states were without authority over on-reservation fishing. On remand from the *Puyallup II* decision, Washington state courts allocated 45 percent of the available natural steelhead run to the treaty fishers' net fishing, but included in the Indian share all fish taken by treaty fishers on the Puyallup reservation.[28] When the case came before the Supreme Court a third time, the Indians maintained that, since the state had no authority to regulate Indian fishing on the reservation, fish caught there should not be counted against the Indian share. In *Puyallup III*, the Supreme Court disagreed.[29]

The Court couched its rationale in *Puyallup III* in terms that appeared to limit the ruling's precedential value. The Court made much of the fact that virtually all of the reservation, including all that abutted the Puyallup River, had since been alienated.[30] Indeed, though noting that whether the reservation even continued to exist was a matter of dispute, the Court declined to resolve it.[31] The Court did say, however, that the very treaty language that assured the Indians a share of the fishery resource also necessarily assured a share to non-Indian fishers and that totally unregulated Indian harvest on the reservation could deprive the latter of their share.[32] Thus, invoking Justice Douglas's admonition in *Puyallup II* that "the treaty does not give the Indians a federal right to pursue the last

[25] *See* Hobbs, *supra* note 15, at 1254.

[26] 31 U.S. (6 Pet.) 515, 557 (1832).

[27] 18 U.S.C. § 1162(b). The quoted language is from the Act of Aug. 15, 1953, ch. 505, Pub. L. No. 83–280, § 2, 67 Stat. 588, 589, often referred to simply as "Public Law 280." Though its provisions relating to state civil jurisdiction over on-reservation Indian activities, 28 U.S.C. § 1360, do not expressly exempt hunting and fishing, this difference between its civil and criminal provisions has not been deemed significant. *See, e.g.*, Menominee Tribe v. United States, 391 U.S. 404 (1968).

[28] Dep't of Game v. Puyallup Tribe, Inc., 86 Wash. 2d 664, 548 P.2d 1058 (1976).

[29] Puyallup Tribe, Inc. v. Dep't of Game of Washington, 433 U.S. 165 (1977).

[30] *Id.* at 174. In Montana v. United States, 450 U.S. 544, 561 (1981), the Court again emphasized this aspect of its Puyallup III decision.

[31] 443 U.S. at 174, n. 11.

[32] *Id.* at 175–76.

living steelhead until it enters their nets,"[33] the Court ruled that fish taken by treaty Indians while on their reservation must be counted against the overall share allocated to them.[34]

Some commentators harshly attacked the *Puyallup III* decision,[35] while even those who agreed with the result expressed doubt whether the same result would obtain for the Stevens treaties that had slightly different language.[36] The two dissenting Justices likewise characterized the majority's opinion as "so narrowly fact specific that it will probably have no significant impact on the Puget Sound Indian fishing rights case still pending in the District Court."[37] That reference was to the still boiling dispute that followed in the wake of the so-called "Boldt decision" of the Western District of Washington in 1974. As it developed, the Supreme Court would itself soon be called upon to resolve that even larger controversy.

The "Boldt Decision" and Its Progeny

One year after the *Puyallup II* decision, Senior District Judge George Boldt of the Western District of Washington translated the Supreme Court's vague and general admonitions into concrete terms. His opinion in *United States v. Washington* attempted to define and assure to the Indians their "fair share" of the fishery resource.[38] The result sent shock waves throughout the Northwest.

The Boldt decision concerned the rights of several tribes under various of the Stevens treaties to fish in the Puget Sound and Olympic Peninsula watersheds and the offshore waters adjacent to those areas. Undoubtedly aware of the enormous implications of his decision, Judge Boldt took great pains to document in detail its legal and factual bases. In sum, he held that

> [e]very regulation of treaty right fishing must be strictly limited to specific measures which before becoming effective have been established by the state, either

[33] 414 U.S. at 49.

[34] 433 U.S. at 178.

[35] *See, e.g.*, Dein, *State Jurisdiction and On-Reservation Affairs: Puyallup Tribe v. Dep't of Game*, 6 Envt'l Aff. 535, 561 (1978) ("The Court in *Puyallup III* ignored the fact that treaties are the supreme law of the land").

[36] *See, e.g.*, Comment, *State Regulation of Indian Treaty Fishing Rights: Putting Puyallup III into Perspective*, 13 Gonzaga L. Rev. 140, 148 n. 62 (1977) (noting that because article III of the Treaty with the Yakimas, June 9, 1855, 12 Stat. 951, 953, guarantees "the exclusive right of taking fish in all the streams, where running through or bordering said reservation," the Yakima Tribe arguably enjoys greater on-reservation fishing rights than does the Puyallup Tribe, whose treaty contains no similar provision).

[37] 433 U.S. at 185 (Brennan, J., dissenting).

[38] 384 F. Supp. 312 (W.D. Wash. 1974), *aff'd*, 520 F.2d 676 (9th Cir. 1975), *cert. denied*, 423 U.S. 1086 (1976).

to the satisfaction of all affected tribes or . . . of this court, to be reasonable and necessary to prevent demonstrable harm to the actual conservation of fish.[39]

Further, he defined "reasonable and necessary" measures as those "essential to conservation" and limited "conservation" to the "perpetuation of a particular run or species of fish,"[40] thus excluding such purposes as assuring a maximum sustained harvest and providing for an orderly fishery.[41] Moreover, he prohibited the state from regulating in any way the fishing rights of those tribes determined to be qualified for self-regulation under criteria set forth in the opinion.

The most far-reaching aspect of the Boldt decision, however, was the determination that the treaty language "in common with" meant "*sharing equally* the opportunity to take fish at 'usual and accustomed grounds and stations.' "[42] Accordingly, Judge Boldt declared that the Indians must be given the opportunity to harvest 50 percent of the fish (not counting those necessary for the minimum conservation purpose of assuring adequate escapement) that "absent harvest en route, would be available for harvest at the treaty tribes' usual and accustomed fishing places."[43]

For a time, it appeared that the Supreme Court would not disturb this result. In 1976, the Court denied Washington's petition for a writ of certiorari to review the court of appeals' decision substantially affirming Judge Boldt's initial decision. Compliance with the Boldt decision, however, did not come easily.

When the Washington Department of Fisheries promulgated regulations to implement the decision, private interests sued in state court to invalidate them. The Washington Supreme Court upheld the challenge, reasoning that Judge Boldt's interpretation of the treaties was erroneous and that compliance with his ruling would violate the equal protection clause of the Fourteenth Amendment to the Constitution and exceed the authority of state agencies under state law.[44] The federal district court responded by entering a series of orders that regulated directly the fisheries at issue, thus displacing the state from any regulatory role.[45] The court of appeals affirmed the district court's action, emphasizing with acerbic comments that state intransigence was largely responsible for the

[39]*Id.* at 342.

[40]*Id.*

[41]*See* 520 F.2d at 686.

[42]384 F. Supp. at 343 (emphasis in original).

[43]*Id.* at 344.

[44]Puget Sound Gillnetters Ass'n v. Moos, 88 Wash. 2d 677, 565 P.2d 1151 (1977); Washington State Commercial Passenger Fishing Vessel Ass'n v. Tollefson, 89 Wash. 2d 276, 571 P.2d 1373 (1977).

[45]Various of the orders of the district court over the period 1974–78 are reported at 459 F. Supp. 1020 (W.D. Wash. 1974–78).

fierce controversy.[46] The Supreme Court agreed to review the Washington Supreme Court's decision and the federal actions it spawned.

The Supreme Court began its analysis in *Washington v. Washington State Commercial Passenger Fishing Vessel Association* by observing that, at the time they negotiated the treaties, neither the tribes nor Governor Stevens "realized or intended that their agreement would determine whether, and if so how, a resource that had always been thought inexhaustible would be allocated between the native Indians and the incoming settlers when it later became scarce."[47] The Court, nevertheless, had to face that issue.

The Court had two choices. The state urged it to interpret the treaties merely to allow the Indians the same opportunity to catch fish "at all usual and accustomed grounds and stations" as non-Indians enjoy. The other choice was to interpret the treaties to allow the Indians a quantifiable share of the fish available for taking at these places. The Court in its earlier decisions had already signalled that the latter choice would prevail. For example, in *Puyallup III* the Court had stated that the states' allocation of 45 percent of the available steelhead run to the Indians was "precisely what we had mandated in *Puyallup II.*"[48]

While acknowledging that the issue before it was "virtually a 'matter decided' by our previous holdings,"[49] the Court undertook an extensive analysis of the treaty language and its historic context. Two considerations, one pragmatic and one formalistic, seemed most important to the Court. The pragmatic consideration was that the "equal opportunity" interpretation urged by the state was likely, "in light of the far superior numbers, capital resources, and technology of the non-Indians," to "net [the Indians] virtually no catch at all."[50] The more formalistic consideration emphasized the treaty language guaranteeing a right to *take* fish, as opposed to a right to *try* to take fish.[51] The right to take fish implied, in the Court's view, a right to take a share of the available fish in a time of shortage.

[46]

> The state's extraordinary machinations in resisting the decree have forced the district court to take over a large share of the management of the state's fishery in order to enforce its decrees. Except for some desegregation cases . . . the district court has faced the most concerted official and private efforts to frustrate a decree of a federal court witnessed in this century. The challenged orders in this appeal must be reviewed by this court in the context of events forced by litigants who offered the court no reasonable choice.

Puget Sound Gillnetters Ass'n v. United States District Court, 573 F.2d 1123, 1126 (9th Cir. 1978).

[47]443 U.S. 658, 669 (1979).
[48]433 U.S. at 177.
[49]443 U.S. at 679.
[50]*Id.* at 677, n. 22.
[51]*Id.* at 678.

To decide that treaty Indians are entitled to a share of available fish does not, however, determine what their share should be. Relying upon reserved water rights cases, the Court reasoned that the treaties guaranteed "so much as, but no more than, is necessary to provide the Indians with a livelihood—that is to say, a moderate living."[52] If that amount were less than 50 percent of the available fish, then an appropriate lower figure should be determined by the district court. The Court concluded, however, that the Indian share could never exceed 50 percent of the available fish because the Indians' right to take fish "in common with" non-Indian settlers implied an equal division.[53]

The Supreme Court thus substantially affirmed Judge Boldt's initial decision. It did, however, adopt a different formula to determine which fish should be counted against the respective Indian and non-Indian shares. Most significantly, it insisted that fish taken by Indian treaty fishers on reservations be counted toward the Indian share. Quietly breaking new ground, the Court declared that "although it is clear that the Tribe may exclude non-Indians from access to fishing within the reservation," *Puyallup III* had "unequivocally rejected the Tribe's claim to an untrammeled right to take as many of the steelhead running through its reservation as it chose."[54] Even Justices Brennan and Marshall, who as dissenters in *Puyallup III* found that decision anything but unequivocal,[55] joined in this sweeping characterization of the *Puyallup III* decision.

The alarming decline of salmon and steelhead populations throughout the Northwest[56] raised a critical question of whether the Indian treaty right to take fish also implies a duty on the part of the government to protect fish habitat from human-caused degradation. Although the courts considered this issue as early as 1980, it remains undecided today. In another of the series of cases entitled *United States v. Washington*, the district court issued a declaratory judgment that the treaties do imply such an "environmental right."[57] The court reasoned that "[t]he most fundamental prerequisite to exercising the right to take fish is the existence of fish to be taken."[58] The Ninth Circuit, however, vacated the judgment on the grounds that declaratory relief was improper in the absence of "concrete facts which underlie a dispute in a particular case."[59] Such a particular

[52]*Id.* at 686.
[53]*Id.* at n. 27.
[54]*Id.* at 683–84.
[55]*See* text accompanying note 37 *supra.*
[56]See Chapter 11 at text accompanying notes 109–131 for a discussion of conflicts over salmon conservation in the Northwest.
[57]506 F. Supp. 187 (W.D. Wa. 1980), *aff'd in part, rev'd in part en banc* in 759 F.2d 1353 (9th Cir.) (1985), *cert. denied,* 474 U.S. 994 (1985).
[58]*Id.* at 203.
[59]759 F.2d at 1357.

case has not yet reached the courts. A determination that the treaties encompass a right to protect fish habitat could have implications even more significant than those of the Boldt decision.

Federal Authority over Native American Hunting and Fishing

Courts reviewing state regulation of Native American hunting and fishing typically focus on the language and intent of the treaties that gave rise to the rights. By contrast, courts reviewing federal regulation of Native American hunting and fishing focus on the language and intent of the legislation under which federal authorities purport to act. The reason for this difference is that Congress can do what no state can do: abrogate or modify treaty obligations. Even if a treaty clearly preserves a right to hunt or fish, a later act of Congress can qualify or extinguish the right.

Congressional power unilaterally to abrogate or modify Indian treaty rights has been recognized since at least the 1903 case of *Lone Wolf v. Hitchcock.*[60] Courts have generally been unwilling, however, to infer congressional intent to abrogate treaty rights, requiring instead clear evidence of such an intent.[61]

With respect to only one of the major federal wildlife conservation laws has the United States Supreme Court resolved the question of its application to activities protected by Indian treaty rights. In 1986, the Court held unanimously in *United States v. Dion*[62] that Congress intended the Bald and Golden Eagle Protection Act to abrogate inconsistent Indian treaty hunting rights.[63]

The defendant, a member of the Yankton Sioux Tribe, shot four bald eagles and a golden eagle on his reservation in South Dakota. He was charged with violating both the Bald and Golden Eagle Protection Act and the Endangered Species Act. The Supreme Court focused almost exclusively on the Bald and Golden Eagle Protection Act violations.

The Court first noted that its precedents "required that Congress' intention to abrogate Indian treaty rights be clear and plain."[64] While the Court prefers that Congress make an "express declaration" of intent to abrogate treaty rights, that preference has never been a *per se* rule.[65]

[60]187 U.S. 553, 564–67 (1903).

[61]*See, e.g.,* Pigeon River Co. v. Cox, Ltd., 291 U.S. 138, 160 (1934); Menominee Tribe of Indians v. United States, 391 U.S. 404, 412–13 (1968).

[62]476 U.S. 734 (1986).

[63]Prior to Dion, there had been a split in the appellate courts on this question. In United States v. White, 508 F.2d 453 (8th Cir. 1974), the Eighth Circuit held that the Bald and Golden Eagle Protection Act did not abrogate treaty hunting rights. In United States v. Fryberg, 622 F.2d 1010 (9th Cir. 1980), the Ninth Circuit held that it did.

[64]476 U.S. at 738 (citations omitted).

[65]The court in White, *supra* note 63, had applied this "express declaration" test.

Where the evidence of congressional intent to abrogate is sufficiently compelling, the Court said, the reviewing court may also infer intent from legislative history.[66] The Court went on to say: "What is essential is clear evidence that Congress actually considered the conflict between its intended action on the one hand and Indian treaty rights on the other, and chose to resolve that conflict by abrogating the treaty."[67]

The Court found that congressional intent to abrogate Indian treaty rights was "strongly suggested" on the face of the statute, since the law, as a result of 1962 amendments, provided for a permit system for taking eagles for the religious purposes of Indian tribes; the permit system would be difficult to explain if the law did not otherwise ban Indians from taking eagles.[68] The Court then examined the legislative history of the Act in detail, noting that the demand for golden eagle feathers for religious ceremonies was one of the threats to eagles that led to the bill, the original bill contained no permit system for takings by Indians for religious purposes, and the bill had been amended in 1962 to include the permit system because without it the impact on Indians would be significant.[69] The 1962 amendments, in the Court's view, "reflected an unmistakable and explicit legislative policy choice that Indian hunting of the bald or golden eagle, except pursuant to permit, is inconsistent with the need to preserve those species."[70]

Having found congressional intent to abrogate Indian treaty rights to take bald eagles, the Court did not consider the further question whether Congress intended the Endangered Species Act to abrogate Indian treaty rights even more broadly. Except with respect to bald eagles, therefore, that issue remained undecided. The following year, however, a federal district court in *United States v. Billie* held that the Endangered Species Act also was intended to abrogate inconsistent Native American rights.[71]

Billie, a Seminole Indian, was charged with violating the Endangered Species Act when he killed a Florida panther on the Big Cypress Seminole Indian Reservation. Billie argued that the Endangered Species Act evinced no congressional intent to abrogate his traditional hunting and fishing rights on the reservation. Those rights did not stem from any treaty, since

[66]This is the test urged by the dissenting judge in White and adopted by the Ninth Circuit in Fryberg.

[67]476 U.S. at 739–40.

[68]*Id.* The defendant argued that the permit system might have been created to allow *non-Indians* to take eagles so Indians could get a sufficient supply of eagle parts for their needs. 752 F.2d at 1270. The Supreme Court rejected this interpretation as "patronizing and strained." 476 U.S. at 744.

[69]*Id.* at 740–43. For an article criticizing the decision in Dion, *see* Laurence, *The Bald Eagle, the Florida Panther and the Nation's Word,* 4. J. Land Use and Envt'l L. 1 (1988). The article contains a discussion of wildlife conservation laws and Indian treaty rights in general.

[70]476 U.S. at 745.

[71]667 F. Supp. 1485 (S.D. Fla. 1987).

the Seminole reservation had been created by Executive Order rather than treaty. The court reasoned that hunting rights nevertheless "were included by implication in the setting aside of the lands as a reservation."[72]

The district court held that the Endangered Species Act abrogates Indian hunting rights, even though the statute does not explicitly so provide. The court relied on Justice Douglas's admonition in *Puyallup II* that Indian treaty rights do not extend to the point of extinction.

> Where conservation measures are necessary to protect endangered wildlife, the Government can intervene on behalf of other federal interests. The migratory nature of the Florida panther gives Indians, the states, and the federal Government a common interest in the preservation of the species. Where the actions of one group can frustrate the others' efforts at conservation . . . reasonable, nondiscriminatory measures may be required to ensure the species' continued existence.[73]

Moreover, the Act's comprehensiveness, its nonexclusion of Indians, and the limited exceptions for certain Alaskan natives, provide "clear evidence that Congress actually considered the conflict between its intended action on the one hand and Indian treaty rights on the other, and chose to resolve that conflict by abrogating" the Indians' rights.[74]

In contrast to the *Billie* case, two federal district courts, in decisions separated by half a century, have held that the Migratory Bird Treaty Act does not abrogate Indian treaty rights. In the most recent of these, *United States v. Bresette,* the court dismissed charges of selling artifacts containing feathers of protected birds against two Chippewa Indians.[75] In the court's view, there was no evidence of congressional intent to abrogate treaty rights, and neither was the Treaty Act a "nondiscriminatory conservation measure" under the *Puyallup I* decision.

In addition to these specific federal conservation statutes, there is one other potential source of broad federal authority to regulate Native American hunting and fishing. The Department of the Interior is charged with managing Native American affairs in accordance with the federal government's trust obligations under federal legislation dating to the 1830s. The contemporary version of that legislation generally authorizes the Secretary of the Interior to manage Indian affairs and to issue regulations as needed.[76] Whether that broad grant of authority authorized the Secretary

[72] *Id.* at 1488.

[73] *Id.* at 1490.

[74] *Id.* at 1491–92, quoting Dion at 740.

[75] 761 F. Supp. 658 (1991). The original case finding no treaty rights abrogation in the Migratory Bird Treaty Act was United States v. Cutler, 37 F. Supp. 724 (D. Id. 1941).

[76] 25 U.S.C. §§ 2 and 9.

to prohibit commercial fishing on the Hoopa Valley Reservation in California was at issue in *United States v. Eberhardt*.[77]

Eberhardt involved Indian fishing in the Klamath River. Since 1933, California had prohibited all commercial fishing in the Klamath. In 1975, however, state courts ruled that the state could not regulate Indian fishing on the Klamath within the Hoopa Reservation.[78] The Department of the Interior responded by beginning to regulate Native American fishing, under its broad authority to manage Indian affairs. Interior eventually imposed a moratorium on all commercial fishing, an action it characterized as temporary and necessary for conservation.

Native Americans charged with violating Interior's regulations challenged them as unauthorized by the statute. The district court held that the regulations abrogated tribal fishing rights, and that abrogation had not been authorized by Congress. The court of appeals, however, reversed. It reasoned that "Interior invokes the general trust statutes only as constituting authority to enact regulations to protect and conserve the fishery resource for the benefit of the Indians, not as power to abrogate reserved tribal rights."[79] The court held that regulations prohibiting temporarily a fishing right that the Indians were conceded to have did not "abrogate" that right because the prohibition was "for the benefit of the Indians." Moreover, the narrow conservation standards that limit state authority over Indian fishing do not restrict federal authority. The court said,

> the validity of these regulations need not be reviewed under standards applicable to state conservation measures either, because this case does not involve the exercise of a state's police power to regulate tribal rights. The challenge to these fishing regulations should be considered under standards generally applicable to an attack on agency rule making.[80]

Thus the *Eberhardt* case, if sound, suggests that the Department of the Interior has far-reaching authority to regulate Native American hunting and fishing without any explicit congressional abrogation of tribal rights.

TRIBAL REGULATION OF ON-RESERVATION HUNTING AND FISHING

In 1979, the Supreme Court in *Washington v. Washington State Commercial Passenger Fishing Vessel Association* observed that it was "clear that the Tribe

[77] 789 F.2d 1354 (1986).

[78] *See* Arnett v. 5 Gill Nets, 48 Cal. App. 3d 454, 121 Cal. Rptr. 906 (1975), *cert. denied*, 425 U.S. 907 (1976).

[79] 789 F.2d at 1360.

[80] *Id.* at 1361.

may exclude non-Indians from access to fishing within the reservation."[81] Just two years later, however, the Court carved out an exception to that broad statement in *Montana v. United States.*[82] The Crow Tribe of Montana sought to prohibit all hunting and fishing within its reservation by nonmembers of the tribe, even though nonmembers owned more than a quarter of the reservation. Addressing first the specific treaty by which the reservation had been created, the Court found that insofar as it conferred upon the tribe the authority to control hunting and fishing on the reservation, that authority did not extend to lands alienated pursuant to later allotment acts.[83]

The Court also considered to what extent tribal authority over reservation hunting and fishing was an element of inherent tribal sovereignty. The test, in the Court's view, was whether tribal authority was "necessary to protect tribal self-government or to control internal relations."[84] As applied to nonmembers hunting or fishing on lands no longer owned by the tribe, the Court concluded that this authority was not necessary to those purposes.[85]

The Court implicitly found that the exercise of authority over nonmembers on tribal lands was necessary to tribal self-government, for it agreed with the court of appeals that the tribe could prohibit nonmembers from hunting or fishing on tribal lands, or condition their entry by charging a fee or establishing bag and creel limits.[86] If, however, the authority to prohibit or condition hunting or fishing by nonmembers is an inherent aspect of tribal sovereignty, then it is fair to ask whether the state may exercise concurrent regulatory authority over nonmembers. The Supreme Court resolved this issue, which was not addressed in *Montana*, two years later in *New Mexico v. Mescalero Apache Tribe.*[87] The Court held that New Mexico could not exercise concurrent authority. In the Court's view, to allow it to do so would undermine Congress's overriding objective of encouraging tribal self-government and economic development.

NATIVE AMERICAN RELIGIOUS PRACTICES AND REGULATION OF HUNTING AND FISHING

Eagle feathers and other wildlife parts or products are commonly used in many Native American religious ceremonies. Conservation laws that

[81]443 U.S. at 683–84. *See* discussion of this case at text accompanying notes 47–55, *supra.*

[82]450 U.S. 544 (1981).

[83]In 1993, the Court extended its *Montana* holding to former reservation lands conveyed to the United States for purposes of constructing a dam and reservoir. *See* South Dakota v. Bourland, 508 U.S. 679 (1993).

[84]450 U.S. at 564.

[85]*Id.* at 566.

[86]*Id.* at 557.

[87]462 U.S. 324 (1983).

prohibit taking or possessing protected wildlife have the potential to affect Native American religious practices. Whether these laws run afoul of the religious freedom guarantees of the First Amendment to the Constitution is a question that only a few courts have addressed.

A brief review of recent developments in First Amendment jurisprudence provides a context for discussion of those cases. Among other things, the First Amendment provides that "Congress shall make no law respecting an establishment of religion, or prohibiting the free exercise thereof." Though simple on its face, this language has given rise to frequent litigation over what it means in practice.

Prior to 1988, federal courts frequently used a balancing test to determine whether governmental actions that incidentally burden the free exercise of religion violate the First Amendment. In general, the balancing test assessed the seriousness of the restraint imposed, whether the regulation advanced a compelling governmental interest, and whether the interest could be achieved through less intrusive means. Using this test, lower courts reached conflicting results as to whether the Bald and Golden Eagle Protection Act's restrictions unduly interfered with Native American religious freedom.

In *United States v. Abeyta*, a district court held that the Act impermissibly constrained Indian religious freedom and dismissed the charge against the defendant for possessing golden eagle parts.[88] The court found that the government's "labyrinthine" procedure whereby Indians could apply for permits to take or possess golden eagle parts was "utterly offensive and ultimately ineffectual," and thus did not cure the constitutional defect.[89] The regulation establishes a depository of eagle parts and feathers, from carcasses of eagles that were killed by accident or died of natural causes. The court found that because of inadequate staffing and inventory, the depository "does not effectively fulfill the [Indians'] need for eagle feathers for ceremonial use."[90] Moreover, the application procedure "is itself unnecessarily intrusive and hostile to religious privacy."[91]

Later the same year, the district court in *United States v. Thirty-Eight Golden Eagles* rejected a defense of religious freedom in a forfeiture case, a decision that was affirmed by the Ninth Circuit without opinion.[92] The

[88]632 F. Supp. 1301 (D.N.M. 1986).

[89]*Id.* at 1307.

[90]*Id.* at 1303.

[91]*Id.* In 1994, President Clinton directed that the process for distributing eagle parts for Native American religious purposes be improved and simplified. *See* Policy Concerning Distribution of Eagle Feathers for Native American Religious Purposes, 59 Fed. Reg. 22953 (1994).

[92]649 F. Supp. 269 (D. Nev. 1986), *aff'd* 829 F.2d 41 (9th Cir. 1987). The forfeiture provision is 16 U.S.C. § 668(b). Ten years earlier, the Ninth Circuit had rejected a religious freedom defense to a conviction for *selling* eagle parts. United States v. Top Sky, 547 F.2d

district court found that the government had a compelling interest in protecting bald and golden eagles, since both are "species whose existence is threatened."[93] The court disregarded the *Abeyta* court's condemnation of the permit system, finding that because the owner of the eagle carcasses had not even applied for a permit, he could challenge only the facial constitutionality of the Act and not its administration.[94]

In major First Amendment cases in 1988 and 1990, however, the Supreme Court refused to apply a balancing test in determining the constitutionality of federal actions that incidentally burden religious practices. In *Lyng v. Northwest Indian Cemetery Protective Association*, several American Indians and allied interests challenged U.S. Forest Service timber harvesting and road-building plans in an area of the Six Rivers National Forest deemed sacred by the Indians.[95] The court of appeals, applying the balancing test, held that the Forest Service activities unconstitutionally burdened the Indians' free exercise of religion.[96]

The Supreme Court reversed. Even though it was undisputed that "the logging and road-building projects at issue in this case could have devastating effects on traditional Indian religious practices,"[97] the Court held that

> incidental effects of government programs, which may make it more difficult to practice certain religions but which have no tendency to coerce individuals into acting contrary to their religious beliefs, [do not] require government to bring forward a compelling justification for its otherwise lawful actions.[98]

For Native American interests who hoped that the *Lyng* reasoning would be limited to federal land management cases, bitter disappointment soon followed with the Supreme Court's decision in *Employment Division, Department of Human Resources of Oregon v. Smith.*[99] At issue in *Smith* was the constitutionality of barring from eligibility for unemployment compensation workers who had been fired for using peyote, an illegal drug that is used in some Native American religious ceremonies. In a sweeping deci-

---

486 (9th Cir. 1976). The court in Top Sky found that selling eagle parts was not part of the Indian religion, and in fact was deplored by Indian religion. *Id.* at 488.

[93]649 F. Supp. at 276. Since the government's interest in protecting threatened species was the basis for the court's holding, query whether the statute could be upheld if bald eagle populations were to recover to a point where they were no longer threatened. Then the question would become whether the government has a compelling interest in protecting the bald eagle (and because of the similarity of immature bald eagles and golden eagles, the golden eagle as well) as a symbol of the nation or for other reasons.

[94]*Id.* at 278.

[95]485 U.S. 439 (1988).

[96]Northwest Indian Protective Ass'n v. Peterson, 795 F.2d 688 (9th Cir. 1986).

[97]485 U.S. at 451.

[98]*Id.* at 450–51.

[99]494 U.S. 872 (1990).

sion, the Court held that if a law is neutral on its face, and not directed specifically at the practice of religion, then no compelling governmental justification for it is necessary, even if it operates to burden the free exercise of religion. In the Court's words, "if prohibiting the exercise of religion . . . is not the object of the [law] but merely the incidental effect of a generally applicable and otherwise valid provision, the First Amendment has not been offended."[100]

Although neither *Lyng* nor *Smith* concerned wildlife conservation laws, their import for these laws is clear. Neutral, generally applicable prohibitions against killing or possessing wildlife would almost certainly not contravene the First Amendment, even if Native Americans could show that the prohibitions impaired their ability to practice their religion. In short, *Abeyta* almost certainly would have been decided differently had it been heard three years later.

The *Smith* decision triggered a political backlash against a ruling that many perceived as a significant erosion of constitutional protection for religious activity, whether of Native Americans or other citizens. As a result, in 1993, Congress enacted the Religious Freedom Restoration Act.[101] Its purpose is to restore the "compelling governmental interest" test in challenges to governmental actions that significantly burden the exercise of religion.

The Religious Freedom Restoration Act likely will lead to more challenges to wildlife laws by Native Americans. Whether these challenges succeed will depend on whether the government can persuade the courts that its interest is compelling. This task is likely to be easier where the law is aimed at species conservation than where its purpose is simply to allocate an abundant resource among competing interests.

In the first case to address this issue under the Act, the Oregon district court did, in fact, hold that prosecution of a Native American for killing bald and golden eagles in violation of the Bald and Golden Eagle Protection Act and the Endangered Species Act did not violate the Religious Freedom Restoration Act.[102] The court found that the government had a compelling interest in protecting golden eagles even though they were not listed as threatened or endangered under the Endangered Species Act, because their population is declining and probably could not withstand serious hunting pressure. The court also found that the government's interest in protecting the bald eagle was compelling despite the fact that the eagle appears to be recovering and the government had proposed reclassifying it from "endangered" to "threatened."

[100]*Id.* at 878.

[101]Pub. L. No. 103–141, 107 Stat. 1488, codified at 42 U.S.C. §§ 2000bb, *et seq.*

[102]United States v. Jim, 888 F. Supp. 1058 (D. Or. 1995). *Accord*, United States v. Lundquist, 932 F. Supp. 1237 (D. Or. 1996).

Chapter 14

INTERNATIONAL WILDLIFE LAW

INTRODUCTION

International wildlife law faces challenges beyond those found in the domestic arena. As noted in Chapter 2, the Constitution provides the ultimate authority for all governmental action in the United States. The federal government derives from the Constitution its authority to enact, enforce, and interpret federal laws, which are binding on all persons subject to United States jurisdiction. The Constitution and federal laws are "the supreme Law of the Land."[1]

In contrast, in the world community there is no "supreme law" under which laws that bind sovereign nations may be developed. There is no elected legislature with authority to enact laws, no universal executive to enforce them, and no supreme judicial tribunal to interpret them. There is instead an ill-defined system for accommodating the competing interests of nations. When that system works smoothly, accommodation takes place through formal agreements between two or more nations in which each commits to restrain the untrammeled pursuit of its own interests. When the system fails, the pursuit of individual interests is restrained only by threat of economic or military sanctions.

Wildlife is a particularly appropriate subject for international regulation. Thousands of bird species migrate between two or more countries during their life cycles. Marine animals, like fish, sea turtles, and most marine mammals, migrate between the waters of different coastal nations or spend part or all of their lives in the high seas, beyond the claim of

[1]U.S. Const. art. VI.

any nation yet subject to exploitation by all. Many terrestrial animals like wolves and large cats routinely move through ranges that cross international borders. Unless limited in the pursuit of its own objectives by agreements or sanctions, each nation may destroy for all nations the value of the species it exploits.[2]

The barriers to establishing an effective means of international restraint are formidable. Most important is the fact that an international agreement can bind only the nations that ratify or accede to it. So long as all the nations affecting a wildlife resource are party to a conservation agreement, the agreement can function effectively. When other nations begin to affect the resource but refuse to be bound by the agreement, its effectiveness is diminished.

The ability of individual nations to thwart international regulation argues for including as many nations as possible in conservation agreements. Doing so, however, may mean diluting substantive conservation measures to the lowest common denominator.

Alternatively, an agreement may contain strong substantive measures, yet allow individual nations to avoid application of measures they deem too stringent. A variety of mechanisms exist to achieve this flexibility. Some agreements require approval of all member nations before any new conservation measure takes effect. Others require approval of only a specified percentage of the member nations but absolve nonconsenting nations from adhering to the measure. The most flexible agreements are those that make a new conservation measure automatically binding when agreed to by some percentage of the member nations but permit objecting nations to escape its applicability.

Even when an agreement encompasses all the nations that it should, and even when they all agree to be bound by each of its conservation measures, enforcement can be a serious problem. The aspect of sovereignty to which nations cling most tenaciously is the exclusive right to enforce the terms of international agreements against their own nationals. Effective enforcement depends not only on officials' ability to investigate and physically observe those who engage in the regulated activity, but also on whether one country's officials trust that their foreign counterparts will diligently enforce the agreement. Lack of trust causes officials to overlook their own citizens' violations to avoid a competitive disadvantage.

International agreements occupy a unique status in United States law. On the one hand, they are like laws enacted by Congress in that the Constitution declares them to be the "supreme Law of the Land." On

[2]Almost thirty years ago, Garrett Hardin observed that each individual or nation exploiting a "common property" resource not only has the potential to destroy it for all others, but has an incentive to do so. *See* Hardin, *The Tragedy of the Commons*, 162 Science 1243 (1968).

the other hand, they are akin to contracts between the United States and other signatory nations. As contracts, they may or may not automatically be effective within the United States without further implementing legislation. Similarly, they may or may not impose mandatory duties or confer enforceable rights on private United States citizens.[3]

A final problem is that courts often are reluctant to interject themselves into disputes arising under international agreements. These disputes involve foreign relations, an area in which the Executive Branch traditionally has been accorded great deference by the courts.[4]

Federal wildlife law has both influenced, and been influenced by, international wildlife law. Early in the century, treaties provided the legal basis for a broad exercise of federal regulatory authority and stimulated a fundamental rethinking of the meaning of the doctrine of state ownership of wildlife. The Convention with Great Britain for the Protection of Migratory Birds is a good example of such a treaty.[5] Later treaties afforded similar bases for even more far-reaching domestic legislation, such as the Endangered Species Act of 1973.[6]

The need to premise exercise of federal authority over wildlife on international treaties has diminished, with broader interpretation of the Constitution's commerce and property clauses.[7] The interplay continues, however, as federal lawmakers seek ways to extend domestic wildlife policies into the international arena.

This chapter does not aim to provide a comprehensive analysis of each of the many wildlife treaties to which the United States is a party. Rather, its objective is to trace the evolution of selected treaties and to compare and contrast their development with the evolution of domestic federal wildlife law. Accordingly, what follows is an examination of some of the more noteworthy international wildlife agreements to which the United States is a party.[8]

[3]The self-executing nature of international treaties and their creation of private rights are addressed in Guilbert, *Wilderness Preservation II: Bringing the Convention into Court*, 3 Envt'l L. Rep. (Envt'l L. Inst.) 50044 (1973), and Comment, *The Migratory Bird Treaty: Another Feather in the Environmentalists' Cap*, 19 S.D.L. Rev. 307 (1974).

[4]In Diggs v. Dent, Civil No. 74–1292 (D.D.C. 1975), the court dismissed a challenge to importation of seal skins from South Africa as a violation of the United Nations Charter, various United Nations Security Council resolutions, and an advisory opinion of the International Court of Justice because the action challenged was "within the foreign policy authority of the President and . . . non-justiciable."

[5]Act of Aug. 16, 1916, 39 Stat. 1702, T.S. No. 628. See Chapter 4.

[6]See Chapter 7.

[7]See Chapter 2.

[8]For the text of international wildlife treaties and agreements to which the United States is a party, see Marine Mammal Commission, Compendium of Selected Treaties, International Agreements and Other Relevant Documents on Marine Resources, Wildlife, and the Environment (1994). A very useful source regarding international wildlife agreements is Simon

The chapter begins by examining treaties pertaining to particular categories of wildlife, including fur seals, whales, polar bears, and endangered species. (Treaties pertaining to migratory birds and ocean fish are covered at length in other chapters and are not included here.[9]) The chapter then examines two significant international efforts to conserve ecosystems and biodiversity on a broad scale, the legal regime for conservation of the Antarctic ecosystem and the 1992 Convention on the Conservation of Biological Diversity.

FUR SEALS

The impetus for passage of the Lacey Act of 1900, the first significant federal wildlife legislation, was the states' inability to protect wildlife against well-organized commercial interests able to take excessive quantities of wildlife and promptly ship them in interstate commerce, out of the reach of the state where they were killed.[10] Similar factors underlay the first effort by the United States to conserve wildlife resources through international agreement. The focus of that effort was the northern fur seal, a marine mammal prized for its fine fur.

Discovery of the Pribilof Islands in the eastern Bering Sea in 1786 revealed an astounding wildlife spectacle. Between 2 and 2.5 million North Pacific fur seals massed there each year to bear their young. About 80 percent of these hauled out on St. Paul Island, one of the four small islands that comprise the Pribilofs. Most of the rest occurred on St. George Island. Although northern fur seals also were found on a few other small islands outside the Pribilofs, the Pribilof population was triple the size of all the others combined.

The Pribilofs' abundant northern fur seals provided the basis for a lucrative fur trade for companies chartered by the Russian czar.[11] This enterprise brought to the Pribilof Islands in the eighteenth century the people from whom many of today's residents are descended.

After the United States purchased Alaska from Russia in 1867, the exclusive right to harvest seals was conveyed to the Alaska Commercial Com-

---

Lyster, International Wildlife Law (Grotius Publications 1985). In addition to many of the agreements discussed in this chapter, Lyster's book includes detailed discussions of a number of other important treaties, including the Convention on Wetlands of International Importance Especially as Waterfowl Habitat ("Ramsar"), 11 I.L.M. 963; T.I.A.S. No 11084, U.K.T.S. no 34 (1976), Cmd.6465, and the Convention on the Conservation of Migratory Species of Wild Animals (the "Bonn Convention"), 19 I.L.M. 15 (1980). An overview of international agreements affecting wildlife conservation may be found in International Law and the Conservation of Biological Diversity (Michael Bowman & Catherine Redgwell eds., Kluwer Law International 1996).

[9]See Chapters 4 (Birds) and 6 (Fish).

[10]See Chapter 3.

[11]Marine Mammal Commission, Annual Report to Congress, 1987 at 25 (1988).

pany, a United States concern. Without access to the principal breeding islands, Russia, Japan, and Great Britain began to take seals on the open ocean, a practice known as pelagic sealing. Pelagic sealing was very wasteful because of the difficulty of retrieving killed or wounded seals and because it often caused the death of pregnant or nursing females.

Pelagic sealing continued despite the great waste because it enabled nations to intercept seals headed for rookeries under the jurisdiction of other countries. Toward the end of the nineteenth century, the United States spearheaded efforts to reach international agreement regarding exploitation of northern fur seals, especially the practice of pelagic sealing.[12] Those efforts finally bore fruit in 1911 when Russia, Japan, Great Britain, and the United States concluded a Treaty for the Preservation and Protection of Fur Seals.[13] By that time, the Pribilof fur seal population had been reduced to about 300,000.

The Fur Seal Treaty of 1911

The Treaty's central provisions were its total ban on pelagic sealing[14] and its requirement that the nation having authority over an island rookery supervise future seal harvests.[15] The *quid pro quo* for agreeing to mutual restraint on pelagic sealing was the requirement that each nation share a specified part of its rookery harvest with the other signatory nations.[16] The 1911 Treaty also prohibited importing sealskins other than those officially marked and certified as having been taken under a government-super-

[12]A brief history of northern fur seal exploitation and of efforts to regulate it can be found in Department of Commerce, National Marine Fisheries Service, Final Environmental Impact Statement, Renegotiation of Interim Convention on Conservation of North Pacific Fur Seals, app. E (1976). A longer and more popular account is Reiger, *Song of the Seal*, 77 Audubon 6 (Sept. 1975). A fascinating part of that history is also contained in an 1898 Supreme Court decision in a case brought by the United States against its exclusive lessee of the right to take seals on the Pribilof Islands for monies owed under the lease. The lessee contended unsuccessfully that restrictions on the size of the permitted take arising from an informal agreement between the United States and Great Britain abrogated the lease. *See* North American Commercial Co. v. United States, 171 U.S. 110 (1898).

[13]July 7, 1911, 37 Stat. 1542, T.S. No. 564.

[14]A special exemption from the pelagic sealing ban was granted to "Indians, Ainos, Aleuts, or other aborigines . . . in canoes . . . propelled entirely by oars, paddles, or sails . . . in the way hitherto practiced and without the use of firearms." *Id.* art. IV.

[15]This aspect of the 1911 Treaty set a precedent that has not been followed in other wildlife treaties. Only with respect to northern fur seals was the United States itself directly involved in the commercial taking of wildlife.

[16]Only Japan was required to share its harvest with all other signatories. The harvests of Russia and the United States were shared with the other two signatories, but not with each other. Since World War II, the former Soviet Union (and now the Russian Federation) has held the island rookeries formerly held by Japan. Under the present agreement, the harvest from those rookeries is shared only with Canada and Japan.

vised harvest and provided a limited means of enforcement by each signatory nation against the others.

The Treaty authorized each nation to seize a vessel of any other nation, except when the offending vessel was in its own territorial waters, if the vessel were engaged in pelagic sealing. However, the seizing nation was obliged to deliver the vessel as soon as possible to the nation having authority over it.[17] Only the latter nation was empowered to try the offense for which the vessel was seized and to impose penalties.[18]

The 1911 Treaty did not articulate any clear policy regarding the level of harvest allowed, nor did it establish any formal mechanism for determining levels. Rather, each country having control over an island rookery was free to establish its own level. It is quite clear, however, that the Treaty contemplated that there would be a substantial harvest.

The Treaty required that the United States make lump sum advance payments of $200,000 to Japan and Great Britain, for which the United States was to reimburse itself by retaining skins from the shares otherwise owing to those nations. Further, the Treaty required that the respective shares for Great Britain and Japan be not less than 1,000 annually, thus contemplating a yearly take on the Pribilofs of at least 6,500 seals. Though the United States could prohibit any harvest in a particular year, it could do so only upon paying $10,000 each to Great Britain and Japan, unless the total seal population on the Pribilofs fell below 100,000.[19] Thus there were significant incentives to continue the commercial seal harvest, wholly apart from the considerable commercial value of the skins. It was clear that the United States' interests were best served by an annual harvest that was as high as could be sustained, though the Treaty did not expressly prescribe this standard.

Japan's withdrawal from the treaty in 1940 led to its termination. From 1941 to 1957, however, a provisional agreement between the United States and Canada governed the Pribilof seal harvest. By the early 1950s the fur seal population had been restored nearly to its historic abundance.[20]

The Interim Fur Seal Convention of 1957

In 1957, a new agreement among Canada, Japan, the Soviet Union, and the United States superseded the 1911 Treaty. Because this agreement was intended to last only six years, it was known as the Interim Convention

[17]Treaty for the Preservation and Protection of Fur Seals, art. I, July 7, 1911, 37 Stat. 1542, T.S. No. 564.

[18]*Id.* art. VI.

[19]*Id.* art. XI.

[20]Marine Mammal Commission, Annual Report to Congress, 1987 at 25–26 (1988).

on the Conservation of North Pacific Fur Seals (hereinafter the Interim Convention).[21] The Interim Convention, implemented domestically through the Fur Seal Act of 1966,[22] changed none of the basic tenets of the original 1911 Treaty.

While the earlier treaty's management goal could only be inferred, the present agreement expressly states a policy of "achieving the maximum sustainable productivity of the fur seal resources . . . so that the fur seal population can be brought to and maintained at the levels which will provide the greatest harvest year after year."[23] Although each signatory retains authority to establish yearly harvest levels on the islands under its control, those levels are based at least in part on the recommendations of the North Pacific Fur Seal Commission, among whose other duties are formulating and coordinating an extensive research program.[24]

Protocols adopted in 1963[25] and 1969[26] extended the Interim Convention through 1976. Its future then became intermingled with Congress's effort to develop a comprehensive federal policy for managing marine mammals in 1972 through passage of the Marine Mammal Protection Act (MMPA).[27] The MMPA sought to incorporate in marine mammal management consideration of the health and stability of the ecosystems of which they are a part, regard for humaneness, and strict compliance with rigorous procedures prior to any taking. The Interim Convention contained no clear authority for any of these. The Act nonetheless created a special exception for taking fur seals pursuant to the Interim Convention by providing that the Act's provisions "shall be deemed to be in addition

[21]Feb. 9, 1957, [1957] 8 U.S.T. 2283, T.I.A.S. No. 3948, 314 U.N.T.S. 105.

[22]Pub. L. 89–702, 80 Stat. 1092 (current version at 16 U.S.C. §§ 1151–1175).

[23]Interim Convention, preamble. This policy is to be tempered by giving "due regard" to the relationship of fur seal populations "to the productivity of other living marine resources of the area." *Id.* Elsewhere, the Convention seems to restrict the range of the "other living marine resources" to be considered by authorizing a cooperative research program to determine "the relationship . . . between fur seals and other living marine resources and whether fur seals have detrimental effects on other living marine resources substantially exploited by any of the Parties." *Id.* art. II, § 1(b).

[24]The original 1911 Treaty did not provide for any commission. The North Pacific Fur Seal Commission established by the Interim Convention consisted of one representative from each of the four member nations. Commission recommendations were to be made by unanimous vote. *Id.* art. V. § 4. However, only those nations sharing the harvest from a particular herd voted on recommendation for the size of that harvest. *Id.* Thus representatives of Japan, Canada, and the United States voted with respect to the recommended harvest level on the Pribilof Islands; the United States had no vote with respect to harvests on islands under the Soviet Union's jurisdiction.

[25]Protocol Amending the Interim Convention, Oct. 8, 1963, [1964] 15 U.S.T. 316, T.I.A.S. No. 5558.

[26]Agreement Extending the Interim Convention, Sept. 3, 1969, [1969] 20 U.S.T. 2992, T.I.A.S. No. 6774.

[27]16 U.S.C. §§ 1361–1421h. See Chapter 5 for a detailed discussion of this law.

to and not in contravention of the provisions of any existing international treaty, convention or agreement, or any statute implementing the same, which may otherwise apply to the taking of marine mammals.''[28]

The MMPA also directed that a ''comprehensive study'' of the Interim Convention be done ''to determine what modifications, if any, should be made . . . to make the Convention and [the Act] consistent with each other.''[29] While directing the Secretary of State to initiate negotiations to modify the Interim Convention in ways the study identified as necessary, the Act also provided that if those efforts failed, the Secretary was to ''take such steps as may be necessary to continue the existing Convention beyond its'' termination date of October 1976.[30]

The ''comprehensive study'' was transmitted to Congress in 1973. It filled two-thirds of a page in the Federal Register, half of which consisted of quotations from the Act and the Interim Convention.[31] It concluded that there was ''no basic incompatibility'' between the Act and the Interim Convention but nevertheless recommended renegotiation of the latter because of ''differences of emphasis'' between the two.

Efforts to renegotiate the Interim Convention began in 1974 and met with very little success. Japan steadfastly resisted all efforts to change the agreement's requirement that harvest levels be established to achieve ''maximum sustainable productivity.'' The United States did, however, persuade its treaty partners to accept amendments authorizing research on the effects of commercial fisheries and human-caused environmental changes on fur seal populations, and requiring use of humane methods in capturing and killing fur seals.[32]

When the Senate ratified the amendments to the Interim Convention, it noted the changed circumstances after the Fishery Conservation and Management Act of 1976 (FCMA).[33] The FCMA amended the MMPA to make its prohibitions on taking marine mammals applicable within a zone extending 200 miles out from the coasts of the United States. The Senate reasoned that the amendment's effect could be to diminish the threat pelagic sealing posed to fur seal populations and urged the Secretary of State to consider negotiating a bilateral treaty with Canada to replace the Interim Convention.[34]

Rather than negotiate a bilateral replacement for the Interim Convention, the United States opted in 1981 to extend the Convention for an-

[28]16 U.S.C. § 1383(a).

[29]*Id.* § 1378(b)(1)(B).

[30]*Id.* § 1378(b)(2).

[31]39 Fed. Reg. 12053 (1974).

[32]Protocol Amending the Interim Convention, May 7, 1976, [1976] 20 U.S.T. 3371, T.I.A.S. No. 8368, arts. II, III, and X.

[33]16 U.S.C. §§ 1801–1882. See Chapter 6 for a discussion of the FCMA.

[34]*See* S. Exec. Rep. No. 36, 94th Cong., 2d Sess. 5 (1976).

other four years.[35] This time, however, the Senate concurred in ratifying that decision only with the understanding that alternative employment for Pribilof Island residents would be studied prior to extending the agreement again.[36] Two years later, Congress enacted comprehensive amendments to the Fur Seal Act that, among other things, terminated federal management of the Pribilofs and established a trust fund to assist residents in making a transition to new economic activities.[37]

Despite the Interim Convention's objective to maintain fur seal populations at levels that would support a maximum sustainable harvest, both fur seal numbers and harvest levels fell significantly. By the mid-1980s the Pribilof population had declined to about 800,000, or less than half the 2 to 2.5 million animals that were believed to have occurred there prior to exploitation and again as recently as the early 1950s.[38] Commercial harvest levels followed a parallel path. In 1968, the commercial harvest on the Pribilofs took 58,908 animals.[39] During the period 1974–1983, the U.S. harvest on the Pribilof Islands declined to an average of about 26,700 annually, and in 1984 was only 22,416.[40] Meanwhile, the Soviets harvested about 7,700 animals annually on Commander and Robben Islands.

The reasons the fur seal population declined were unknown. Commercial harvest levels did not appear to account for the decline, since it was evident on both St. Paul and St. George Islands, even though no commercial harvest had occurred on St. George since it was halted for experimental purposes in 1973.[41] Entanglement in lost or discarded fishing gear, disease, changes in prey availability, and other factors were among other possible explanations. Another possible cause, entanglement in Asian squid driftnets, was eliminated as a result of the global moratorium on high seas driftnet fisheries that took effect December 31, 1992. Earlier estimates indicate that these fisheries may have killed more than 2,000 fur seals annually.[42]

The unexplained dramatic decline in fur seals made extension of the Interim Convention beyond the fourth protocol's 1984 termination no longer a sure thing. A fifth protocol was negotiated that would have extended the Interim Convention through 1988. Animal welfare interests in the United States, committed to ending the commercial harvest and armed with the fact of the seal's steep decline, fought vigorously to block

[35]Protocol Amending the Interim Convention on Conservation of North Pacific Fur Seals, Oct. 14, 1980, [1981] 32 U.S.T. 5881, T.I.A.S. No. 10020.

[36]*See* 127 Cong. Rec. S6078–87 (June 11, 1981).

[37]16 U.S.C. §§ 1151–1175.

[38]Marine Mammal Commission, Annual Report to Congress, 1987 at 26 (1988).

[39]Marine Mammal Commission, Annual Report to Congress, 1988 at 46 (1989).

[40]*Id.*

[41]Marine Mammal Commission, Annual Report to Congress, 1987 at 26 (1988).

[42]Marine Mammal Commission, Annual Report to Congress, 1990 at 47 (1991).

U.S. ratification. Under pressure from these and other quarters, the Senate failed to ratify the protocol, bringing the Interim Convention to an end.

Opposing ratification of the fifth protocol was only one element of the campaign to terminate the harvest. Animal welfare groups also brought an unsuccessful lawsuit and petitioned to list the fur seal as a threatened species.[43] In 1985, the National Marine Fisheries Service (NMFS) determined against listing the fur seal as threatened.[44] Meanwhile, the Marine Mammal Commission urged NMFS to designate the seal as depleted under the MMPA, a designation that was clearly appropriate in light of the animal's documented population decline.[45] A formal determination that the seal was indeed depleted did not occur, however, until May 18, 1988.[46]

The Fur Seal Commission met early in 1985, anticipating that all parties would eventually ratify the protocol. However, as the summer harvest season approached that year, the United States had still not done so. Without an international agreement, fur seal management on the Pribilofs was subject exclusively to the domestic law of the United States—the MMPA and the Fur Seal Act. Because the MMPA prohibits taking marine mammals for commercial purposes without a waiver of its moratorium, for the first time since 1942, there was no commercial harvest of northern fur seals on the Pribilofs in 1985.[47]

The MMPA does, however, allow Alaskan natives to take marine mammals for subsistence. The local natives' subsistence needs had been met with the meat and other by-products of commercial harvests when those harvests occurred.[48] Without a commercial harvest, meeting the natives' needs would require a subsistence harvest. The National Marine Fisheries Service published emergency regulations to assert control over the 1985 harvest by the Aleut residents of the Pribilofs.[49] By 1986, with the prospects for U.S. ratification of the protocol waning, the Fur Seal Commission no longer had any reason or authority to exist. The transition to U.S. management was essentially complete with the publication of permanent subsistence harvest regulations on July 9, 1986.[50]

The former parties to the Convention met in 1987 to explore the possibility of a new agreement, but had little success.[51] The National Marine

[43]Marine Mammal Commission, Annual Report to Congress, 1984 at 44–49 (1985).

[44]50 Fed. Reg. 9232 (1985).

[45]53 Fed. Reg. 17888 (1988).

[46]*Id.*

[47]There had been earlier brief periods without commercial harvest in the early 1800s, and between 1912 and 1916. *See* Marine Mammal Commission, Annual Report to Congress, 1988 at 45 (1989).

[48]Marine Mammal Commission, Annual Report to Congress, 1990 at 44 (1991).

[49]50 Fed. Reg. 27914 (1985).

[50]51 Fed. Reg. 24828 (1986).

[51]Marine Mammal Commission, Annual Report to Congress, 1987 at 35–36 (1988).

Fisheries Service drafted a proposed agreement in 1988, but before the year was out the agency suspended efforts to stimulate international interest in the draft.[52] Meanwhile, annual fur seal harvest on the Pribilofs, now done solely for the subsistence needs of local natives, has averaged only about 1,600 animals since 1986.[53] Since the end of commercial harvests, the Pribilof population of fur seals has remained stable or slightly increased. In 1994 the population was estimated to be about one million.[54]

Ultimately, however, the fate of the Pribilof population of northern fur seals may be determined by factors other than harvest. The end of commercial seal hunting in 1984 caused Pribilof residents to seek new sources of revenue. Several seafood processing facilities have begun operating there, and ship traffic has greatly increased. A sharp increase in the number of seals with tar in their coats was noted in 1994. While overall seal numbers have remained fairly constant, populations at the rookeries closest to the new industrial facilities and their marine outfalls have declined. According to the Marine Mammal Commission, these new developments may pose "a potentially significant threat to fur seals and certain rookeries on the Pribilof Islands."[55]

The 1988 MMPA amendments directed that a fur seal conservation plan be prepared by the end of 1989. The plan was not actually completed, however, until 1993.

WHALES

The Era of Commercial Exploitation

Probably the best known of the many international wildlife agreements to which the United States is a party are those pertaining to whaling. Efforts to bring under international control the heedless destruction of whale stocks began at least as early as the 1920s.[56] Early efforts culminated in 1931 in the largely ineffectual Convention for the Regulation of Whaling.[57] This first of several international agreements to regulate whaling

[52]Marine Mammal Commission, Annual Report to Congress, 1989 at 48 (1990).

[53]Marine Mammal Commission, Annual Report to Congress, 1994 at 50 (1995).

[54]*Id.*

[55]*Id.* at 51.

[56]A good early source for a comprehensive treatment of historical, biological, and legal aspects of whaling and its regulation is The Whale Problem: A Status Report (W. Schevill ed. 1974). *See also* R. Michael M'Gonigle, *The "Economizing" of Ecology: Why Big, Rare Whales Still Die*, 9 Ecol. L. Q. 119 (1980).

[57]Sept. 24, 1931, [1934] 49 Stat. 3079, T.S. No. 880. While nearly fifty nations were party to the 1931 Convention, including the United States, they did not include the world's major whaling nations, Japan and the former Soviet Union.

was limited substantively to prohibiting the most egregious forms of waste, such as "killing of calves or suckling whales, immature whales, and female whales which are accompanied by calves (or suckling whales)."[58] These specific directives were accompanied by the more general directive to make the "fullest possible use" of all whales taken.[59]

In the decade and a half after this convention was concluded, there were several attempts to improve it. The outbreak of World War II cut short these efforts, just as it interrupted most commercial whaling. In the year following the war's end, however, the world's major whaling nations concluded the agreement that has since governed most whaling activity, the International Convention for the Regulation of Whaling.[60] Only about one-third the number of countries that had signed the 1931 Convention signed the 1946 agreement. The signatories, however, for the first time included the major whaling nations of Japan and the former Soviet Union.

The 1946 Convention retains the major substantive provisions of the 1931 agreement. In addition, it establishes an International Whaling Commission (IWC), comprised of one representative from each signatory, with the power to designate species of whales as protected species, fix open and closed seasons or areas, and specify size limits, overall catch limits, and methods of whaling.[61] The IWC's authority extends to all oceans in which whales occur.

The IWC exercises its powers through periodic amendments to a part of the Convention known as the Schedule. To amend the Schedule, however, requires the concurrence of three-fourths of the members. In addition, any member not wishing to be bound by a proposed amendment to the Schedule may, by registering an objection within ninety days of its adoption, escape its applicability.[62]

The Convention's preamble sets forth the agreement's policies and objectives in ambiguous and arguably conflicting terms. The preamble states that the nations of the world have an interest "in safeguarding for future generations the great natural resources represented by the whale stocks." The Convention proposes to achieve "increases in the size of whale stocks," thus permitting the attainment of "the optimum level of whale

[58] *Id.* art. 5.

[59] *Id.* art. 6.

[60] December 2, 1946, [1948] 62 Stat. 1716, T.I.A.S. 1849. The convention is implemented in the United States by the Whaling Convention Act of 1949, 16 U.S.C. §§ 916–916*l*.

[61] *Id.* art. V, § 1. One of the powers expressly denied the IWC is the power to allocate catch quotas among the signatory nations. Nonetheless, the practice of national allocations has arisen outside formal IWC mechanisms.

[62] *Id.* § 3. If any nation registers an objection within the prescribed 90-day period, an additional period of up to 120 days is allowed for other nations to withdraw their previously registered approval.

stocks as rapidly as possible.'' On the other hand, the preamble recites an objective to ''make possible the orderly development of the whaling industry.''

The duality of purposes reflected in the 1946 Convention, conservation of the living resource and ''development'' of the industry that exploits it, resulted in a long history of decisions sacrificing conservation for development. For example, although the Convention preamble declares that to achieve its objectives ''whaling operations should be confined to those species best able to sustain exploitation,'' the IWC in fact promptly adopted and utilized for two and a half decades the ''blue whale unit'' as its yardstick for annual catch limits. This measure treated all species of baleen whales as fungible commodities and negated any effort to develop conservation measures that took into account each species' unique status.[63] Similarly, although the Convention recited an undefined goal of attaining ''optimum levels'' of whale stocks, the actual management standard used was ''maximum sustainable yield.''[64]

With no clear mandate for whale conservation, a directive to consider the needs of the whaling industry, and a regulatory structure that enables individual nations to frustrate conservation actions by registering objections to them, the IWC has for most of its life served as overseer of successive depletions of individual whale stocks.[65] In recent years, however, prompted in part by the United States, the IWC has moved dramatically to fulfill its conservation mandate.[66]

[63]The oil of baleen whales is essentially interchangeable. The blue whale unit served as a device for regulating the annual catch in terms of the gross measure of oil capable of being produced from the catch, rather than in terms of the effects on individual species. For example, in blue whale units, one fin whale was equivalent to one and a fourth humpback whales or three sei whales. *See* Christol, Schmidhauser & Totten, Law and the Whale: Current Developments in the International Whaling Controversy 4 (working paper presented at the Seventh Conference on the Law of the World, 1975).

[64]Department of Commerce, National Oceanic and Atmospheric Administration, International Whaling Commission and Related Activities 8 (July 1976). Maximum sustainable yield or MSY is defined as follows:

> MSY is a concept often employed by managers of living resources to provide a goal towards which exploited populations would be managed. At MSY level, the population is thought to be capable of producing on a continuing basis the largest amount (either in weight or numbers) of harvestable surplus. In its simplest application, the MSY concept is limited to a single species or stock and does not consider ecosystem relationships or external factors such as economics or esthetics.

Id. at 3. MSY and its relationship to ''optimum'' population levels are discussed at greater length in Chapter 5.

[65]Overexploitation caused the IWC to prohibit all commercial taking of blue and humpback whales in 1966, to prohibit commercial taking of fin and sei whales in the North Pacific in 1975, and to extend those prohibitions to still other areas and other species in 1976.

[66]The change in IWC effectiveness was also influenced greatly by world opinion, which by 1972 had turned strongly against the whaling policies of the former Soviet Union and Japan.

The Era of Whale Conservation

After the last vestige of its own commercial whaling industry had disappeared, the United States became an aggressive advocate for an end to commercial whaling worldwide. The ultimate aim of U.S. policy was to persuade the IWC to embrace a total ban. The U.S. was instrumental in securing as part of the 1972 "Stockholm Declaration" of the United Nations Conference on the Human Environment a recommendation for a ten-year moratorium on commercial whaling.[67] The more immediate objectives, however, were simply to secure compliance with IWC whaling quotas by member nations and to gain some leverage over nonmember nations whose citizens were engaged in whaling.

A principal tool used to secure compliance was a 1971 law commonly called the Pelly Amendment to the Fishermen's Protective Act of 1967.[68] The Pelly Amendment was enacted in response to a dispute among the United States, Denmark, Germany, and Norway over harvest of Atlantic salmon. The latter three nations, though members of the International Convention for the Northwest Atlantic Fisheries (ICNAF), had exempted themselves from certain ICNAF harvest regulations by exercising their right to file objections. To discourage these objections, the Pelly Amendment authorized the President to restrict importation of fish products from any nation whose citizens were certified by the Secretary of Commerce to be "conducting fishing operations in a manner or under circumstances which diminish the effectiveness of an international fishery conservation program."[69] Though prompted by a dispute over salmon and framed in terms of "fishery" conservation programs, the Pelly Amendment clearly was intended to apply to the Whaling Convention and other agreements.[70]

In practice, the mere threat of trade sanctions often proved sufficient to force compliance with IWC decisions. Between 1971 and 1979, the Secretary of Commerce twice certified nations under the Pelly Amendment, but on each occasion the President secured commitments of future com-

In fact, one of the recommendations of the 1972 Stockholm Declaration, *infra* note 67, was a ten-year moratorium on commercial whaling.

[67]June 16, 1972, 67 Dep't of State Bull. 116 (1972), 11 Int'l Leg. Mats. 1416–69 (1972).

[68]22 U.S.C. § 1978.

[69]The Pelly Amendment's substantive standard ("diminish the effectiveness") can be traced to 1962 amendments to the Tuna Convention Act of 1950, 64 Stat. 777, 16 U.S.C. § 951 *et seq.* The President's authority to impose trade sanctions extends only "to the extent that such prohibition is sanctioned by the General Agreement on Tariffs and Trade." 22 U.S.C. § 1978(a).

[70]The Pelly Amendment defines the term "international fishery conservation program" to include not only programs for the conservation of fish, but also those pertaining to any "living resource of the sea." *Id.*, § 1978(h)(3).

pliance without actually imposing sanctions.[71] Some in Congress saw this reluctance to impose sanctions as a significant shortcoming. That perception led in 1979 to enactment of the Packwood Amendment to the Magnuson Fishery Conservation and Management Act.[72] The Packwood-Magnuson Amendment, as it is called, directed that any nation certified pursuant to the Pelly Amendment as diminishing the effectiveness of the Whaling Convention must have its fishery allocation within the U.S. fishery conservation zone reduced by at least 50 percent. Certification thus would lead automatically to sanctions; no discretion to withhold them was allowed (though the Packwood-Magnuson Amendment did not alter the President's Pelly Amendment discretion to restrict importation of fish products from a certain nation). The issue of what discretion, if any, the Secretary of Commerce has to withhold certification of a nation that has disregarded IWC requirements eventually found its way to the Supreme Court.[73]

The Moratorium

The IWC at its July 1982 meeting finally agreed to a moratorium on commercial whaling, ten years after the Stockholm Declaration had called for one. The moratorium was not to take effect, however, until 1986. The moratorium's duration was indefinite, but the resolution imposing it provided that by no later than 1990 the IWC was to undertake a comprehensive assessment of the moratorium's effects upon whale stocks and to consider modifying it.

The 25 to 7 vote to approve the moratorium (with five abstentions)[74] was made possible by the addition of eight non-whaling nations to the IWC since its 1981 meeting. Seven voted in favor of the moratorium, while the eighth abstained. Japan, the Soviet Union, and Norway filed objections. The United States applied a variety of pressures to gain acceptance of the moratorium. For example, the United States responded to Japan's

[71]The first time a President considered imposing Pelly Amendment sanctions occurred in 1975, when President Ford declined to impose sanctions against Japan and the Soviet Union. He justified his decision on the grounds that he believed Japan and the Soviet Union would abide by future IWC quotas and that sanctions would cause economic disruption in the United States. See President's Message to Congress Submitting a Report on International Whaling Operations and Conservation Programs, 11 Pres. Doc. 55 (Jan. 16, 1975). For a more detailed account, *see* Comment, *Not Saving Whales: President Ford Refuses to Banish Imports from Nations Which Have Violated International Whaling Quotas,* 5 Envt'l L. Rep. (Envt'l L. Inst.) 10044 (1975).

[72]22 U.S.C. § 1978(a)(3) and 16 U.S.C. § 1821(e)(2).

[73]*See* text accompanying note 80 *infra.*

[74]Amendments to the IWC's Schedule require the approval of three-fourths of those voting.

objection by reducing Japan's allocation of fish in U.S. waters by over 100,000 metric tons.[75]

Not all whaling nations resisted the moratorium, however. Peru, for example, initially filed an objection but withdrew it shortly thereafter. Chile opted to prohibit whaling by its citizens as of July 1983.[76] The Soviet Union, despite filing an objection, announced in 1985 that it would cease commercial whaling in 1987, a year after the moratorium was to take effect.[77] The year it made this announcement, however, the Soviet Union took more than its share of the Southern Hemisphere minke whale quota, causing the overall quota to be exceeded.

The Secretary of Commerce promptly certified the Soviet Union under the Pelly Amendment and the Packwood-Magnuson Amendment. The Secretary's certification was the first ever against any nation under the Packwood-Magnuson Amendment for whaling practices. It caused the Soviet Union's allocation of fish from the U.S. fishery conservation zone to be reduced by 50 percent.[78] When the Soviet Union later honored its pledge to cease whaling after 1987, the Secretary withdrew certification.[79]

Even before the commercial whaling moratorium took effect, Japan also found itself embroiled in controversy with the United States. The dispute stemmed from Japan's plans to continue killing sperm whales in the western North Pacific in 1984, notwithstanding that a zero quota on that stock was to take effect that year. Japan had filed a formal objection to this quota, but the United States threatened to impose sanctions under the Pelly and Packwood-Magnuson Amendments. Seeking to head off sanctions, Japan negotiated an executive agreement with the United States in 1984 whereby it agreed to withdraw its objection to the moratorium and cease all commercial whaling by April 1, 1988, two years after the moratorium was scheduled to take effect. In addition, Japan agreed to limit its sperm whale kill prior to the moratorium taking effect. The United States, in turn, agreed not to certify Japan for sanctions under the two statutes. Animal welfare groups challenged in court the U.S. failure to certify. The case eventually made its way to the Supreme Court, which held, in a 5–4 decision, that the law allowed the government to accept the negotiated arrangement with Japan in lieu of certification.[80]

As of 1986, when the commercial whaling moratorium was scheduled to begin, only Japan, the Soviet Union, and Norway were still whaling, and the first two of these had committed to end their whaling shortly. On June 9, 1986, after learning that Norwegian whalers had taken minke whales in

[75] *See* Marine Mammal Commission, Annual Report to Congress, 1983 at 22 (1984).
[76] *Id.*
[77] Marine Mammal Commission, Annual Report to Congress, 1985 at 33 (1986).
[78] Marine Mammal Commission, Annual Report to Congress, 1986 at 38 (1987).
[79] Marine Mammal Commission, Annual Report to Congress, 1988 at 104 (1989).
[80] Japan Whaling Ass'n v. American Cetacean Society, 478 U.S. 221 (1986).

the northeast Atlantic in accordance with Norway's formal objection to the moratorium, the Secretary of Commerce certified Norway for possible trade sanctions under the Pelly Amendment. In less than a month, Norway relented, agreeing to cease commercial whaling at the end of the 1987 season. This agreement was comparable to the agreement by which Japan had avoided U.S. sanctions. As a result, the President chose not to impose sanctions.[81]

The ultimate acceptance of the moratorium by the major whaling nations appeared to set the stage for an end to whaling. However, the parties had agreed in 1982 to a moratorium, not a permanent end to whaling. It was possible that commercial whaling would resume after 1990, when the assessment of the moratorium's effects was completed, under quotas the IWC judged scientifically sound. The requirement that three-quarters of the voting members approve any change to the IWC's schedule of quotas, however, is a significant practical barrier to the resumption of commercial whaling.

"Research Whaling" Proposals

In order to gather data for the assessment of the moratorium's effects on whales, some of the whaling nations proposed to begin "research whaling." They pointed to Article VIII of the Whaling Convention, which allows any member nation to issue permits authorizing taking whales for scientific research. The provision further allows any whales taken under these permits to be processed and sold as the member nation directs. When the governments of Iceland, Japan, and Korea proposed to issue research permits, and Iceland's proposal involved taking about half the number of whales that its commercial whaling had taken, those who had celebrated the moratorium became concerned that Article VIII provided an end-run around it.

The IWC's response to the research initiatives was to develop procedures under which proposals would be submitted to the IWC's Scientific Committee, and the Committee would offer its advice on them. At its 1987 meeting, the Scientific Committee reviewed the proposed research and, based on the review, the IWC adopted a resolution recommending that Japan, Iceland, and Korea not carry out the proposed research until they had addressed the Committee's concerns.[82]

At later annual meetings of the IWC, still more research proposals were put forward by Norway, the Soviet Union, Japan, and Iceland. No proposal has ever received the unqualified endorsement of the IWC. Instead, with regard to every proposal a resolution has expressed the view that the pro-

[81]Marine Mammal Commission, Annual Report to Congress, 1986 at 39–40 (1987).
[82]Marine Mammal Commission, Annual Report to Congress, 1987 at 99 (1988).

posed research was deficient in some manner and urged that it be reconsidered. Resolutions are nonbinding, but the United States took the view that failure to comply with a nonbinding resolution could be the basis for imposing sanctions under the Pelly and Packwood-Magnuson Amendments. The United States used the threat of sanctions to persuade Korea to abandon its research plans and Iceland to limit its take of whales in 1987 and to abide by the Scientific Committee's recommendations after that year.[83]

Japan, however, was not so easily dissuaded. Notwithstanding the IWC's resolution requesting that it refrain from carrying out its planned research, Japan began taking minke whales in the Antarctic in early 1988. This action forced the United States' hand. In February 1988, the Secretary of Commerce certified Japan under the Pelly and Packwood-Magnuson Amendments.

The sanctions imposed on Japan were rather modest. Its pending request for permission to harvest sea snails and Pacific whiting in the U.S. fishery conservation zone was denied, as provided by the Packwood-Magnuson Amendment, but the President declined to impose immediately any trade sanctions under the Pelly Amendment.[84] To maintain pressure on Japan (and very likely to placate domestic environmental interests) the President held out the possibility of future Pelly Amendment sanctions if Japan continued on its course. Japan never wavered, however, and no Pelly Amendment sanctions have yet been imposed.

Opposition to whaling has been the policy of both Democratic and Republican administrations since the 1970s. The absence of any domestic whaling industry has made that policy a politically painless choice. The possible ramifications of imposing trade sanctions on our major trading partner and an important geopolitical ally, however, have been too great for any President yet to invoke sanctions against Japan.

In 1988, Norway also proposed to begin taking whales for scientific research purposes. Although the IWC approved a resolution criticizing the Norwegian research proposal, the United States and Norway negotiated an understanding that avoided certification and sanctions. The understanding, however, proved short-lived. Norway in each of the following two years went forward with new proposals that had been disapproved by the IWC.

In October 1990, the Secretary of Commerce certified Norway under the Pelly and Packwood-Magnuson Amendments. The Packwood-Magnuson certification was of no practical consequence, since Norway had no allocation of fish in the U.S. fishery conservation zone. The Pelly Amendment certification also was for nought, as the President declined

[83]Marine Mammal Commission, Annual Report to Congress, 1987 at 100–101 (1988).

[84]Marine Mammal Commission, Annual Report to Congress, 1988 at 63 (1989).

to impose any trade sanctions. The same ritual was repeated again in 1992: The IWC called upon Norway to refrain from issuing a research permit; Norway issued the permit and continued taking whales; the Secretary of Commerce certified Norway; and the President declined to impose trade sanctions.[85]

Post-1990 Developments

As part of the comprehensive assessment of the moratorium's effects on whale stocks, the IWC Scientific Committee developed "revised management procedures" to determine biologically acceptable catch limits in the event whaling were resumed. The IWC adopted these procedures in 1992. They provided that no harvest should be carried out on a whale stock that is below 54 percent of its carrying capacity or preexploitation size, and that long-term management efforts should be aimed at not reducing any stock below 72 percent of its carrying capacity.

The revised management procedures appeared to set the stage for approval of new quotas under which commercial whaling could resume. When the IWC elected instead to delay implementing its revised management procedures and to ask the Scientific Committee to report the following year on how the procedures should be implemented, Iceland had had enough. Its commissioner to the IWC, citing this decision and the disapproval of every one of its research proposals, announced that it would withdraw from the IWC.

Iceland perceived that the moratorium, though indefinite on its face, might in fact be permanent, even though some of the whale stocks were abundant enough to permit whaling under the new management procedures. That perception was reinforced by policy discussions taking place in the United States.

In 1991, the U.S. Marine Mammal Commission carried out its own comprehensive review of the International Convention for the Regulation of Whaling.[86] It concluded that the Convention was "outdated and in need of revision."[87] It urged the United States to renegotiate the Convention to seek recognition of the nonconsumptive value of cetaceans and to clarify the Convention's authority over small cetaceans (*e.g.*, dolphins, narwhals, and beluga whales). The Commission argued that although a resumption of whaling might be justified purely on scientific grounds, there were moral and ethical considerations as well. The Commission concluded that "[u]ntil such time as the Whaling Convention is amended to take account of the above points, . . . the United States position should be

[85]Marine Mammal Commission, Annual Report to Congress, 1992 at 126 (1993).
[86]The Commission's conclusions were published at 57 Fed. Reg. 4603–15 (1992).
[87]*Id.* at 4604.

to continue to oppose any resumption of commercial whaling."[88] The Commission's recommendations were largely affirmed by a nonbinding resolution of the U.S. House of Representatives in May 1992,[89] and by both the House and Senate in 1993.[90] These resolutions called upon the United States to work to maintain the moratorium.

Meanwhile, in April 1992, Iceland, Norway, Greenland, and the Faroe Islands entered into an "Agreement on Research, Conservation and Management of Marine Mammals in the North Atlantic."[91] The agreement set up a parallel and rival structure to the IWC. Its purpose, at least in part, was to send a signal that unless the moratorium were lifted, some of its members were prepared to abandon the IWC. Norway, though not withdrawing from the IWC, did something no less dramatic. In May 1993, it announced that it would resume commercial whaling by killing up to 160 minke whales in the North Atlantic that year, notwithstanding the moratorium. Norway was already subject to a Pelly and Packwood-Magnuson Amendment certification because of its research whaling, though no trade sanctions had been imposed upon it. Its unilateral resumption of commercial whaling forced the President to reexamine the issue of possible trade sanctions. Though he advised Congress that he had directed that a list of possible sanctions be drawn up, none has yet been imposed.

Unless the IWC disbands, Japan is unlikely to follow either Iceland's or Norway's lead, both because of the formal agreement it reached with the United States in 1984 and because it continues to take whales under its own research permits. Nevertheless, in late 1995, the Secretary of Commerce once again certified Japan under the Pelly and Packwood-Magnuson Amendments. The most recent certification stemmed from Japan's taking of whales for research purposes in an area designated by the IWC as a whale sanctuary. As the year ended, the President was again faced with the delicate question whether to impose trade sanctions against a close geopolitical ally and major trading partner. If past experience is any guide, it seems unlikely that he will do so.

The commercial whaling moratorium remains in place nearly a decade after it was to begin, though ignored by Norway and undercut by the "research" whaling activities of several nations. Despite adoption of new procedures for determining whaling quotas, the IWC has failed to authorize any resumption of commercial whaling. The IWC decided that in addition to the new procedures for fixing catch limits, it also had to put in place new systems for enforcing compliance with the limits and for monitoring whale stocks before it could approve resumption. Its inability

[88] Marine Mammal Commission, Annual Report to Congress, 1991 at 118 (1992).
[89] H. Concurrent Res. 177, 102d Cong., 2d Sess.
[90] H. Concurrent Res. 34, 103d Cong., 1st Sess.
[91] Marine Mammal Commission, Annual Report to Congress, 1992 at 120 (1993).

to agree on these new systems has meant that the moratorium continues and looks more and more like a permanent arrangement.

With the moratorium in place and prospects for its termination slight, the IWC has become an institution in search of a mission. Some of the nations that led the fight against whaling have sought to focus the IWC on entirely new issues, such as regulation of whale watching and incidental taking of whales in commercial fisheries, and the plight of small cetaceans.[92] In recent years, many nations simply seem less interested in IWC affairs, a fact reflected by lapsed payments of assessed contributions and declining attendance at IWC meetings.[93]

Regulation of Aboriginal Subsistence Whaling

While the IWC was reducing commercial whale harvests in the 1970s, aboriginal subsistence whaling in Alaska and elsewhere was unregulated and growing. Restrictions on subsistence hunting of caribou, and money from settlement of land claims and the booming oil economy that made possible the purchase of motorboats are among the factors suggested as possible explanations for the increased whaling activity.[94]

In 1976, Alaskan Eskimos landed forty-eight endangered bowhead whales and struck but lost forty-three others. The following year, they struck more whales, and landed a smaller percentage, than ever previously documented: only 26 landed out of 108 struck.[95] The inefficiency of the Eskimo whaling, and the growing number of whales being struck, raised serious conservation concerns because the western Arctic stock of bowheads was then thought to number only 600 to 2,000 animals.

In 1977, the IWC eliminated the exemption for aboriginal subsistence whaling from the zero harvest quota for bowheads. Later that same year, however, under pressure from the United States, the IWC reversed its position and authorized sharply limited quotas for subsistence harvesting. The initial quota for 1978 allowed only twelve whales to be landed or eighteen struck, whichever came first. The IWC later approved annual bowhead quotas for 1979 and 1980, and multi-year "block quotas" for subsequent years.

[92]Although the Whaling Convention does not state which whale species are subject to its jurisdiction, in practice the IWC has never sought to regulate harvest of smaller cetaceans. The IWC's authority to do so is disputed within the IWC itself and has never been resolved.

[93]For example, in 1990, eight nations failed to pay their assessed contributions and six had their voting rights suspended because of non-payment (three of which were among the non-whaling nations that had joined the IWC in the year preceding the 1982 vote in favor of the moratorium). Only thirty of the forty-one member nations attended that year's IWC meeting. Marine Mammal Commission, Annual Report to Congress, 1990 at 123, 126–27 (1991).

[94]Marine Mammal Commission, Annual Report to Congress, 1990 at 65 (1991).

[95]Marine Mammal Commission, Annual Report to Congress, 1984 at 102 (1985).

At the 1982 IWC meeting, when the commercial whaling moratorium was approved, the IWC established a special management scheme for aboriginal subsistence whaling. Under that scheme, to take effect in 1984, whale stocks below MSY levels could be harvested for aboriginal subsistence purposes, so long as catch limits permitted the stocks to increase to MSY levels. For stocks already at or above MSY levels, aboriginal subsistence harvest could be permitted so long as total removals did not exceed 90 percent of MSY.

In 1982, the IWC Scientific Committee estimated the bowhead population at 3,857 animals, or between 21 and 43 percent of its historic abundance.[96] Later IWC estimates have pushed the bowhead numbers steadily upward, to 4,417 in 1985,[97] to 7,200 in 1987, roughly twice the estimate only five years earlier, and finally to 7,992 in 1993.[98] Reflecting these higher estimates, the quota of allowable whale strikes increased, eventually reaching 54 in 1992, triple the limit set in 1978 and half the number of strikes that occurred in 1977.

With the exception of 1980 and 1991, when the number of whales struck exceeded permissible limits, the Eskimo harvest has complied with the quotas. Compliance has been aided by the fact that in March 1981, the National Oceanic and Atmospheric Administration signed a cooperative agreement with the Alaska Eskimo Whaling Commission under which the latter assumed significant responsibility for regulating the subsistence bowhead harvest, including allocating the harvest among native villages, monitoring compliance, and conducting research. Self-regulation probably has contributed to the harvest's improved efficiency as well as compliance with quotas. Since 1985, the percentage of struck whales that were successfully landed has never fallen below 60 percent and has been as high as 79 percent. For six of the twelve years prior to 1985, the percentage of struck whales successfully landed was 48 percent or less; the highest it ever reached was only 67 percent.

POLAR BEARS

Polar bears are found in the north polar region. There are believed to be six relatively discrete populations, two of which occur in Alaska. One is the western Alaska (Bering-Chukchi Seas) population, shared with the Russian Federation. The other is the northern (Beaufort Sea) population, shared with Canada. The total number of polar bears off Alaska is estimated at 3,000–5,000.[99]

[96]Marine Mammal Commission, Annual Report to Congress, 1982 at 61–62 (1983).
[97]Marine Mammal Commission, Annual Report to Congress, 1985 at 109 (1986).
[98]Marine Mammal Commission, Annual Report to Congress, 1994 at 72 (1995).
[99]*Id.* at 92.

Before the 1940s, polar bears were taken primarily by natives for subsistence and for hides. In the middle of the twentieth century, trophy hunters began shooting bears from airplanes. Hunting pressures on bears increased significantly, and Alaska adopted sport hunting regulations to restrict the hunting season in 1961. The state banned all sport hunting of polar bears in 1972.[100]

In the same year, Congress took two major steps that would aid in conserving polar bears. First, it passed the MMPA.[101] The MMPA established a moratorium on taking marine mammals, including polar bears, although the moratorium allows for numerous exceptions.[102] Second, in yet another example of its stepped-up efforts to influence international wildlife policy, Congress passed a joint resolution directing the executive branch to seek an international agreement to protect polar bears.[103]

The Agreement on the Conservation of Polar Bears

The next year the United States, Norway, Canada, Denmark (for Greenland), and the Soviet Union concluded an Agreement on the Conservation of Polar Bears (Polar Bear Agreement).[104] The Agreement's three main purposes are to restrict the killing and capturing of polar bears, to protect polar bear habitat, and to coordinate research efforts among nations with polar bears.[105]

Prohibition on Taking Polar Bears

The Agreement, like the MMPA, prohibits taking polar bears.[106] Also like the MMPA, however, the Agreement's take prohibition is riddled with

[100] *Id.* at 92–93.

[101] 16 U.S.C. §§ 1361–1421h.

[102] See Chapter 5 for a discussion of the take prohibition and the general exceptions to it. Alaskan natives may take polar bears and other marine mammals for subsistence purposes and for use in making and selling traditional handicrafts and clothing. 16 U.S.C. § 1371(b).

[103] H.J. Res. 1268, 92d Cong., 2d Sess. (1972). The United States was not alone in its concern for the future of polar bears. The former Soviet Union had banned all hunting of the bears in 1955, and the countries in whose territories polar bears occur had first met to discuss the bear's decline in 1965.

[104] Nov. 15, 1973, [1976] 27 U.S.T. 3918, T.I.A.S. No. 8409. The Russian Federation has succeeded to the obligations of the Soviet Union under the Agreement. In addition, at the end of 1995 discussions were underway between the United States and the Russian Federation on a bilateral agreement regarding management and conservation of the shared Bering-Chukchi Sea polar bear population. Marine Mammal Commission, Annual Report to Congress, 1995 at 144–47 (1996).

[105] Art. VII requires the parties to conduct research programs on polar bears, "particularly research relating to the conservation and management of the species." Parties are to coordinate research programs and exchange information, research results, and data on bears taken. For a discussion of research carried out after the Agreement was signed, see Lyster, *supra* note 8, at 56, and annual Reports to Congress by the Marine Mammal Commission.

[106] Art. I. "Taking" includes "hunting, killing and capturing."

exceptions. The most extensive exception authorizes a country to allow the taking of polar bears "wherever polar bears have or might have been subject to taking by traditional means by its nationals."[107] This exception is ambiguous and is susceptible to widely differing interpretations, depending on whether the phrase "by traditional means by its nationals" is read to limit how and by whom taking is to be allowed in the future, or simply to define the areas where taking is allowed. It thus could mean that (1) taking is allowed only by a country's own nationals, and only by traditional means, in areas where nationals have or might have taken bears in the past by traditional means, or (2) taking is allowed by anyone using any means, as long as it is done in the described areas.

Soon after the Agreement took effect, the Fish and Wildlife Service appeared to adopt the former view.[108] Canada favors the latter view; it allows limited sport hunting by non-nationals. The U.S. Marine Mammal Commission now agrees with Canada and has suggested that the Fish and Wildlife Service consider the more liberal interpretation.[109]

Neither view imposes any restrictions as to area or means for taking pursuant to the Agreement's other exceptions. These include taking for scientific or conservation purposes, "to prevent serious disturbance of the management of other living resources," and "by local people using traditional methods in the exercise of their traditional rights."[110] Despite these exceptions, in practice most parties allow very limited killing of polar bears.[111] In the 1980s, Canada's subsistence hunt accounted for over 75 percent of the number of bears killed each year.[112]

Article IV's ban on the use of aircraft or large motorized vessels to take polar bears appears to apply to any form of taking permitted under the agreement.[113] However, the ban does not apply where it would be "in-

[107]Art. III(1)(e).

[108]Department of the Interior, Fish and Wildlife Service, Environmental Assessment: Ratification of the Agreement on the Conservation of Polar Bears 6 (April 1975).

[109]*See* Marine Mammal Commission, Annual Report to Congress, 1995 at 148 (1996).

[110]Art. III. It is never explained whether there is any difference between "traditional methods" and "traditional means." Canada interprets this exception to allow local people to sell polar bear permits to non-native hunters. Marine Mammal Commission, Annual Report to Congress, 1995 at 148 (1996). Trade in skins and other polar bear products lawfully taken under Art. III is regulated by the CITES, since the polar bear is listed on CITES Appendix II. See discussion of CITES, *infra.*

[111]The former Soviet Union and Norway have banned all taking of polar bears since 1955 and 1973 respectively. Lyster, *supra* note 8, at 58. The parties to the agreement adopted a resolution in 1973 banning the hunting of female polar bears with cubs and their cubs, and banning hunting bears in denning areas. Marine Mammal Commission, Annual Report to Congress, 1995 at 149 (1996).

[112]*Id.* The Canadian polar bear population is estimated at about 13,000 bears, about half the estimated worldwide population of 21,000 to 28,000 bears. Marine Mammal Commission, Annual Report to Congress, 1995 at 148 (1996).

[113]The Fish and Wildlife Service interprets the prohibition to be inapplicable to takings for scientific purposes. *Supra* note 108, at 6.

consistent with domestic laws." One commentator notes that these words "empty [the ban] of any real meaning."[114]

Protection of Polar Bear Habitat

Article II of the Polar Bear Agreement obligates each party to "take appropriate action to protect the ecosystems of which polar bears are a part, with special attention to habitat components such as denning and feeding sites and migration patterns." Although the duty is mandatory, the choice of "appropriate action" is discretionary,[115] and the exhortation is nowhere translated into concrete directives.

The habitat provision is significant, nevertheless, because of growing interest in exploration of the Arctic for oil and gas. Worried about the effects of mineral exploration on polar bears, the parties at their first consultative meeting in 1978 stated that it would be desirable to establish protected zones around denning areas where human disturbance might otherwise occur. Several parties have established these protected zones.[116]

The most important polar bear on-land denning area in Alaska is in the coastal plain of the Arctic National Wildlife Refuge.[117] The coastal plain is one of the areas the United States oil and gas industry is most eager to explore, and designation as a wildlife refuge does not prohibit development or establish a protected zone.[118] Congress closed the area to oil exploration and development in 1980 to allow time to study the potential environmental impacts of oil and gas development, but this decision is quite controversial and could be reversed.[119]

In 1989, the Marine Mammal Commission issued a report noting that polar bears and their habitat could be affected in many ways by Arctic oil and gas development.[120] The Commission observed that in some cases,

[114]Lyster, *supra* note 8, at 58. Both Canada and the Fish and Wildlife Service have expressed the view that using aircraft to aid in the hunting of polar bears can be consistent with the Agreement. 60 Fed. Reg. 36389 (1995). *But see* Marine Mammal Commission, Annual Report to Congress, 1995 at 149 (1996).

[115]*See* Lyster, *supra* note 8, at 59. Lyster notes, however, that "the underlying obligation is firm, and the Agreement makes no provision for any exceptions to be made to it." *Id.*

[116]*See* Lyster, *supra* note 8, at 60.

[117]U.S. Department of the Interior, Arctic National Wildlife Refuge, Alaska, Coastal Plain Resource Assessment: Report and Recommendation to the Congress of the United States and Final Legislative Environmental Impact Statement 5, 30 (1987). The Arctic National Wildlife Refuge is "the only remaining relatively undisturbed on-land polar bear denning area in Alaska." Marine Mammal Commission, Annual Report to Congress, 1991 at 50 (1992). Reproductive success seems to be greater in on-land dens than in pack ice dens. *Id.*

[118]See Chapter 8 for a discussion of the National Wildlife Refuge System and the laws applicable to refuges. Legislation to designate the coastal plain as wilderness under the Wilderness Act has been introduced in several sessions of Congress but has not been acted upon.

[119]See Chapter 8.

[120]*See* Marine Mammal Commission, Annual Report to Congress, 1991 at 48 (1992).

development might be inconsistent with the Polar Bear Agreement's directive to protect polar bear habitat. In 1991, in response to a letter from the Commission regarding the potential conflict, the Fish and Wildlife Service stated that "extensive measures to protect polar bears" would be instituted if Congress were to authorize drilling in the refuge, and it was "presumptuous to speculate about potential exploration or development scenarios before Congress acted."[121]

Oil and gas exploration has begun, however, in the Chukchi and Beaufort Seas. The Fish and Wildlife Service, pursuant to the small-take provisions of the MMPA, has authorized the take of walruses and polar bears incidental to oil and gas exploration in the Chukchi and Beaufort Seas.[122] The rule specifically excluded areas within the Arctic National Wildlife Refuge.

Also in 1993, the Secretary of the Interior directed the Fish and Wildlife Service to develop and begin implementing a Polar Bear Habitat Conservation Strategy to further the Polar Bear Agreement's habitat protection goals.[123] The draft Conservation Strategy was issued in early 1995 and received extensive public comment. The Marine Mammal Commission expressed doubt that the conservation measures the Service proposed in its draft strategy would be effective in protecting polar bear habitat. The Commission suggested that specific measures to prevent habitat degradation and other adverse impacts on the bear be incorporated into regulations or letters of authorization for incidental take of polar bears. Rather than adopt specific protective measures, however, the final conservation strategy provided for addressing habitat conservation on a case-by-case basis.[124]

The Polar Bear Agreement and Domestic Law

Congress has never enacted implementing legislation for the Polar Bear Agreement. Rather, the MMPA is the principal domestic law pertaining to polar bear conservation. In 1991, both the Marine Mammal Commission and the Fish and Wildlife Service expressed their belief that regula-

[121] *Id.* at 51. For a detailed discussion of oil development in the Arctic and U.S. obligations under the Polar Bear Agreement and other wildlife agreements, see Walker, *Oil Development in the Arctic National Wildlife Refuge and its Impact on United States International Wildlife Commitments,* 4 Int'l Legal Perspectives 1 (1992).

[122] 50 C.F.R. § 18.111 *et seq.* (1995).

[123] This conservation strategy is distinct from the polar bear conservation plan the Fish and Wildlife Service prepared pursuant to 1988 amendments to the MMPA. The conservation plan was completed in 1994. *See* Marine Mammal Commission, Annual Report to Congress, 1994 at 93–94 (1995).

[124] Marine Mammal Commission, Annual Report to Congress, 1995 at 202 (1996).

tions or implementing legislation were required in order to give full effect to the Agreement.[125]

In 1992, the Commission contracted for a comprehensive legal assessment to determine whether the United States was meeting its obligations under the Agreement and whether additional legislation was necessary. The contractor's report, *Reconciling the Legal Mechanisms to Protect and Manage Polar Bears under United States Law and the Agreement for the Conservation of Polar Bears,* was completed in 1993. It noted that

> while the goals and provisions of the [MMPA] and the Polar Bear Agreement are similar, there are some inconsistencies. For example, the Polar Bear Agreement prohibits, but the [MMPA] does not prohibit, the use of aircraft and large motorized vessels to hunt polar bears, and the taking of polar bears from the wild for purposes of public display. Also, the [MMPA] authorizes the take of polar bears incidental to industrial activities if certain conditions are met, whereas the Polar Bear Agreement does not. Further, the Polar Bear Agreement obligates the parties to protect denning areas, feeding areas, and other areas of similar importance to polar bears, but neither the [MMPA] nor other domestic statutes provide explicit authority for doing so.[126]

Congress adopted extensive amendments to the MMPA in 1994. It did not, however, amend the Act to eliminate the inconsistencies with the Polar Bear Agreement noted in the legal assessment. Instead, Congress required the Secretary of the Interior to conduct two reviews. First, in consultation with the contracting parties, the Secretary was to review the Polar Bear Agreement's effectiveness.[127] Second, in consultation with the Secretary of State and the Marine Mammal Commission, the Secretary was to review the effectiveness of U.S. implementation, particularly with respect to the Agreement's habitat protection mandates.[128]

The Fish and Wildlife Service convened a meeting of interested groups to review U.S. implementation of the Agreement in June 1995. A draft report was circulated to the participants after the meeting, and a final report will be submitted to Congress.[129]

By the end of 1995, the Service had not convened a meeting of the parties to review the Polar Bear Agreement's effectiveness. Independent of the review directed by the MMPA amendments, however, a task force of the Arctic Environmental Protection Strategy has initiated a review of the Agreement as it pertains to sustainable development in the Arctic.[130]

[125]Marine Mammal Commission, Annual Report to Congress, 1991 at 49 (1992).
[126]Marine Mammal Commission, Annual Report to Congress, 1994 at 158 (1995).
[127]16 U.S.C. § 1383(b).
[128]*Id.* § 1383(c).
[129]Marine Mammal Commission, Annual Report to Congress, 1995 at 144 (1996).
[130]*Id.* In 1991 the eight Arctic countries adopted the Arctic Environmental Protection Strategy. One of the Strategy's program areas is conservation of Arctic flora and fauna. The

The 1994 MMPA amendments also allow the Secretary of the Interior to issue permits to import sport hunted polar bear trophies from Canada.[131] Permits may be issued provided that the Secretary, in consultation with the Marine Mammal Commission, has made certain determinations with regard to Canada's sport hunting program. The Fish and Wildlife Service issued a proposed rule in 1995.[132] The Marine Mammal Commission's review of the proposed rule concluded that some of the Service's findings could be better explained or justified, and the final rule had not been published by the end of 1995.[133]

THE CONVENTION ON INTERNATIONAL TRADE IN ENDANGERED SPECIES OF WILD FAUNA AND FLORA

The Endangered Species Conservation Act of 1969[134] directed the Secretary of the Interior, in conjunction with the Secretary of State, to seek to convene an international meeting for the purpose of concluding "a binding international convention on the conservation of endangered species."[135] The meeting took place in early 1973. The agreement it produced, the Convention on International Trade in Endangered Species of Wild Fauna and Flora (CITES) is, as its name implies, limited solely to matters of international trade and is not a truly comprehensive "convention on the conservation of endangered species" as the 1969 Act seemed to contemplate.[136] CITES nonetheless is quite important, not only because of its substantive restrictions, but also because of the conceptual underpinnings that it provided for later domestic legislation.

CITES' most significant innovation was its recognition that some species are more vulnerable to extinction than others and its provision for regulatory measures that reflect this recognition. Each species[137] protected under CITES is assigned to one of three Appendices. Those listed in

---

Strategy contains no legally binding obligations, but the signatory nations have committed themselves to implementing it. For a discussion of the Strategy's purposes and implementation, see *id.* at 139–42.

[131] 16 U.S.C. § 1374(c)(5)(A).

[132] 60 Fed. Reg. 70 (1995) and 60 Fed. Reg. 36382 (1995).

[133] Marine Mammal Commission, Annual Report to Congress, 1995 at 147–52 (1996). The Commission's report includes a detailed discussion of the proposed rule, the Service's findings with regard to polar bear population trends and illegal trade in bear parts, and other pertinent issues.

[134] Pub. L. No. 91–135, 83 Stat. 275. See Chapter 7 for a discussion of this law.

[135] *Id.* § 5(b).

[136] March 3, 1973, [1975] 27 U.S.T. 1087, T.I.A.S. No. 8249. For a detailed discussion of how CITES works, see International Wildlife Trade: A CITES Sourcebook (Ginette Hemley, ed. 1994).

[137] CITES defines "species" to include "any species, subspecies, or geographically separate population thereof."

Appendix I are the most vulnerable. They include "all species threatened with extinction which are or may be affected by trade" and are made subject to the most stringent protections.[138] The most significant is that these species may not be traded for primarily commercial purposes.

Species listed in Appendix II are somewhat less vulnerable and include:

> (a) all species which although not necessarily now threatened with extinction may become so unless trade in specimens of such species is subject to strict regulation in order to avoid utilization incompatible with their survival; and
>
> (b) other species which must be subject to regulation in order that trade in specimens of certain species referred to in subparagraph (a) of this paragraph may be brought under effective control.[139]

The species referred to in (b) include any that are so similar in appearance to those referred to in (a) as to be nearly indistinguishable from them.[140] Appendix II species may be traded for commercial purposes, subject to several restrictions, the most important of which is that the trade not be detrimental to the species' survival.

The parties to CITES agreed upon the initial Appendix I and II lists at the time it was adopted, and they have been frequently amended.[141] Appendix III, on the other hand, includes species unilaterally designated by any party "as being subject to regulation within its jurisdiction for the purpose of preventing or restricting exploitation, and as needing the cooperation of other parties in the control of trade."[142] Changes in Appen-

[138]CITES, art. II, § 1. Trade is defined broadly in CITES to mean "export, re-export, import and introduction from the sea." *Id.* art. I(c). At the first Conference of the Parties in 1976, the parties adopted criteria for adding species to Appendix I. The criteria provide that if a particular species is, as a biological matter, "currently threatened with extinction," it should be included on Appendix I as long as it "might be expected to be traded for any purpose, scientific or otherwise." Since any species threatened with extinction presumably would be of interest to scientific investigators, it would seem that the biological determination alone is sufficient for listing. *See* Report of the U.S. Delegation to the Conference of the Parties, app. at 11 (hereinafter U.S. Delegation Report).

[139]CITES, art. II, § 2.

[140]*Compare* the "similarity of appearance" authority under Section 4(e) of the Endangered Species Act of 1973, discussed in Chapter 7.

[141]Appendices I and II may be amended by a two-thirds majority of parties voting at the biennial Conferences of the parties, or between Conferences by a two-thirds majority of those voting in a "postal vote" conducted under article XV. Any party making a timely reservation to a particular amendment will be treated as not a party to CITES with respect to trade in the species concerned. Similarly, any nation may, upon ratifying or acceding to CITES, enter a reservation as to any species on any of the Appendices or any parts or derivatives specified in relation to a species included in Appendix III. *See* CITES, art. XXIII, § 2.

[142]CITES, art. II, § 3.

dix III can come about only as a result of the withdrawal of a particular species by the party responsible for placing it there.

CITES' principal substantive provision is its prohibition of trade in "specimens" of species included in any of the Appendices except in accordance with CITES's terms.[143] Trade in Appendix I species requires both an export permit (or a re-export permit if export is from a country other than the country of origin) from the exporting country and an import permit from the importing country.[144] If an Appendix I species is taken in a marine environment not under the jurisdiction of any nation, only a certificate authorizing "introduction from the sea" is required.[145]

Granting of permits and certificates is done by each nation's "Management Authority" on the advice of its "Scientific Authority."[146] Export permits for Appendix I species may be granted only if the Management Authority determines that the specimen has been lawfully obtained, that it will be "so prepared and shipped as to minimize the risk of injury, damage to health or cruel treatment," and that an import permit has been granted. In addition, the Scientific Authority must determine "that such export will not be detrimental to the survival of that species."[147] Re-export permits require similar determinations.[148] Import permits require determinations by the Management Authority "that the specimen is not to be used for primarily commercial purposes," and by the Scientific Authority "that the proposed recipient of a living specimen is suitably equipped to house and care for it" and "that the import will be for purposes which are not detrimental to the survival of the species involved."[149]

[143]The term "specimen" has a different meaning in CITES depending on whether the species concerned is a plant or an animal and on which appendix the species is listed. "Specimen" always includes the whole animal or plant, whether living or dead. It also includes "any recognizable part or derivative thereof," provided that such part or derivative is specified in Appendix III or, in the case of plants only, Appendix II. The parties decided at their conference in 1976, as a temporary measure, that any additions to Appendix III should specify that all readily recognizable parts and derivatives are also to be included. U.S. Delegation Report, app. at 11.

[144]CITES, art. III, § 2–4.

[145]*Id.* § 5. CITES offers no clue as to whether the 200-mile fishery conservation zone created by the Fishery Conservation and Management Act of 1976, or similar zones established by other coastal nations, will be deemed to be "under the jurisdiction" of these countries for purposes of CITES. If so, a country claiming such a zone may allow the introduction for commercial or other purposes of otherwise protected species taken in the zone without the need for a determination of the effect of that introduction on the species' survival.

[146]CITES, art. IX. In 1979, the United States designated the Secretary of the Interior as both its Management and Scientific Authorities. 16 U.S.C. § 1537A. Earlier, an Executive Order vested the Scientific Authority's responsibilities in an independent "Endangered Species Scientific Authority." *See* 41 Fed. Reg. 15683 (1976), revoked by Exec. Order No. 12,608, 52 Fed. Reg. 34617 (1987).

[147]CITES, art. III, § 2.

[148]*Id.* § 4.

[149]*Id.* § 3. This last determination's function is not readily apparent. Presumably, if the

With respect to certificates for introduction from the sea, the Management Authority has the same duties as it does with respect to import permits and the Scientific Authority must advise that introduction will not be detrimental to the species' survival.[150]

CITES operates in much the same way for Appendix II species, except that no import permits are required. Accordingly, no determination as to whether the purpose of the trade is commercial or noncommercial is required. However, the Scientific Authority of each nation must monitor the export of Appendix II species and recommend to its Management Authority "suitable measures to be taken to limit the grant of export permits" for any Appendix II species whenever it "determines that the export of specimens of any such species should be limited in order to maintain that species throughout its range at a level consistent with its role in the ecosystems in which it occurs and well above the level at which that species might become eligible for inclusion in Appendix I."[151]

Finally, for Appendix III species, export from any nation responsible for its inclusion in that Appendix requires an export permit, which may be granted only upon a determination by that nation's Management Authority that the specimen was lawfully obtained and that it will be shipped in a humane manner.[152] An importing nation must require a certificate of origin and, if the state of origin is responsible for the species' inclusion in Appendix III, an export permit.[153]

Scientific Authority of the exporting nation is acting in good faith, its determination that export will not be detrimental to survival should make irrelevant any consideration of the purpose of importation. However, the requirement that the Management Authority of the importing nation determine that the specimen is not to be used for primarily commercial purposes seems to constitute a conclusive presumption that importation for commercial purposes is detrimental to the survival of any Appendix I species, notwithstanding the determination of the Scientific Authority of the exporting nation. The further required determination of the Scientific Authority of the importing nation that the importation be for purposes not detrimental to survival compounds the confusion inherent in these overlapping standards. A seemingly more sensible approach would be to require that the Scientific Authority of the importing nation determine that the purposes of the importation will affirmatively enhance the prospects for the species' survival, such as by artificial propagation or similar mechanisms.

[150] *Id.* § 5.

[151] CITES, art. IV, § 3. *Compare* the language in the Marine Mammal Protection Act that marine mammal "species and population stocks should not be permitted to diminish beyond the point at which they cease to be a significant functioning element in the ecosystem of which they are a part." 16 U.S.C. § 1361(2).

[152] CITES, art. V, § 2. Appendix III is, in effect, an internationalized Lacey Act. For a discussion of CITES' origins, see Comment, *Commentary Upon the IUCN Draft Convention on the Export, Import and Transit of Certain Species of Wild Animals and Plants,* 21 Cath. U. L. Rev. 665 (1972).

[153] CITES, art. V, § 3.

CITES also provides a means of gaining some leverage over nonparty countries. "Comparable documentation" may be required for import from, or export to, these countries.[154]

CITES' requirements are subject to several carefully circumscribed exceptions. The most important of these pertain to specimens acquired before CITES applied to them, specimens that are "personal or household effects," and specimens loaned, donated, or exchanged by scientists or scientific institutions registered by the Management Authority of their nation.[155]

A potential threat to CITES' success is its omission of any effective enforcement mechanism against states that give it only perfunctory attention. While many countries have joined CITES,[156] the quality of their implementation and enforcement varies markedly.[157] Even the United States' enforcement of CITES has been subject to criticism.[158]

At least one commentator has portrayed CITES as "perhaps the most successful of all international treaties concerned with the conservation of wildlife,"[159] and implementation of CITES has expanded significantly

[154]CITES, art. X. The Parties decided at their 1976 Conference to require documentation for all non-party countries, reducing the problem of "laundry" countries (those that import from one country and re-export to another). *See* U.S. Delegation Report, app. at 1.

[155]CITES, art. VII. The first of these, the so-called "pre-Convention exemption," has been subject to varying interpretations by CITES parties, particularly as it applies to parties that ratified or acceded after the effective date of a species' listing or that initially entered a reservation with respect to a species and later withdrew it. Since CITES applies to all specimens of a species as of the date of its listing, regardless of the specimens' location, the relevant date should be the listing date rather than the later date of a party's ratification or withdrawal of reservation. This view is compatible with CITES' language. Moreover, it serves CITES' purposes by discouraging hoarding in anticipation of ratification or withdrawal of a ratification, and by promoting uniform implementation.

[156]As of January 1, 1995, 132 countries were parties to CITES. The United States, which was the first country to ratify CITES, did not designate its Management and Scientific Authorities until April 1976, Exec. Order No. 11911, 41 Fed. Reg. 15683 (1976), or propose regulations to implement CITES until June 1976. *See* 41 Fed. Reg. 24367 (1976). The final regulations were promulgated in February 1977 and are codified at 50 C.F.R. Part 23 (1995). *See* 42 Fed. Reg. 10465 (1977).

[157]*See* William C. Burns, *CITES and the Regulation of International Trade in Endangered Species of Flora: A Critical Appraisal*, 8 Dick. J. Int'l L. 203 (1990) (discussion of how enforcement of CITES is skewed in favor of fauna over flora); John B. Heppes and Eric J. McFadden, *The Convention on International Trade in Endangered Species of Wild Fauna and Flora: Improving the Prospects for Preserving Our Biological Heritage*, 5 B.U. Int'l L.J. 229, 232–45 (1987) (discussing problems with record-keeping, enforcement, trade with non-parties, and linguistics); Kathryn S. Fuller, Ginette Hemley and Sarah Fitzgerald, *Wildlife Trade Law Implementation in Developing Countries: The Experience in Latin America*, 5 B.U. Int'l L.J. 289 (1987); Eric McFadden, *Asian Compliance with CITES: Problems and Prospects*, 5 B.U. Int'l L.J. 311 (1987).

[158]Meena Alagappan, Comment, *The United States' Enforcement of the Convention on International Trade in Endangered Species of Wild Fauna and Flora*, 10 Nw. J. Int'l L. & Bus. 541 (1990).

[159]Lyster, *supra* note 8, at 240.

since it entered into force in 1975. CITES' value as a wildlife conservation device is limited by the fact that the international trade that it seeks to control is but one of several causes of wildlife endangerment.[160]

THE ANTARCTIC ECOSYSTEM

Twelve nations concluded the Antarctic Treaty in 1959.[161] The Treaty provided the initial framework for an international legal regime applying to the Antarctic continent and the Southern Ocean. It is the centerpiece of a system of agreements that has evolved over the last four decades, called the Antarctic Treaty System.[162] Of particular importance for wildlife are agreements pertaining to seals, living marine resources, and protection of the environment.

The 1959 Treaty's principal aims were to bar the use of Antarctica for military purposes, to prohibit nuclear explosions and the disposal of radioactive waste there, and to foreclose assertions of claims to territorial sovereignty. Although Antarctica represented the world's last great wilderness, the Treaty itself said virtually nothing about conservation of the continent's wildlife. It merely delegated to each of its signatory nations the authority to make recommendations at periodic meetings for the "preservation and conservation of living resources."[163]

More than a dozen species of whales, seals, porpoises, and dolphins are residents or seasonal visitors to the seas surrounding Antarctica. Regional populations of several of the whale species are severely depleted, as are two of the six resident seal species.[164] In addition to these marine mammals, finfish, squid, and birds are abundant. In fact, the Southern Ocean is "biologically among the most productive on earth."[165] Concern for the integrity of this ecosystem has prompted a number of recommendations and agreements to conserve Antarctica's animals and plants.

[160]*See, e.g.*, Heppes and McFadden, *supra* note 157, at 230, n. 6 ("The gravest danger to wildlife species is the loss of wildlife habitat because of population pressures, urbanization and major projects such as dam and road building").

[161]Dec. 1, 1959, [1961] 12 U.S.T. 794, T.I.A.S. No. 4780, 402 U.N.T.S. 71. The nations were Argentina, Australia, Belgium, Chile, France, Japan, New Zealand, Norway, South Africa, the Soviet Union, Great Britain, and the United States. There now are fourteen additional "consultative parties," for a total of twenty-six nations that claim decision-making power over Antarctica.

[162]In addition to the Antarctic Treaty System, the Southern Ocean is subject to the 1982 Law of the Sea (LOS) Convention. For a discussion of the relationship between the Treaty System and the LOS Convention, see Christopher C. Joyner, *The Antarctic Treaty System and the Law of the Sea—Competing Regimes in the Southern Ocean?* 10 Int'l J. of Marine and Coastal L. 301 (1995).

[163]Antarctic Treaty, art. IX, § 1(f).

[164]Marine Mammal Commission, Annual Report to Congress, 1995 at 126 (1996).

[165]Joyner, *supra* note 162, at 311.

The signatories to the Antarctic Treaty first promulgated a set of Agreed Measures for the Conservation of Antarctic Fauna and Flora in 1964, under the authority to preserve and conserve living resources.[166] In the same year, a Norwegian company began an exploratory sealing expedition in the Antarctic. This expedition, and the fact that neither the Agreed Measures nor later recommendations regarding sealing were binding, led the parties to conclude that there was a need for a new treaty addressing sealing. The result was the 1972 Convention for the Conservation of Antarctic Seals, which was ratified by the Senate in 1976 and became effective in 1978.[167] Although Norway did not initiate commercial sealing and there has been none in the Antarctic since the 1950s, this agreement provides a mechanism for regulating sealing should it resume.

This Convention provides that yearly harvest levels not exceed "optimum sustainable yield," prohibits pelagic sealing, and fixes other harvest restrictions. It does not, however, provide for any enforcement mechanism or any permanent commission. The State Department was criticized for the secrecy that surrounded negotiations leading to the Convention. The Department frankly acknowledges that the Convention falls short of the Marine Mammal Protection Act's standards for marine mammal management.[168] Nonetheless, to prevent commercial sealing from taking place without any international regulation, the Senate consented to ratify the Convention.

The Convention on the Conservation of Antarctic Marine Living Resources (CCAMLR)

In 1977, the Antarctic parties recommended establishing a "definitive regime" for the conservation of Antarctic marine living resources.[169] That recommendation led to the 1980 Convention on the Conservation of Antarctic Marine Living Resources (CCAMLR).[170]

The major impetus for CCAMLR was the likelihood of a greatly expanded, unregulated commercial harvest of krill, a shrimp-like crustacean

[166]Recommendations of the Third Antarctic Consultative Meeting, June 2–13, 1964, [1966] 17 U.S.T. 991; T.I.A.S. No. 6058, app. The Agreed Measures were implemented in the United States by the Antarctic Conservation Act of 1978, Pub. L. No. 95–541, 92 Stat. 2048, which prohibits the taking of mammals or birds in Antarctica and the taking of native plants in protected areas except under permit from the Director of the National Science Foundation. *See* 16 U.S.C. §§ 2401–2412.

[167]June 1, 1972, 29 U.S.T. 441, T.I.A.S. No. 8826.

[168]*See* S. Exec. Rep. No. 35, 94th Cong., 2d Sess. 5 (1976).

[169]Recommendations of the Ninth Antarctic Consultative Meeting, Sept. 19–Oct. 7, 1977, [1983] T.I.A.S. No. 10735, Recommendation IX-2.

[170]May 20, 1980, 33 U.S.T. 3476, T.I.A.S. No. 10240. Initially there were fifteen parties to the Convention. By 1995 the number had grown to twenty-seven. *See* Joyner, *supra* note 162, at 303 n.13 for a list of the current signatories.

that is abundant in the Southern Ocean and is the central link in the Antarctic marine food chain.[171] The main concern was not so much for conservation for its own sake, but rather for sustaining commercially viable populations of marine mammals that depend on the krill, particularly baleen whales. A legal regime fostered by the desire to sustain resource exploitation has, however, evolved to encompass broader concerns, as has been the case with international regulation of whaling itself.

The Convention is remarkable in two respects. First, it was signed before a major Antarctic krill fishery had emerged. It is thus "one of the few international treaties concerned with wildlife conservation to be concluded prior to heavy commercial pressure on the species it was designed to protect." [172] Second, as noted above, CCAMLR obliges the parties to adopt an ecosystem approach to the exploitation of Antarctic marine living resources. The Convention is an attempt to insure that krill harvesters consider impacts of the harvest on other species and thus to protect the stability of the Antarctic marine ecosystem.[173]

Although regulating the krill harvest is the Convention's main purpose, it is not the only one. The Convention applies to all Antarctic living marine resources, a term defined to include all species of living organisms, including birds, found south of the Antarctic Convergence.[174] Article II of the Convention sets forth the following three "principles of conservation" to govern living marine resources and their taking:

(a) prevention of decrease in the size of any harvested population to levels below those which ensure its stable recruitment. For this purpose its size should not be allowed to fall below a level close to that which ensures the greatest net annual increment;

(b) maintenance of the ecological relationships between harvested, dependent and related populations of Antarctic marine living resources and the restoration of depleted populations to the levels defined in sub-paragraph (a) above; and

(c) prevention of changes or minimization of the risk of changes in the marine ecosystem which are not potentially reversible over two or three decades, taking

[171] *See* Joyner, *supra* note 162, at 311. In 1981, an estimated 200–600 million metric tons of krill were present in Antarctic waters during the summer. National Academy of Science, An Evaluation of Antarctic Marine Ecosystem Research (1981).

[172] Lyster, *supra* note 8, at 157. *See also* Christopher C. Joyner, *Fragile Ecosystems: Preclusive Restoration in the Antarctic*, 34 Nat. Resources J. 879 (1994). Prof. Joyner examines how the agreements in the Antarctic Treaty System aim to prevent damage to the fragile Antarctic ecosystem while the ecosystem is still in good health.

[173] It is especially noteworthy that an ecosystem approach "is likely to mean much lower harvest levels than would result from the traditional approach," since harvest must be kept to levels that will allow recovery of depleted predator species. Lyster, *supra* note 8, at 158.

[174] Art. 1, ¶ 2. The Antarctic Convergence is a seasonally shifting oceanographic boundary between Antarctic waters to the south and warmer, sub-Antarctic waters to the north. Article 1, ¶ 4 defines it arbitrarily by reference to a set of coordinates roughly approximating the actual Convergence.

into account the state of available knowledge of the direct and indirect impact of harvesting, the effect of the introduction of alien species, the effects of associated activities on the marine ecosystem and of the effects of environmental changes, with the aim of making possible the sustained conservation of Antarctic marine living resources.[175]

Subparagraph (a) and the reference to "depleted populations" in the second clause of subparagraph (b) appear to require that harvest and associated activities not cause any population in the Antarctic ecosystem, whether or not harvested itself, to be reduced below the level of greatest net annual increment.[176] Because this requirement applies to *all* species, effective implementation requires careful selection of vulnerable species and indicator species for research and monitoring.

The third conservation principle, requiring the "[p]revention of changes or minimization of the risk of changes in the marine ecosystem which are not potentially reversible over two or three decades," was an important innovation in international conservation law. It addresses the resilience of the ecosystem to harvest and associated activities, rather than just desired states of an ecosystem, such as attainment of particular population levels.

To implement these principles, the Convention establishes regulatory and scientific bodies with the authority to impose a full panoply of conservation measures and the flexibility to adopt or to revise the measures expeditiously. The Convention's principal regulatory body is the Commission for the Conservation of Antarctic Marine Living Resources. All signatories and certain other parties are represented on the Commission,[177] which makes decisions by consensus on matters of substance.[178]

The Convention directs the Commission to hold regular annual meetings and to provide for the conservation of Antarctic marine living resources in accordance with the principles of Article II. The Commission is to achieve this objective by collecting and analyzing necessary data and adopting conservation measures. The measures may include, but are not limited to, limitations on the kind, quantity, place, duration, and methods of harvest, and designation of special areas for protection and for scientific study.[179] The designation provision authorizes the Commission to establish a system of marine reserves paralleling the terrestrial system of protected areas and sites of special scientific interest

[175]Art. II, ¶ 3.

[176]The term "greatest net annual increment" is likely to be interpreted to mean "maximum productivity" as that term is used in the Marine Mammal Protection Act's definition of "optimum sustainable population." *See* Chapter 5 at text accompanying notes 21–35.

[177]Art. VII.

[178]Art. XII, ¶ 1.

[179]Art. IX.

established by the Agreed Measures for the Conservation of Antarctic Fauna and Flora and adopted at various meetings of the Antarctic Treaty Consultative Parties.

Congress enacted implementing legislation for CCAMLR in 1984. The Antarctic Marine Living Resources Convention Act[180] makes it unlawful to engage in harvesting or other associated activities in violation of provisions of the Convention or of a conservation measure in force, to violate any regulation promulgated under the Act, or to transport, sell, purchase, or possess any Antarctic living marine resource that was harvested in violation of any of the above.[181] The Act provides for civil and criminal penalties.[182]

After fourteen years of management under CCAMLR, it remains unclear how effective the Convention has been in achieving its goals.[183] Some general observations, however, may be made. First, while our knowledge of the Antarctic ecosystem has grown during the time CCAMLR has been in force, we still know far too little to make accurate predictions as to impacts of particular harvest levels of krill, finfish, or other species on predators or on the ecosystem as a whole. Collecting data has been difficult,[184] and the complexity of managing an ecosystem rather than species-by-species adds to the problem. Thus, while the Convention's aspiration to avoid adverse ecosystem impacts is innovative and laudable, in practice it is extremely difficult to determine how to move toward this goal and to what extent it has been achieved.[185]

The lack of scientific data figured prominently in the Commission's failure to adopt any conservation measures pertaining to krill prior to 1991, despite the fact that concern for ecosystem impacts of krill harvest was the main impetus for CCAMLR's creation. Russia and Japan, the major krill fishing nations, objected to conservation measures on the grounds that there was no data on the Southern Ocean that would justify imposing harvest limits.[186] In 1991, however, the Commission adopted a harvest

[180]Pub. L. No. 98–623, 98 Stat. 3401, codified at 16 U.S.C. §§ 2431–2442.

[181]*Id.* § 2435. Regulations are found at 50 C.F.R. § 380 (1995).

[182]16 U.S.C. §§ 2437, 2438.

[183]For a detailed discussion of how CCAMLR has operated and how effective it has been, see Stuart B. Kaye, *Legal Approaches to Polar Fisheries Regimes: A Comparative Analysis of the Convention for the Conservation of Antarctic Marine Living Resources and the Bering Sea Doughnut Hole Convention*, 26 Cal. W. Int'l L.J. 75, 87–97 (1995).

[184]The Commission has no independent ability to collect data but instead must rely on the parties for information, particularly those involved in fishing. There is a lack of long-term data on the fisheries and no comprehensive data bank on Southern Ocean marine life. *Id.* at 88.

[185]*See id.* at 88. *See also* William T. Burke, The New International Law of Fisheries 114–15 (1994) (noting that CCAMLR remained the only marine living resource convention based on an ecosystem management approach).

[186]Christopher C. Joyner, Antarctica and the Law of the Sea 234 (1992). The Article XII consensus requirement allows a single state to block action by the other parties.

quota of 1.5 million metric tons (mt) for the area in which most krill are taken.[187] Harvest levels have never come close to this figure, the highest being about 528,000 mt in 1982.[188]

Considering that the "precautionary" harvest limit is almost three times the highest catch ever reported, and that there are no meaningful quotas in particular subareas or locales, it is difficult to conclude that CCAMLR has contributed significantly to ecosystem protection through management of the krill harvest. The Commission will need to address krill harvest on a much more localized and sophisticated level in future years if CCAMLR's promise of an ecosystem approach is to be realized.[189]

In recent years the Commission has, however, adopted conservation measures that address incidental catch of seabirds in longline gear, bycatch levels, exploratory fisheries, and finfish catch limits and restricted fisheries.[190] In addition, the Commission has set up the CCAMLR Ecosystem Monitoring Program and prepared special site management plans for important research areas.[191] Of course, in light of how little we understand of the Antarctic ecosystem, the cautious approach would be to make areas open to harvest the exception rather than the rule. Designation of protected areas where commercial harvest would be prohibited and where ecological relationships can be maintained would assure that relatively

[187]Conservation Measure 32/X (1991) (applies to Area 48, the Atlantic sector). In 1992, the Commission adopted a conservation measure specifying subarea quotas if the catch in three subareas of the Atlantic sector were to exceed 620,000 mt. Conservation Measure 46/XI (1992).

[188]See Kaye, *supra* note 183, at 94, Table I, for information on krill harvest from 1971 to 1994. The krill catch has fluctuated widely. There was a dramatic decline from about 300,000 mt in 1992 to 89,000 in 1993. *Id.* The former Soviet Union historically accounted for the vast majority of the harvest (for example, Soviet vessels took over 75 percent of the harvest in 1990), and the sharp decline in 1993 was due primarily to reduction in fishing effort by Russia and the Ukraine. Marine Mammal Commission, Annual Report to Congress, 1993 at 141 (1994). See Kaye, *supra* note 183, at 95–96 for further discussion of the possible reasons for decline in the krill harvest.

[189]It appears from biomass and recruitment data that the krill stock as a whole is in no danger. However, distribution as well as abundance is important to predators, as is timing of harvest. In 1995 the CCAMLR Commission's working group on ecosystem monitoring and management noted the need to

> (a) ensure that krill catches are not concentrated in time and space to an extent that local populations of dependent species may be affected adversely and (b) take into account relevant biological and environmental variables, when determining precautionary catch limits and subdividing limits for broad areas.

Marine Mammal Commission, Annual Report to Congress, 1995 at 134 (1996).

[190]See Kaye, *supra* note 183, at 90–92, for a discussion of these measures. *See also* Marine Mammal Commission, Annual Report to Congress, 1995 at 131–35 (1996).

[191]Conservation Measures 18/IX (1992) and 82/XIII (1994). *See also* Kaye, *supra* note 183, at 91.

pristine areas in Antarctica are maintained. Moreover, it would buffer harvested areas against overexploitation until our knowledge permits us to make more precise determinations of critical areas for protection.

As with most wildlife treaties, compliance with measures adopted under CCAMLR remains a problem. Fishing in closed areas and illegal longline fishing are two kinds of violations that have been documented,[192] and the parties have not reached consensus on suggestions for improving compliance. For example, the United States proposed in 1994 and 1995 that satellite monitoring be used to detect the presence of vessels in closed areas. Some parties objected on the grounds that satellite monitoring would infringe on either high seas rights or national jurisdiction.[193]

The Protocol on Environmental Protection

In 1988, the parties to the Antarctic Treaty concluded a convention to regulate mineral exploration and development, the Convention on the Regulation of Antarctic Mineral Resource Activities (CRAMRA). The Convention never came into force, however, because of environmental concerns about the impacts of mineral development on the fragile Antarctic ecosystem. In 1989, Australia and France announced that they would not sign CRAMRA. Because all parties to the Antarctic Treaty must sign any convention before it is adopted, this declaration meant that CRAMRA likely would never be implemented.[194]

In place of CRAMRA, in 1991 the parties concluded a Protocol on Environmental Protection to the Antarctic Treaty in Madrid, Spain, known as the Madrid Protocol.[195] The Protocol bans all mineral exploration and development in the Antarctic for fifty years.[196] It has not yet come into force, however, because it has been ratified by only nineteen of the twenty-six consultative parties.[197] The United States Senate gave its consent to the Protocol in 1992, but no legislation has been passed providing authority to implement its provisions and the United States has not deposited its instrument of ratification.[198]

[192] *See, e.g.*, Marine Mammal Commission, Annual Report to Congress, 1995 at 133 (1996).

[193] *Id.* Despite the problems, however, the Marine Mammal Commission "believes that efforts to implement the Convention have been innovative and generally successful." *Id.* at 135.

[194] *See* Joyner, *supra* note 186, at 304 ("[L]argely out of environmental concerns, [CRAMRA] has been set aside and relegated to the status of legal dormancy."). See Marine Mammal Commission, Compendium of Selected Treaties, International Agreements, and Other Relevant Documents on Marine Resources, Wildlife, and the Environment at 432–73 (1994) for a text of CRAMRA.

[195] Sen. Treaty Doc. 102–22, 102d Congress, 2d Sess. (1992) (hereinafter "Protocol").

[196] *Id.*, art. 7.

[197] Marine Mammal Commission, Annual Report to Congress, 1995 at 127 (1996).

[198] *Id.*

Although the Madrid Protocol's most widely known provision is the ban on mining, the Protocol and its five annexes in fact provide a comprehensive regime of environmental protection for the Antarctic.[199] Article 2 sets forth the Protocol's objective: "The parties commit themselves to the comprehensive protection of the Antarctic environment and dependent and associated ecosystems and hereby designate Antarctica as a natural reserve, devoted to peace and science." Article 3 provides that "protection of the Antarctic environment and dependent and associated ecosystems, including its wilderness and aesthetic values . . . shall be fundamental considerations in the planning and conduct of all activities" in the Antarctic. The five annexes deal with environmental impact assessment, conservation of Antarctic Fauna and Flora, waste disposal and waste management, prevention of marine pollution, and area protection and management.[200]

Annex II, Conservation of Antarctic Fauna and Flora, essentially restates the 1964 Agreed Measures for the Conservation of Arctic Fauna and Flora. A significant innovation in the Annex, however, is a prohibition on introducing any species of animal or plant not native to the Antarctic without a permit.[201]

Annex V, Area Protection and Management, is an effort to improve the existing system of Antarctic protected areas and has substantial potential to aid wildlife conservation. It allows any area in Antarctica, including marine areas, to be designated as an Antarctic Specially Protected Area or an Antarctic Specially Managed Area.[202] A Specially Protected Area may be designated to protect "outstanding environmental, scientific, historic, aesthetic or wilderness values . . . or ongoing or planned scientific research."[203] Management plans adopted under the annex may prohibit, restrict, or manage activities in these areas.[204] Any party, the Committee for Environmental Protection, the Scientific Committee on Antarctic Research, or the Commission for the Conservation of Antarctic Marine Living Resources may nominate an area for designation.[205]

Article 11 of the Protocol establishes a Committee for Environmental Protection. The committee is not meant to be an independent body to oversee compliance with the Protocol; rather, "the functions of the Com-

[199]The Protocol has been called "the most comprehensive multilateral document ever adopted on the international protection of the environment." S.K.N. Blay, *New Trends in the Protection of the Antarctic Environment: The 1991 Madrid Protocol, 86 Am. J. Int'l L. 377 (1992).* It is not meant to supplant previous agreements, however. Article 4 cautions that the Protocol supplements the Antarctic Treaty and does not derogate from the rights and obligations of the parties under the other instruments in force under the Antarctic Treaty System.

[200]For a detailed discussion of the Protocol and its annexes, see Blay, *supra* note 199.

[201]Protocol Annex II, art. 41.

[202]*Id.*, Annex V, arts. 2 and 3.

[203]*Id.*, art. 2.

[204]*Id.*, art. 5, ¶ 3(f).

[205]*Id.*, art. 5, ¶ 1.

mittee shall be to provide advice and formulate recommendations to the Parties in connection with the implementation of this Protocol, including the operation of its Annexes, for consideration at Antarctic Treaty Consultative Meetings."[206]

Although like CRAMRA, the Madrid Protocol has not yet entered into force, it appears more likely to be ratified by all parties than was CRAMRA because of today's heightened level of international environmental consciousness. Parties who have yet to ratify the Protocol may be moved to do so by the desire to be recognized as environmentally responsible.[207]

Conclusion

For the moment, whaling and sealing do not pose major threats to the Antarctic ecosystem.[208] Further, if the Madrid Protocol comes into force, it will go a long way toward protecting the ecosystem from impacts of mineral development.[209] The CCAMLR Commission is beginning to monitor krill harvest on a more sophisticated level. Finally, there have been encouraging developments within the Commission to deal with other threats, such as new fisheries, bycatch, incidental mortality of seabirds, and the effects of increased tourism on the fragile environment.[210]

It is clear that the international community recognizes the value of one of the world's last, unspoiled natural areas. Whether CCAMLR and the Protocol together will live up to their great potential of preserving the Antarctic through their comprehensive ecosystem approach remains to be seen.[211]

CONVENTION ON BIOLOGICAL DIVERSITY

Global concern for the environment in the face of rapid development led the United Nations in 1983 to establish an independent body called

[206]Protocol, art. 12.

[207]*See* Blay, *supra* note 199, at 397.

[208]In addition to the moratorium on whaling, the IWC in 1994 designated the Southern Ocean a sanctuary for all whale species found there. Thus, even if the moratorium were lifted, whaling would not resume in the Southern Ocean unless this designation were removed.

[209]Even if the Protocol comes into force, however, problems remain. *See* Comment, *A Certain False Security: The Madrid Protocol to the Antarctic Treaty*, 4 Colo. J. Int'l Envt'l L. & Pol'y 458 (1993) (noting potential problems with non-party nations, state succession to the Antarctic Treaty System, and offshore jurisdiction).

[210]See Marine Mammal Commission, Annual Report to Congress, 1995 at 129–30 (1996) for a discussion of recent efforts to reduce impacts of the growing Antarctic tourist industry.

[211]For an examination of the Antarctic regime's ecosystem approach and its relationship to the Convention on Biological Diversity, discussed below, see Catherine Redgwell, "The Protection of the Antarctic Environment and the Ecosystem Approach," in Bowman and Redgwell, *supra* note 8, at 109–28.

the World Commission on Environment and Development. In 1987, the Commission issued an influential report entitled *Our Common Future.*[212] The Commission's report, also commonly called the "Brundtland Report" for the Prime Minister of Norway who chaired the Commission, sought to highlight the vital link between economic development and environmental protection. The report was among the first to promote the concept of "sustainable development."

The report's recommendations triggered several international initiatives. Among them was the decision of the United Nations General Assembly in December 1989 to hold a United Nations Conference on Environment and Development in 1992. The resulting UNCED conference, also known simply as the "Rio Conference," for its location in Rio de Janeiro, Brazil, produced international agreements on a wide range of subjects. One was the Convention on Biological Diversity, the result of nearly four years of effort by an intergovernmental negotiating committee sponsored by the United Nations Environment Program. The Convention addressed several of the major causes of global species decline that CITES had not, including habitat destruction and introduction of alien species.

The Convention on Biological Diversity was adopted on May 22, 1992,[213] and within three weeks 153 nations and the European Community had signed it. Significantly, however, the United States was not among the signatories. This was no small irony, since the United States had served as host of and acknowledged world leader in the meeting that produced the CITES treaty in 1972. Just twenty years later, the United States stood virtually alone among the major nations of the world as a nonsignatory to this major Convention.

When the Clinton administration took office in 1993, it promptly signed the Convention and sought the Senate's consent to ratification. By the end of 1996, however, the Convention remained unratified, with little prospect for favorable action in the foreseeable future. Meanwhile, the Convention has entered into force, the parties to it have met twice and made several important decisions, key administrative positions have been filled,[214] and the Convention has become fully operational. Because of the Convention's significance, its key provisions and the concerns underlying U.S. reluctance to become a party to it are discussed below.[215]

[212]World Commission on Environment and Development, Our Common Future (1987).

[213]For a text of the Convention, see Marine Mammal Commission, *supra* note 194, at 1054–74.

[214]Dr. Calestous Juma, a prominent Kenyan conservationist, was appointed Executive Secretary of the Convention and Peter Schei, a highly regarded Norwegian scientist, heads the important Subsidiary Body on Scientific, Technical and Technological Advice. *See* Convention, Article 25.

[215]For an in-depth examination of the Convention and its relationship to other international agreements that contribute to biodiversity protection, see Bowman and Redgwell, *supra* note 8.

The Convention on Biological Diversity has a much broader set of objectives than those of CITES. CITES concerns only species that are considered to be at risk of extinction, and it seeks to achieve its conservation goal solely through regulating international trade in those species. The Convention on Biological Diversity seeks to conserve "biological diversity," a term that it broadly defines as "the variability among living organisms from all sources, including, *inter alia*, terrestrial, marine and other aquatic ecosystems and the ecological complexes of which they are part; this includes diversity within species, between species and of ecosystems."[216] To achieve this goal, the Convention provides that each party is to take a wide range of actions "as far as possible and as appropriate."[217]

Traditional conservation measures, aimed at conserving biological resources both in natural habitats (*in situ* conservation) and otherwise (*ex situ* conservation), were not the source of the United States' concern. In fact, the United States had already demonstrated considerable leadership with respect to most of these measures. For example, the Convention calls upon the member nations to "[e]stablish a system of protected areas or areas where special measures need to be taken to conserve biological diversity," to "[r]egulate or manage biological resources important for the conservation of biological diversity . . . with a view to ensuring their conservation and sustainable use," to "[p]romote the protection of ecosystems, natural habitats and the maintenance of viable populations of species in natural surroundings," to "[p]revent the introduction of, control or eradicate those alien species which threaten ecosystems, habitats or species," and to "[d]evelop or maintain necessary legislation and/or other regulatory provisions for the protection of threatened species and populations."[218] With respect to each of these actions, the United States already had legislation or programs that could serve as models for much of the rest of the world.

The United States' concerns stemmed not from these provisions, but rather from those aimed at adjusting the balance between the less developed nations of the world that harbor much of the remaining biological diversity, and the developed nations that often capture the economic benefit of that diversity. Developed nations profit from biological resources by, for example, patenting new medicines, developing new crop varieties, and creating entirely new organisms through biotechnology. The Convention is premised on the beliefs that biodiversity can be conserved only if less developed nations benefit economically from its conservation (a view that is widely embraced), and that to achieve this objective, developed or "user" nations should share equitably the economic value of biodiversity

[216]Art. 2.
[217]Art. 9.
[218]Art. 8.

with less developed or "source" nations. The latter view is, in practical application at least, highly controversial.

The Convention's first article describes the agreement's objectives as "the conservation of biological diversity, the sustainable use of its components and the fair and equitable sharing of the benefits arising out of the utilization of genetic resources, including by appropriate access to genetic resources and by appropriate transfer of relevant technologies, taking into account all rights over those resources and to technologies, and by appropriate funding."[219] The qualifying phrase in this passage ("taking into account all rights over those resources and to technologies") reflects the efforts of the Convention's negotiators to accommodate the concerns of the United States and other developed nations for the protection of "intellectual property rights." Intellectual property rights include patents and similar rights that confer upon those who discover or create a new medicine, crop variety, or genetically engineered organism an exclusive right to produce and market their product.

The conflict over the Convention, and particularly over whether the United States should sign and ratify it, was in large part an argument about intellectual property rights. One camp regarded the system of intellectual property rights as part of the problem, contributing to the disparity between the developed and undeveloped world by allowing developed nations to usurp the genetic resources and traditional knowledge of less developed nations without compensation. The other regarded protection of intellectual property rights as a *sine qua non* not only for economic progress but also for conservation, contending that without this protection, the incentive to make discoveries that reveal the value of biodiversity vanishes.

As so often happens in negotiation, the final product represented not so much a triumph of one camp over the other, but rather an ambiguous accommodation that appeared to give both sides much of what they wanted. On one hand, the Convention recognizes "the sovereign rights of States over their natural resources" and acknowledges that "authority to determine access to genetic resources rests with the national governments and is subject to national legislation."[220] On the other hand, a member state cannot simply refuse to provide access to its genetic resources, for the Convention also requires that "[e]ach Contracting Party shall endeavor to create conditions to facilitate access to genetic resources for environmentally sound uses by other Contracting Parties and not to impose restrictions that run counter to the objectives" of the Convention.[221] Lest the latter provision be assumed to mean that a nation har-

[219]Art. 1.
[220]Art. 15, ¶ 1.
[221]Art. 15, ¶ 2.

boring potentially valuable genetic resources must simply open its doors to all who would seek to exploit them, however, the Convention also provides that "[a]ccess, where granted, shall be on mutually agreed terms."[222]

A similar back-and-forth quality characterizes the Convention's provisions governing transfer of technologies that make use of genetic resources, including biotechnology. The Convention states that each contracting party is to "provide and/or facilitate access for and transfer to other Contracting Parties" of these technologies.[223] Further, access to and transfer of technologies to developing countries shall be "under fair and most favourable terms, including on concessional and preferential terms where mutually agreed."[224] These requirements would appear to undermine, if not be entirely inconsistent with, intellectual property rights. However, the Convention goes on to provide that "[i]n the case of technology subject to patents and other intellectual property rights, such access and transfer shall be provided on terms which recognize and are consistent with the adequate and effective protection of intellectual property rights."[225] The latter provision does not necessarily freeze in place the existing system of intellectual property rights, however, for yet another provision states that "[t]he Contracting Parties, recognizing that patents and other intellectual property rights may have an influence on the implementation of this Convention, shall cooperate in this regard subject to national legislation and international law in order to ensure that such rights are supportive of and do not run counter to its objectives."

When President Bush refused to sign the Biodiversity Convention at the Rio Conference in 1992, the concerns he cited were being voiced principally by the U.S. biotechnology industry. At best, that industry saw in the ambiguous and apparently conflicting provisions of the Convention the opportunity for endless wrangling and uncertainty about what the Convention required. At worst, they feared erosion of the patent protection that provides the essential underpinning for this new industry.

If the choice were ever between having a treaty or having no treaty, that is clearly not the case now. Rather, the choice for the United States is between being part of a treaty that nearly every other nation of the world participates in and that will greatly influence how conservation and intellectual property rights goals are reconciled, and being forever on the outside. American businesses that once expressed concern about the Convention's potential interpretation now recognize that many of the principles it embraces are being implemented through access and other

[222]Art. 15, ¶ 4.
[223]Art. 16, ¶ 1.
[224]Art. 16, ¶ 2.
[225]*Id.*

commercial agreements. Thus, many business interests now conclude that the United States will have more influence over how those principles are put into practice if it joins the Convention. Just as commercial concerns overrode conservation interests in the United States' initial decision not to sign the Convention, the same concerns probably will cause reversal of that decision in the future.

TABLE OF AUTHORITIES

FEDERAL CASES

Biderman v. Secretary of Interior, 7 Envt'l Rep. Cas. (BNA) 1279 (E.D.N.Y. July 19, 1974), *aff'd sub nom. Biderman v. Morton*, 5 Envt'l L. Rep. (Envt'l L. Inst.) 20027 (2d Cir. 1974), **294n**

Big Hole Ranchers Ass'n, Inc. v. Forest Service, 686 F. Supp. 256 (D. Mont. 1988), **317n**

Bishop v. United States, 126 F. Supp. 449 (Ct. Cl. 1954), **91n**

Bob Marshall Alliance v. Hodel, 852 F.2d 1223 (9th Cir. 1988), **319–20**

Bogle v. White, 61 F.2d 930 (5th Cir. 1932), **70n**

Boteler v. National Wildlife Federation, 429 U.S. 979 (1976), **222n, 245n**

Brown v. Anderson, 202 F. Supp. 96 (D. Alas. 1962), **23n, 30n**

C & W Fish Co. v. Fox, 931 F.2d 1556 (D.C. Cir. 1991), **164n, 171n, 176–77**

Cabinet Mountains Wilderness v. Peterson, 685 F.2d 678 (D.C. Cir. 1982), **319n**

California v. Bergland, 483 F. Supp. 465 (E.D. Cal. 1980), *aff'd in part, rev'd in part sub nom. California v. Block*, 690 F.2d 753 (9th Cir. 1982), **317**

California v. Watt, 520 F. Supp. 1359 (C.D. Cal. 1981), **248n**

California v. Watt, 683 F.2d 1253 (9th Cir. 1982), **416n**

Calipatria Land Co. v. Lujan, 793 F. Supp. 241 (S.D. Cal. 1990), **75**

Cameron Parish Police Jury v. Hickel, 302 F. Supp. 689 (W.D. La. 1969), **289n**

Camfield v. United States, 167 U.S. 518 (1897), **385**

Cape Henry Bird Club v. Laird, 359 F. Supp. 404 (W.D. Va.), *aff'd*, 484 F.2d 453 (4th Cir. 1973), **410n**

Cappaert v. United States, 426 U.S. 128 (1976), **286n, 309**

Carpenter v. Andrus, 485 F. Supp. 320 (D. Del. 1980), **228–29**

Carson-Truckee Water Conservancy District v. Watt, 549 F. Supp. 704 (D. Nev. 1982), *aff'd sub nom. Carson-Truckee Water Conservancy District v. Clark*, 741 F.2d 257 (9th Cir. 1984), *cert. denied*, 470 U.S. 1083 (1985), **238n**

Catron County Board of Commissioners v. U.S. Fish and Wildlife Service, 75 F.3d 1429 (10th Cir. 1996), **210, 251, 252, 262**

Cayman Turtle Farm, Ltd. v. Andrus, 478 F. Supp. 125 (D.D.C. 1979), *aff'd without opinion* (D.C. Cir. Dec. 12, 1980), **92n, 226–27, 230n, 273**

Cerritos Gun Club v. Hall, 96 F.2d 620 (9th Cir. 1938), **24n, 72**

Chalk v. United States, 114 F.2d 207 (4th Cir. 1940), **20, 21**

Choctaw Nation v. United States, 119 U.S. 1 (1886), **449n**

Christy v. Hodel, 857 F.2d 1324 (9th Cir. 1988), **37, 92, 232**

Chrysler Corp. v. Brown, 441 U.S. 281 (1979), **415n**

Citizens Interested in Bull Run, Inc. v. Edrington, 781 F. Supp. 1502 (D. Or. 1991), **76n, 81**

Citizens to End Animal Suffering and Exploitation, Inc. v. New England Aquarium, 836 F. Supp. 45 (D. Mass. 1993), **141n**

City of Las Vegas v. Lujan, 891 F.2d 927 (D.C. Cir. 1989), **208**

City of Tenakee Springs v. Block, 778 F.2d 1402 (9th Cir. 1985), **318n**

Clemons v. United States, 245 F.2d 298 (6th Cir. 1957), **73n**

Cochrane v. United States, 92 F.2d 623 (7th Cir. 1937), **24n, 72**

Colorado River Water Conservation District v. Andrus, Civ. No. 78-A-1191 (D. Colo. Dec. 28, 1981), **203n, 242n, 266n**

Committee for Humane Legislation, Inc. v. Richardson, 414 F. Supp. 297 (D.D.C. 1976), *aff'd*, 540 F.2d 1141 (D.C. Cir. 1976), **120n, 124**

Commonwealth Edison Co. v. Montana, 453 U.S. 609 (1981), **35n**

Complete Auto Transit v. Brady, 430 U.S. 227 (1977), **35n**

Confederated Tribes and Bands of the Yakima Indian Nation v. FERC, 746 F. 466 (9th Cir. 1984), **418–19, 423**

Connor v. Andrus, 453 F. Supp. 1037 (W.D. Tex. 1978), **90, 237**

Conner v. Burford, 836 F.2d 1521 (9th Cir. 1988), **319**

Conner v. Burford, 848 F.2d 1441 (9th Cir. 1988), *cert. denied sub nom. Sun Exploration & Product Co. v. Lujan*, 489 U.S. 1012 (1989), **248–49**

Conservation Law Foundation v. Andrus, 623 F.2d 712 (1st Cir. 1979), **247–48**
Conservation Law Foundation v. Clark, 590 F. Supp. 1467 (D. Mass. 1984), *aff'd*, 864 F.2d 954 (1st Cir. 1989), **309n**
Conservation Law Foundation v. Franklin, 989 F.2d 54 (1st Cir. 1992), **155n, 177n**
Conservation Law Foundation v. Mosbacher, 966 F.2d 39 (1st Cir. 1992), **177–78**
Conservation Law Foundation v. Mosbacher, Civil Action No. 91–11759-MA, 1991 WL 501640 (D. Mass. Aug. 28, 1991), **177–78**
Coupland v. Morton, 5 Envt'l L. Rep. (Envt'l L. Inst.) 20504 (E.D. Va.) *aff'd*, 5 Envt'l L. Rep. (Envt'l L. Inst.) 20507 (4th Cir. 1975), **293**
County of Bergen v. Dole, 620 F. Supp. 1009 (D. N.J. 1985), **412n**
County of Trinity v. Andrus, 438 F. Supp. 1368 (E.D. Cal. 1977), **410n, 411n, 415n**
Craft v. National Park Service, 34 F.3d 918 (9th Cir. 1994), **334–35**
Cresenzi Bird Importers, Inc. v. State of New York, 658 F. Supp. 1441 (S.D.N.Y. 1987), *aff'd*, 831 F.2d 410 (2d Cir. 1987), **271**
Dahl v. Clark, 600 F. Supp. 585 (D. Nev. 1984), **395n, 400–401**
Defenders of Wildlife v. Andrus, 428 F. Supp. 167 (D.D.C. 1977), **90, 236–37, 238**
Defenders of Wildlife v. Andrus, 11 Envt'l Rep. Cas. (BNA) 2098 (D.D.C. July 14, 1978), **293, 294–95**
Defenders of Wildlife v. Andrus, 455 F. Supp. 446 (D.D.C. 1978), **293, 294–95**
Defenders of Wildlife v. Andrus, 627 F.2d 1238 (D.C. Cir. 1980), **393**
Defenders of Wildlife v. Endangered Species Scientific Authority, 659 F.2d 168 (D.C. Cir.), *cert. denied sub nom. International Ass'n of Fish and Wildlife Agencies v. Defenders of Wildlife*, 454 U.S. 963 (1981), **272**
Defenders of Wildlife v. Endangered Species Scientific Authority, 725 F.2d 726 (D.C. Cir. 1984), **273**
Defenders of Wildlife v. EPA, 688 F. Supp. 1334 (D. Minn. 1988), *modified*, 882 F.2d 1294 (8th Cir. 1989), **71n, 76n, 78, 221**
Defenders of Wildlife v. Lujan, Civil No. C92–1643 (W.D. Wash. April 11, 1994), **292**
Delbay Pharmaceuticals, Inc. v. Department of Commerce, 409 F. Supp. 637 (D.D.C. 1976), **228n, 230n**
Diamond v. Chakrabarty, 447 U.S. 303 (1980), **5**
Didrickson v. U.S. Department of Interior, 982 F.2d 1332 (9th Cir. 1992), **142n**
Diggs v. Dent, Civil No. 74–1292 (D.D.C. 1975), **470n**
Dorothy Thomas Foundation v. Hardin, 317 F. Supp. 1072 (W.D.N.C. 1970), **346**
Douglas County v. Babbitt, 48 F.3d 1495 (9th Cir. 1995), *cert. denied*, 116 S.Ct. 698 (1996), **210, 251, 252, 266**
Douglas v. Seacoast Products, Inc., 431 U.S. 265 (1977), **23, 30–31**
Edwards v. Leaver, 102 F. Supp. 698 (D.R.I. 1952), **30n**
Employment Division, Department of Human Resources of Oregon v. Smith, 494 U.S. 872 (1990), **466–67**
Endangered Species Committee of the Building Industry Ass'n of Southern California v. Babbitt, 852 F. Supp. 32 (D.D.C. 1994), **208–9**
Environmental Defense Fund v. Alexander, 501 F. Supp. 742 (N.D. Miss. 1980), *aff'd in part and rev'd in part on other grounds sub nom. Environmental Defense Fund v. Marsh*, 651 F.2d 983 (5th Cir. 1981), **411n, 415n**
Environmental Defense Fund v. Corps of Engineers, 325 F. Supp. 749 (E.D. Ark. 1971), **409–410**
Environmental Defense Fund v. Froehlke, 473 F.2d 346 (8th Cir. 1972), **410**
Environmental Defense Fund v. Watt, 18 Envt'l Rep. Cas. (BNA) 1336 (E.D.N.Y. Oct. 22, 1982), **292n**
Escondido Mutual Water Co. v. La Jolla Band of Mission Indians, 466 U.S. 765 (1984), *rehearing denied*, 467 U.S. 1267, **421**
Fallini v. BLM, 92 IBLA 200 (1986), **402n**
Fallini v. Hodel, 725 F. Supp. 1113 (D. Nev. 1989), **402n**

Fallini v Hodel, 783 F.2d 1343 (9th Cir. 1986), **401–3**
Fallini v. Hodel, 963 F.2d 275 (9th Cir. 1992), **401–3**
Flood v. Kuhn, 407 U.S. 258 (1972), **35**
Florida Key Deer v. Stickney, 864 F. Supp. 1222 (S.D. Fla. 1994), **239**
Forest Conservation Council v. Espy, 835 F. Supp. 1202 (D. Idaho 1993), **368n**
Forest Conservation Council v. Rosboro Lumber Co., 50 F.3d 781 (9th Cir. 1995), **217n**
Foster-Fountain Packing Co. v. Haydel, 278 U.S. 1 (1928), **23, 29, 72n**
Fouke Co. v. Mandel, 386 F. Supp. 1341 (D. Md. 1974), **143, 197n**
Friends of Animals v. Hodel, 1988 WL 236545 (D.D.C. 1988), **293n, 297n**
Friends of Endangered Species, Inc. v. Jantzen, 760 F.2d 976 (9th Cir. 1985), **235**
Friends of the Boundary Waters Wilderness v. Robertson, 978 F.2d 1484 (8th Cir. 1992), *cert. denied*, 113 S. Ct. 2962 (1993), **322**
Fund for Animals, Inc. v. Andrus, 11 Envt'l Rep. Cas. (BNA) 1189 (D. Minn. July 14 and Aug. 30, 1978), **266n**
Fund for Animals, Inc. v. Espy, 814 F. Supp. 142 (D.D.C. 1993), **313–14**
Fund for Animals, Inc. v. Florida Game and Fresh Water Fish Commission, 550 F. Supp. 1206 (S.D. Fla. 1982), **221n**
Fund for Animals, Inc. v. Frizzell, 530 F.2d 982 (D.C. Cir. 1975), *aff'g* 402 F. Supp. 35 (D.D.C. 1975), **87, 88**
Fund for Animals, Inc. v. Lujan, 962 F.2d 1391 (9th Cir. 1992), **313**
Fund for Animals, Inc. v. Morton, Civil No. 74–1581 (D.N.J. 1974), **83n**
Geer v. Connecticut, 161 U.S. 519 (1896), **8n, 14–15, 16, 17, 20, 23, 32, 33–34, 37, 90, 450**
Glover River Org. v. Department of Interior, 675 F.2d 251 (10th Cir. 1982), **266n**
Glover River Org. v. Department of Interior, Civ. No. 78–202-C (E.D. Okla. Dec. 12, 1980), **209n**
Greenpeace Action v. Franklin, 982 F.2d 1342 (9th Cir. 1992), **155n**
Greenpeace Action v. Franklin, 14 F.3d 1324 (9th Cir. 1993), **246n**
Greenpeace U.S.A. v. Evans, 688 F. Supp. 579 (W.D. Wash. 1987), **139n**
Haavik v. Alaska Packers' Association, 263 U.S. 510 (1924), **27–28, 29**
Hallstrom v. Tillamook County, 493 U.S. 20 (1989), **267n**
Hanson v. Klutznick, 506 F. Supp. 582 (D. Alaska 1981), **171n**
Headwaters v. Bureau of Land Management, 914 F.2d 1174 (9th Cir. 1990), **387n**
Heckler v. Chaney, 470 U.S. 821 (1985), **85**
Hill v. TVA, 419 F. Supp. 753 (E.D. Tenn. 1976), **240n**
Hill v. TVA, 549 F.2d 1064 (6th Cir. 1977), **241n**
Hoffman Homes, Inc. v. EPA, 999 F.2d 256 (7th Cir. 1993), **26–27**
Humane Society of the United States v. Babbitt, 849 F. Supp. 814 (D.D.C. 1994), **107n**
Humane Society of the United States v. Clark, Civil Action No. 84–3630 (D.D.C. July 25, 1986), **296n**
Humane Society of the United States v. Hodel, 840 F.2d 45 (D.C. Cir. 1988), **296n**
Humane Society of the United States v. Lujan, 768 F. Supp. 360 (D.D.C. 1991), **97n, 293n, 297n**
Humane Society of the United States v. Watt, 551 F. Supp. 1310 (D.D.C. 1982) *aff'd* 713 F.2d 865 (D.C. Cir. 1983), **71n, 89–90**
Hunt v. United States, 278 U.S. 96 (1928), **19–20, 21, 342**
Idaho Farm Bureau Federation v. Babbitt, 58 F.3d 1392 (9th Cir. 1995), **208–9**
Idaho Public Utilities Commission v. ICC, 35 F.3d 585 (D.C. Cir. 1994), **243n**
Inland Empire Public Lands Council v. Schultz, 992 F.2d 977 (9th Cir. 1993), **346n**
Izaak Walton League of America v. St. Clair, 353 F. Supp. 698 (D. Minn. 1973), *rev'd on other grounds*, 497 F.2d 849 (8th Cir. 1974), **316n**
J.H. Miles & Co. v. Brown, 910 F. Supp. 1138 (E.D. Va. 1995), **171n, 176n**
Japan Whaling Ass'n v. American Cetacean Society, 478 U.S. 221 (1986), **483n**
Jensen v. United States, 743 F. Supp. 1091 (D.N.J. 1990), **171n**
Jones v. Gordon, 792 F.2d 821 (9th Cir. 1986), **139n**

Katelnikoff v. U.S. Department of Interior, 657 F. Supp. 659 (D. Alas. 1986), **142n**
Kidd v. United States Department of the Interior, Bureau of Land Management, 756 F.2d 1410 (9th Cir. 1985), **402n**
Kisner v. Butz, 350 F. Supp. 310 (N.D.W. Va. 1972), **348n**
Kleppe v. New Mexico, 426 U.S. 529 (1976), **21–22, 23, 312, 325, 394n**
Kokechik Fishermen's Ass'n v. Secretary of Commerce, 839 F.2d 795 (D.C. Cir. 1988), *cert. denied sub nom., Verity v. Center for Environmental Educ.*, 488 U.S. 1004 (1989), **132–33**
Kramer v. Mosbacher, 878 F.2d 134 (4th Cir. 1989), **171, 171n**
Krichbaum v. Kelley, 844 F. Supp. 1107 (W.D. Va. (1994), **367n**
Lane County Audubon Society v. Jamison, 958 F.2d 290 (9th Cir. 1992), **250**
Lansden v. Hart, 168 F.2d 409 (7th Cir.), *cert. denied*, 335 U.S. 858 (1948), **36, 91n**
Lansden v. Hart, 180 F.2d 679 (7th Cir.), *cert. denied*, 340 U.S. 824 (1950), **91n**
Leo Sheep Co. v. United States, 440 U.S. 668 (1979), **22n**
Leslie Salt Co. v. United States, 896 F.2d 354 (9th Cir. 1990), **26**
Libby Rod & Gun Club v. Poteat, 457 F. Supp. 1177 (D. Mont. 1978), *aff'd in part, rev'd in part on other grounds*, 594 F.2d 742 (9th Cir. 1979), **266n**
Lone Wolf v. Hitchcock, 187 U.S. 553 (1903), **460**
Louisiana ex rel. Guste v. Verity, 853 F.2d 322 (5th Cir. 1988), **223–24, 226–27**
Louisiana v. Baldridge, 538 F. Supp. 625 (E.D. La. 1982), **171n, 178–79**
Lucas v. South Carolina Coastal Council, 505 U.S. 1027 (1992), **38**
Lujan v. Defenders of Wildlife, 504 U.S. 555 (1992), **263n**
Lyng v. Northwest Indian Cemetery Protective Ass'n, 485 U.S. 439 (1988), **466, 467**
Maine v. Kreps, 563 F.2d 1043 (1st Cir. 1977), **158n, 172, 174**
Maine v. Taylor, 477 U.S. 131 (1986), **34n**
Man Hing Ivory and Imports, Inc. v. Deukmejian, 702 F.2d 760 (9th Cir. 1983), **271**
Manchester v. Commonwealth of Massachusetts, 139 U.S. 240 (1891), **13–14, 27**
Marbled Murrelet v. Pacific Lumber Co., 880 F. Supp. 1343 (N.D. Cal. 1995), **217n, 224n**
Marsh v. Oregon Natural Resources Council, 490 U.S. 360 (1989), **368n, 386n**
Martin v. Waddell in 1842, 41 U.S. (16 Pet.) 367 (1842), **10–12, 13**
Massey v. Apollonio, 387 F. Supp. 373 (D. Me. 1974), **30n**
McCready v. Virginia, 94 U.S. 391 (1876), **13, 27, 29**
Menominee Tribe of Indians v. United States, 391 U.S. 404 (1968), **450n, 460n**
Michigan United Conservation Clubs v. Lujan, 949 F.2d 202 (6th Cir. 1991), **306n, 311**
Minnesota v. Bergland, 499 F. Supp. 1253 (D. Minn. 1980), **22n**
Missouri v. Holland, 252 U.S. 416 (1920), **18, 23, 28, 63n, 64**
Montana v. United States, 450 U.S. 544 (1981), **464**
Morissette v. United States, 342 U.S. 246 (1952), **99n**
Morrill v. Lujan, 802 F. Supp. 424 (S.D. Ala. 1992), **222**
Mountain States Legal Foundation v. Andrus, 499 F. Supp. 383 (D. Wyo. 1980), **318–19**
Mountain States Legal Foundation v. Andrus, 16 Envt'l Rep. Cas. (BNA) 1351 (D. Wyo. April 21, 1981), **400n**
Mt. Graham Red Squirrel v. Madigan, 954 F.2d 1441 (9th Cir. 1992), **361n**
Mullaney v. Anderson, 342 U.S. 415 (1952), **28n, 30n**
National Audubon Society v. Babbitt, Civil No. C92–1641 (W.D. Wash. Oct. 20, 1993), **292n, 293n**
National Audubon Society v. Forest Service, 4 F.3d 832 (9th Cir. 1993), *modified*, 46 F.3d 1437 (1994), **318n**
National Audubon Society v. Hodel, 606 F. Supp. 825 (D. Alaska 1984), **293n, 302–3**
National Fisheries Institute v. Mosbacher, 732 F. Supp. 210 (D.D.C. 1990), **164n, 179, 180n**
National Parks and Conservation Ass'n v. Federal Aviation Administration, 998 F.2d 1523, **376n**
National Rifle Ass'n v. Kleppe, 425 F. Supp. 1101 (D.D.C. 1976), **71n, 82**
National Rifle Ass'n v. Potter, 628 F. Supp. 903 (D.D.C. 1986), **310–11**

STATE CASES

FEDERAL STATUTES

Act for the Preservation of American Antiquities, 16 U.S.C. §§ 431–433, **286n, 310n**

Administrative Procedure Act, 5 U.S.C. §§ 551–559 and 701–706, **71, 155n, 197, 203, 376, 415**

Agriculture Appropriation Act of 1907, 34 Stat. 1269, **342**

Airborne Hunting Act, 16 U.S.C. § 742j–l, **24**

Alaska Game Law, Act of Jan. 13, 1925, 43 Stat. 739, **84**

Alaska National Interest Lands Conservation Act of 1980, Pub. L. No. 96–487 (codified in scattered sections of 16, 43 U.S.C.), **146n, 290n, 299–305**

Alaska Native Claims Settlement Act, Pub. L. 92-203, Dec. 18, 1971, 85 Stat. 688, **303**

Alaska Statehood Act, Pub. L. 85–508, § 8(d), 72 Stat. 339 (1958), 48 U.S.C. notes prec. § 21, **84n**

American Fisheries Promotion Act, Pub. L. No. 96–561, Title II, Part C, 94 Stat. 3296, **184–85**

American Indian Religious Freedom Act, 42 U.S.C. § 1996, **100**

Anadromous Fish Conservation Act of 1965, Pub. L. 89–304, § 2, 79 Stat. 1125 (current version at 16 U.S.C. § 757b), **408n**

Animal Welfare Act, 7 U.S.C. §§ 1231–2159, **140**

Antarctic Conservation Act of 1978, Pub. L. No. 95–541, 92 Stat. 2048, **501n**

Antarctic Marine Living Resources Convention Act, Pub. L. No. 98–623, 98 Stat. 3401, codified at 16 U.S.C. §§ 2431–2442, **504**

Arizona-Idaho Conservation Act, 16 U.S.C. § 431 note and scattered sections, **362n**

Bald Eagle Protection Act, 16 U.S.C. §§ 668–668d, **39–40, 77, 78, 79, 80, 87, 93–102, 231, 460n, 465, 467**

Bartlett Act, 16 U.S.C. §§ 1081–1086 2nd 1091–1094 (1976) (repealed as of March 1, 1977), **149–50**

Black Bass Act of 1926, Act of May 20, 1926, Ch. 346, 44 Stat. 576 (codified at 16 U.S.C. §§ 851–856 (1976)) (repealed 1981), **39, 42–43, 45, 196, 198**

Boundary Revision Act, Act of Jan. 3, 1983, Pub. L. No. 97–405, 96 Stat. 2028, 16 U.S.C. § 160a–1, **312, 313**

California Desert Protection Act, Pub. L. No. 103–433, 108 Stat. 4471, **391**

California Wilderness Act of 1984, 16 U.S.C. § 1132 and scattered sections, **316**

Classification and Multiple Use Act of 1964, 43 U.S.C. §§ 1411–1418, **373–75**

Clean Water Act, 33 U.S.C. §§ 1251–1387, **166, 356, 408, 413, 439–40**

Coastal Barrier Resources Act, 16 U.S.C. §§ 3501–3510, **442–43**

Coastal Zone Management Act of 1972, 16 U.S.C. §§ 1451–1464, **155n**

Commercial Fisheries Research and Development Act of 1982, Pub. L. No. 97–389, § 201, 96 Stat. 1949, **132n**

Conservation Act of August 12, 1949, ch. 421, § 2, 63 Stat. 600, **296**

Cooperative Forestry Assistance Act of 1978, Pub. L. 95–313, 92 Stat. 365, **438**

Department of Interior Appropriations Act for Fiscal Year 1988, Pub. L. No. 100-202 § 314, 101 Stat. 1329-254 (1987), **387**

Dep't of Interior and Related Agencies Appropriations Act for Fiscal Year 1993, Pub. L. No. 102–381, 106 Stat. 1419, 16 U.S.C. § 1612 note, **356n**

Dolphin Protection Consumer Information Act, Pub. L. 101–627, § 901, 104 Stat. 4465 (1990), codified at 16 U.S.C. § 1385, **128**

Electric Consumers Protection Act, Act of October 16, 1986, Pub. L. No. 99–495, 100 Stat. 1243, **419**

Emergency Supplemental Appropriations and Rescissions Act, Act of July 27, 1995, Pub. L. No. 104–19, **365, 388**

Endangered Species Act Amendments of 1978, Pub. L. No. 95–632, 92 Stat. 3751 (1978), **254–58**

INDEX

About the Authors

MICHAEL J. BEAN is Chair of the Environmental Defense Fund's Wildlife Program and is well known for his expertise on endangered species conservation.

MELANIE J. ROWLAND has been a faculty member at the University of Washington and Senior Counsel with The Wilderness Society. She is now an attorney with the National Oceanic and Atmospheric Administration's Office of General Counsel in Seattle.

ISBN 0-275-95988-0
90000>
EAN
9 780275 959883
HARDCOVER BAR CODE